Applied Hydrogeochemistry

Groundwater flow systems can be evaluated by solute and isotopic methods, both powerful tools for students and professionals in hydrogeology. This accessible textbook provides foundational knowledge and promotes a comprehensive understanding users need to fully evaluate both the solute and isotopic aspects of groundwater. It focuses on groundwater chemistry and discusses the most important solutes, stable and radiogenic isotopes, and groundwater age dating – topics often inadequately treated or ignored in other texts. The reader will find chapters on equilibria, gases, kinetics, organic compounds, and the most important species: carbon, sulfur, nitrogen, silica, noble gas, and chloride, and stable isotopes, groundwater dating, graphical methods, and geothermometers.

Features

- Presents a balanced approach between solutes and isotopes, the first book of its kind to do so.
- Explains in detail how to apply stable isotopic data to help solve groundwater problems.
- Describes the appropriate groundwater dating tools for both young and old groundwater and how to apply the tools.
- Provides the reader with the background knowledge necessary to conceptualize the geochemical evolution of a groundwater system before using a geochemical modeling program.
- Examines how geothermometers and graphical methods can be used to illustrate chemical evolution.

This textbook is an excellent resource for upper-level undergraduates and graduate students taking courses in groundwater chemistry that are taught in departments of geology, physical geography, environmental science, and environmental engineering, as well as for researchers and working professionals involved with water and groundwater.

Applied Hydrogeochemistry

Fundamentals, Solutes, and Isotopes

Alan L. Mayo

CRC Press
Taylor & Francis Group
Boca Raton London New York

CRC Press is an imprint of the
Taylor & Francis Group, an **informa** business

Designed cover image: Linda D. Mayo, Mammoth Terrace, Yellowstone National Park

First edition published 2026
by CRC Press
2385 NW Executive Center Drive, Suite 320, Boca Raton FL 33431

and by CRC Press
4 Park Square, Milton Park, Abingdon, Oxon, OX14 4RN

CRC Press is an imprint of Taylor & Francis Group, LLC

ISBN: 978-1-041-08588-1 (hbk)
ISBN: 978-1-041-08589-8 (pbk)
ISBN: 978-1-003-64612-9 (ebk)

DOI: 10.1201/9781003646129

Typeset in Times
by Apex CoVantage, LLC

Access the Instructor and Student Resources: www.routlege.com/9781041085881

To my wife, Linda, for her encouragement and enduring patience while writing this book.

In memory of Anthony B. Muller. Mentor, colleague, and dear friend who introduced me to isotope geochemistry.

Contents

Preface xiii
About the Author xv
Acknowledgments xvii

PART I Solutes

Chapter 1 Solutes and Solutions 3

1.1 Definition of a Solution 3
1.2 Units and Solute Concentrations 3
1.2.1 Length, Mass, Volume, Time, Temperature, and Pressure 3
1.2.2 Solute Concentrations 5
1.3 Chemical Bonding 9
1.4 Physical and Chemical Properties of Water 10
1.4.1 The Water Molecule 10
1.4.2 Water as an Electrolyte Solution 13
1.4.3 Speciation 14
1.4.4 Salinity 15
1.4.5 Density and Temperature 16
1.5 Inorganic Chemical Reactions 17
1.5.1 Mineral Dissolution–Precipitation 18
1.5.2 Acid–Base Reactions 18
1.5.3 Hydrolysis 22
1.5.4 Reduction–Oxidation (Redox) 25
1.5.5 Oxidizing Environments 29
1.5.6 Reducing Environments 31
1.6 Mineral Surface Reactions 31
1.6.1 Sorption Isotherms 33
1.6.2 Ion Exchange 35
1.6.3 Ion Hydration, Ion Pairs, and Ion Complexation 36
1.6.4 Speciation Stability Constants 38
1.7 Complex Ion Speciation 40
References Cited 47

Chapter 2 Chemical Equilibria 49

2.1 Chemical Equilibrium 49
2.1.1 Law of Mass Action 49
2.1.2 Equilibrium Defined 50
2.1.3 Chemical Activity 51
2.2 Thermodynamics 59
2.2.1 Enthalpy and Entropy 59
2.2.2 Gibbs Free Energy 60
2.2.3 Phase Rule 64
2.3 Reduction–Oxidation (Electrochemical Equilibrium) 66
2.4 Mineral Dissolution 70

2.4.1 Saturation Index (SI) ... 70
2.4.2 Ionic Strength Effect ... 72
2.4.3 Common Ion Effect ... 72
2.4.4 Congruent and Incongruent Dissolution ... 73
References Cited ... 75

Chapter 3 Gases ... 76

3.1 Nature of Gas ... 76
3.2 Gas Laws ... 78
3.3 Gases and Solutions – Gas Exchange Processes ... 80
3.3.1 Ideal Solutions (Gas–Liquid Systems) ... 80
3.3.2 Nonideal Solutions ... 86
3.4 CO_2 System ... 86
References Cited ... 89

Chapter 4 Kinetics ... 90

4.1 Introduction ... 90
4.2 Equilibrium Thermodynamics and Kinetics ... 91
4.3 Classification of Kinetic Reactions ... 91
4.4 Kinetic Reaction Rates ... 94
4.5 Kinetic Precipitation and Dissolution Mechanisms ... 99
4.5.1 Mineral Precipitation ... 100
4.5.2 Weathering and Other Dissolutions ... 104
References Cited ... 107

Chapter 5 Organic Chemistry ... 108

5.1 Introduction ... 108
5.2 Properties of Organic Compounds ... 108
5.3 The Carbon Cycle ... 113
5.4 Hydrocarbon Framework ... 114
5.5 Organic Carbon Functional Groups ... 117
5.6 Humic Substances ... 120
References Cited ... 121

Chapter 6 Major and Trace Aqueous Species ... 123

6.1 Introduction ... 123
6.2 Carbonate System ... 123
6.2.1 CO_2 and pH Control ... 125
6.2.2 Calcite and Dolomite ... 128
6.2.3 Open and Closed Groundwater Systems ... 129
6.2.4 Other CO_2–Carbonate Considerations ... 133
6.3 Chloride-Bromide ... 138
6.4 Sulfur ... 141
6.4.1 Atmospheric Sulfur ... 142
6.4.2 Biogenic Sulfur ... 143
6.4.3 Sulfur in Groundwater ... 145
6.5 Iron and Manganese ... 148

6.6 Silica ... 150
6.7 Nitrogen ... 156
6.8 Noble Gas ... 159
6.9 Minor Constituents ... 163
6.9.1 Alkali Metals ... 164
6.9.2 Alkali Earth Metals ... 165
6.9.3 Transition Metals ... 166
6.9.4 Other Metals ... 169
6.9.5 Nonmetallic Elements ... 172
References Cited ... 175

PART II Isotopes

Chapter 7 Isotopic Definitions and Standards ... 183
7.1 Introduction ... 183
7.2 Isotopic Fractionation and Isotopic Exchange ... 184
7.3 Fractionation Factors ... 186
7.4 Isotopic Equilibrium ... 190
7.5 Radioactive Isotopes ... 193
7.6 Isotope Reporting Standards ... 193
7.6.1 Oxygen ... 194
7.6.2 Hydrogen ... 195
7.6.3 Carbon ... 195
7.6.4 Sulfur ... 196
7.6.5 Nitrogen ... 196
7.6.6 Chlorine ... 196
References Cited ... 198

Chapter 8 Hydrogen and Oxygen Isotopes ... 199
8.1 Introduction ... 199
8.2 Global Meteoric Water Line (GMWL) ... 200
8.3 Meteoric and Groundwater ... 201
8.3.1 Isotopic Trends in Meteoric Water ... 204
8.3.2 Thermal Groundwater ... 219
8.3.3 Magmatic and Connate Water ... 222
8.4 Chapter Notation ... 224
References Cited ... 224

Chapter 9 Carbon-13 ... 229
9.1 Introduction ... 229
9.2 Carbon Isotopic Fractionation ... 231
9.2.1 Atmospheric, Oceanic, and Biologic Carbon ... 231
9.2.2 Aqueous Carbon Exchange Processes ... 234
9.3 Sources of Groundwater Carbon ... 234
9.3.1 Groundwater CO_2 ... 235
9.4 Soil Zone and Aquifer Carbon Processes ... 237
9.5 Other Inorganic Carbon Sources ... 240

9.5.1 Magmatic Carbon 240
9.5.2 Metamorphic Carbon 241
9.5.3 Sulfate Reduction Carbon 242
9.5.4 Connate Water 242
References Cited 243

Chapter 10 Sulfur-34 and Nitrogen-15 246

10.1 Sulfur Isotope Fractionation and the Sulfur Cycle 246
10.1.1 Atmospheric Sulfur 248
10.1.2 Calcium Sulfate and Calcium Carbonate Minerals 249
10.1.3 Sulfate-Sulfide Redox and Biologic Fractionation 250
10.1.4 Anthropogenic Sulfur 253
10.2 Nitrogen Cycle and Isotopic Fractionation 255
10.2.1 Nitrogen Sources 255
10.2.2 Nitrogen Isotopic Fractionation 259
10.2.3 Nitrogen in Groundwater 262
References Cited 266

Chapter 11 Groundwater Dating 272

11.1 Introduction 272
11.2 Tritium 274
11.2.1 Atmospheric Tritium 274
11.2.2 Tritium Groundwater Dating 278
11.3 $^3H/^3He$ Dating 279
11.4 Radiocarbon 280
11.4.1 Radiocarbon Dating 283
11.4.2 Models for Estimating A_o 288
11.4.3 A_o Numerical Instabilities and Model Sensitivity 295
11.5 Chlorine-36 298
11.6 Chlorofluorocarbons and Sulphur Hexafluoride 300
References Cited 305

PART III Data Manipulation and Geothermometers

Chapter 12 Project Design, Data Management, and Data Manipulation: Computer and Graphical Methods 311

12.1 Project Design and Data Management 311
12.1.1 Project Design 311
12.1.2 Data Management 312
12.2 Computer Analytical Methods 315
12.3 Graphical Methods 319
12.3.1 Shape Diagrams 320
12.3.2 Hydrogeochemical Facies 331
12.3.3 Scatter Plots 333
12.3.4 Tie Line Diagrams 334
12.4 Contour Maps 345
12.5 Equilibrium Diagrams 345

12.5.1 E_h-pH Diagram 349
12.5.2 Congruent Solubility Diagrams 358
12.5.3 Incongruent Solubility Diagrams 360
12.5.4 Partial Pressure and Fugacity Diagrams 365
12.5.5 Problems 371
References Cited 373

Chapter 13 Geothermometers 375

13.1 Introduction 375
13.2 Qualitative Geothermometers 376
13.3 Quantitative Solute Geothermometers 378
13.3.1 Solubility Geothermometers 379
13.3.2 Exchange Geothermometers 380
13.3.3 Giggenbach Na–K–Mg Triangular Diagram 385
13.3.4 Mixing Geothermometer Models 387
13.4 Gas Geothermometers 388
13.5 Isotopic Geothermometers 389
13.5.1 Carbon Geothermometers 390
13.5.2 Hydrogen Geothermometers 392
13.5.3 Oxygen Geothermometers 392
13.5.4 Sulfur Geothermometers 392
13.5.5 Problems 392
References Cited 393

Appendix **397**

Index **415**

Preface

The mysterious nature of water moving below the ground holds a fascination for many observers of the natural world. My interest began as a child when I first saw geysers in Yellowstone. For most of human history the physical behavior of groundwater was little understood and its chemical nature was completely unknown. The physical aspects of groundwater flow are based on the three elegant yet easily understood principles: Darcy's law, fluid potential, and the idea that the earth is the matrix through which water flows. From these concepts physical hydrogeology developed and is widely used.

Using chemical hydrogeology can be more daunting because our understanding of fluid flow must be coupled with knowledge of often complex solute and isotopic rock–water interactions. This complex nature can inhibit the application of the powerful tools that hydrogeochemistry has to offer. However, hydrogeochemistry can add greatly to the knowledge gleaned from physical hydrogeology. While physical hydrogeology will suggest what potentially can happen with a water, hydrogeochemistry interrogates the water chemistry. Those answers will tell us what has happened.

Hydrogeochemistry can shed light on where and when groundwater recharge occurred, what actual rock packages the water has encountered along the flow path, how deep the water has flowed, what other waters it may have mixed with, how long the water has been in the ground, and what are the source and likely fate of the groundwater contaminants. In many investigations both physical and chemical hydrogeology need to be coupled to provide a complete picture. For example, ^{14}C groundwater ages may be the only reliable tool to validate particle tracks in a groundwater flow model.

As I wrote this text, I tried to keep in mind the book that I wish had been available when I embarked on my career as a hydrogeologist and the book I wish had been available to use with my own students. So the purpose of this book is twofold: 1) to provide students with the necessary background for more advanced study and 2) to provide students and working professionals with the theoretical and practical background necessary to confidently tackle and solve gaseous, inorganic and organic solute, and stable and radiogenic isotopic aqueous geochemical problems.

This book is organized in sections that describe the most widely used solute and isotopic components of groundwater chemistry. Solving real-world problems requires the careful selection and integration of solute and isotopic techniques. In my own research and consulting I try to think of the encounters an individual water molecule has with gases, organic matter, contaminants, and minerals as it first falls from the sky and then interacts with the soil zone and the aquifer skeleton as it travels along the flow path. This conceptual understanding is useful for helping sort out which physical and chemical tools are best suited to help evaluate the groundwater flow system.

Although geochemistry is computational by nature, my hope is that readers will use the theoretical background given here to understand each aspect of hydrogeochemistry when solving groundwater problems – and then let computer codes such as PHREEQC do the computational heavy lifting.

About the Author

Alan L Mayo received B.S. and M.S. degrees in geology from San Diego State University and his Ph.D. in hydrogeology from the University of Idaho. He has been a senior environmental planner with San Diego County and a senior hydrogeologist with Law Engineering in Atlanta, Georgia, and has held teaching and research positions at the University of Idaho, the University of Colorado at Colorado Springs, and Brigham Young University, where he originated the water lab. He was an HEW Fellow and a Fulbright Scholar. His research interests are in mountain and basin hydrogeology, geothermal systems, and groundwater chemistry. He has conducted research in North, Central, and South America, as well as Central Europe, has presented research results at conferences worldwide, and has published 100+ research papers.

While in Utah, Alan founded Mayo and Associates, LC, consultants in hydrogeology. Later, he was vice president and chief technical officer of HydroGeo Group, a company that specializes in hydrogeology, aquifer analysis, aquifer hydrodynamics, and solute and isotopic hydrogeochemistry. His consulting work has included coal and hard rock mining operations, gas and oil fields, mine and other permitting, environmental analysis, water supplies, groundwater contamination, and litigation. Earlier in his career Alan was the hydrogeologist for the plaintiffs on the so-called Erin Brockovich case.

He has been married to Linda (his partner in adventure) for 47 years and has two children and four grandchildren. In addition to California, where he is a registered geologist, Alan has lived in Idaho, Georgia, Colorado, Arizona, Utah, and the Czech Republic. He currently summers in Colorado and winters in Arizona. He has traveled in all of western and central Europe and parts of the Middle East, China, southern Africa, Australia and New Zealand, Iceland, and Central and South America.

Acknowledgments

No author can claim independent credit for their work. In my case my research and this book are the result of the opportunities afforded by my public education and my professions experiences and encounters. Over the course of my career, I have received generous personal and financial support from so many people and institutions. I would like to thank the citizens of the states of California and Idaho who financially supported San Diego State University and the University of Idaho, where I obtained my bachelor's, master's, and doctoral degrees in geology and hydrogeology. I am indebted to the US Department of Human Health Education and Welfare for my doctoral fellowship, the US Fulbright program, and the University of Idaho, University of Colorado at Colorado Springs, and Brigham Young University for my teaching–research positions. I also thank the National Science Foundation for the support in establishing our solute and isotopic laboratory. Special thanks go to Brigham Young University, which supported my research and attendance at numerous domestic and foreign conferences, and to the dozens of anonymous colleagues who peer-reviewed my research. I am particularly indebted to David Tingey, who organized and ran our laboratory and was a critical research colleague and resourceful field investigator. My research would not have been possible without the interaction with my colleagues, students, student researchers, and teaching assistants. I wish to thank the many commercial interests who allowed me to ply my trade and who gave me the freedom to use new-to-them isotopic techniques, particularly JBR Environmental in Salt Lake City, Utah, the coal mining industries in Colorado, Wyoming, Utah, and Canada, the gold mining industries in Nevada and Utah, the petroleum industries in Wyoming, the phosphate industries in Idaho, the energy industries in Nevada and Utah, and the refining and mineral processing industries. I am grateful to the many law firms that allowed me to represent their clients in local, state, and federal litigation across the United States.

Part I

Solutes

1 Solutes and Solutions

1.1 DEFINITION OF A SOLUTION

Hydrogeochemical investigations require the study and understanding of chemical reactions between aqueous solutions and gases, solids, and other liquids. Typical reactions include the acquisition of a gas such as CO_2, the dissolution of soil zone and aquifer minerals, the precipitation of minerals in the aquifer and at the land's surface, and interactions between groundwater with organic contaminants such as gasoline. In order to effectively undertake hydrogeochemical investigations it is important to understand the nature of solutions and the general types of reactions that may have occurred.

A solution may be thought of as a homogenous system composed of a mixture of components whose relative abundances may vary. By using the term system to describe any portion of the universe, we are able to define the solution selected for study as narrowly or as broadly as is necessary. Most solutions of interest to hydrogeology are binary, meaning they consist of a solvent, the substance in greater proportion, and one or more solutes, the substances in lesser proportions. Obviously, water is the solvent of interest in groundwater investigations. Neither the solvent nor the solutes need to be pure substances.

It is convenient to classify solutions in terms of the physical states of the solute and solvent. Examples of binary solutions are listed in Table 1.1.

1.2 UNITS AND SOLUTE CONCENTRATIONS

1.2.1 Length, Mass, Volume, Time, Temperature, and Pressure

In science the metric system, also known as the International System of Units, is the universally accepted system of measurement for length, mass, volume, and time. This system is often referred to as the SI system (System International) of units. Length is commonly expressed in meters (m), mass in grams (g), volume in liters (L), and time in seconds (s) or minutes (min). In the United States, the English system of measurement is often used where measurements of groundwater flow are involved. Commonly used English units for volume include gallons (gal), cubic feet (ft^3), and acre-feet (af). For flow per unit time, English units include gallons per minute (gpm), cubic feet per second (ft^3/sec), and acre-feet per year (af/yr).

The SI system of units has the fundamental advantage over the English system in that SI units are easily manipulated by decimal factors of 10. These decimal relationships are indicated by a series of commonly used prefixes for multiples and fractions (Table 1.2). In hydrogeochemistry the most frequently used units for measurement of length, mass, volume, time, and temperature are listed in Table 1.3.

The ease of using SI units is evident by considering mass, where a milligram is 1/1,000 of a gram and a kilogram is 1,000 grams. Times (*t*) in both the SI and English units are the same. Short time intervals are usually reported in seconds or minutes and longer time intervals are often reported in hours, days, and years. In international science the most frequently used measure of temperature (*T*) is degrees centigrade (°C), although degrees Kelvin (°K), also known as absolute temperature, is sometimes used, particularly in equilibrium calculations. Zero °K is the temperature at which a quantity of gas will occupy the minimum volume and may be thought of as the temperature where molecular motion ceases. In the United States and some Caribbean countries, degrees Fahrenheit (°F) is commonly used to report temperature in everyday life and in scientific reports that are not

DOI: 10.1201/9781003646129-2

TABLE 1.1
Solutions of Interest in Hydrogeologic Investigations

Solute	Solvent	Examples
gas	gas	air, all mixtures of gases
gas	liquid	CO_2 in water
gas	solid	fluid inclusion in rock matrix
liquid	liquid	gasoline in water
liquid	solid	fluid inclusion in rock matrix
solid	liquid	dissolved and particulate substances in water
solid	solid	igneous rock

TABLE 1.2
Prefixes for SI Units

Prefix	Factor	Symbol
exa	10^{18}	E
peta	10^{15}	P
tera	10^{12}	T
giga	10^{9}	G
mega	10^{6}	M
kilo	10^{3}	k
hecto	10^{2}	h
decka	10^{1}	da
deci	10^{-1}	d
centi	10^{-2}	c
milli	10^{-3}	m
micro	10^{-6}	μ
nano	10^{-9}	n
pico	10^{-12}	p
femto	10^{-15}	f
atto	10^{-18}	a

TABLE 1.3
Commonly Used Measurements of Length, Mass, Volume, Time, and Temperature in Hydrogeochemistry

Length	Length	Volume	Time	Temperature
kilometer (km)	kilogram (kg)	liter (L)	second (s)	centigrade (°C)
meter (m)	gram (g)	milliliter (ml)	minute (min)	Kelvin (°K)
centimeter (cm)	milligram (mg)		hour (h)	
millimeter (mm)	microgram (μg)		day (d)-year (yr)	

intended for international use. The centigrade scale is based on the freezing and boiling points of pure water at a pressure of 1 atmosphere (atm). At 1 atm pure water freezes at 0 °C and boils at 100 °C, which are equivalent to 273.16 °K and 373.16 °K and 32 °F and 212 °F, respectively. Conversions between centigrade, Kelvin, and Fahrenheit may be performed using the following equations:

$$°K = 273.16 + °C \tag{1.1}$$

$$°C = -273.17 + °K \tag{1.2}$$

$$°F = \frac{9}{5}°C + 32 \tag{1.3}$$

$$°C = \frac{5}{9}°F - 32 \tag{1.4}$$

Pressure has little effect on most low temperature, near surface solute and aqueous isotopic reactions. However, pressure may profoundly influence reactions involving gases and may be important in deeper thermal groundwater systems. The force of the atmosphere pressing down at sea level is the nominal reference pressure for measurements. Atmospheric pressure has historically been measured using a simple mercury barometer consisting of an 80 cm long tube sealed at one end and filled with mercury. When the open end of the tube is carefully placed in an open container full of mercury, the mercury in the tube will fall, creating a vacuum. Changes in atmosphere pressure will cause the height of mercury in the tube to fall and rise. Changes in the height of the column of mercury may be measured as inches (in Hg), centimeters of mercury (cm Hg) or millimeters of mercury (mm Hg).

One standard atmosphere is 76 cm Hg or about 29.9 in Hg. Pressure can also be reported as bars (bar), where 1 atm approximately equals 1 bar (1 atm = 1.01325 bar). Although the bar is a metric unit that is not an approved SI unit, it has been recognized by the European Union. The SI system uses pascal (Pa) units for pressure. One bar is equivalent to 100,000 Pa or 100 kPa. Other common pressure equivalents for one bar are 0.987 atm, 29.53 in Hg, 750.06 mm Hg, 750.06 torr, and 100,000 N/m^2 ($Newton/m^2$).

In the deep portions of the ocean, in deep groundwater systems, and in metamorphic systems, hydrostatic and lithostatic pressures greatly exceed 1 atm. As a rule of thumb, there is about a 300-bar pressure increase for each km increase in depth (Turner, 1968). Thus, thermal groundwater circulating at depth of 3.3 km would be under a pressure of about 1 kilobar (kbar).

Results of many laboratory experiments and equilibrium constants are reported at standard temperature and pressure (STP), which is 25 °C and 1 atm, respectively. Except for gases and unusual groundwater conditions, corrections seldom need to be made for pressure variations. Because few groundwater systems occur at the standard 25 °C, equilibrium constants for geochemical reactions must be recalculated for actual aquifer temperatures before they can be meaningfully used. Procedures for calculating temperature-adjusted equilibrium constants are described in Chapter 2. Fortunately, most groundwater geochemical programs calculate equilibrium constants for the measured water temperature. It is commonly assumed that field measured water temperatures represent the in situ aquifer temperature.

1.2.2 Solute Concentrations

A variety of concentration units are used for reporting water quality analysis and for use in chemical calculations. These units belong to three general categories:

1. laboratory reporting units,
2. geochemical units, and
3. chemical equivalence units.

Dissolved substances in groundwater are almost entirely dissociated species. Dissociated species are two or more ions or neutral molecules formed by the dissolution of one or more solids. Laboratories usually report the solute composition of solutions in weight per weight units. Laboratory analyses have historically been reported as parts per million (ppm), parts per thousand (ppt), and parts per billion (ppb). In 1967 the U.S. Geological Survey adopted the milligram per liter (mg/L) standard. In low salinity water, ppm and mg/L are essentially the same. The former ppm standard is:

$$ppm = \frac{g\ solute}{g\ solution \times 10^6} \tag{1.5}$$

Measurements in mg/L are strictly correct only at the laboratory analytical temperature due to temperature dependent volumetric changes of water. Under most circumstances such temperature dependent changes are of little consequence and may be ignored. In practice mg/L is considered equivalent to milligrams per kilogram (mg/kg) for groundwater having solute concentrations less than about 7,000 mg/L. The concentration of trace metals and many hazardous substances are often reported in micrograms per kilogram (μg/kg) or in micrograms per liter (μg/L), which corresponds to parts per trillion.

Units developed by laboratory chemists for the manipulation of aqueous chemical data include:

Molality (*m*) – moles of solute per kg of solvent,

Formality (*f*) – mass of solute per liter of solution, and

Molarity (M) – moles of solute per liter of solvent.

Molality is measured relative to the solvent mass, whereas molarity is measured relative to the space a solvent occupies. Formality is the concentration of the compound in the solution and does not consider how much of the compound is dissolved. A mole, the standard unit for a mass of substance, is the formula weight of the substance in grams and is often referred to as the gram formula weight (gfw). Atomic weights are measured relative to the atomic weight of carbon-12, which has been assigned an atomic weight of 12. The gfw of an element is approximately equal to the atomic number (i.e., the number of protons) of the element. Most elements have a gfw that is slightly greater than the atomic number because the gfw is the average mass of all of the element's isotopes. Isotopes are defined by the number of protons plus neutrons. For example, carbon-12 has 12 protons and no neutrons. Carbon-13 has 12 protons and one neutron. The number of atoms in a mole of a solid substance is Avogadro's number (6.023×10^{23}) molecules. One mole of a gas at STP occupies a volume of 22.4 L and also contains one Avogadro's number of atoms of the gas.

Although chemists commonly use molality when performing chemical calculations, in dilute solutions, including typical groundwater, differences between *m*, *f*, and M are so small that they can generally be neglected because one liter of pure water weighs1 kg at 3.89 °C. When the solute concentration is less than about 7,000 mg /L density, corrections may be omitted, as errors introduced by the density of water will usually be less than other analytical errors. During calculations of free energies the differences between *m*, *f*, and M are commonly less than are the errors in published free energy values.

When weight is in g and volume is in ml the units of molality, formality, and molarity may be calculated as follows:

$$m = \frac{\text{g solute} \times 1{,}000}{\text{gfw solute} \times \text{g solute}} \tag{1.6a}$$

$$f = \frac{\text{g solute} \times 1{,}000}{\text{gfw solute} \times \text{g solution}} \tag{1.6b}$$

$$\mathrm{M} = \frac{\text{g solute} \times 1{,}000}{\text{gfw solute} \times \text{ml solution}} \tag{1.6c}$$

$$m = f \times \frac{\text{g solution}}{\text{g solution} - \text{total g solutes}} \tag{1.6d}$$

$$m = M \frac{\text{g solution}}{(\text{g solution} - \text{total g solutes}) \times \text{density}} \tag{1.6e}$$

$$M = f \times \text{densityofsolution} \tag{1.6f}$$

Milligrams per liter may be converted to molarity by dividing by the atomic weight of the constituent:

$$M = \frac{\text{mg}}{\text{L}} \times \frac{10^{-3}}{\text{gfw of solute}} \tag{1.7}$$

In most groundwater, M may therefore be used as an approximation for *m*.

Chemists also use the unit mole fraction for water and for the partial pressure of a gas. Partial pressure is recorded in atmospheres. Mole fraction (N) is the ratio of the moles of a particular constituent relative to the total moles of all constituents. Mole fraction may be calculated for solid and liquid substances by the following relationship:

$$N = \frac{\text{moles}^1}{\text{moles}_1\ \text{moles}_2 + \cdots \text{moles}_x} \tag{1.8}$$

where the superscript 1 is the sum of the total moles and the subscripts are the moles of each species.

In most groundwater investigation we are more concerned with the reacting equivalence between ionic species than we are with laboratory results reported in mass per unit volume (i.e., mg/L). Chemical equivalence or equivalence per liter (eq/L) is the number of moles of solute in a liter of solution multiplied times the valence of the ion. Moles of a species and eq/L may be determined as follows:

$$\text{moles} = \frac{\text{g solute}}{\text{gfw of species}} \tag{1.9}$$

$$\text{mmoles} = \frac{\text{mg solute}}{\text{gfw of species}} \tag{1.10}$$

$$\frac{\text{eq}}{\text{L}} = \frac{\text{moles} \times \text{valance}}{\text{L}} \tag{1.11}$$

where valance is the charge on the species.

Most groundwater only contains fractions of a mole of each ionic species in a liter of water. Therefore, in hydrogeology chemical equivalence is usually reported in milligram equivalents per liter (milliequivalents/L or meq/L). Meq/L may be calculated as follows:

$$\frac{\text{meq}}{\text{L}} = \frac{\text{mmoles} \times \text{valance}}{\text{L}} \tag{1.12}$$

where mmoles are the milligrams of solute per gfw. When ionic analyses are expressed in meq/L, the concentrations of the cations and anions associated with chemical reactions are chemically equivalent. What this means is that 1 meq/L of Ca^{2+} will react with 1 meq/L of SO_4^{2-}, forming

1 mmole of $CaSO_4$, and 1 mmole of $CaSO_4$ will dissociate into 1 meq/L of Ca^{2+} and 1 meq/L of SO_4^{2-}. Conversion factors for converting chemical analyses reported in mg/L to meq L and to mmoles L are listed in Appendix A. Less commonly, chemical equivalence may be reported as milligram-equivalents per kilogram of solution (epm). Not all solutes are amenable to the meq/L convention. For example, laboratory analyses do not always report the valence for multivalent species, and silica is reported as having a charge of zero. It is common practice to report such species in mg/L. Mg/L conversion equivalents for 1 meq/L for major ions are shown in Table 1.4. From the conversion equivalents it is apparent that heavy anion molecules such as HCO_3^- contribute disproportionately to total dissolved solids (TDS) relative to their reacting cation counterparts, such as Ca^{2+} .

Chemical equivalency is one of the great tools available to hydrogeologists when trying to sort out groundwater flow and rock–water interactions. The advantage of using meq/L units over mg/L units when analyzing water chemistry data is apparent in Table 1.5, where the results of four laboratory analyses reported in mg/L have been converted to meq/L. The mg/L results provide little insight into possible rock–water interactions and possible groundwater flow paths other than that there appears to be lots of bicarbonate and varying amounts of cations and other anions. A brief interrogation of the Table 1.5 meq/L conversions provides a quick and easy starting point for subsequent analysis of the potential rock–water interactions. In Water 1, the Na^+ and Cl^- are equal, suggesting halite dissolution, and the Ca^{2+} plus Mg^{2+} are equal to the HCO_3^- , suggesting dolomite dissolution. Water 2 is little more complex in that the Cl^- does not offset the Na^+, and the Mg^{2+} is four times as great as the Ca^{2+}. A plausible explanation is that groundwater encountered dolomite and then underwent Ca-Na ion exchange by contact with clay minerals. At first glance, Water 3 appears to be confusing. The Ca^{2+} and HCO_3^- are equal suggesting calcite dissolution, but there are no common $MgSO_4$ minerals to offset the SO_4^{2-}. The SO_4^{2-} balances one half of the Ca^{2+}, suggesting gypsum dissolution, leaving dolomite dissolution to account for the sum of the remaining Ca^{2+} plus Mg^{2+} to offset the HCO_3^-. Water 4 is quite interesting in that no cation has an equivalent anion concentration. The Cl^- concentration is quite low, which makes ion exchange a possible source for the elevated Na^+. Assuming ion exchange, it is possible to add the 3 meq/L of Na^+ to the 3 meq/L of Ca^{2+}, resulting in 6 meq/L of Ca^{2+}, which offsets both the HCO_3^- and SO_4^{2-} for calcite and gypsum dissolution.

Important Takeaway – Always convert major ion laboratory results to reacting equivalents, and then perform a preliminary analysis of the data by considering common rock–water reactions.

TABLE 1.4
Mg/L Conversion Equivalents for Major Ions in meq/L

	meq/l	mg/l
Ca^{2+}	1	20.0
Mg^{2+}	1	12.2
Na^+	1	23.0
K^+	1	39.1
HCO_3^-	1	61.0
SO_4^{2-}	1	48.0
Cl^-	1	35.5

TABLE 1.5
Comparison of mg/L and meq/L in Laboratory Water Analysis for Four Waters

	Water 1		Water 2		Water 3		Water 4	
	mg/L	meq/L	mg/L	meq/L	mg/L	meq/L	mg/L	meq/L
Ca^{2+}	30.06	1.5	10.02	0.5	80.16	4	80.16	4.00
Mg^{2+}	18.23	1.5	24.30	2	24.30	2	1.22	0.10
Na^{+}	34.48	1.5	34.48	1.5	2.30	0.1	45.98	2.00
K^{+}	3.91	0.1	3.91	0.1	3.91	0.1	3.91	0.10
HCO_3^-	183.04	3	244.05	4	244.05	4	183.04	3.00
SO_4^{2-}	4.80	0.1	4.80	0.1	96.06	2	144.09	3.00
Cl^-	53.17	1.5	3.54	0.1	3.54	0.1	3.54	0.10

Example E1.1 Concentration conversion calculation

Problem: Calculate the conversion factors for converting SO_4^{2-} concentrations in mg/L, as determined by laboratory analysis into mmole/L and into meq/L.

Solution: From the periodic table, the gram formula weight (gfw) of S and O are 32 and 16 g, respectively, and the gfw of SO_4^{2-} is 96 g (32 g of S and 16 g of O). The factor for converting mg/L to mmole/L may be calculated by substituting the gfw of SO_4^{2-} in Equation 1.10 and adding liters to both sides of the equation:

$$\text{mmole } SO_4^{2-} = \frac{\text{mg } SO_4^{2-}}{96\text{g } SO_4^{2-}} \tag{E1.1a}$$

which may be rearranged into the form:

$$\text{mmole } SO_4^{2-} = \frac{0.0104 \text{ mg } SO_4^{2-}}{1} \tag{E1.1b}$$

The factor for converting mg /L into meq /L may be calculated by substituting Equation E1.1a into Equation1.12:

$$\text{meq/L} = \frac{\text{mg } SO_4^{2-}\,(2)}{1(96\text{ g})SO_4^{2-}} \tag{E1.1c}$$

which may be arranged into the form:

$$\text{meq/L} = \frac{0.0208 \text{ mg } SO_4^{2-}}{1} \tag{E1.1d}$$

when the valance of SO_4^{2-} is 2.

1.3 CHEMICAL BONDING

Soil zone and aquifer minerals, organic matter, gases, and other substances encountered by surface and groundwater are composed of elements that are chemically bonded together. The kind of chemical bond is of great importance in controlling the solubility of a substance with respect to water and other solvents. It is convenient to classify chemical bonds as ionic, covalent, or metallic. Many substances are held together by a combination of bond types.

Except for some noble gases, such as He, Ne, Ar, Kr, Xe, and Rn, all elements have an outer electron shell that is not full. Noble gases, also known as rare gases, are effectively nonreactive because of their full outer electron shell. Atoms are most stable when they attain the outer shell electron configuration of the noble gases. One of the easiest ways for atoms to acquire a full outermost electron shell is by either giving away or by accepting electrons from another atom by a process known as electron transfer. Ionic bonds form by the transfer of electrons from one species to another. The resulting molecule is held together by electrostatic or coulombic attraction. Because ionic bonds are held together by the weak force of electrostatic attraction between positively and negatively charged species, such as Na^+ and Cl^-, minimal energy is required to form or break such bonds. Ionic bonds are therefore commonly soluble in dipolar liquids such as water. Typical examples of ionically bonded soil zone and sedimentary aquifer minerals include most carbonate, sulfate, and chloride minerals.

Cations are species that give up one or more electrons and acquire a positive. Anions are species that receive one or more electrons and acquire a negative charge. The number of excess or deficient electrons is known as the ionic valence or charge of the ion. Most ions in groundwater are either monovalent or divalent, although some elements such as sulfur, iron, and chromium have two or more oxidation or valence states. For example, both sodium and chloride lose or gain one electron and are monovalent, whereas calcium and sulfate lose or gain two electrons and are divalent.

Covalent bonds share electrons to acquire the rare gas electron configuration. Gases such as O_2, N_2, and CH_4, organic substances, silicate minerals, and very importantly water molecules utilize this type of bond. Covalent bonds generally form uncharged species; however, some ionized species such as NH_4^+ are also bonded covalently. Each bond type behaves differently when in contact with water. Water's tendency toward hydrogen bonding and its dipolar character makes it a powerful solvent for electrovalent solids, liquids, and gases and also gives it unusual physical properties. The high dielectric constant (Dk) of water (78.25 at 25 °C), with positive and negative molecular ends allows it to readily break apart or ionize ionically bonded minerals such as calcite, gypsum, and halite. Aluminosilicate minerals are ubiquitous in sedimentary, igneous, and metamorphic terrains, but they are very insoluble in water because they are held together by covalent rather than by ionic bonds.

1.4 PHYSICAL AND CHEMICAL PROPERTIES OF WATER

1.4.1 The Water Molecule

The water molecule is formed by the covalent bonding of two oxygen and one hydrogen atoms. Hydrogen has two stable and one radioactive isotope, and oxygen has three stable isotopes (Table 1.6). There are 18 possible water molecule configurations and 12 ionic species configurations. The rarer isotopes ^2H, ^3H, ^{17}O, and ^{18}O usually combine with the common isotopes ^1H and ^{16}O; thus, hydrogen atoms in the water molecule have an average atomic weight of 1.00797 and oxygen atoms have an average atomic weight of 15.9994. Water molecules have an average atomic weight of 18.01534, although each molecular isotopic configuration has a specific atomic weight.

The two hydrogen atoms of the water molecule are asymmetrically and covalently bonded to the oxygen nucleus at an angle of 104.5°, forming a tetrahedron molecular structure with the oxygen atom in the center (Figure 1.1a). At any particular moment the two hydrogen atoms occupy two corners of the tetrahedron and two electron pairs occupy the other two corners. The four electrons of the two electron pairs are located as far away from the oxygen nucleus as possible. Four of the remaining six electrons of the molecule participate in the covalent bond, which holds the water molecule together, and the remaining two electrons remain close to the oxygen nucleus. This asymmetrical

TABLE 1.6
Average Terrestrial Abundance of Water Molecule and Other Isotopes of Hydrogeologic Interest

Element	Isotope	Terrestrial abundance (%)	
Hydrogen	^{1}H	99.9885	
	^{2}H	0.0115	
	^{3}H	10^{-14} to 10^{-16}	radioactive, $t_{1/2}$ = 12.32 years
Oxygen	^{16}O	99.757	
	^{17}O	0.038	
	^{18}O	0.205	
Helium	^{3}He	99.99986	
	^{4}He	0.000137	
Carbon	^{12}C	98.93	
	^{13}C	1.07	
	^{14}C	10^{-14} to 10^{-16}	radioactive, $t_{1/2}$ = 5730 years
Nitrogen	^{14}N	99.632	
	^{15}N	0.368	
Chlorine	^{35}Cl	75.78	
	^{36}Cl	10^{-13}	radioactive, $t_{1/2}$ = 301,300 years
	^{37}C	24.22	
Sulfur	^{32}S	94.93	
	^{33}S	0.76	
	^{34}S	4.29	
	^{36}S	0.02	

Source: Data from Fritz and Fontes, 1980; Rossman and Taylor, 1998

arrangement and the distance between the centers of gravities of the positive charges (hydrogen atoms) and the negative charges (two electron pairs) results in a dipolar electrical moment. What this means is that water molecules have a positive and negative end.

a) Covalently bonded water molecule
b) Hydrogen bonded water molecule

Water molecules are held together in groups by hydrogen bonding (Figure 1.1b). It is hydrogen bonding, where a positively charged hydrogen atom in one water molecule is attached to one of the unshared electrons in another water molecule, that gives water its unusual physical properties. In other words, it is an electrostatic bond. Hydrogen bonds are 10 to 50 times weaker than are the covalent bonds, which hold each individual water molecule together (Berner and Berner, 1987). The groups or clusters of water molecules consist of approximately 90 molecules at 0 °C and 25 molecules at 70 °C. Breaking the intermolecular bonds of water requires a great amount of energy, which explains its high surface tension (71.97×10^{-3} N m^{-1} at 25 °C) and latent heat of vaporization (2282.2 J g^{-1} at 1,013.24 mbar). The high surface tension, due to hydrogen bonding holding water molecules together, plays a major role in the formation of rain droplets. The heat of vaporization allows evaporating water to convey considerable amounts of heat without appreciably cooling the oceans and surface water bodies.

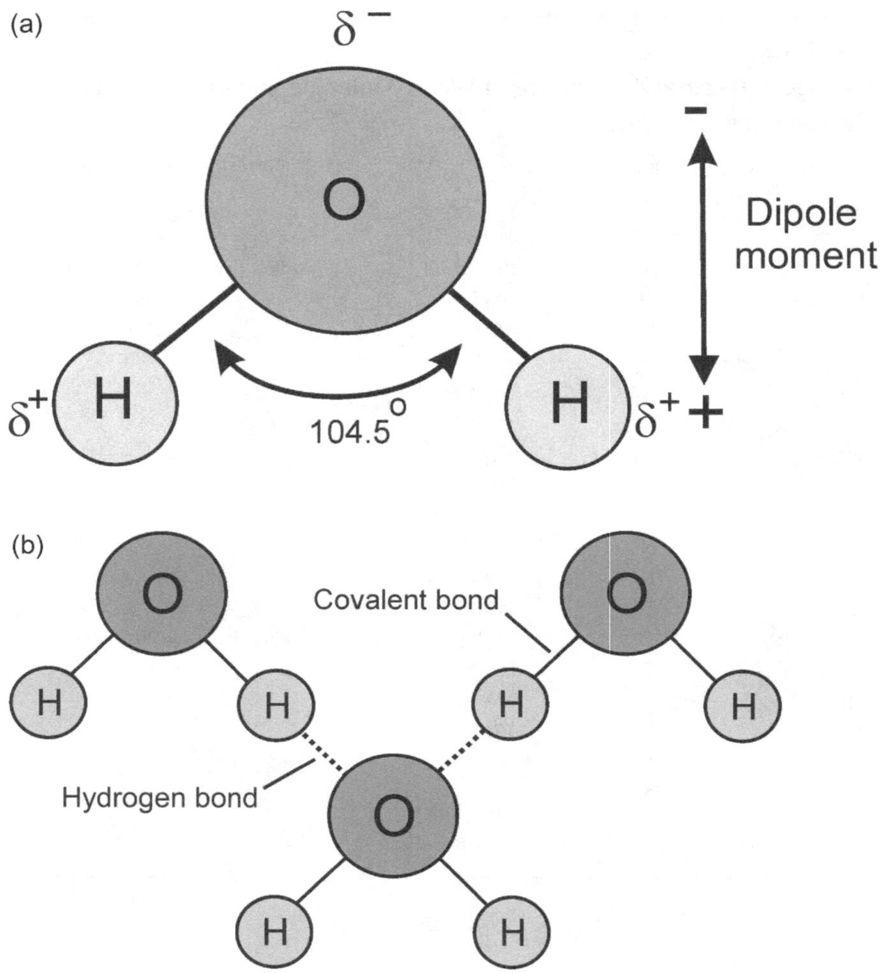

FIGURE 1.1 Models of water molecules. The molecule is held together by a) covalent bonding and b) weaker hydrogen bonding. The δ sign indicates a partial positive or negative charge.

Many of the important physical properties of water such as phase (i.e., gas, liquid, or solid), vapor pressure, density, viscosity, and compressibility are temperature dependent (Figure 1.2).

Physical properties greatly influence the movement of water, and temperature effects may have to be considered when analyzing groundwater migration patterns and pump testing results. Phase changes have a profound influence on the mobility of groundwater. For example, the formation of near-surface ice lenses may prevent groundwater recharge, and water vapor in geothermal systems may move under concentration and density gradients. Temperature-driven density gradients, particularly in high temperature geothermal systems, and salinity-driven density gradients may be the driving force determining the direction of groundwater flow.

Important Takeaway – The asymmetrical arrangement of water molecule atoms, the distance between the centers of gravities of the positive charges (hydrogen atoms) and the negative charges (two electron pairs), results in a dipolar molecule. This dipolar electrical moment with positive and negative ends allows water to readily break apart or ionize ionically bonded minerals such as calcite, gypsum, and halite.

Important Takeaway – Many of the physical properties of water are temperature dependent.

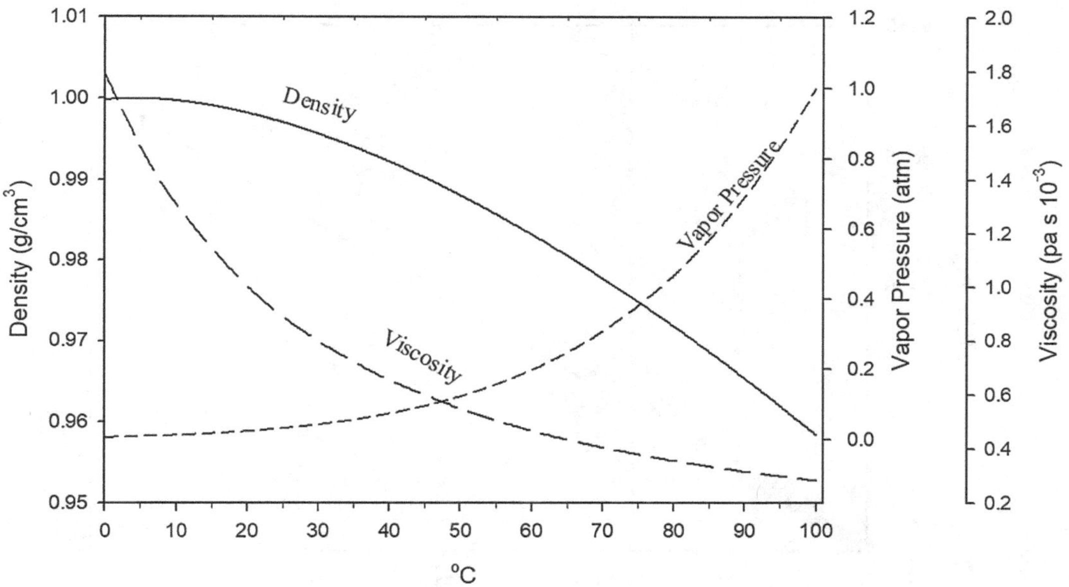

FIGURE 1.2 Temperature dependent characteristics of water. Modified after Matthess, 1982.

1.4.2 Water as an Electrolyte Solution

Surface and groundwater may contain a variety of dissolved and non-dissolved inorganic and organic substances. These substances include ionic and non-ionic species, colloidal material, gases, and organic compounds. Solutions are either true or colloidal. True solutions, such as groundwater, transmit light because they contain dissolved polar substances that have dissociated into ions and charged molecules. Distances of 10^{-6} to 10^{-8} cm separate these charged particles from each other. The dissolved substances, known as solutes, are either electrolytes or nonelectrolytes. Electrolytes, such as acids, bases, salts, and other dissociated ions and molecules, have either a positive or a negative charge. Inorganic electrolyte compounds are commonly called salts. Each positive and negative ionic charge has a corresponding electrical charge of 1.6016×10^{-19} coulombs.

The number of ionic charges equals the ionic valance of the species. For example, Na^+ and Cl^- have one electrical charge each, whereas Ca^{2+} and SO_4^{2-} have two electrical charges each. Nonelectrolytes are nonionic compounds such as sugar and urea. The electrical charge of electrolyte solutions permits the solution to conduct an electrical current, whereas nonelectrolyte solutions cannot. The electrolyte strength of a solution may be measured by the electrical conductivity of the solution. Electrical conductivity measurements, which may be rapidly made in the field, often provide a powerful and simple method to estimate the relative TDS of a surface or groundwater. There is a general linear relationship between conductivity and solute content:

$$S = KA \tag{1.13}$$

where S is the TDS (total dissolved solids), usually reported as mg/L, K is specific conductance in micromhos per cm (μmhos/cm) and A is a linear regression correlation factor. The A parameter generally has a value 0.55 to 0.75, although values as great as 0.98 have been reported (Hem, 1970). A plot of conductivity vs. TDS is shown in Figure 1.3.

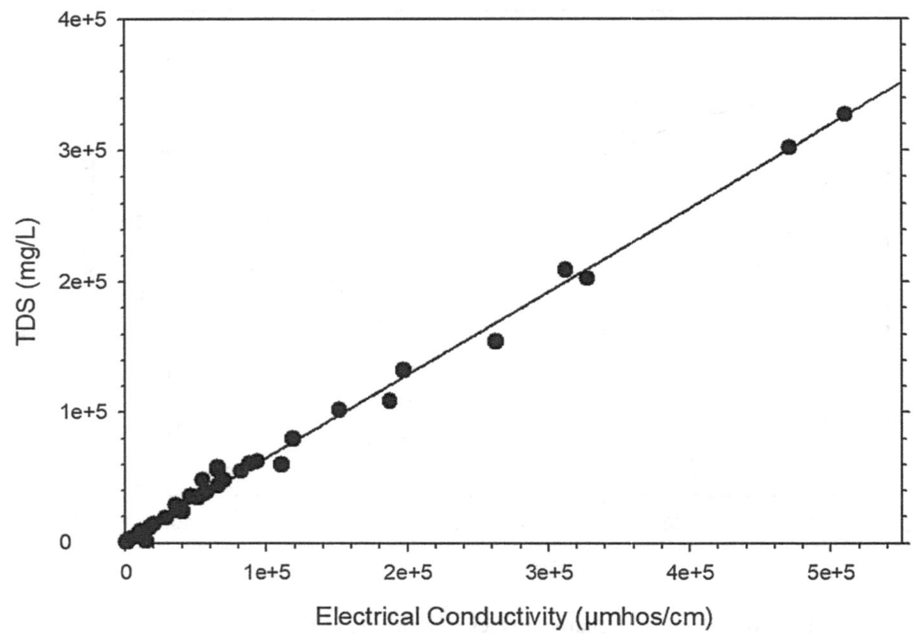

FIGURE 1.3 TDS vs. specific conductance of fresh to brine groundwater in Pilot Valley, Utah-Nevada. The calculated linear correlation A parameter is 0.63, $R^2 = 0.995$.

Electrolytes may be classified as either strong or weak. Strong means that the electrolyte has a large equilibrium constant, is highly soluble, and has a high electrical conductance at elevated concentrations. Strong electrolytes completely dissociate in a solvent. Weak electrolytes have a small equilibrium constant and a low electrical conductance.

Surface and groundwater contain both non-associated and associated electrolytes. Non-associated electrolytes only dissociate into simple ions and their hydrated forms. In other words, the dissociation products tend to remain as ionic species. An important distinction between simple ions and other molecular forms, such as ion pairs and ion complexes, is described in Section 1.4.3. Strong electrolytes in aqueous solutions include HCl, HBr, HNO_3, H_2SO_4, H_3PO_4, NaOH, and most soluble salts such as NaCl and $CaCO_3$. Associated electrolytes dissociate into a few simple ions, ion pairs, complex ions, and undissociated covalent molecules. They include most acids and bases, H_2S, and a few soluble salts. For example, carbon dioxide in water may occur as undissociated H_2CO_3 molecules, SO_4^{2-}, and CO_3^{2-} ions.

1.4.3 Speciation

Groundwater contains a variety of dissolved and non-dissolved inorganic and organic substances. These substances include simple ions, complex ions, uncharged molecules, colloidal material, gases, and organic compounds. Dissolved substances, regardless of phase, may be designated with the subscript (aq) for aqueous. Undissolved and exsolved gas may be designated with the subscript (g), and solid particles may be designated with the subscript (s). In most equations the phase is readily apparent and the subscripts are omitted. Neutral or uncharged inorganic substances that remain aqueous and do not precipitate are often distinguished from solids by the superscript o. The actual form that the substance takes is known as its species. For example, a few of the possible carbon species found in aqueous solutions include $CO_{2(g)}$, $CO_{2(aq)}$, $H_2CO_3^o$, HCO_3^-, CO_3^{2-}, $CaCO_3^o$, $CaCO_{3(s)}^o$, and organic compounds.

Dissolved substances in aqueous solutions can occur as elemental, free ionic, ion pair, inorganic and organic complex, neutral molecular, colloidal, and sorbed forms. In most groundwater almost all of the mass of aqueous species are ionic. An ion is any charged particle. In the strict sense, a simple ion is the charged form of a single element and an ion pair is the charged form of the combination of two more elements. Following these definitions, examples of simple ions include Ca^{2+}, Mg^{2+} and Cl^-, and examples of ion pairs include SO_4^{2-} and HC. The strength of the ion pair bonding ranges from week electrostatic attraction, where the cation and anion are solvent separated, to bonds that have covalent-like characteristics, such as SO_4^{2-}.

In the general sense an aqueous complex is the chemical combination of two or more different atoms in a solution. A complex may be an ionic or neutral species. According to this definition, common ion pairs such as SO_4^{2-} and HCO_3^- are complexes. It is important here to make a distinction between common ion pairs, uncommon ion pairs, and complexes. For practical purposes, uncommon ion pairs and complexes may be distinguished from common ion pairs by the fact that uncommon ion pairs and complexes tend to dissociate, whereas simple ions and common ion pairs do not. In other words, simple ions and common ion pairs such as SO_4^{2-} and HCO_3^- are treated as thermodynamic entities, and ion pairs such as $CaSO_4^o$ and $NaCO_3^-$ are treated as a dissolved substance. There are also structural differences between complexes and ion pairs, which are discussed later in this chapter.

In this book the term ion refers to simple ions and common ion pairs; ion pair is restricted to uncommon inorganic, ionic molecules, and neutral species; and the term aqueous complex is reserved for some uncommon inorganic and organic ion pairs and neutral molecular species.

1.4.4 Salinity

The salinity of water is the total concentration of dissolved inorganic material and has been referred to as total solids, total dissolved solids (TDS), and dissolved solids. In hydrogeology TDS is the preferred acronym. It is often convenient and useful to classify water according to the following common salinity-reporting classifications (Table 1.7).

Groundwater in regions containing insoluble rock and high rainfall may have a TDS as small as 25 mg /L, whereas saturated sodium chloride hypersaline water at 25 °C will have a TDS of about 300,000 mg /L. Seawater has a TDS of about 35,000 mg /L. In groundwater investigations involving highly saline water it is often useful to consider TDS greater than 200,000 mg/L as hypersaline (Figure 1.4).

Two general approaches are used to determine TDS: 1) direct measurement of the weight of the evaporation residue of a filtered aliquot of water, and 2) computational procedures. Evaporation involves the drying an aliquot of water for 1 hour at a specified temperature and then weighing the dried residue. Computational methods include calculations using specific conductance

TABLE 1.7
Salinity Classification of Water

Classification	mg/L
Fresh	0 to 1,000
Brackish	1,000 to 10,000
Saline	10,000 to 100,000
Brine	100,000 to 200,000
Hypersaline	>200,000

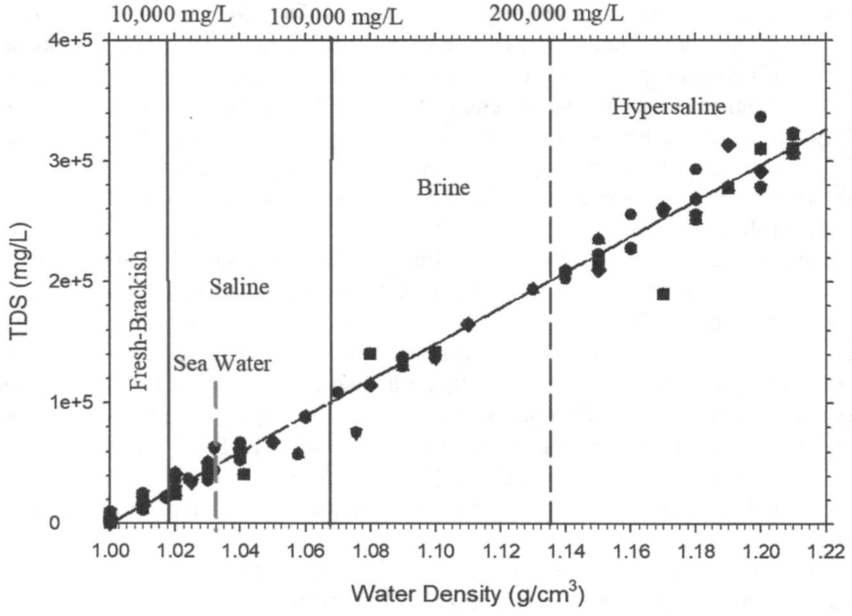

FIGURE 1.4 Classification of highly saline groundwater in Pilot Valley, Utah-Nevada, where brine has been subclassified as brine and hypersaline. Modified after Mayo et al., 2020.

measurements, specific gravity of the solution, and summation of the concentrations of laboratory-reported dissolved species. Results of TDS determinations often vary greatly depending upon the method used, and there is no chemically correct method. Fortunately, TDS is usually used as a gross indicator of water quality, and the precise knowledge of TDS is not necessary or required.

Most regulatory agencies require TDS to be determined by direct measurement; however, determining the exact concentration by evaporation is often difficult because 1) organic matter may be only partially volatile, 2) some inorganic anions such as nitrate and boron are partially volatile, 3) at 100 °C, half of the bicarbonate ions (HCO_3^-) are converted to $CO_{2(g)}$ and escape, and 3) some inorganic ions, such as Ca^{2+} and SO_4^{2-} which forms gypsum, may incorporate water in their crystal lattice during evaporation. The specific conductance method is very useful for waters where the relationships between salinity and conductivity have been previously established, and the specific gravity method is most useful in highly concentrated solutions. The accuracy of the summation method is dependent upon the completeness of the water quality analysis.

Important Takeaway – TDS is often a required analyte, but the results vary by analytical method, and TDS has limited use other than as a screening tool.

1.4.5 Density and Temperature

Because the molecular movement of most molecules increases with temperature, most substances are denser in the solid phase than in the liquid phase. Fresh water is unusual in that it does not follow this pattern (Figure 1.5). At zero degrees centigrade water undergoes a profound change in physical state. Below 0 °C water is ice, and above 0 °C water is liquid. At 0 °C the density of pure water changes from about 0.917 to about 0.9995 gm/cm³ at about 4 °C. Fresh water has a minimum density at 0 °C and a maximum density at 3.94 °C of about 1 gm/cm³. Above about 4 °C the density of water decreases. This behavior is due to the hydrogen bonding that holds the water molecules

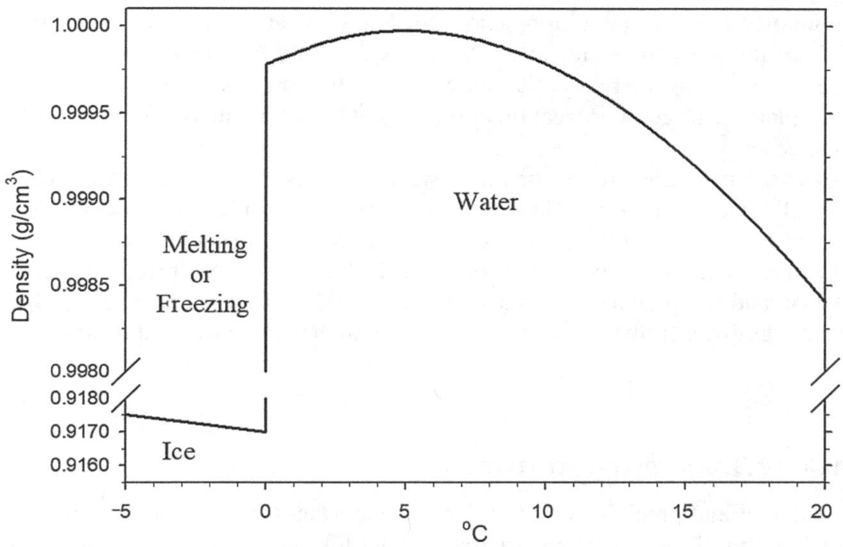

FIGURE 1.5 Density of water and ice as a function of temperature. Maximum density occurs at about 4 °C. See Figure 1.2 for the density–temperature relationship above 20 °C.

together. Frozen water has complete hydrogen bonding, and the structure of the linked molecules is open. When heated between 0 °C and 4 °C many of the hydrogen bonds break and bend, allowing water molecules to move closer together. Above about 4 °C water begins to behave like most other liquids in that the density decreases with increases in temperature. This normal behavior is due to the random motion of molecules that overshadow the effects of hydrogen bonding. The density characteristics of water at low temperature permit the freezing of surface water from the top down. The net result of freezing from the top down is that the floating ice layer in a lake or pond becomes a thermal blanket.

The density of water increases as the salinity increases. Seawater, with a salinity of about 35,000 mg /L and a density of 1.025, behaves as a normal liquid; however, at extremely elevated salinity groundwater behaves as a more viscous fluid. In most groundwater systems density does not need to be considered. In systems with elevated density, corrections are required for determining the freshwater equivalent hydraulic head when measuring groundwater levels in monitoring wells. Density also plays a critical role in the evaluation of the extent of sea water intrusion in coastal areas.

1.5 INORGANIC CHEMICAL REACTIONS

Most aqueous chemical reactions of hydrogeologic interest occur or near at the hydrosphere–lithosphere and hydrosphere–atmosphere interfaces. Hydrosphere–lithosphere processes include weathering, mineral dissolution and precipitation, and ion exchange. Hydrosphere–atmosphere processes include evaporation of water and volatile organic compounds and the absorption of atmospheric, soil zone, and other gases. A few aquatic chemical reactions such as ion complexation and dissolution of soluble organic compounds occur at liquid–liquid interfaces. Inorganic aquatic chemical reactions may be broadly classified as

1. mineral dissolution–precipitation reactions,
2. mineral surface reactions, and
3. aqueous complexation reactions.

Mineral dissolution and precipitation reactions are transformations from the aqueous to the solid phase and from the solid to the aqueous phase, respectively. Aqueous phases include ionic and dissolved neutral species, whereas solid phases are neutral species. Dissolution and precipitation entail the complete destruction or creation of the crystal lattice or amorphous structure of solid and semisolid molecules.

Mineral surface reactions are responses between aqueous–solid phases that occur only at the surface of a solid phase. These reactions result in chemical modifications to existing solid phases and include adsorption, desorption, and exchange of ions and other solutes on solid surfaces such as crystal faces and clay particles. Complexation reactions are aqueous reactions that do not entail a phase change, and the products and reactants can include both ionic and neutral species. Complexation reactions differ from normal mineral precipitation reactions in that neutral species remain aqueous.

1.5.1 Mineral Dissolution–Precipitation

Mineral dissolution and precipitation are the most important types of chemical reactions in most groundwater systems. These reactions usually account for the greatest exchange of mass between solid and liquid phases, and they are the easiest to quantify. Dissolution and precipitation reactions occur because they improve the stability of the outer shell electrons of the species involved. These dissolution and precipitation reactions belong to two general classes: acid–base and reduction–oxidation.

Acid–base reactions, broadly defined as proton transfer reactions, are not accompanied by changes in the oxidation state of either the products or reactants. Acid–base reactions are responsible for the regulation of the pH of most groundwater and include the congruent and incongruent dissolution of most common aquifer minerals. During congruent dissolution all of the products are aqueous species, whereas during incongruent dissolution one or more of the products are a solid or colloidal species. Incongruent weathering is a primary factor in soil formation and the abundance of clays in many geologic environments.

Reduction–oxidation reactions, known as redox, are a second great class of chemical reactions. Redox reactions differ from acid–base reactions in that they involve the transfer of negative charges (electrons) between reactants and products. The net result of redox is a change in the oxidization state of both reactants and products.

1.5.2 Acid–Base Reactions

The concept of acid–base reactions can be confusing because acids and bases may be defined in two different ways. In the simplest of terms, we should think of acid–base reactions as proton transfers. Acids and bases may be either of the Bronsted-Lowry or Lewis types. Bronsted-Lowry acids are acids and bases as commonly envisioned. When ionized, this class of acids yields hydrogen ions (H^+), and the bases yield hydronium (OH^-) ions. Lewis acids are more encompassing in that they include all species that, when ionized, serve the same chemical function as Bronsted-Lowry acids.

In Bronsted-Lowry type acid–base reactions, proton donors are acids, proton acceptors are bases, and a proton donor–acceptor pair is known as a conjugate acid–base pair. According to the Bronsted-Lowry theory the electrical field around bare protons (i.e., H^+ ions) is so strong that they cannot exist in a solvent, and they must complex with molecules of the solvent. In water H^+ reacts with water molecules forming hydronium ions:

$$H^+ + H_2O = H_3O^+ \tag{1.14}$$

Hydronium ion (H_3O^+) formation is pervasive in aqueous system and is usually omitted in written equations. H^+ is considered to be synonymous with H_3O^+ in most reactions. For example, the dissolution of calcite, which is usually dependent on the presence of H_3O^+, is correctly represented as:

$$CaCO_3 + H_3O^+ = Ca^{2+} + HCO_3^- + H_2O \tag{1.15}$$

but is commonly represented as:

$$CaCO_3 + H^+ = Ca^{2+} + HCO_3^- \tag{1.16}$$

Water can be omitted from Equation 1.15 because it is present as a reactant and as a product in equal proportions.

Because many compounds such as $CaSO_4$ and NaCl dissociate and precipitate independently of hydronium and hydroxide ions (i.e., they are pH independent), it is sometimes convenient to think of acid and bases in terms of the Lewis concept. According to the Lewis view, an acid is any species that acts as an electron-pair acceptor (i.e., any cation), and a base is any species that acts as an electron-pair donor (i.e., any anion). It is important to notice that the Lewis definition of acids and bases includes all Bronsted-Lowry acids and bases. The advantage of using the Lewis definition is that it allows non-hydrogen-bearing substances, such as Ca^{2+}, which perform the same function as a Bronsted-Lowry acid, to be considered as acids.

Although the Lewis concept is meaningful and allows us to classify many precipitation and dissolution reactions, unless otherwise stated, acids and bases are herein discussed in terms of the Bronsted-Lowry concept. Adoption of the Bronsted concept is meaningful in the context of natural water systems because most hydrogeologists, engineers, and non-scientists have an intuitive idea of acids and bases that is consistent with the Bronsted-Lowry theory.

Acids and bases may be either strong or weak. The concepts of strong and weak are used in a fashion similar to their usage regarding electrolytes. Strong acids and bases have large equilibrium constants and tend to completely ionize in solution. In other words, in water they usually are reaction products. Weak acids and bases have smaller equilibrium constants and only partially ionize in solution. Many laboratory procedures utilize strong acids such as hydrochloric (HCl) and nitric (HNO_3) and the strong base sodium hydroxide (NaOH). Strong acids, such as hydrosulfuric (H_2S), sulfuric (H_2SO_4), and hydrochloric (HCl), are commonly associated with acid mine drainage, acid rain, and volcanic and magmatic emanations.

The majority of acids and bases in surface and groundwater systems are not highly ionized. Common weak acids in natural waters include carbonic (H_4CO_3), siliceous (H_4SiO_3), phosphoric (H_3PO_4), boric ($B(OH)_3$), and humic. The equilibrium constants of many acids and their conjugate bases are listed in Table 1.8. Equilibrium constants are described in Chapter 2. Strong acids have weak conjugate bases, and strong bases have week conjugate acids. Because water will dissociate to a limited extent, it is also a weak acid.

The dissociation of water may be expressed as:

$$H_2O = H^+ + OH^- \tag{1.17}$$

Expressing Equation 1.17 in terms of the law of mass action and thermodynamic equilibrium, we obtain:

$$k_w = \frac{H^+OH^-}{H_2O} \tag{1.18}$$

TABLE 1.8
Equilibrium Constants of Some Acids and Conjugate Bases of Hydrogeochemical Interest

Acid (–log $K_a = pK_a$)		p*K*a	Conjugate base (–log $K_b = pK_b$)		pK_b
perchloric acid	$HClO_4$	–7	perchlorate	ClO_4^-	21
hydrochloric acid	HCl	–3	chloride ion	Cl^-	17
sulfuric acid	H_2SO_4	–2.8	hydrogen sulfate	HSO_4^-	17
nitric acid	HNO_3	–1.3	nitrate ion	NO_3^-	15
hydronium ion	H_3O^+	0	water	H_2O	14
bisulfate	HSO_4^-	1.9	sulfate ion	SO_4^{2-}	12.1
Phosphoric acid	H_3PO_4	2.2	dihydrogen phosphate ion	$H_2PO_4^-$	11.9
hydrofluoric acid	HF	3.2	fluoride ion	F^-	10.8
acetic acid	CH_3CO_2H	4.8	acetate ion	CH_3COO^-	9.3
carbonic acid	H_2CO_2	6.3	bicarbonate	HCO_3^-	7.7
hydrosulfuric	H_2S	7.1	hydrogen sulfide ion	HS^-	6.9
hydrogen cyanide	HCN	9.2	cyanide ion	CN^-	4.7
boric acid	H_2BO_3	9.7	borate ion	$B(OH)_4^-$	4.7
bicarbonate	HCO_3^-	10.3	carbonate	CO_3^{2-}	3.7
hydrogen peroxide	H_2O_2	11.8	peroxide	HO_2^-	2.3
orthosilicic acid	H_4SIO_4	9.8	trihydrogen silicate	$SiO(OH)_3^-$	4.5
water	H_2O_2	14	hydroxide	OH^-	0
ammonium ion	NH_4^+	23	ammonia	NH_3	4.7
ammonia	NH_3	~23	amide ion	NH_2^-	-10
methane	CH_4	~50	methane ion	CH_3^-	-20

Source: Data from Stumm and Morgan, 1981.

where K_w is the equilibrium constant of water at STP. Except for highly saline water, the activity of water is nearly equal to 1. The activity of water decreases to 0.98 in a 3 percent NaCl solution (i.e., seawater) and to 0.84 in a 20 percent NaCl solution (Freese and Cherry, 1979). Pressure has a negligible effect on the activity of water in most groundwater. At STP the activity and fugacity of water are 1 and 0.03125, respectively. The pressure on groundwater systems rarely exceeds 100 bar, where the activity of water at 25 °C is 1.0757. The concepts of activity and fugacity are discussed in detail in Chapters 2 and 3. Briefly, activity is thermodynamic concept that describes the effective concentration of a species in a mixture, and fugacity is the effective partial pressure of a gas in a mixture of gases.

Because the activity of water is approximately 1 for most groundwater, Equation 1.18 may be written as:

$$k_w = [H^+][OH^-] \quad (1.19)$$

H^+ is a master variable in acid–base equilibria relationships. Introduction or consumption of H^+ in one chemical reaction in a solution will influence the concentrations of products and reactants of all other acid–base reactions in the same solution. The concentration of this master variable is designated by the parameter pH, which is defined as the negative log of the hydrogen ion concentration (–log [H^+]). From Equation 1.19, it is apparent that in neutral solutions [H^+] = [OH^-] at 25 °C K_w = 1×10^{-14}, and the pH of a neutral solution is therefore 7. Neutral solutions have a pH >~7 at lower temperatures and a pH <~7 at higher temperatures. Values of K_w and pH for water temperatures between 0 and 100 °C are listed in Table 1.9.

TABLE 1.9
K_w and Neutral pH of Water at Various Temperatures

°C	Kw × 10^{-14}	Neutral pH
0	0.1139	7.47
5	0.1846	7.37
10	0.2920	7.27
15	0.4505	7.17
20	0.6809	7.08
25	1.008	7.00
30	1.469	6.92
35	2.089	6.84
40	2.919	6.77
45	4.018	6.70
50	5.474	6.63
100	51.3	6.15

Source: Data from Garrels and Christ, 1965; Shabunya et al., 2019.

Because the H^+ and OH^- ions are equal at equilibrium, a solution that only contains water cannot have a pH greater than 7 at STP. Basic solutions require the dissociation of other species in order to consume H^+ ions and to produce a surplus of OH^- ions.

Example E1.2 Calculate the pH of neutral of water.

Problem: Calculate the pH of ion free water at 30 °C.
Solution: From Equation 1.19 and Table 1.9, the equilibrium point of water at 30 °C is:

$$K_w = 1.469 \times 10^{-14} = [H^+][OH^-] \tag{E1.2a}$$

$$[H^+] = [OH^-] = 1.212 \times 10^{-7} \tag{E1.2b}$$

$$-\log [H^+] = 6.92 \tag{E1.2c}$$

therefore

$$pH = 6.92 \text{ at neutrality} \tag{E1.2d}$$

Due to the abundance of $CO_{2(g)}$ in the soil zone and the abundance of carbonate minerals in most soil and in many aquifers, carbonic acid (H_2CO_3) is the most plentiful and important acid in groundwater. The dissociation of carbonic acid is the principal factor regulating the pH of most groundwater. Except for unusual conditions, the contribution of other acids to the solute chemistry and the pH of groundwater is minor. Typical concentrations of acids in fresh and seawater are listed in Table 1.10.

Carbonic acid has more than one hydrogen ion dissociation, and it belongs to a class of acids known as polyprotic acids. H_2CO_3 dissociates into H^+ and HCO_3^- that can in turn dissociate into H^+ and CO_3^{2-}. The formation and dissociation of carbonic acid and its relationship to carbonate minerals are described in Chapter 6. Other polyprotic acids of interest in groundwater investigations include siliceous, phosphatic, and hydrosulfuric.

TABLE 1.10
Mean Concentrations of Weak Acids and Bases in Natural Water

	Freshwater	Seawater		
		Warm surface	Deep Atlantic	Deep Pacific
Carbonate	0.96 mM	2.1 mM	2.3 mM	2.4 mM
Silicate	220 μM	220 μM	<3 μM	150 μM
Ammonia	0–10 μM	<0.5 μM	<0.5 μM	<0.5 μM
Phosphate	220 μM	<0.2 μM	0.4 mM	0.4 mM
Borate	<0.1 μM	0.4 mM	0.4 mM	0.4 mM
	Typical Anoxic Hypolimnion	**Black Sea (Deep Water)**	**Cariaco Trench**	
Sulfide	50–150 μM	330 μM	20 μM	
Ammonia	10–40 μM	53μM	10 μM	

Source: Modified after Morel, 1983.

Interactions between H^+ and OH^- and the speciation of H^+ and OH^- donors and acceptors are complex and very important. The speciation of several groundwater acids relative to pH is graphically illustrated in Figure 1.6. Ignoring the effects of water dissociation, the pH of water containing only a single acid will be regulated by the equilibrium constant of that acid. However, such simple aqueous systems seldom naturally occur. In most groundwater, where the concentration of carbonic acid greatly exceeds the concentrations of other acids, the carbonic acid concentration regulates the pH of the most water and controls the speciation of lessor occurring acids. In systems where the quantity of a second acid is also great, both acids contribute H^+, and the two acids may mutually affect the speciation of each other.

1.5.3 Hydrolysis

Hydrolysis is an important type of acid–base dissolution reaction. Hydrolysis differs from other acid–base reactions in that during hydrolysis, water is a reactant. When water dissociates into H^+ and OH^-, these ions either go into solution or are incorporated into the reaction products. Water must therefore be written into equations representing hydrolysis. Recall that water may be omitted from reaction equations representing most other types of reactions because water is not altered during the reaction.

The very slow dissolution of relatively insoluble silicate and aluminosilicate minerals is accomplished by hydrolysis. Silicate and aluminosilicate minerals constitute about 58 percent of the earth's crust. Granitic rocks are almost entirely composed of silicate and aluminosilicate minerals, and these minerals are important constituents in many metamorphic assemblages. Because most aluminosilicate minerals dissolve incongruently (Chapter 2) the solid phase reaction products constitute the bulk of many soils and of clastic sedimentary rocks. The dissolution of albite is an example of incongruent hydrolysis:

$$\underset{\text{(albite)}}{2NaAlSi_3O_8} + 2H^+ + 9H_2O = \underset{\text{(kaolinite)}}{Al_2Si_2O_5(OH)_4} + 4H_4SiO_4 + 2Na^+ \tag{1.20}$$

The presence of H^+ as a reactant means that albite dissolution is pH dependent. Because soil zone carbonic acid is the most common source of H^+ in the shallow subsurface, the bulk of pH-dependent hydrolysis weathering reactions take place in the soil zone and bedrock located immediately below the soil zone, where the decay of organic matters yields abundant $CO_{2(g)}$. Soil zone $CO_{2(g)}$ is responsible for the thick weathered horizons often found between the soil veneer and unweather granitoid

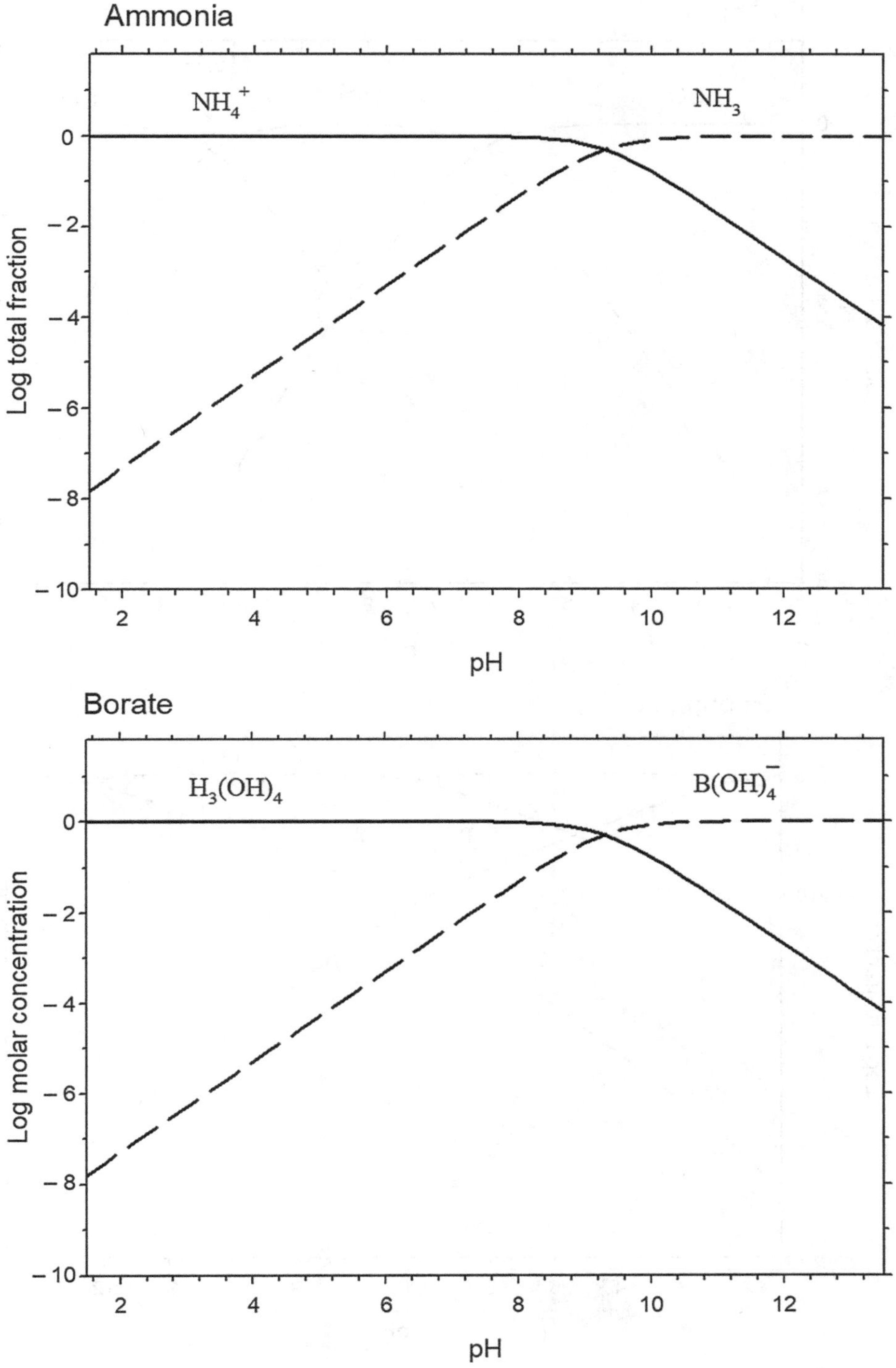

FIGURE 1.6 Concentration vs. pH of weak acids that commonly occur in natural waters.

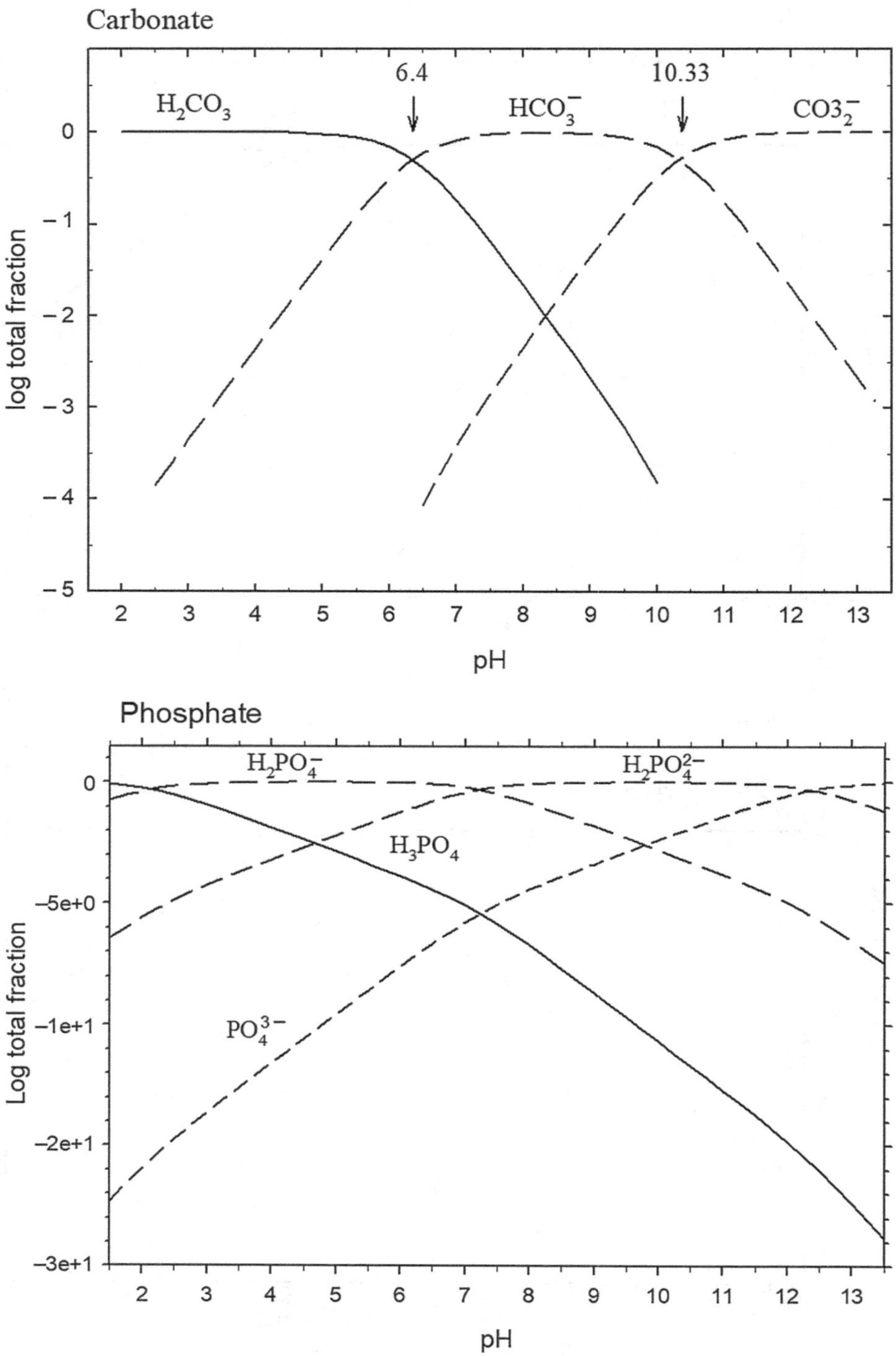

FIGURE 1.6 (Continued)

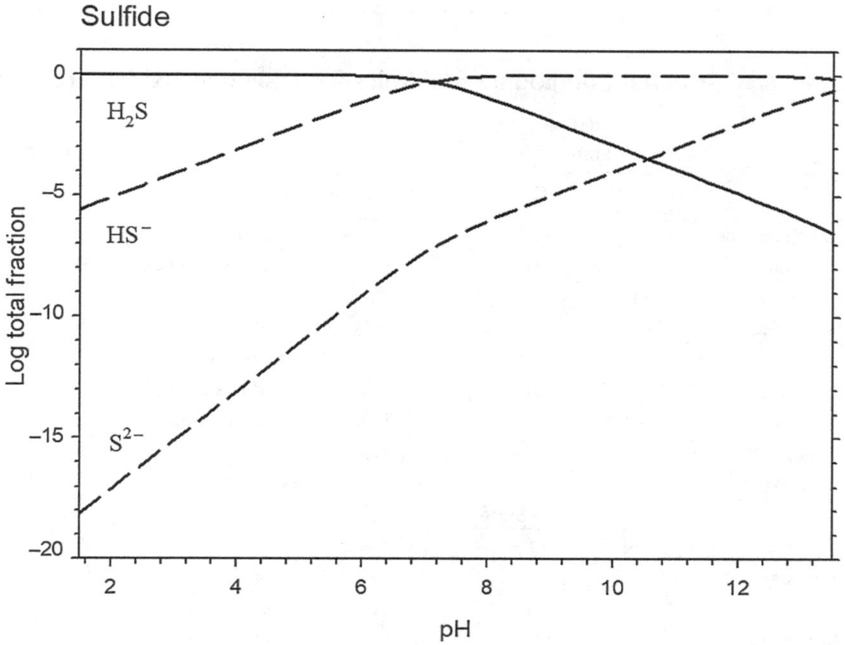

FIGURE 1.6 (Continued)

and metamorphic bedrock. The abundance of soil zone $CO_{2(g)}$ is readily apparent by comparing the partial pressures of $CO_{2(g)}$ in the atmosphere and typical soils zone. Typical CO_2 partial pressures $\left(P_{CO_2}\right)$ in these two environments are $10^{-3.5}$ and 10^{-2} atm, respectively.

1.5.4 Reduction–Oxidation (Redox)

Recall that redox is accomplished by the transfer of electrons and results in oxidation state changes of both reactants and products. The oxidization state is the theoretical charge a species would have if it were to dissociate. Oxidation states are represented by Roman numerals, whereas Arabic numbers represent valance states. Elements that participate in redox (O, C, N, S, Fe, Mn, Cr, U, etc.) have multiple oxidation states. Oxidation states of several substances of importance to groundwater are listed in Table 1.11.

Rules for assigning oxidation states.

1. The oxidation state of free elements, whether atomic or molecular is zero.
2. The oxidation state of an element in its simple ionic form is equal to the valance of the species. For example, Fe^{2+} has an oxidation state of +II.
3. The sum of the oxidation sates for molecules is zero and is equal to the charge of the charge for ions.
4. The oxidation state for oxygen in an oxygen-bearing compound is –II, except for O_2 and O_3, where it is 0. In peroxides such as H_2O_2 it is –I.
5. The oxidation sate of hydrogen is +I, except for H_2 and compounds where it is combined with a less electronegative element such as CaH_2.

TABLE 1.11
Oxidation States of Selected Compounds and Rules for Assigning Oxidation States

Species		Oxidation State C	Oxidation State O
CO_2	carbon dioxide	+IV	–II
HCO_3^-	bicarbonate	+IV	–II
CO_3^{2-}	carbonate	+IV	–II
CH_2O	Formaldehyde	0	–II
$C_6H_{12}O_6$	sugar	0	–II
CH_3OH	methanol	–II	–II
CH_4	methane	–IV	
H_2O_2	hydrogen peroxide		–II

Species		Oxidation State S	O	Fe
S	sulfur	0		
H_2S	hdyrogen sulfide	–II		
SO_3^{2-}	sulfite	+IV	–II	
SO_4^{2-}	sulfate	+VI	–II	
$S_2O_3^{2-}$	thiosulfate	+II	–II	
HS^-	hydrogen sulfide	–II		
FeS^2	iron disulfide	–I		+II
FeS	iron sulfate	–II		+II

Nitrogen Compounds

Species		Oxidation State N	O	C
N	nitrogen	0		
NO_3^-	nitrate	+V	–II	
NO_2^-	nitrogen dioxide	+III	–II	
NH_4^-	ammonium	+III		
HCN	hydrogen cyanide	–III		+II
SCN^-	thiocyanate	–III		+IV

Iron Compounds

Species		Oxidation State Fe	O	C
Fe	iron	0		
FeO	iron (II) oxide	+II	–II	
$Fe(OH)_2$	iron hydroxide	+II	–II	
$FeCO_3$	iron carbonate	+II	–II	+IV
FeO_3	feric (III) oxide	+VI	–II	
$Fe(OH)_3$	iron (III) oxide-hydroxide	+III	–II	
FeOOH	geothite	+III	–II	

Data from Stumm and Morgan, 1981, and Freeze and Cherry, 1979

Redox reactions differ from acid–base reactions in three basic ways:

1. H^+, either as hydronium ions or in some other hydrated form, exist in solution, whereas free electrons (e^-) do not. Free electrons can be represented as e^- or as e.
2. Once initiated, acid–base reactions are typically fast, whereas redox reactions are relatively slow and are usually mediated by microorganisms. Microorganisms act as catalysts that speed up the reactions.
3. Relative to acid–base equilibrium constants, equilibrium constants for redox reactions are often extremely large or small due to the great energy required for the reactions. The large differences in energies are primarily due to the fact that most acids in groundwater are weak, and the proton transfer reactions only involve one or two protons, whereas redox reactions often involve the transfer of several electrons.

Oxidation, originally defined as the chemical combination of a substance with oxygen, has taken on a broader meaning with the advent of modern atomic theory. Oxidation is now defined as the loss of electrons (i.e., the oxidation state becomes more positive) and reduction is defined as the gain of electrons (i.e., the oxidation state becomes more negative). It is significant to note that this behavior is fundamentally different from the behavior of acids and bases. During redox, electrons are physically removed from one ionic or molecular species and are conveyed to another ionic or molecular species. An oxidant is a substance that causes oxidation to occur while it is simultaneously reduced. Because free electrons cannot exist in a solution, oxidation and reduction must occur concurrently.

It is conventional to write an overall redox reaction as two half-reactions in the form

$$\text{oxidized state} + ne = \text{reduced state} \tag{1.21}$$

where *ne* is the number of electrons.

Let us consider the oxidization of ferrous iron:

$$4Fe^{+2} + O_2 + 4H^+ = 4Fe^{+3} + 2H_2O \tag{1.22}$$

where Fe^{2+} is oxidized and O_2 is reduced. The oxidization of iron, like many other redox reactions that are accompanied by the transfer of protons, is pH dependent. The relationship between pH and redox reactions is discussed in Chapter 12. The two half-reactions for Equation 1.22 are:

$$O_2 + 4H^+ + 4e = 2H_2O \qquad \text{(reduction)}$$

$$\frac{4Fe^{2+} = 4Fe^{3+} + 4e}{O_2 + 4H^+ + 4Fe^{+2} = 4Fe^{+3} + 2H_2O} \qquad \text{(oxidation)}$$

where e is transferred electrons. The fundamental electron changes accompany Equation 1.22 are the oxidation of ferrous iron to ferric iron:

$$Fe^{2+} = Fe^{3+} + e$$

and the reduction of molecular oxygen:

$$O_2 + e = O_2^{2-}$$

In hydrogeology we are interested in redox reactions because they can profoundly affect groundwater systems and the environments they interact with in three important ways (Table 1.12). Oxygen rich groundwater oxidizes sulfur, iron, nitrogen, and manganese ions and minerals. In organic-rich environments, organic matter is consumed by O_{2g} oxidation and is involved in denitrification, methane fermentation, Mn(IV), Fe(III), and sulfate reduction, and methane fermentation. Redox reactions are also responsible for the mobilization of many trace metals.

Many important half-reactions are listed in Table 1.13. From the table it is apparent that water participates in redox in four possible ways. Water may be:

1. an inert solvent, and it does not appear in the redox equation,
2. oxidized as in reaction 1,
3. reduced as in reaction 3, and
4. an acid or base and appears as H^+ or OH^- on one side of the equation.

In most groundwater systems the supply of reducing agents greatly exceeds the supply of oxidizing agents. Important oxidizing agents in groundwater are dissolved oxygen, nitrate, sulfate, and sometimes water. The primary source of dissolved oxygen and nitrate is recharge water that acquires O_2 and NO_3^- as it passes through the atmosphere and the soil zone. Once these atmospheric and soil gases are consumed, additional redox will not occur unless the groundwater encounters

TABLE 1.12
Some Redox Processes of Importance in Groundwater Systems

	Inorganic process that consumes oxygen	
Iron oxidation	$4Fe^{2+} + O_2 + 4H^+ = 4Fe^{3+} + 2H_2O$	
Iron sulfide oxidation (step 2)	$FeS_2 + \frac{7}{2}O_2 + H_2O = Fe_2 + +2SO_4^{2-} + 2H^+$	
Iron sulfide oxidation (step 3)	$Fe^{2+} + H_2O = Fe(OH)_{2(s)} + 2H^+$	
Nitrification step 1 (oxidization of ammonium to nitrite)	$2NH_4^+ + 3O_2 = 2NO_2^- + 4H^+ + 2H_2O$	
Nitrification step 2 (oxidization of nitrite to nitrate)	$2NO_{2^-} + O2 = 2NO_3^-$	
Manganese oxidation	$2MN^{2+} + 4OH^- + O_2 = 2MNO_{2(s)} + 2H_2O$	
Sulfide oxidation	$2O_2 + HS^- = H^+ + SO_4^{2-}$	
	Organic process in systems containing O_2, NO_3^-, SO_4^{2-} and CO_2	**ΔG (kJ at pH = 7)**
Aerobic respiration	$CH_2O + O_{2(g)} = CO_{2(g)} + H_2O$	–502.4
Denitrification	$5CH_2O + 4NO_3^- + 4H^+ = 7H_2O + 2N_{2(g)} + 5CO_{2(g)}$	–476.9
Mn(IV) reduction	$CH_2O + 3CO_2 + 3MNO_{3(s)} = 2MN^{2+} + 3H_2O + CO_{2(g)}$	–340.4
Fe(III) reduction	$CH_2O + 4Fe(OH)_{3(s)} + 8H^+ = 4Fe^{2+} + CO_{2(g)} + H_2O$	–116
Sulfate reduction	$2CH_2O + SO_4^{2-} + H^+ = HS^- + H_2O + CO_{2(g)}$	–104.7
Methanogenic	$2CH_2O = CH_4 + CO_2$	–92.9
Nitrogen fixation	$CH_2O + H_2O + \frac{1}{2}N_{2(g)} + \frac{4}{3}H^+ = \frac{4}{3}NH_4^+ + CO_{2(g)}$	–80.4

Mn^{2+} can oxidize to MnO_2, Mn_2O_3, Mn_3O_4 and other species.
Organic matter is represented as CH_2O

TABLE 1.13
Redox Half-Reactions for Some Constituents Found in Groundwater Environments

1) $\frac{1}{4}O_2 + H^+ + e = \frac{1}{2}H_2O$
2) $H^+ + e = \frac{1}{2}H_{2(g)}$
3) $H_2O + e = \frac{1}{2}H_{2(g)} + OH^-$
4) $\frac{1}{5}NO_3^- + \frac{6}{5}H^+ + e = \frac{1}{10}N_{2(g)} + \frac{3}{5}H_2O$
5) $\frac{1}{2}NO_3^- + H^+ + e = \frac{1}{2}NO_2^- + \frac{1}{2}H_2O$
6) $\frac{1}{8}NO_3^- + \frac{5}{4}H^+ + e = \frac{1}{8}NH_4^+ + \frac{3}{8}H_2O$
7) $\frac{1}{6}NO_2^- + \frac{4}{3}H^+ + e = \frac{1}{6}NH_4^+ + \frac{1}{3}H_2O$
8) $\frac{1}{4}NO_2^- + \frac{5}{4}H^+ + e = \frac{1}{8}N_2O_{(g)} + \frac{5}{8}H_2O$
9) $\frac{1}{2}NO_2^- + \frac{3}{2}H^+ + e = \frac{1}{4}N_2O_{(g)} + \frac{3}{4}H_2O$
10) $\frac{1}{6}N_{2(g)} + \frac{4}{3}H^+ + e = \frac{1}{3}NH_4^+$
18) $\frac{1}{8}SO_4^{2-} + \frac{9}{8}H^+ + e = \frac{1}{8}HS^- + \frac{1}{2}H_2O$
19) $\frac{1}{2}S_s + H^+ + e = \frac{1}{2}H_2S_{(g)}$
20) $Fe^{3+} + e = Fe^{2+}$
21) $Fe(OH)_{3(s)} + HCO_3^- + 2H^+ + e = Fe(CO_3)_{(s)} + 3H_2O$
22) $Fe(OH)_{3(s)} + 3H^+ + e = Fe^{2+} + 3H_2O$
23) $Fe(OH)_{3(s)} + H^+ + e = Fe(OH)_{2(s)} + H_2O$
24) $\frac{1}{2}FeS_{3(s)} + 2H^+ + e = \frac{1}{2}Fe^{2+} + H_2S_{(g)}$
25) $\frac{1}{2}Fe^{2+} + S_{(s)} + e = \frac{1}{2}FeS_{2(s)}$
26) $\frac{1}{16}Fe^{2+} + \frac{1}{8}SO_4^{2-} + e = \frac{1}{16}FeS_{2(s)} + \frac{1}{2}H_2O$
27) $\frac{1}{14}Fe(OH)_{2(s)} + \frac{1}{7}SO_4^{2-} + \frac{9}{7}H^+ + e = \frac{1}{14}FeS_{2(S)} + \frac{5}{7}H_2O$

TABLE 1.13 (*Continued*)
Redox Half-Reactions for Some Constituents Found in Groundwater Environments

11) $\frac{1}{4}CH_2O + H^+ + e = \frac{1}{4}CH_{4(g)} + \frac{1}{4}H_2O$

12) $\frac{1}{4}CO_{2(g)} + H^+ + e = \frac{1}{4}CH_2O + \frac{1}{4}H_2O$

13) $\frac{1}{2}CH_2O + H^+ + e = \frac{1}{2}CH_3OH + \frac{1}{4}H_2O$

14) $\frac{1}{8}CO_{2(g)} + H^+ + e = \frac{1}{8}CH_{4(g)} + \frac{1}{4}H_2O$

15) $\frac{1}{2}CH_3OH + H^+ + e = \frac{1}{2}CH_{4(g)} + \frac{1}{2}H_2O$

16) $\frac{1}{6}SO_4^{2-} + \frac{4}{3}H^+ + e = \frac{1}{6}S_{(s)} + \frac{3}{2}H_2O$

17) $\frac{1}{8}SO_4^{2-} + \frac{5}{4}H^+ + e = \frac{1}{8}H_2S_{(g)} + \frac{1}{2}H_2O$

28) $\frac{1}{14}Fe(CO)_{3(s)} + \frac{1}{7}SO_4^{2-} + \frac{17}{14}H^+ + e = \frac{1}{14}FeS_{2(S)} + \frac{4}{7}H_2O + \frac{1}{14}HCO_3^-$

29) $\frac{1}{2}MnO_{2(s)} + \frac{1}{2}HCO_3^- + \frac{3}{2}H^+ + e = \frac{1}{2}MnCO_{3(s)} + \frac{3}{8}H_2O$

30) $Mn^{2+} + 2e = Mn_{(s)}$

31) $\frac{1}{2}MnCO_3 + \frac{1}{2}H^+ + e = \frac{1}{2}Mn_{(s)} + \frac{1}{2}HCO_3^-$

32) $MnOOH_{(s)} + HCO_3^- + 2H^+ + e = MnCO_3 + 2H_2O$

33) $MnO_2 + H^+ + e = MnOOH$

Source: After Freeze and Cherry, 1979.

additional gas or other oxidizing agents and some groundwater pollutants. What this means is that these reactions primarily occur in the recharge area of groundwater systems, in open in-mine environments, and in some discharge locations. Important reducing agents include organic compounds such as carbohydrates, humic substances, and hydrocarbons, sulfide minerals such as pyrite, and iron(II) silicate minerals.

Redox is of great importance because it regulates the consumption of both dissolved oxygen and organic matter in the soil zone and in the aquifer and the production of acid mine drainage, and redox is the key to bioremediation of organically contaminated groundwater. Iron-rich mining environments and mine waste piles that are open to atmospheric gas, particularly O_2, account for most acid mine drainage. Redox reactions involving the transfer of one or two electrons, such as Fe^{2+} / Fe^{3+} and Mn^{2+} / Mn^{3+}, are rapid, whereas reactions that involve multiple electrons, such as SO_4^{2-}/S^{2-}, NO^{3-} / NH_3 and CO_3^{2-}/CH_4, are slow and kinetically impaired. The rapid oxidization of Fe^{2+} is readily apparent by how quickly bare metal rusts and iron-rich mine drainage becomes acidic. Kinetically impaired reaction rates tend to be very slow in the absence of biological mediation. Non-photosynthetic microorganisms do not oxidize organic matter or reduce inorganic compounds; instead they mediate redox reactions by facilitating the transfer of electrons. In the process the microorganisms use some of the energy that is stored in inorganic chemical bonds or is released by the oxidation of unstable products of photosynthesis.

1.5.5 Oxidizing Environments

Oxygen consumption (i.e., O_2 reduction) is the results of the oxidation of inorganic substances and the oxidation of reduced organic matter. Not only do these two types of processes compete for available O_2, but they also have large ΔG (change in Gibbs free energy) differences between some inorganic oxidation processes that establish a preferential oxidation order. ΔG is a measure of the amount of energy available to be released during a chemical reaction. For example, when both sulfide and Mn(II) are available, the more negative ΔG of sulfide oxidation favors sulfide oxidation over manganese oxidation. Gibbs free energy is discussed in Chapter 2. Common aqueous inorganic oxidation reactions include sulfide oxidation, iron oxidation, nitrification, manganese oxidation, and iron sulfide oxidation (Table 1.12). In addition to these common reactions, numerous inorganic substances of economic and environmental interest such as S, Se, As, Cr, Cd, V, Mo, W, and U can be readily oxidized to soluble anion complexes. After mobilization, species of economic interest often precipitate as reduced species at the aerobic–anaerobic boundary. Such deposits of economic minerals are known as secondary enrichment. Secondary enrichment was responsible for the vast silver and copper deposits that were mined in the western United States in the late 1800s and early 1900s.

Organic matter, the primary reducing substance in most groundwater systems, is ubiquitous in most soils and is common in many reduced sedimentary rocks. Natural organic matter may be in many forms including polysaccharides, saccharides, fatty acids, amino acids, and phenols. Oxidation of reduced organic matter is dependent on bacterial mediation. Bacteria may be aerobic, anaerobic, or facultative (i.e., they can be either aerobic or anaerobic). Facultative bacteria will preferentially utilize their aerobic metabolic pathway when free molecular O_2 is present. Aerobic reduction of organic compounds (i.e., aerobic respiration) is more energy favorable than is anaerobic reduction, and aerobic bacteria have a metabolic advantage over anaerobic bacteria. The significance of this is that organic degradation tends to be aerobically mediated in the soil zone and in the aquifer near the recharge area, and it tends to be anaerobically mediated elsewhere in the aquifer.

There must be a continual supply of O_2 or groundwater will rapidly become anaerobic because 1) molecular oxygen is readily reduced by inorganic and organic compounds and 2) the solubility of O_2 in water is low and decreases with increased temperatures (Figure 1.7). Water in contact with the atmospheric can contain 15.4 mg/L of oxygen at 0 °C and only 9.6 mg/L at 20 °C. Anaerobic conditions may be thought of as the absence of free molecular oxygen. The actual boundary between aerobic and anaerobic conditions depends on the oxygen reduction process and the oxygen demand of various bacterial species. The threshold boundary lies between 0.7 and 0.1 mg/L of O_2 at 8 °C for bacterially mediated oxygen consumption.

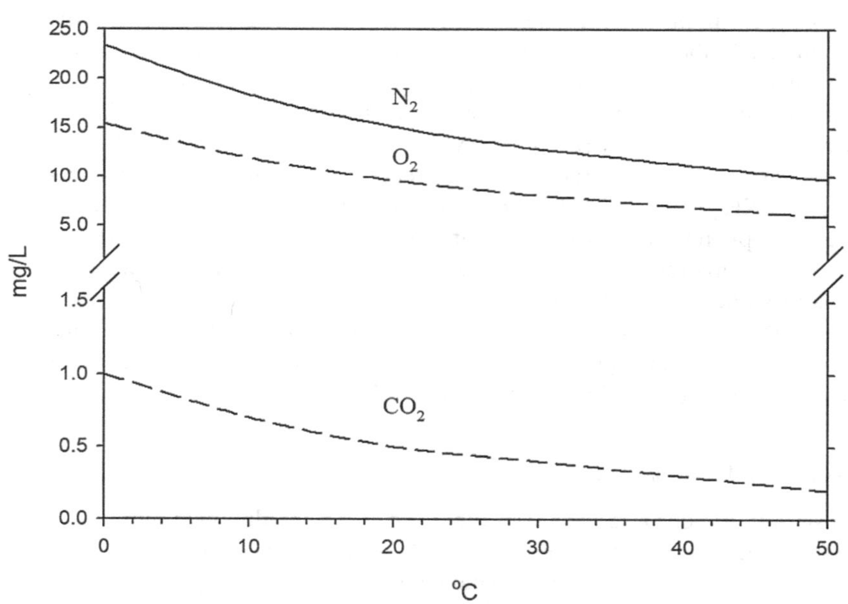

FIGURE 1.7 Temperature-dependent solubility of atmospheric gases in pure water. Data from Matthess, 1982.

In recharge areas, where open system conditions prevail, a continual supply of oxygen from the atmosphere and root respiration keeps the groundwater oxic. It is not surprising that the reduction of free molecular O_2, accompanying the oxidation of organic matter in the soil zone and in the open system portion of aquifers, plays such a critical role in the geochemical evolution of most groundwater. In the presence of free molecular O_2, organic matter, which for illustrative purposes may be represented by the simple carbohydrate (CH_2O), is oxidized to CO_2:

$$O_{2(g)} + CH_2O = CO_{2(g)} + H_2O \qquad (1.23)$$

Aerobic liberation of $CO_{2(g)}$ in the soil zone and open portion of aquifers is the primary source of most carbonic acid in groundwater. In organic rich aquifers substantial $CO_{2(g)}$ is also released by anaerobic processes such as denitrification, sulfate reduction, and methanogenesis (Table 1.12). The partial pressure of atmospheric $CO_{2(g)}$ is so low $P_{CO_2} = 10^{-3.5}\,atm)$ that, except in areas generally devoid of vegetation and in other areas during torrential precipitation recharge events, atmospheric CO_2 has little influence on the carbonic acid content of most groundwater. The partial pressure of soil zone $CO_{2(g)}$ in most recharge areas is considerably elevated ($P_{CO_2} = 10^{-2}$ to $10^{-2.5}$ atm). In many groundwaters, H^+ ions liberated by the dissociation of carbonic acid do not have an appreciable effect on groundwater pH because these ions are consumed by mineral dissolution in the soil zone or aquifer. The consumption of the H^+ ions accompanying the dissolution of carbonate minerals tends to make the groundwater basic. In forested granitic rock terrains and other environments where carbonate or other H^+ ion consuming minerals are scarce, the oxidation of organic matter often leads to acidic groundwater.

1.5.6 Reducing Environments

Based on the stoichiometric relationship in Equation 1.23, only 8.4 mg/L of CH_2O is needed to consume 9 mg/L of O_2. When a groundwater flow path becomes closed to atmospheric the supply of oxygen is rapidly depleted and the groundwater becomes anaerobic. Champ et al. (1979) proposed a thermodynamic model for the sequential oxidation of excess organic matter by the progressive reduction of oxidized species. The idea behind this model is that oxidized species with more negative ΔG values (Table 1.12) will preferentially oxidize organic matter relative to oxidized species with less negative ΔG values. According to this model nitrate reduction, also known as denitrification, begins at low O_2 concentrations. As the nitrate is consumed the groundwater becomes progressively more reduced. Nitrate reduction is followed by Mn(IV) reduction, which in turn is followed by Fe(III) reduction, and so forth. The actual redox processes that occur are dependent on the organic content of the aquifer and on the availability of oxidized species such as nitrate and sulfate.

Because oxygen consumption and nitrate reductions readily occur, large concentrations of O_2 and N_2 are often good evidence of close proximity of a groundwater to its recharge source. Groundwaters that have been reduced contain little or no O_2 and NO_3^-, may have diminished SO_4^{2-}, and may have elevated Mn^{2+}, Fe^{2+}, H_2S, N_2, NO_2^-, and NH_4^+ contents.

1.6 MINERAL SURFACE REACTIONS

Reactions between aqueous species and the surfaces of colloidal and solid particles greatly influence the geochemical evolution of many groundwaters. Surface reactions may soften groundwater and retard and even immobilize the spread of some organic and inorganic contaminants, are largely responsible for determining the fate of trace metals and radioactive elements in solution, and are responsible for the ion exchange in clay-rich environments.

Surface reactions include adsorption and desorption. There is no practical way to distinguish between the several types of adsorption processes, so these processes are collectively known as sorption. These processes include:

1. Physical adsorption is the result of nonspecific forces of attraction involving the entire electron shells of elements in solution and of the adsorbing substrate. This process is rapid and reversible and is independent of temperature and the ionic strength of the solution. The presence of complexing agents and the pH of the solution have a large influence on physical adsorption. The uptake of neutral metal complexes is an example of physical adsorption.
2. Electrostatic adsorption is the result of coulombic forces of attraction between charged solute species and the absorbate. This process is rapid, largely reversible, somewhat dependent on temperature, and strongly dependent on the composition of the sorbent as well as the ionic strength and composition of the solution. Ion exchange, a significant process in many groundwaters, is an example of electrostatic adsorption. Ion exchange is discussed in later in this chapter.
3. Specific adsorption occurs when sorbate ions are adsorbed and thus induce the formation of surface sites of opposite charge. Specific adsorption also occurs when the sorbate ion, as a trace species, occupies some of the small number sites of opposite charge that exist on a surface regardless of the net surface charge. These processes may be either slow or rapid, are highly dependent on the composition of the sorbent as well as the concentrations of solutes, and are temperature dependent.
4. Chemical substitution (co-precipitation or solid solution) is due to the action of chemical forces involving chemical bonding. This process may be slow, only partially reversible, highly solute selective, temperature dependent, and highly dependent on the concentration of the solute. Chemical substitution is not a true form of adsorption but is difficult to distinguish from adsorption under experimental conditions.

Desorption is the release of an adsorbed substance into solution from an adsorbing substrate. Surface reactions occur because all solid particles have electrically charged surfaces that attract and hold ions of the opposite charge. Minerals composed of large crystal lattices (e.g., carbonate, igneous silicate minerals) have small surface affects relative to the overall electrical state of the crystal. Surface charge affects are more pronounced on very fine-grained substances where the surface area is great relative to the total volume of the particle. Of particular importance to hydrogeology are colloid particles that have equivalent spherical diameters of 2 μm or less. Colloids, which include most clay minerals and iron and manganese oxides, are so small that their electrochemistry is dominated by surface affects. Other aquifer materials that often undergo important surface reactions include zeolites, hydrous iron, aluminum oxides, organic substances and humus, glauconite, micas, feldspars, and hornblende.

Charges on crystal surfaces are due to unsatisfied valences on boundary atoms at silicate surfaces and to imperfections or ionic substitutions in crystal lattices. Charged surfaces attract polar water molecules, H^+ and OH^- ions, and dissociated and undissociated organic and inorganic species and solids. Adsorbed substances are held to the active surface by varying degrees of force ranging from weak Van der Waals forces to the strong forces of covalent bonds. Van der Waals forces only exert a week electrostatic attraction, whereas the adsorbed substances may be strongly held to the crystal surface by incorporation into the crystal lattice accompanying covalent bonding.

The net positive or negative surface charge of the solid surface is offset by an equivalent number of counter ions of the opposite charge that have been adsorbed on the solid surface. Counter ions occupy a region known as the electrical double layer. The double layer (Figure 1.8) is envisioned as a fixed layer of oppositely charged ions known as the Stern layer, and a diffuse layer of counter ions is also known as the Gouy layer (Sparnaay, 1972). Although the diffuse layer contains both positive and negative charges, the overall charge of the double layer is opposite but of equal magnitude to

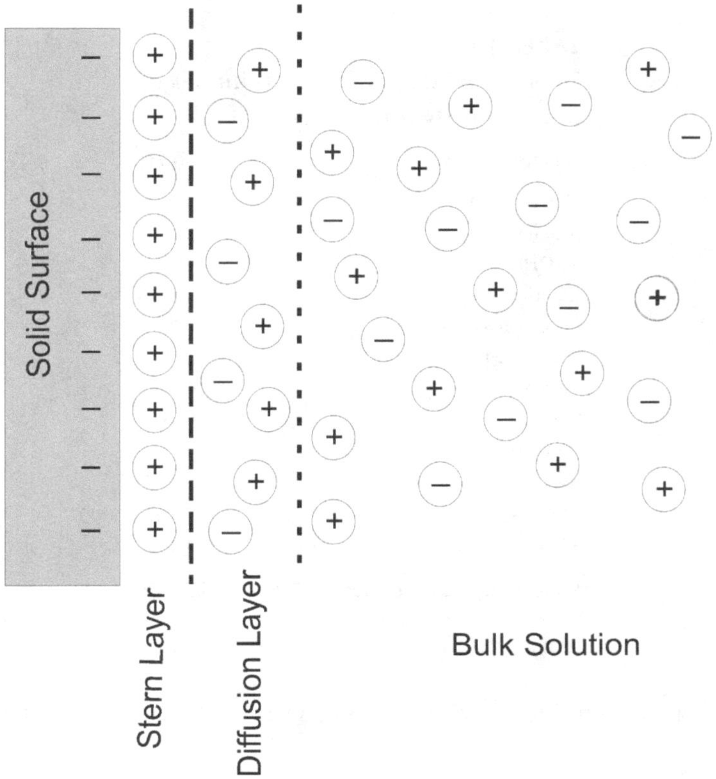

FIGURE 1.8 The distribution of charges in the electric double layer. The Stern layer has a fixed charge and is solidly attached to the crystal surface, whereas the Gouy or diffusion layer contains both positive and negative charges.

surface charge of the solid phase. Ions and charged particles that make up the double layer are often colloidal and remain separated from each due to electrostatic repulsion.

The fact that the surface charge is pH dependent is very important. The surface charge is positive in acidic solutions and negative in alkaline solutions. At an intermediate pH, known as the zero or isoelectric point, the surface charge is neutral. The pH of the zero point is substance dependent and varies from less than 3 to 9 (Table 1.14). In most groundwater, colloids of iron, manganese, and aluminum attract cations or anions, whereas crystalline minerals and clays tend to attract cations.

1.6.1 Sorption Isotherms

The quantitative relationship between the mass of substance absorbed on a surface (S) and the mass of the solid absorbate (*sm*) may be represented as:

$$S = \frac{(C_1 - C)\text{ solution volume}}{sm} \tag{1.24}$$

where C_l and C are equilibrium concentrations in mg/L, *sm* is in g, and S is in mg/L. Because equilibrium is temperature dependent, Equation 1.24 is usually expressed in terms of equations known as sorption isotherms. Sorption isotherms describe the mass absorbed relative to the concentration of absorbent in solution at a specific temperature. Isotherms have no set shape; however, several

TABLE 1.14
Isoelectric Points of Some Minerals of Interest in Hydrogeology

Species	pH
SiO_2 (quartz)	2.0–3.5
Aluminum oxide	8.5
$Al(OH_3)$ (Gibbsite)	~9
Chlorite	<3
$Fe3O_4$ (magnetite)	6.5
$Fe2O_3$ (hemitite)	5–9
Feldspar	<3
Goethite	6–8
Hydrous ferric oxide	8.7
Illite	<2
Kaolinite	<3
Talc	<3

Source: Data from Drever, 1997, Kosmulski, 2020, and Parks, 1965.

equations have been developed to describe classes of shape. Two common equations are the exponential Freundlich isotherm:

$$S = KC^n \tag{1.25}$$

and the hyperbolic Langmuir isotherm:

$$S = Qo\frac{KC}{1 + KC} \tag{1.26}$$

where *K* and *n* are empirical coefficients specific to particular substances, and Qo is the maximum sorptive capacity of the surface. The Freundlich and Langmuir isotherm equations are based on empirical and theoretical considerations, respectively. Equations 1.25 and 1.26 verify our intuitive idea that an increase in solute concentration results in an increase in adsorption, and that a decrease in solute concentration results in desorption. Some possible shapes of Freundlich and Langmuir isotherms are illustrated in Figure 1.9.

Sorption isotherms are used in solute mass transport models to predict the fate of aqueous substances that are affected by surface interactions. A common form of the Freundlich isotherm often used in solute transport modeling is:

$$S = KdC \tag{1.27}$$

This equation is a modified version of Equation 1.25 where $n = 1$ and where *Kd* is a distribution coefficient relating S to *C*. *Kd* describes the linear slope of the isotherm and may be determined by laboratory experiments for site-specific substances absorbed and solid absorbates. Because sorption retards the migration of radionuclides, a major effort has been undertaken to determine *Kd* values for the host rock types found at potential radioactive waste repositories. Extensive literature and tables of *Kd* values are readily available on the internet.

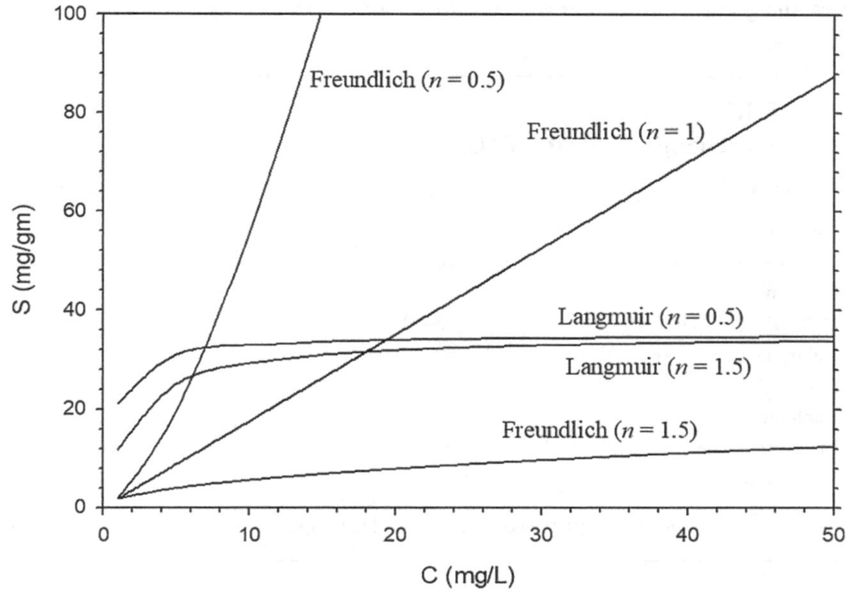

FIGURE 1.9 Examples of Freundlich and Langmuir isotherms. Freundlich isotherms have a K value of 1.75 and n values of 0.5, 1.0, and 1.5. Langmuir isotherms have K values of 0.5 and 1.5 and a Qo value of 25 mg/gm.

1.6.2 Ion Exchange

The stoichiometric exchange of ions in a solution with ions adsorbed on solid substances is known as ion exchange. Ion exchange is possible because many counter ions are relatively mobile due to the weak electrostatic forces that bind them to crystal surfaces. Although both anion and cation exchange are possible, cation exchange is most prevalent in groundwater due to the fact that most clay surfaces have a negative charge in the normal pH range of groundwater. Cation exchange capacity (CEC) and anion exchange capacity (AEC) are determined in the laboratory by measuring the quantity of NH_4^+ uptake or release from a 1 M solution of ammonium acetate at pH 7.0. CEC and AEC are usually reported as the mg/L of exchangeable ions per 100 g of dry solid.

Exchange capacity is not rigorously quantifiable. In some mineral species CEC increases with increases in pH above the zero-point pH. CEC also varies as a function of the species that make up the double layer. Typical ranges of CEC for clay minerals are listed in Table 1.15. Ion exchange is not a random process. The affinity for specific ions to exchange for other ions is well understood. The Gouy theory correctly predicts that multivalent ions will preferentially concentrate in the double layer; however, this preference decreases as the ionic strength of the solution increases. The preferential concentration of multivalent ions means that during exchange, multivalent ions from the solution tend to exchange with univalent ions absorbed on the double layer. In many clay-rich groundwater systems, divalent Ca^{2+} ions replace absorbed monovalent Na^+ on clay surfaces. The net effect is to flood the solution with Na^+ ions and deplete the solution in Ca^{2+} ions. Because the water remains electrically neutral, twice as many Na^+ ions are released as Ca^{2+} ions are absorbed. Na–Ca exchange is the chemical basis for common water softeners where the substrate becomes Ca^{2+} saturated by exchanging for Na^+ ions. The process is reversed by flooding the system with Na^+ (NaCl).

Not only do multivalent ions tend to displace univalent ions, but ions with larger hydrated radius tend to displace ions with smaller hydrated radius. The affinity for sorption of most clays, known as the Hofmeister series, is:

$$Cs^+ > K^+ > Na^+ > Li^+ \text{ and } Ba^{2+} > Sr^{2+} > Ca^{2+} > Mg^{2+} \quad (1.28)$$

Some zeolites and glasses exhibit a reversed affinity for exchange.

TABLE 1.15
Cation Exchange Capacity (CEC)

Clay mineral	CEC (meq/100 g)	Surface area (m^2/g)
Allophane	70	
Chlorites	10–40	80
Glauconite	11–20	
Halloysite ($2H_2O$)	5–10	
Halloysite ($4H_2O$)	40–50	
Illite	10–40	80
Kaolinite	3–15	15
Montmorillonite	70–100	800
Vermiculite	100–150	

Source: Data from Carroll, 1959, and Domenico and Schwartz, 1990.

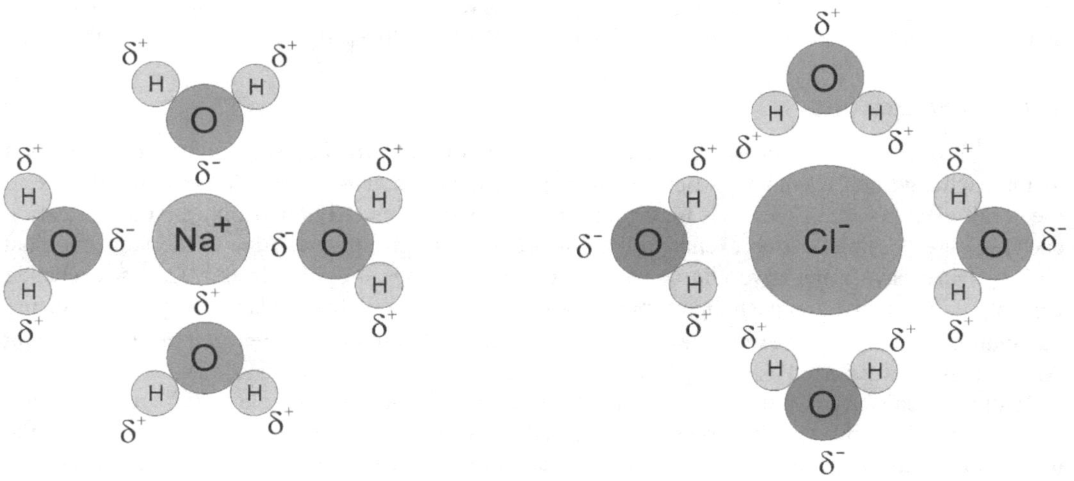

FIGURE 1.10 Hydration of Na^+ and Cl^-.

1.6.3 Ion Hydration, Ion Pairs, and Ion Complexation

Bare simple ions, such as Na^+, Ca^{2+}, and Cl^-, do not exist in aqueous solutions. Instead, they form aqua-molecules by a process known as hydration, where the ion is bonded with bipolar H_2O and with OH^- ligands. Each cation is surrounded by water molecules with the oxygen atoms positioned inward, and each anion is surrounded by water molecules with the oxygen atoms positioned outward (Figure 1.10). The number of hydration ions (n) can vary greatly. For example, the hydration of Na^+ may be represented as $Na(H_2O)_n$ where $n = 1–6$. Hydration occurs because of the dipolar force of water molecules. The NaH_2O molecule is held together by the electrostatic force between the dipolar water molecules and the positive charge of Na^+ ion. Ion hydration stabilizes the solution by reducing the attractive forces between oppositely charge ions.

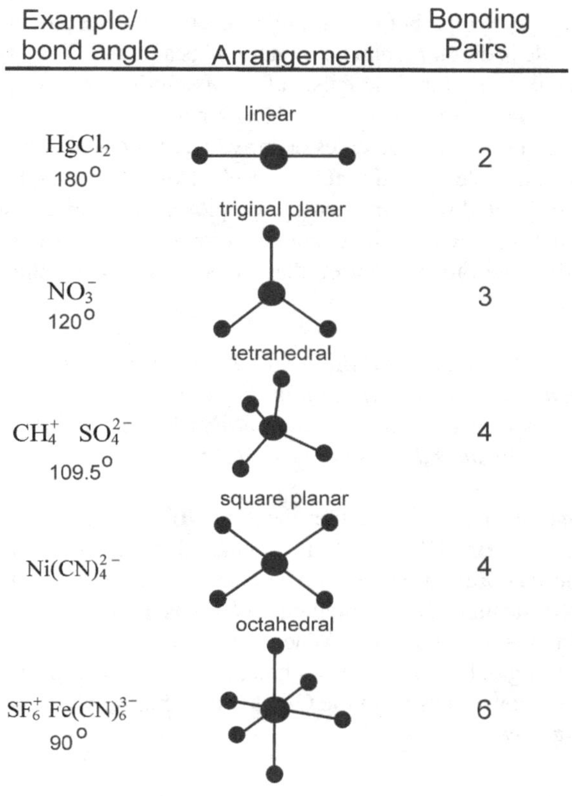

FIGURE 1.11 Arrangement of ligands with respect to the central ion in ion pairs and complex ions.

Coordination chemistry is only briefly introduced here as an aid to understanding the underlying principles that govern simple ions, ion pairs, and aqueous complexation. Ion pairing and complexation may be thought of as Lewis acid–base reactions. It is sometimes convenient in the cases of simple pairing, complex ion pairing, and complexation to think of acid-base reactions as changes in coordinative relationships rather than in terms of the more familiar thermodynamic relationships, which are discussed in Chapter 2.

The molecular structure of most ionic molecules, ion pairs, and complexes of interest to hydrogeology are mononuclear in that they consist of a central atom surrounded by anions or other negatively charged molecules, known as ligands. Many organic compounds, including most hydrocarbons, that contribute to groundwater contamination are polynuclear in that they contain more than one central atom. A benzene ring is an example of a polynuclear atom. Ligand comes from the Latin word ligare meaning "to bond." Ligands donate or share electrons with the central cation. The number of ligand atoms bonded to the central atom defines what is known as the coordination number of the central atom. The central atom is commonly a transition metal, but some anions can be the central atom. H^+ has a coordination number of 1; most other central atoms have coordination numbers of 2, 4, 6, or sometimes 8, and in some situations, such as CO_3^{2-}, where carbon has a coordination number of 3. The most common coordination number is 6. When H^+ is the central atom it only bonds with a single ligand, and when it is a central atom with a coordination number of 2 it bonds with two ligands. When more than one ligand is involved, the ligands may be the same or different species. The coordination number determines the physical locations of the ligands with respect to the central ion, which in turn controls the shape of the ion pair or complex. Ligands form linear structures with central ions having a coordination number of 2 and form more complex geometric shapes with central ions having higher coordination numbers (Figure 1.11).

Reactions of single cations with other anions and molecules can be thought of as ligand exchange; because the water molecule is dipolar it can be thought of as a ligand. The overwhelming abundance of unpaired water molecules in all but the most saline of solutions gives water its incredible solute capabilities. In coordination chemistry H^+ (i.e., H_3O^+) and OH^- ions are considered metals and ligands, respectively. Common ionic molecules occur when the central atom is in direct electrostatic contact with the ligand(s). Simple ion pairs also have electrostatic bonds, but the central ion and or the ligand(s) retain a coordinated water molecule. Complexes are more stable than simple ion pairs, because the electron donating ligands of inner sphere complexes are covalently bonded to the metal. It is important to note that laboratory reported data does not specify if the ions are simple ion, ion pairs, or ion complexes.

Important Takeaway – A fundamental difference between a simple ion pair and a complex is that a hydration ion forms a critical bonding link in an ion pair, whereas complexes form by the direct bonding of a central ion and ligands. The abundance of unpaired water molecules give water its incredible solute capabilities.

A ligand can occupy one or more coordination locations. If each base occupies a single coordination location, such as Cl^-, F^-, Br^-, CN^- and NH_3, the ligand is known as a monodentate ligand. If a base occupies two or more coordination locations it is known as a multidentate ligand. Multidentate ligands, which bond more strongly than monodentate ligands, may be bidentate, tridentate, etc. Ions such as SO_4^{2-} can be mono- and bidentate. When a ligand occupies more than one coordination location the multidentate ligand is known as a chelate, and it bonds to the metal ion by a process known as chelation. Chelation comes from the Greek word chelos, which means claw and is used in reference to two binding claws.

1.6.4 Speciation Stability Constants

Except for polynuclear compounds, aqueous complexes attain thermodynamic equilibrium relatively fast. The thermodynamic ability of aqueous complexes to remain complexed is represented by the stability constant of the complex. Equilibrium constants are, on the other hand, a quantitative measure of the temperature-dependent tendency of substance to dissolve or otherwise dissociate (Chapter 2). Large stability constants indicate a great tendency to remain complexed, whereas large equilibrium constants indicate a great tendency to dissociate. Stability constants are calculated in a manner similar to the calculation of equilibrium expressions. Expressed in terms of the law of mass action (Equation 2.9), complexation may be represented as:

$$mM + lL + hH = M_mL_lH_h \tag{1.29}$$

and the general form for calculating stability constants is:

$$\beta_{mlh} = \frac{[M_mL_lH_h]}{[M]^m[L]^l[H]^h} \tag{1.30}$$

where M, L and H are metal, ligand, and proton (i.e., hydrogen ion), respectively, and the super and subscripts are the number of moles of each substance. If $m = 1$ the super and subscripts are omitted. β is the stability constant. For many aqueous complexes several different formation and dissociation reaction pathways are possible. Selected stability constants are listed in Appendix B.

Important Takeaway – Stability constants differ from equilibrium constants in that stability constants are a quantitative measure of the temperature-dependent tendency of a complex to remain complexed.

Example E1.3 Aluminum speciation

Problem: Calculate the speciation of Al among the various hydroxide complexes and Al^{3+} for a groundwater with a pH of 6 discharging from granitic terrain contains a trace amount of Al ($Al_T = 10^{-5}$ M).

Solution: The stability constants, to the base 10 (Appendix B), are 9.0, 18.7, and 33.0 for $AlOH^{2+}$, $Al(OH)_2^+$, and $Al(OH)_4^-$, respectively. Other aluminum hydroxide complexes are ignored in this example. Assume the pH remains constant and the neutral species remains in solution.

The mole balance for Al_T is:

$$(Al)_T = (Al^{3+}) + (AlOH^{2+}) + (Al(OH)_2^+) + (Al(OH)_4^-) \quad \text{(E1.3a)}$$

By substituting mass balance equations:

$$(Al)_T = (Al^{3+}) + \beta_1(Al^{3+})(OH^-) + \beta_2(Al^{3+})(OH^-)^2 + \beta_3(Al^{3+})(OH^-)^4 \quad \text{(E1.3b)}$$

$$(Al)_T = (Al^{3+})\left[1 + \beta_1(OH^-) + \beta_2(OH^-)^2 + \beta_3(OH^-)\right] \quad \text{(E1.3c)}$$

OH^- is 10^{-8} because the pH remains constant. The concentration of Al^{3+} may be determined by substituting known values of β:

$$Al^{3+} = \frac{(Al)_T}{\left[1 + \beta_1(OH^-) + \beta_2(OH^-)^2 + \beta_3(OH^-)^4\right]} \quad \text{(E1.3d)}$$

$$Al^{3+} = \frac{10^{-5}}{\left[1 + 10^9(10^{-8}) + 10^{18.7}(10^{-8})^2 + 10^{33}(10^{-8})^4\right]} \quad \text{(E1.3e)}$$

and

$$Al^{3+} = \frac{10^{-5}}{\left[1 + 10 + 10^{18.7}(10^{-16}) + 10^{33}(10^{-32})\right]} \quad \text{(E1.3f)}$$

$$= \frac{10^{-5}}{\left[1 + 10 + 10^{2.7} + 10^1\right]} \quad \text{(E1.3g)}$$

$$= 10^{-7.7}\ \text{M} \quad \text{(E1.3h)}$$

The concentration of Al complexes may be calculated by substituting the concentrations of Al^{3+} and OH^- into mass balance equations:

$$AlOH^{2+} = \beta_1(Al^{3+})(OH^-) \quad \text{(E1.3i)}$$

$$= 10^9(10^{-7.7})(10^{-8})$$

$$= 10^{-6.7}\ \text{M} \quad \text{(E1.3j)}$$

$$Al(OH)_2^+ = \beta_2(Al^{3+})(OH^-)^2$$

$$= 10^{18.7}(10^{-7.7})(10^{-8})^2$$

$$= 10^{-5.0}\ \text{M} \quad \text{(E1.3k)}$$

About 30 percent of the total Al is free Al^{3+} ion, and the remainder is in complex forms.

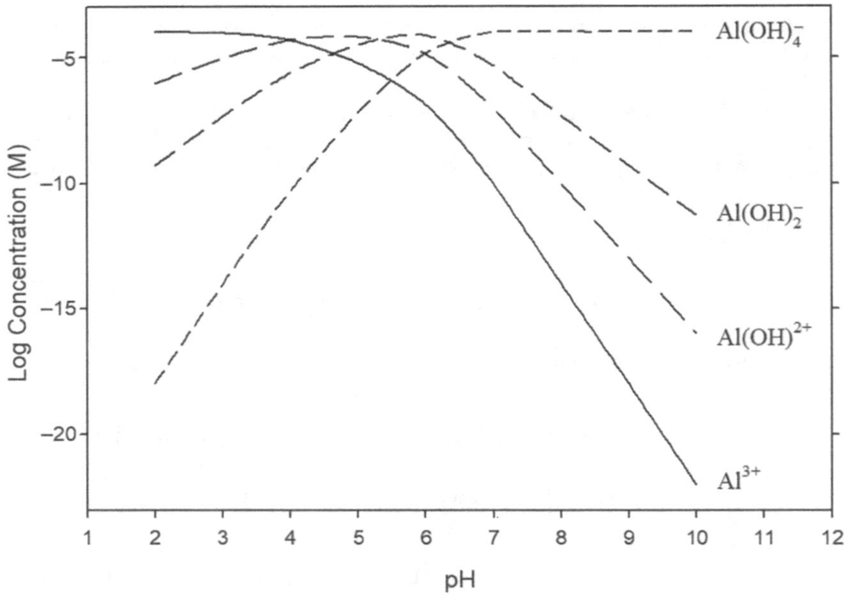

FIGURE 1.12 Speciation of Al^{3+} and Al-complexes as a function of pH at 25 °C.

$$Al(OH)_4^- = \beta_3\left(Al^{3+}\right)\left(OH^-\right)^4 \tag{E1.3l}$$

$$= 10^{33}\left(10^{-7.7}\right)\left(10^{-8}\right)^4$$

$$= 10^{-6.7}\ M \tag{E1.3m}$$

Because the reactions in Example E1.3 involve OH^- is it evident that the speciation of Al^{3+} and Al-hydroxide complexes are pH dependent (Figure 1.12). Speciation can be further complicated by the existence of mixed complexes. For example, trace amounts of Fe^{2+} can be tied up in a mixture of hydroxide, carbonate, and fluoride complexes. Speciation calculations for mixed complexes are treated in a manner similar to non-mixed complexes. Most groundwater contains an abundance of major and trace constituents that interact with each other. Speciation calculations for such real-world solutions are complex and tiresome. Fortunately, several groundwater chemistry modeling computer codes exist that can readily perform the needed calculations (Chapter 12).

1.7 COMPLEX ION SPECIATION

Prior to the growth of the field of contaminant hydrology, our understanding of aqueous complexes in groundwater systems was limited and largely restricted to saline solutions such as seawater. In recent years a better understanding of the role of complexes in the geochemical evolution of groundwater has evolved. Our interest in complexation is ultimately one of speciation. Aqueous complexation may affect the speciation of 1) major inorganic ions, 2) trace metals in the presence of major inorganic ions, and 3) organic compounds. Complexation involving major ions may reduce the concentrations or activities of free ions in solution relative to the total concentration of ions in solution. Complexation of trace metals may facilitate or retard trace metal transport. Organic complexation may affect the concentration of nutrients essential to biota and of toxic substances that are deleterious to biota.

Only 10 elements – H, Na, K, Mg, Ca, Cl, O, S, and C, and Si – make up 99 percent of the total dissolved material in most natural waters. These elements originate from the dissolution and dissociation of weak acids and the weathering and dissolution of sedimentary, metamorphic, and crystalline rocks. In dilute solutions, ion pairing of major dissolved species is quantitatively unimportant (Garrels and Christ, 1965), and most speciation may be accounted for by a few monatomic ions and common ion molecules (i.e., H^+, Na^+, K^+, Mg^{2+}, Ca^{2+}, HCO_3^-, Cl^-, SO_4^{2-}, and $H_3SiO_4^-$). What this means is that complexes have a minimal effect on the ionic strength of dilute solutions, and therefore they do not influence solubility of aquifer minerals and gases. This list, excluding H^+ and $H_3SiO_4^-$, are commonly known as the major ions, and H^+ is accounted for by pH measurements.

The tendency for major ions to form ion pairs and complexes increases with increased salinity. Two factors contribute to this. First, with increased salinity, particles in solution are, on the average, forced closer together so that bond formation takes place more often. Second, increases in ionic strength destabilize hydrated ions by affecting the structure of their attached water molecules. The introduction of additional solutes destabilizes water molecules attached to simple ions and exposes them to complexation.

Chemical analysis of two samples collected from a groundwater in central Utah illustrates the effects of complexing in relatively fresh water (Table 1.16). Only about 1.5 percent of the dissolved species occurs as complexes in the recharge water that has a TDS of about 10.4 meq /L. After the groundwater had the opportunity to interact with calcite, dolomite, and some gypsum, the TDS increased to nearly 30 meq /L, and approximately 5.2 percent of the dissolved species are in complex form. Table 1.16 also highlights the tendency for sulfate to preferentially complex relative to other inorganic ligands. In more concentrated waters, particularly saline to hypersaline solutions, and in waters containing relatively large concentrations of trace metals, the formation of complexes consumes considerably larger quantities of dissolved ions (Table 1.17).

Complex formation affects the solubility of aquifer minerals by two mechanisms. Complexation changes the activity coefficients of major free ions and ion pairs by increasing the ionic strength of

TABLE 1.16

Comparison of the Speciation into Free Ions and Ion Pairs vs. Complexes Due to Increased Salinity of the Geochemical Evolution of a Groundwater in Central Utah that Encountered Calcite, Dolomite, and Gypsum along the Flow Path

		Free ions and ion pairs (molarity × 10^{-3})								
Aquifer	**TDS (meq/L)**	**Ca^{2+}**	**Mg^{2+}**	**Na^+**	**K+**	**HCO_3^-**	**CO_3^{2-}**	**Cl^-**	**SO_4^{2-}**	**Species %**
Inflow	0.38	1.32	0.81	0.57	0.04	4,2	0.002	0.31	0.29	98.51
Discharge	27.98	2.15	2.09	3.33	0.14	5.57	0.007	0.51	3.19	94.8

	Complexes (molarity × 10^{-3})			complex
	HCO_3^-	CO_3^{2-}	SO_4^{2-}	%
Inflow	0.066	0.002	0.046	1.49
Discharge	0.157	0.012	0.76	5.2

Inflow: 12.8 °C, pH 7.0, log SIc = –0.476, log SId = –1.212, log SIg = –2.109

Discharge: 13.8 °C, pH 7.5; log SIc = 0.178, log SId = 0.317, log SIg = –1.057

SO_4^{2-} complexes include $CaCO_3^0$, $MgCO_3^0$, $NaCO_3^-$, KSO_4^-

HCO_3^- complexes include $CaHCO_3^-$, $MgHCO_3^-$, $NaHCO_3^0$,

CO_3^{2-} complexes include $CaCO_3^0$, $MgCO_3^0$, $NaCO_3^-$

Molarity, ion complexes and saturation indices (SI) were calculated using the computer code WATEQF: SIc (calcite), SId (dolomite), SIg (gypsum)

TABLE 1.17
Sulfur, Carbon and Chloride Complexes in Highly Saline Water

	Percent Complexed		
Species	**Seawater[1]**	**Sea water[2]**	**Brine spring[2]**
SO_4^{2-}	60	46	24
HCO_3^-	36	31	33
CO_3^{2-}	–	81	87
Cl^-	0	0	0

[1] *Source:* Morel (1983).
[2] *Source:* Garrels and Christ (1965).

the solution, and it increases the overall solubility of aquifer minerals by reducing the concentration of hydrated ionic species. An accounting of the major dissolved constituents in concentrated solutions must include the speciation of complexes as well as free ions.

The speciation of trace metals complexes is somewhat simpler to evaluate than is the speciation of major ions complexes. Trace metal speciation can be evaluated one metal at a time, whereas the speciation of all major ions must be evaluated simultaneously. This is because trace metal concentrations are small relative to the concentrations of the inorganic ligands that they bind. Important inorganic ligand groups that bind trace metals include OH^-, CO_3^{2-}, HCO_3^-, O_3^-, SO_4^{2-}, SO^{2-}, and H^+. Other inorganic ligands are usually in too low of concentrations to be important. Except for hydroxides, few inorganic trace metal complexes form in oxic waters. The major exceptions are carbonate complexes with copper in alkaline solutions and chloride complexes with cadmium, silver, and mercury in chloride-rich solutions. Sulfide, thiosulfate, and polysulfide complexes may be important in anoxic waters (Boulegue et al., 1979).

Trace metal speciation is pH dependent, because bare metals exist as aqua complexes whose formation is favored by acidic conditions. Speciation of hydroxide and carbonate complexes are strongly pH dependent as the concentration of these ligands are also pH dependent. At low pH levels, trace metals are highly mobile, and metal cations are the dominant species. In near neutral to alkaline solutions, concentrations of bare metals decline, and trace metal complexes, some of which are solids, become dominant.

Trace metal complexation of metal M between ligands A^- and B^- may be represented as:

$$M^{2+} + mA^- = M\left(A_m^{m\text{-}2}\right)^- ;\ \beta_m \tag{1.31}$$

$$M^{2+} + nB^- = M\left(B_n^{n-2}\right)^- ;\ \beta_n$$

where m and n are moles of ligands A and B, respectively. The total concentration of the trace metal (TOTM) as bare metal and trace metal complexes may be represented as:

$$TOTM = \left(M^{2+}\right) + M\left(A_m^{m-2}\right)^- + M\left(B_n^{n-2}\right)^- + \ldots. \tag{1.32}$$

By substituting appropriate stability constants and rearranging, Equation 1.32 becomes:

$$TOTM = \left(M^{2+}\right)(1 + \beta_m\left(A^-\right)^m + \beta_n\left(B^-\right)^n + \ldots.) \tag{1.33}$$

A fairly simple method to calculate the inorganic complexation of trace metals is described by Morel (1983) and is summarized here:

1. List all possible species. Many trace metal species are listed in Table 1.18.
2. Calculate free ligand concentrations using an ion interaction model, and calculate the speciation of acids and bases using acidity equilibrium constants.
3. Calculate the bracketed terms of Equation1.33 and use those species that are in significant concentrations.
4. Calculate the concentration of the bare metal by rearranging Equation 1.33.

TABLE 1.18
Inorganic Trace Metal Species of Potential Interest in Hydrogeology

Ag

$Ag^+, AgCl, AgCl_2^-, AgS^-, AgHS, AgHS_2^{2-}, Ag(HS)_2^-$

$Ag_{2(s)}, AgCl_{(s)}, AgBr_{(s)}$

Al

$Al(OH)_3, AK(OH)_4^-, AlF^{2+}, AlF_2^+$

$Al(OH)_{3(s)}, Al_2O_{3(s)}, Al_2Si_2O_5(OH)_{4(s)}, Al_{silicates}$

Ba

Ba^{2+}

$BaSO_{4(s)}, BaCO_{3(s)}$

Cd

$Cd^{2+}, CdCl^+, CdCl_2, CdCl_3^-, CdOH^+, CdS, CdHS^+, Cd(HS)_2, Cd(HS)_3^-$

$Cd(HS)_4^{2-}$

$Cd_{(s)}, CdCO_{3(s)}, Cd(OH)_{2(s)}$

Co

$Co^{2+}, CoCl^+, CoSO_4$

$CoS_{(s)}, Co(OH)_{2(s)}, CoCO_{3(s)}, Co(OH)_{2(s)}$

Cu

$Cu^{2+}, CuCO_3, CuOH^+$

$CuS_{(s)}., CuFeS_{2(s)}, Cu_2CO_3(OH)_{2(s)}, Cu_3(CO_3)_2(OH)_{2(s)}, Cu(OH)_{2(s)}$

$CuO_{(s)}$

Cr

$Cr(OH)_2^+, Cr(OH)_3, Cr(OH)_4^-, HCrO_4^-. CrO_4^{2-}$

$Cr(OH)_{3(s)}$

Fe

$Fe^{2+}, FeCl^+, FeSO_4, Fe(OH)_2^+, Fe(OH)_4^-$

$FeS_{(s)}, Fe_{2(s)}, FeCO_{3(S)}, Fe_2, O_{3(s)}, Fe_sO_{4(s)}$

$FePO_{4(s)}. Fe_3(PO_{4(s)}, Fe_{silicates}$

(Continued)

TABLE 1.18 (*Continued*)
Inorganic Trace Metal Species of Potential Interest in Hydrogeology

Hg
$Hg^{2+}, HgCl^{+}, HgCl_2, HgCl_3^{-}, HgOHCl, Hg(OH)_2, HgS_2^{2-}, HgOHS^{-}$
$HgS_{(s)}, Hg_{(liq)}, Hg(OH)_{2(s)}$
Mn
$Mn^{2+}, MnCl^{+}$
$MnS_{(s)}, MnCO_{3(s)}, Mn(OH)_{2(s)}, MnO_{2(s)}$
Ni
$Ni^{2+}, NiCl^{+}, NiSO_4$
$NiS_{(s)}, Ni(OH)_{2(s)}$
Pb
$Pb^{2+}, PbCl^{+}, PbCl_2, PbCl_3^{-}, PbOH^{+}, PbCO_3$
$PbS_{(s)}, PbCO_{3(s)}, Pb(OH)_{2(s)}, PbO_2$
Zn
$Zn^{2+}, ZnCl^{+}, ZnSO_4, ZnOH^{+}, ZnCO_2, ZnS$
$ZnS_{(s)}, ZnCO_{3(s)}, ZnSiO_{3(s)}$

Sources:
[1] Freeze and Cherry, 1979
[2] Helgeson et al., 1978
[3] Smith and Martell, 1976
[4] Bates and Mesmer, 1976
[5] Parkhurst and Appelo, 2013

Organic complexing of trace metals is briefly described here, and organic hydrogeochemistry is further discussed in Chapter 5. Surface and most groundwater contains organic compounds. Aqueous organic compounds are derived directly or indirectly from photosynthetic process and are either the product of biosynthesis or biodegradation. The concentration of dissolved organic carbon (DOC) ranges from 10^{-1} to 10^{1} mmoles/L. The chemical character of organic matter in aqueous systems is poorly understood. Humic and fluvic acids account for 40 percent to 99 percent of DOC in most natural waters. Humic and fluvic acids are general terms used to describe broad mixtures of compounds. The fraction of organic matter that is insoluble at all pH values is known as humin. *Humin* is the insoluble component of soil organic matter (SOM) that remains after extraction of the other components. Humin, which may constitute up to 60 percent of organic ligand binding sites, often irreversibly fixes major cations and trace metals (Matthess, 1982). Once fixed, these ions may be removed from surface water by the sedimentary deposition of humin. In groundwater systems containing organic rich layers, fixation may reduce the trace metal content of the groundwater. Humic acid is the fraction that is insoluble and precipitates at pH values of 2.0 or less. Fluvic acid is the fraction that remains soluble at all pH values. All classes of organic matter are structurally similar, but they differ by molecular weight and by the structure of their functional groups. Fluvic acids generally are smaller molecules with lower molecular weights but tend to be more hydrophilic.

In most natural water, organic molecule complexation of trace metals is of minor importance. Free protons and major cations, such as Ca^{2+} and Mg^{2+}, are usually in greater abundance than are organic ligands and can out-compete trace metals for the available binding sites on organic ligands. Major cations do not actually compete for binding sites; instead, they decrease the affinity of organic ligands for trace metals. Trace metal complexation is more important in fresh water than in saline water due to the elevated cation content of saline water relative to fresh water. Organic complexation of trace metals is only quantitatively significant in fresh water where organic ligands have a high affinity for trace metals or where organic ligand concentrations are very high (Morel, 1983; Morel et al., 1973). These conditions commonly occur in lakes choked by algal growth and in aqueous systems with a high organic content (e.g., bogs, organic rich lacustrine deposits). The affinity for trace metal-organic complexation varies greatly from one trace metal to another.

Chapter 1 Problems

1. Calculate the conversion factors for converting HCO_3^- into mole/L and meq/L.
2. Calculate the molarity of each species for Water 1 in Table 1.5.
3. Calculate the mole fraction of sulfate for Water 1 in Table 1.5.
4. Calculate the neutral pH of ion free water at 22 °C.
5. A water has a measured electrical conductance of 400,000 umohs/cm. What is a reasonable estimate of the TDS, what are the likely major ions, and how did this water acquire its TDS?
6. The analytical error (E) of a laboratory reported water analysis can be calculated using the equation

$$E = \Sigma \frac{\text{anion} - \text{cations}}{\text{anions} + \text{cations}} \times 100$$

   where the laboratory data have been converted to meq/L. Laboratory errors should not exceed 5% and usually are considerably less. Using the error balance equation, calculate the error balances for each of the waters in Table 1.5. Then recalculate the errors after increasing the mg/L of SO_4^{2-} by 40 percent and decreasing the mg/L of Ca^{2+} by 10 percent. Assuming that all of the significant ion concentrations are listed in the table, are there any analyses that could be rejected from further analysis?
7. Most groundwater has a pH in the range of 6.4 to 8.3. Assuming that water contains carbonate, phosphate, borate, sulfide, sulfate, and nitrogen species, what are the likely dominant dissolved forms of each species, and what are the fractional concentrations of each?
8. Discuss the chemical impact on a groundwater as it circulates in a) a granitic terrain and b) in a carbonate terrain as it moves from the recharge zone to confided aquifer conditions. Consider dissolved O_2 and CO_2 concentrations, oxidizing conditions, TDS, anticipated dissolved ions and their concentrations, and pH as the water moves from the recharge zone to confined conditions. Consider the effects of organic matter, mineral dissolution, and temperature.
9. A groundwater only containing Ca^{2+} and HCO_3^- ions encounters a Na-clay and undergoes complete ion exchange. Assume the initial Ca^{2+} and HCO_3^- concentrations of each species are 2 meq/L. What is the final concentration of each species in meq/L? What is the initial and final TDS of the solution in mg/L?
10. A water containing 4.7 meq/L of Ca^{2+} encounters a Na-clay and undergoes complete ion exchange. What is the net TDS change (in mg/L) of the water after the exchange?

Chapter notation

$>$ greater than
$<$ less than
$=$ equal

subscripts

aq aqueous
g gas

l liquid
s solid

prefix

E	exa (10^{18})
P	peta (10^{15})
T	tera (10^{12})
G	giga (10^{9})
M	mega (10^{6})
k	kilo (10^{3})
h	hecto (10^{2})
da	decka (10^{1})
d	deci (10^{-1})
c	centi (10^{-2})
m	milli (10^{-3})
μ	micro (10^{-6})
n	nano (10^{-9})
p	pico (10^{-12})
f	femto (10^{-15})
a	atto (10^{-18})
o	uncharged or neutral species
°C	temperature in degrees Centigrade
°K	temperature in degrees Kelvin
°F	temperature in degrees Fahrenheit
ΔG	Gibbs free energy of formation
μg	microgram
μhos	microohm
A	linear regression correlation factor
AEC	anion exchange capacity
atm	atmosphere (unit of pressure)
CEC	cation exchange capacity
cm	centimeter
d	day
DIC	dissolved inorganic carbon
DOC	dissolved organic carbon
e^- or e	electron
eq	equivalence
epm	milligram-equivalents per kilogram of solution
f	formality
g	gram
gfw	gram formula weight
h	hour
J	joule (J = 0.239 calorie)
K	specific conductance
K	empirical coefficient for Freundlich isotherm
kb	kilobar
Kd	distribution coefficient
Kg	kilogram
km	kilometer
K_w	equilibrium constant for water

L	liter
Log	logarithm to the base 10
lL	moles of ligand L
m	meter
m	molality
M	molarity
mA^-	moles of ligand A^-
mbar	millibar
meq	milliequivalent
mg	milligram
ml	milliliter
mmole	millimole
n	empirical coefficient for Freundlich isotherm
nB^-	moles of ligand B^-
N	mole fraction
P_{CO_2}	partial pressure of $CO_{2(g)}$
pH	negative log of H^+ concentration
ppm	parts per million
ppt	parts per thousand
s	second
sm	mass of solid absorbate
S	solute concentration
S	mineral surface
STP	standard conditions (25 °C or 298°K, 1 atm pressure)
t	time
TDS	total dissolved solids
TOTM	total concentration of trace metal
y	year

REFERENCES CITED

Alvarez-Silva, M., et al., 2010, Point of zero charge, isoelectric point and aggregation of phyllosilicate minerals, *Canadian Metallurgical Quarterly*, 49(4), 405–410.

Berner, E.K., and R.A. Berner, 1987, *The global water cycle*, Prentice-Hall.

Boulegue, J.P., et al., 1979, Field titrations of dissolved sulfur species in anoxic environments – geochemistry of Puzzichello waters (Corsica, France), *Chemical Geology*, 25, 19–29.

Carroll, D., 1959, Ion exchange in clays and other minerals, *Geological Society of America Bulletin*, 70, 749–779.

Champ, D.R., et al., 1979, Oxidation reduction sequences in ground water flow systems, *Canadian Journal of Earth Science*, 16, 12–23.

Domenico, P.A., and F.W. Schwartz, 1990, *Physical and chemical hydrogeology*, John Wiley & Sons.

Drever, J.A., 1997, *The geochemistry of natural waters*, 3rd ed, Prentice Hall.

Freeze, J.A., and J.A. Cherry, 1979, *Groundwater*, Prentice-Hall.

Fritz, P., and J.C. Fontes, 1980, *Introduction in handbook of environmental isotope geochemistry*, Elsevier, 1.

Garrels, R.M., and C.L. Christ, 1965, *Solutions, minerals, and equilibria*, Freeman, Cooper & Company.

Hem, J.D., 1970, *Study and interpretation of the chemical characteristics of natural water: U.S. Geological Survey Water-supply paper 1473.*

Kosmulski, M., 2020, The pH dependent surface charging and point of zero charge. VIII. Update, *Advances in Colloid and Interface Science*, 275.

Matthess, G., 1982, *The properties of groundwater*, translated by J.C. Harvey, Wiley and Sons.

Mayo, A.L., et al., 2020, Shallow ground water flow and inverted fresh/saline water interface in a hypersaline endorheic basin – Great Basin, USA, *Hydrogeology Journal*, 28, 2877–2902.

Morel, F.M.M., 1983, *Principles of aquatic chemistry*, Wiley and Sons.

Morel, F.M.M., et al., 1973, Interactions and chemostasis in aquatic chemical systems: Role of pH, p*E*, solubility, and complexation, in *Trace metals and metal-organic interactions in natural waters*, P.C. Singer (ed.), Ann Arbor Science, 157–200.

Parks, G.A., 1965, The isoelectric points of solid oxides, solid hydroxides, and aqueous hydroxo complex systems, *Chemical Review*, 65(2), 177–198.

Rossman, K.J.R., and P.D.P. Taylor, 1998, Isotopic compositions of the elements 1997, *Pure and Applied Chemistry*, 70(1), 217–235.

Shabunya, S., et al., 2019, Modeling of pH of concentrated aqueous solutions of sodium metaborate, *Russian Chemical Bulletin*, *68*, 1183–1189.

Sparnaay, M.J., 1972, *The electrical double layer*, Pergamon.

Stumm, W., and J.J. Morgan, 1981, *Aquatic chemistry*, Wiley & Sons.

Turner, F.J., 1968, *Metamorphic petrology*, McGraw Hill.

2 Chemical Equilibria

2.1 CHEMICAL EQUILIBRIUM

Chemical equilibrium is the condition when the concentrations of both reactants and products in a chemical reaction no longer change. In groundwater systems we are usually concerned with the processes involved in reaching equilibrium and the state of chemical equilibrium between water and the soil zone and aquifer minerals and gases.

2.1.1 Law of Mass Action

It is readily apparent that mineral solubility varies greatly from one mineral species to another and that solubility is temperature dependent. For example, in three beakers of water containing equal proportions of powdered halite (NaCl), gypsum ($CaSO_4$), and granite considerably more halite will dissolve in the halite beaker than will gypsum in the gypsum beaker, and it will be extremely difficult to measure the quantity of granite, or for that matter the quantity of the glass beaker that has dissolved. Heating the beaker will increase the solubility of each mineral but will not make gypsum more soluble than halite. Differences in mineral solubility and their temperature dependence may be described in terms of thermodynamic equilibrium. It is convenient to represent chemical reactions by the general law of mass action equation:

$$aA + bB = cC + dD \tag{2.1}$$

$$\text{(reactants) (products)}$$

where a is the moles of substance A, b is the moles of B, etc. By convention, substances on the left side of the equation are known as reactants and substances on the right side of the equation are known as products. Reactants chemically interact and yield new substances known as products. In groundwater systems reactants are typically minerals and gases that dissolve into ionic species and precipitate as minerals such as clay. Both reactants and products may contain an unlimited number of species and may be any combination of solids, gases, ionic species, and liquids. For example, there may be three reactants and four products, in which case A, B, and C would be reactants and D, E, F, and G would be products.

The symbols $\leftrightarrow$, =, and $\rightarrow$ are commonly used to separate the reactants from the products in chemical equations. These symbols indicate the direction of the reaction. The symbols $\leftrightarrow$ and = show that the reaction is reversible, and the symbol $\rightarrow$ indicates that the reaction only goes in one direction, yielding products.

In terms of Equation 2.1 the dissolution of halite and gypsum would be written respectively as:

$$NaCl = Na^+ + Cl^- \tag{2.2}$$

and

$$CaSO_4 = Ca^{2+} + SO_4^{2-} \tag{2.3}$$

Water has been omitted from the equations but could be added to both reactants and products. We omit water because there is no change in the volume of water or the isotopic composition of the

DOI: 10.1201/9781003646129-3

water in either the reactants or products. If the water isotopically fractionated during the reaction, water would be included and the isotopic composition of the fractionating hydrogen and/or oxygen would be symbolically represented. Both Equations 2.2 and 2.3 are single step reactions. Sometimes it is convenient to represent multiple step reactions as a single step. For example, the formation of bicarbonate from soil zone CO_2 gas involves the sequential reactions:

$$CO_{2(g)} + H_2O = CO_{2(aq)} + H_2O \quad (2.4)$$

$$CO_{2(aq)} + H_2O = H_2CO_3 \quad (2.5)$$

$$H_2CO_3 = HCO_3^- + H^+ \quad (2.6)$$

which are often simplified to the single equations:

$$CO_{2(g)} + H_2O = HCO_3^- + H^+ \quad (2.7)$$

or

$$CO_{2(aq)} + H_2O = HCO_3^- + H^+ \quad (2.8)$$

2.1.2 Equilibrium Defined

Chemical processes tend to occur spontaneously in the direction that will lead to equilibrium. Equilibrium is the state of lowest energy, and equilibrium processes of hydrogeologic interest include dynamic, static, and non-dynamic. Dynamic equilibrium is the condition where the reaction continues to occur after equilibrium is achieved, but the rate of the forward reaction equals the rate of the reverse reaction. For example, after NaCl dissolution reaches equilibrium the rate of continued NaCl dissolution equals the rate of NaCl precipitation. Once dynamic equilibrium is attained there is no additional change in the absolute mass of either the reactants or the products. Static equilibrium is the situation where no further forward or reverse reactions occur after equilibrium is achieved. Non-dynamic equilibrium is the condition where a reaction is not reversible and can only proceed toward products. In equations representing non-dynamic equilibrium, the arrow → is used to separate reactants from products. Many incongruent reactions, such as the dissolution of feldspar to clay, and most biological reactions are non-dynamic. During incongruent reactions the initial mineral is leached and modified and converted into a secondary solid product. In other words, one or more of the products is a solid. During congruent reactions the total reactant (i.e., the mineral phase) has the potential to completely dissolve and become a product in the solution. The actual amount of the product is determined by the equilibrium constant of the reactant.

Dynamic equilibrium may be represented in terms of the thermodynamic equilibrium constant, denoted as K, K_{eq}, or K_{sp}, where:

$$Keq = \frac{[C]^c\,[D]^d}{[A]^a\,[B]^b} \quad (2.9)$$

where the brackets [ ] are the effective molar concentration and the superscript is the moles of each species. K_{eq} is temperature dependent and must be calculated for the water temperature of the reaction. In practice the water temperature is generally assumed to be the temperature of the surface or groundwater at the sampling location. The relationship between effective molar concentration and molar concentration is discussed later in this chapter. The effective molar concentration of solid mineral species, such as $CaSO_4$, is considered to be 1. Equation 2.9 implies that chemical equilibrium occurs in all groundwater systems where the availability of reactants (e.g., aquifer minerals) is not a limiting factor. As discussed later, this assumption is not always satisfied. Equation 2.9 also provides a measure of the ability of water to both dissolve and precipitate mineral species. K_{eq} is a numerical value that describes how far the reaction will proceed from left to right and describes the completeness of the chemical reaction. K_{eq} tells us how much of the reactant will dissolve and what will be the concentration of dissolved substances in the water. The larger the K_{eq}, the greater the concentration of products relative to concentration of reactants. It is important to keep in mind that the abundance of major aquifer minerals is essentially unaffected by groundwater dissolution because the mass of minerals greatly exceed the mass of water. In other words the large mass of aquifer minerals buffers the reaction. Equilibrium constants for selected minerals at standard conditions are listed in Table 2.1. From the table it is apparent that halite is extremely soluble and explains why the oceans, as well as saline and hypersaline lakes and groundwater, mostly contain Na^+ and Cl^- ions. The table also demonstrates that the relative solubility of mineral classes may be represented as:

chlorides > carbonates > oxides > silicate and aluminosilicate

and that the relative solubility of common rock-forming minerals may be represented as:

gypsum >> calcite > dolomite >>> silicate and aluminosilicate minerals.

2.1.3 Chemical Activity

Laboratory analysis of a water sample is usually reported as the absolute mass of constituent per unit weight or unit volume of water (i.e., ppm, mg/L). However, most surface and groundwaters are nonideal solutions in that the interaction of ions with each other and with water molecules makes the ions chemically behave as if they are less concentrated than are their laboratory reported abundances. The effective molar concentration of an ion or molecule is known as chemical activity or simply activity. Concentration units used in equilibrium calculations such as Equation 2.9 are activity rather than the laboratory-reported abundance. Chemical activity may be thought of as the effective molar concentration and is symbolically represented by the prefix *a* or is enclosed in brackets [ ]. For example, the activity of Ca^{2+} may be represented as *a* or $[Ca^{2+}]$. Laboratory-reported values are often represented in parentheses ( ). The chemical activity of a gas is reported as the partial pressure of the gas in atmospheres. For dilute solutions, such as most groundwater, the activity of solid mineral phases and the solvent water are assigned a value of 1.

Chemical activity may be measured or calculated from other known parameters. The chemical activity of many ions may be measured directly by means of specific ion electrodes. The most common ion that is directly measured is H^+. The activity of H^+ is reported as the familiar pH unit. The p in pH symbolically represents the negative log, and H represents hydrogen ion activity. Consequently, pH actually stands for the negative log of the hydrogen ion activity. For reasons of convenience, the chemical activity of most ions is calculated rather than directly measured. Chemical activities of individual ions may be calculated using laboratory-determined molality and the activity coefficient (y) for the ion:

$$a_i = m_i y_i \tag{2.10}$$

or

$$[\ ]_i = (\ )_i y_i \tag{2.11}$$

TABLE 2.1
Equilibrium Constants for Selected Reactions at 25 °C and 1 Bar Total Pressure

Mineral	Dissociation Reaction	Keq	Source
Silicate			
Albite	$NaAlSi_3O_8 + 8H_2O = Na^+ + Al(OH)_4^- + 3H_4SiO_4$	9.55×10^{-10}	5
Anorthite	$CaAl_2Si_2O_8 + 8H_2O = Ca^{2+} + 2Al(OH)_4^- + 2H_4SiO_4$	9.55×10^{-19}	5
K-feldspar	$KAlSi_3O_8 + 8H_2O = K^+ + Al(OH)_4^- + 3H_4SiO_4$	1.93×10^{-20}	5
Carbonate			
Aragonite	$CaCO_3 = Ca^{2+} + CO_3^{2-}$	6.03×10^{-9}	3
Azurite	$Cu3(CO_3)_2(OH)_2 = 3CU^{2+} + 2CO_3^{2-}$	10^{-46}	3
Calcite	$CaCO_3 = Ca^{2+} + CO_3^{2-}$	4.67×10^{-9}	3
Cerussite	$PbCO_3 = Pb^{2+} + CO_3^{2-}$	7.94×10^{-14}	3
Dolomite	$CaMg(CO_3)_2 = Ca^{2+} + Mg^{2+} + 2CO_3^{2-}$	10^{-17}	3
Magnesite	$MgCO_3 = Mg^{2+} + CO_3^{2-}$	3.47×10^{-8}	3
Malachite	$Cu_2CO_3(OH)_2 = 2Cu^{2+} + CO_3^{2-} + 2OH^-$	1.59×10^{-34}	3
Nesquehonite	$MgCO_3 \bullet 3H_2O = Mg^{2+} + CO_3^{2-} + 2H_2O$	2.14×10^{-5}	3
Rhodochrosite	$MnCO_3 = Mn^{2+} + CO_3^{2-}$	3.98×10^{-11}	3
Siderite	$FeCO_3 = Fe^{2+} + CO_3^{2-}$	1.96×10^{-11}	3
Strontianite	$SrCO_3 = Sr^{2+} + CO_3^{2-}$	10^{-9}	3
Chloride			
Chlorargyrite	$AgCl = Ag^+ + Cl^-$	$10^{-9.74}$	3
Halite	$NaCl = Na^+ + Cl^-$	$10^{+1.54}$	3
Sylvite	$KCl = K^+ + Cl^-$	$10^{+0.98}$	1
Fluorite			
Fluorite	$CaF_2 = Ca^{2+} + 2F^-$	1.59×10^{-10}	1
Hydroxide and Oxide			
Calcium Hydroxide	$Ca(OH)_2 = Ca^{2+} + 2OH^-$	6.46×10^{-6}	3
Cupric Hydroxide	$Cu(OH)_2 = Cu^{2+} + 2OH^-$	3.98×10^{-20}	3
Brucite	$Mg(OH)_2 = Mg^{2+} + 2OH^-$	7.94×10^{-12}	3
Ferris Hydroxide	$Fe(OH)_2 = Fe^{2+} + 2OH^-$	7.94×10^{-16}	4
Goethite	$FeOOH + H_2O = Fe^{3+} + 3OH^-$	3.16×10^{-42}	3
Gibbsite	$Al(OH)_3 = Al^{3+} + 3OH^-$	3.16×10^{-34}	3
Hematite	$Fe2O_3 + 3/2H_2O = Fe^{3+} + 3OH^-$	15.85×10^{-12}	4
Litharge (red)	$PbO + H_2O = Pb^{2+} + 2OH^-$	5.01×10^{-16}	4
Massicot (yellow)	$PbO + H_2O = Pb^{2+} + 2OH^-$	7.94×10^{-16}	4
Manganous Hydroxide	$Mn(OH)_2 = Mn^{2+} + 2OH^-$	1.59×10^{-13}	4
Tenorite	$CuO + H_2O = Cu^{2+} + 2OH^-$	3.98×10^{-20}	4
Sulfide			
Anhydrite	$CaSO_4 = Ca^{2+} + SO_4^{2-}$	4.93×10^{-5}	5
Gypsum	$CaSO_4 \bullet 2H_2O = Ca^{2+} + SO_4^{2-} + H_2O$	2.62×10^{-5}	5
Celestite	$SrSO_4 = Sr^{2+} + SO_4^{2-}$	2.34×10^{-7}	3

Sources:

[1] Freeze and Cherry, 1979
[2] Helgeson et al., 1978
[3] Smith and Martell, 1976
[4] Bates and Mesmer, 1976
[5] Parkhurst and Appelo, 2013

where the subscript i represent any ion or molecule and m or () represent laboratory measured or theoretical molality of the ion or molecule. Activity coefficients are dimensionless parameters, which are related to the ionic strength of the solution and to the radius and charge of the ion or molecule. The ionic strength (I) of a solution may be calculation according to the equation:

$$I = \frac{1}{2}\Sigma m_i z_i \tag{2.12}$$

Where Σ is the sum of all species, m_i is the molality of species I, and z_i is the valance or charge of species i. For example, to calculate the ionic strength of a solution only containing Ca^{2+} and HCO_3^- you would determine the molarity of both ions in the laboratory and use 2 and 1 for the Zi of Ca^{2+} and HCO_3^-, respectively.

Example E2.1 Calculate ionic strength

Problem: Calculate the ionic strength for the following waters:

				-------------------------- meq/L --------------------------							
System	**Samples**	**°C**	**pH**	**Ca^{2+}**	**Mg^{2+}**	**Na^+**	**K^+**	**HCO_3^-**	**SO_4^{2-}**	**Cl^-**	**Total**
Island crystalline system	12	14.0	8.2	2.53	1.42	4.06	0.07	3.1	1.52	3.19	15.98
Island sedimentary system	8	13.6	8.2	4.6	2.21	10.4	0.42	4.6	1.97	10.99	45.06

Solution: Use the conversion factors in Appendix A to convert the concentrations to mg/L by dividing each concentration by the appropriate F_1. If the concentrations are in mg/L, this step is not necessary. The converted concentrations are:

	------------------------------------ mg/L ------------------------------------							
System	**Ca^{2+}**	**Mg^{2+}**	**Na^+**	**K^+**	**HCO_3^-**	**SO_4^{2-}**	**Cl^-**	**Total**
Island crystalline system	50.70	17.26	93.33	2.74	194.64	73.01	113.08	544.75
Island sedimentary system	92.18	26.86	239.08	16.42	280.66	94.62	389.58	1139.40

Using Equation 1.7

$$M = \frac{mg}{L} \times \frac{10^{-3}}{\text{gfw of solute}}$$

where the gfw of each species is from the periodic table. The calculated molarity of each species is:

System	**Ca^{2+}**	**Mg^{2+}**	**Na^+**	**K^+**	**HCO_3^-**	**SO_4^{2-}**	**Cl^-**
Island crystalline system	0.00127	0.00075	0.00406	0.00007	0.00319	0.00076	0.00319
Island sedimentary system	0.00230	0.00117	0.01040	0.00042	0.00460	0.00099	0.01100

The ionic strength of a solution may be calculated using Equation 2.12:

$$I = \frac{1}{2}\Sigma m_i z_i$$

Where Σ is the sum of all species, m_i is the molality of species i and z_i is the valance or charge of species I. For the Island crystalline system, the ionic strength is:

$$I = (0.00127 \times 2 + 0.00075 \times 2 + 0.00406 + 0.00007 + 0.00319 + 0.00076 \times 2 + 0.00319) \times \tfrac{1}{2} = 0.016$$

The ionic strength of the Island sedimentary system is 0.035.

Chemical activities approach laboratory-measured solute concentrations in very dilute solutions. With increasing ionic strength, activities first decrease and then increase when the solute concentrations are extremely high (Figure 2.1). Activity coefficients differ greatly between the dissolved species in the same solution. This nonideal behavior of ions and molecules is primarily the result of the long-range electrostatic effects of the ionic charges. Ionic charges inhibit randomness in a solution because like charges are repelled from each other and dissimilar charges are attracted to each other.

It is very difficult if not impossible to precisely determine chemical activities in surface and groundwater. Fortunately, in most investigations it is possible to estimate activity coefficients within an acceptable level. Numerous mathematical models have been developed to calculate ion activity coefficients for individual species and molecules. It is seldom necessary to manually calculate activity coefficients because one or more activity coefficient models are incorporated into most geochemical modeling computer codes. The five most commonly used models for calculating activity coefficients are listed in Table 2.2, and their equation constants are listed in Table 2.3. Additional models include the empirical mean salt (Harned et al., 1959), refer to Eqs 2.19 to 2.22, and the B-dot (Helgeson, 1969) which is an extended version of the Debye-Hückel. Considerable variations in the constants used in activity coefficient models are reported in the literature, and for critical analytical work the original references should be consulted.

It is important to select the appropriate activity coefficient model because the models yield significantly different results with varying ionic strengths (Figure 2.2). In very dilute solutions of univalent

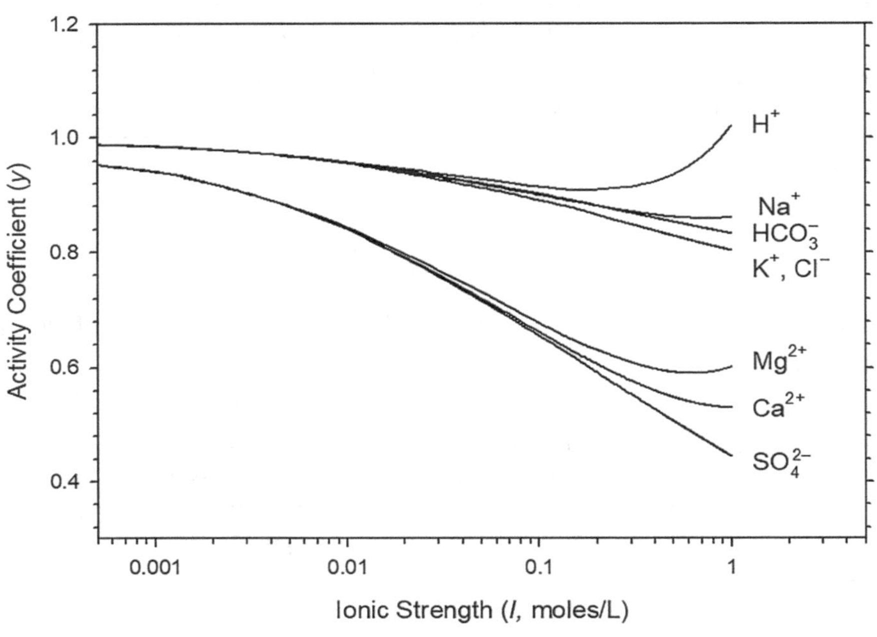

FIGURE 2.1 Activity coefficients vs. ionic strength of selected ions. Activity coefficients are based on the Truesdell-Jones activity model.

TABLE 2.2
Commonly Used Models for Calculating Activity Coefficients of Individual Ions or Charged Molecules Based on the Total Ionic Strength of the Solution

Acceptable range of method	Equation	Ionic strength (mole/L)	
Debye-Hückel	$\log y_i = -Az_i^2 \sqrt{I}$	$<10^{-2.3}$	(2.13)
Extended Debye-Hückel	$\log y_i = -Az_i^2 \dfrac{\sqrt{I}}{1 + a_i B\sqrt{I}}$	0 to 10^{-1}	(2.14)
Davies[1]	$\log y_i = -Az_i^2 \left(\dfrac{\sqrt{I}}{1+\sqrt{I}} - 0.3I \right)$	0 to 0.6	(2.15)
Truesdell-Jones[2]	$\log y_i = -Az_i^2 \left(\dfrac{\sqrt{I}}{1+Ba_i\sqrt{I}} \right) + b_i I$	0 to 4	(2.16)
Pitzer[3]	Use computer code such as PHREEQC	0 to 20	(2.17)

y_i = activity coefficient
a_i = the radius of the hydrated sphere of the ion or molecular (see Table 2.3); Truesdell-Jones model uses different values
I = ionic strength of solution
A and B = pressure and temperature dependent constants related to the solvent. A depends on the dielectric constant for water at a given temperature. B (see Table 2.3).

Sources:
[1] Davies, 1962
[2] Truesdell and Jones, 1974
[3] Pitzer, 1973, 1979

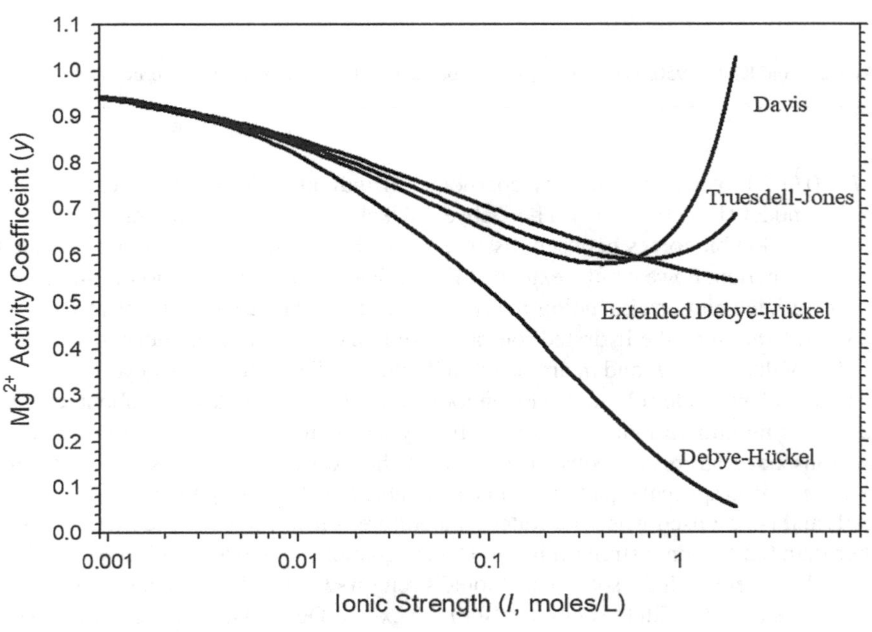

FIGURE 2.2 Magnesium activity coefficients vs. ionic strength calculated using different activity coefficient models. The models yield significantly different results at varying ionic strengths.

TABLE 2.3
Equation Constants for Use in the Activity Coefficient Models (See Table 2.2)

				Extended Debye-Hückel	Truesdell-Jones	
°C	*A*	*B* (10^8)		a_i (10^{-8})	a_i (10^{-8})	*b*
0	0.4883	0.3241	Al^{+3}	9.0	6.65	0.19
5	0.4921	0.3249	Ba^{2+}	5.0	4.55	0.09
10	0.4960	0.3258	Ca^{2+}	6.0	4.86	0.15
15	0.5000	0.3262	Cd^{2+}	5.0	5.8	0.10
20	0.5042	0.3273	Cl^-	3.0	3.71	0.01
25	0.5085	0.3281	CO^{2+}	6.0	6.17	0.22
30	0.5130	0.3290	CO_3^-	4.5	5.4	2.00
35	0.5175	0.3297	F^-	3.5	3.46	0.08
40	0.5221	0.3305	Fe^{2+}	6.0	5.08	0.16
45	0.5287	0.3334	H^+	9.0	4	0.24
50	0.5319	0.3321	HCO_3^{2-}	4.0	5.4	0.00
55	0.5399	0.3360	K^+	3.0	3.71	0.01
60	0.5425	0.3338	Mg^{2+}	8.0	5.46	0.22
			Mn^{2+}	6.0	7.04	0.22
			Na^+	4.0	4.32	0.06
			NH_4^+	2.5	–	–
			Ni_4^+	6.0	5.51	0.22
			OH^-	3.5	10.65	0.21
			Pb^{2+}	4.5	4.8	0.01
			SO_4^{2-}	4.0	5.31	–0.07
			Sr^{2+}	5.0	5.48	0.11
			Zn^{2+}	6.0	4.87	0.24

Source: Data are from Klotz, 1950, Manov et al., 1943, Robinson and Stokes, 1970, and Truesdell and Jones, 1974.

ions (i.e., $I < 0.005$ moles/L) the activity coefficient of individual ions may be calculated using the Debye-Hückel model (Equation 2.13). The Debye-Hückel equation is pioneering work that considers the long-range coulombic forces between dissolved ions. Equation 2.13 is seldom used due to its low ionic strength limitation; however, the extended Debye-Hückel model (Equation 2.14) has widespread application for dilute solutions including many groundwaters. The extended Debye-Hückel equation includes a consideration of the hydrated ion radius (a_i), and the temperate dependent solutions constants A and B. Values of A, B and a_i are listed in Table 2.2. The extended Debye-Hückel equation is applicable to univalent dilute solutions with an ionic strength <0.10 moles/L or about <5,800 mg/L of NaCl. The usable maximum ionic strength is slightly less than 0.10 moles/L in most water because multivalent ions have stronger electrical attractions than do univalent ions. The Davies equation (Equation 2.15) is an empirical equation that is applicable to solutions up to an ionic strength of about 0.60 moles/L and is best used when the water temperature is near 25 °C. Some literature uses a factor of 0.2 rather than 0.3 for ionic strength in the Davis equation. The extended Debye-Hückel equation provides slightly better result for solutions of ionic strength <0.10 moles/L than does the Davies equation and should be used for dilute solutions. Both the extend Debye-Hückel and the Davies equations are included in many geochemical modeling codes.

The extended Debye-Hückel and Davies equations do not reliably predict the individual activity coefficients for brackish, saline, and hypersaline solutions, because ionic behavior in these solutions is affected by the formation of ion pairs and complex ions in addition to long range electrostatic effects. At ionic strengths greater than about 0.5 moles/L the activity coefficients of many ions tend to increase with increased ionic strength (Figure 2.1).

Several methods have been developed to calculate activity coefficients for ion pairs and highly concentrated solutions. The activity of ion pairs can be calculated using the ion pair approach. This method is based on the idea that single ion activity coefficients may be determined using the Debye-Hückel theory and that ion pair activity coefficients may be determined using single ion activity coefficients and ion pair stability constants. For example, the activity coefficient of the ion pair $CaSO_4^o$ is related to the following stability constant (Appendix B):

$$K_{\text{stab}} = \frac{\text{aCaSO}_4^o}{\left(\text{aCa}^{2+})(\text{aSO}_4^{2-}\right)} \tag{2.17}$$

and

$$a\text{CaSO}_4^o = \frac{\left(a\text{Ca}^{2+}\right)\left(\text{aSO}_4^{2-}\right)}{K_{stab}} \tag{2.18}$$

where K_{stab} is the stability constant of the neutral ion pair. Sometimes stability constants are listed in the reciprocal form and are known as instability constants. The effect of ion pair formation is to decrease the concentration of the ions that participate in ion pairing and to a lesser extent to decrease the ionic strength of the solution.

The activity of ion pairs can also be calculated using the mean-salt method, which assumes that the $a_{\text{cation}} = a_{\text{anion}}$ in each univalent–univalent electrolyte pair for the ionic strength of interest (Garrels and Christ, 1965). The behavior of KCl, which may be determined experimentally, is the standard electrolyte for determining individual ion activity coefficients:

$$a\text{KCl} = \left\{\left[a\text{K}^+\right]\left[a\text{Cl}^-\right]\right\}^{½} = a\text{K}^+ = a\text{Cl}^- \tag{2.19}$$

Using the relationship in Equation 2.19, a table of values for other ions may be prepared. For monovalent chloride the activity coefficients of monovalent cations are calculated according to:

$$a\text{MCl} = \left[\left(a\text{M}^+\right)\left(a\text{Cl}^-\right)\right]^{½} = \left[\left(a\text{M}^+\right)\left(a\text{KCl}\right)\right]^{½} \tag{2.20a}$$

and

$$\text{aM}^+ = \frac{a^2\text{MCl}}{\text{aKCl}} \tag{2.20b}$$

where M is any monovalent cation. The activity coefficient of divalent cations is calculated according to:

$$\text{aMCl}^2 = \left[\left(\text{aM}^{2+}\right)\left(\text{aCl}^-\right)^2\right]^{1/3} = \left[\left(\text{aM}^{2+}\right)\left(\text{aKCl}^2\right)\right]^{1/3} \tag{2.21a}$$

and

$$\text{aM}^{2+} = \frac{a^3\text{MCl}^2}{a^2\text{KCl}} \tag{2.21b}$$

For anions such as SO_4^{2-}, activity coefficients may be calculated according to:

$$aK_2SO_4^{2-} = [(aK^+)^2 (aSO_4^{2-})]^{1/3} = [(aK^+)^2 (aSO_4^{2-})]^{1/3} \quad (2.22a)$$

and

$$aSO_4^{2-} = \frac{a^3 k_2SO_4^{2-}}{a^2\ KCl} \quad (2.22b)$$

The three most common methods for calculating activity for high salinity solutions are the mean-salt, Truesdell-Jones, and Pitzer. The Truesdell-Jones equation is based expansion of the Debye-Hückel using empirical data and includes both a_i and b parameters. The Truesdell-Jones a_i parameter is not applicable to other models. Calculation of activity coefficients in more concentrated solution such as brines requires more sophisticated models such as proposed by Pitzer (1973, 1979, 1980), and Pitzer and Kim (1974). The Pitzer model is more difficult to use but has provided accurate estimates for up to 20 M solutions.

Example E2.2 Calculate activity coefficients

Problem: Use the ionic strengths calculated in Example E2.1, the extended Debye-Hückel method (Equation 2.14), and the temperature dependent coefficients (Table 2.3) to calculate the activity coefficient Ca^{2+} in the Island crystalline and Island sedimentary waters.

Solution: Based on a water temperature of approximately 14 °C reasonable estimates of A and B coefficients are 0.499 and 0.3261 × 10⁻⁸, respectively. The activity coefficients for Ca^{2+} are calculated as:

$$\log y_i = -Az_i^2 \frac{\sqrt{I}}{1 + a_iB\sqrt{I}}$$

$$\log \gamma_{Ca^{2+}} = \frac{-0.499 \times 2^2 \times 0.016^{0.5}}{1 + 6 \times 10^{-8} \times 0.3261 \times 10^{-8} \times 016^{0.5}} = -0.252$$

$$\gamma_{Ca^{2+}} = 10^{-0.252} = 0.56$$

and

$$\log \gamma_{Ca^{2+}} = \frac{-0.499 \times 2^2 \times 0.035^{0.5}}{1 + 6 \times 10^{-8} \times 0.3261 \times 10^{-8} \times 035^{0.5}} = -0.373$$

$$\gamma_{Ca^{2+}} = 10^{-0.373} = 0.424$$

Important Takeaway – Laboratory analysis of water chemistry must be converted to effective molar concentration, also known as activity, before the data are used for geochemical modeling. Laboratory reported chemical analysis are converted to activity by multiplying the laboratory data by the activity coefficient adjusted for the ionic strength (i.e., salinity) of the surface or groundwater. Several models are available for calculating activity, and care must be taken when selecting the activity coefficient model. The Debye-Hückel and Truesdell-Jones methods are most commonly used in groundwater investigations. Raw laboratory data may be used for many graphical representations such as Stiff and Trilinear plots (Chapter 12).

2.2 THERMODYNAMICS

All geochemical reactions and transformations are accompanied by changes in energy. The ultimate state of energy is heat. Other energy forms include mechanical, electrical, radiant, and chemical. Commonly used energy measurement units include calorie (cal) and joule (J) for heat and electrical energy, and erg for mechanical energy. One cal equals 4.184 J. The study of the various forms of energy and their interactions is known as thermodynamics. A system is any portion of the universe that is isolated from the rest of the universe by a real or an imagined container or barrier. A system may be as large as the whole universe or as small as the dissociation of a single molecule of water in to H^+ and OH^-. Thermodynamic investigations are usually concerned with time independent changes in the energy of systems. The study of time dependent changes is known as kinetics (Chapter 4).

Thermodynamic calculations are an integral part of many hydrogeochemical investigations. They are necessary to determine if a water will dissolve, is in equilibrium with, or will precipitate a mineral and are useful for evaluating potential geochemical reaction pathways and the fate of mixed groundwater systems. Knowledge of reaction pathways is an extremely useful tool in that it can help identify which bedrock lithologies a groundwater has or has not been in contact with and help sort out the resulting groundwater chemistry when two or more surface and/or groundwaters mix and what the result will be when a solution is injected into a bedrock formation. Reaction pathway calculations are easily and routinely performed using a variety of readily available computer models (Chapter 12).

Working hydrogeologists seldom, if ever, need to manually perform thermodynamic calculations, and many hydrogeochemical investigations and reaction pathway calculations can be performed without an understanding of the principles of thermodynamics. Nonetheless, a basic understanding of the concepts of enthalpy, entropy, and Gibbs free energy will greatly enhance your understanding and interpretation of the geochemistry in groundwater systems. It is also necessary to manually calculate the equilibrium constants for unusual mineral phases that are not included in the thermodynamic libraries of popular geochemical modeling codes. Once calculated, the constant can be included in the model code library.

The thermodynamic datasets containing enthalpy, entropy, and Gibbs free energy values lie at the heart of thermodynamic calculations. Over the years numerous compilations of these data have been developed and are published in numerous texts and online. The dataset must be internally consistent, which means the data are based on common scales for temperature, energy, atomic mass, and physical constraints (Lu et al., 2022). Each datum should also have an appropriate reference. Thermodynamic data for some species of hydrogeochemical interest are compiled in Appendix C.

2.2.1 Enthalpy and Entropy

Geochemical reactions obey the first and second laws of thermodynamics. The first law of thermodynamics, the conservation of energy, states that energy cannot be created or destroyed. However, during chemical reactions energy is often transformed into heat. Such transfers are measured by changes in enthalpy. Enthalpy (H) is a measure of the heat of a system and may be defined as:

$$H = E + PV \tag{2.23}$$

where E is the total energy of the system and P and V are pressure and volume, respectively. Units of H are kilojoules per mole (kJ/mol) or calories per mole (cal/mol). In most groundwater systems the approximation $H = E$ is reasonable because P and V are generally constant and may be neglected. The total energy of the system includes both kinetic and potential energy. Kinetic energy is due to the molecular motion of the system and potential energy is due to the structure or form of the system. The absolute value of energy may be calculated from Einstein's equation $E = mc^2$, where m is mass and c is the velocity of light. In aqueous thermodynamics, where the energy involved in chemical reactions is extremely small, such a calculation would be extremely difficult and is not necessary. Instead, it is easier and considerably more practical to calculate energy from the quantity known as the heat of the reaction (ΔH) which is the change in H associated with the chemical reaction according to:

$$\Delta H = H_2 - H_1 \tag{2.24}$$

where H_2 and H_1 are the enthalpy of the products and reactants of a chemical reaction, respectively. It is instructive to think of the heat of the reaction as the change in heat content of each substance during the chemical reaction. Most reactions are exothermic; that is, heat is liberated by the reaction. The burring of gasoline or diesel fuel in your car is exothermic, which is why internal combustion auto engines require a radiator and radiator fluid to manage the engine heat. Some reactions are endothermic, where heat is consumed by the reaction. Instant cold compresses used in first aid are based on endothermic chemical reactions. The heat of a reaction may be determined experimentally by measuring the change in heat accompanying the reaction or by calculating the heat using known heats of reactions.

The enthalpy for 1 mole of a substance at standard pressure and temperature is designated as ΔH^o. The enthalpy of all elements in their standard states is 0. Standard heats of reactions have been calculated for the heat of formation of compounds, solutions, and ions, the heat of combustion, and the heat of fusion, vaporization, sublimation, dissociation, etc. Heats of reactions for most chemical reactions are readily available from the literature and the internet or can be calculated from other known heats of reactions. Selected ΔH values are in Appendix C.

Entropy is a measure of the randomness of the system. The second law of thermodynamics states that natural processes are accompanied by an increase in entropy and that the direction of change is always toward increasing entropy (S). For example, when electrical or mechanical energy is used to perform work, heat is produced, and the chaos, disorder, or mixed-upness of the energy in the system increases. It is helpful to remember that heat energy is due to the random motion of atoms and molecules, and the greater the heat energy the greater the random motion. The entropy of a system is dependent upon pressure, volume, and temperature and is usually expressed at standard entropy conditions (S^o). By convention, standard entropy is usually reported at 25 °C, but S^o may be reported for other temperatures. Units of S are joules per mole degree (J/deg-mol). As in the case of enthalpy, we are usually concerned with the change in entropy of a chemical reaction or process rather than the absolute entropy of the system. The change in entropy (ΔS) is:

$$\Delta S = S_2 - S_1 \tag{2.25}$$

where S_2 and S_1 are the entropies of the final and initial states of the system, respectively.

2.2.2 Gibbs Free Energy

The measure of the net energy available for a chemical reaction at a constant temperature and pressure is described by Gibbs free energy (Gibbs, 1948):

$$G = H - TS \tag{2.26}$$

where
G = Gibbs free energy of formation
H = enthalpy in calories
T = temperature in degrees Kelvin (°K)
S = entropy in calories/°K

G may be used to evaluate the potential for a chemical reaction to occur spontaneously. For the reaction:

$$a\mathrm{A} + b\mathrm{B} = c\mathrm{C} + d\mathrm{D} \tag{2.27}$$

the Gibbs free energy of the reaction may be calculated from the free energy of formation of the products minus the free energy of formation of the reactants:

$$\Delta G^{\circ}_{\mathrm{R}} = \left(\mathrm{c}\Delta G^{\circ}_{\mathrm{C}} + \mathrm{d}\Delta G^{\circ}_{\mathrm{D}}\right) - \left(\mathrm{a}\Delta G^{\circ}_{\mathrm{A}} - \mathrm{b}\Delta G^{\circ}_{\mathrm{B}}\right) \tag{2.28}$$

and

$$\Delta G^{o}_{R} = \Delta G^{o}_{p} - \Delta G^{o}_{r} \tag{2.29}$$

where
$\Delta G^{\circ}_{\mathrm{R}}$ = Gibbs free energy of formation of the reaction,
$\Delta G^{\circ}_{\mathrm{p}}$ = Gibbs free energy of formation of the products, and
$\Delta G^{\circ}_{\mathrm{r}}$ = Gibbs free energy of formation of the reactants, for standard conditions.

The standard free energy of formation is the free energy needed to yield 1 mole of a substance or compound from a stable element. Common units for G are J/mole and kcal/mole. The ΔG° of elemental forms is taken as 0 at 25°C (298°K) and 1 atmosphere of pressure. At a constant temperature and pressure, the change in free energy for a reaction is:

$$\Delta G^{\circ}_{\mathrm{R}} = \Delta \mathrm{DH} - T\Delta \mathrm{S} \tag{2.30}$$

The sign (positive, neutral, or negative) of a calculated Gibbs free energy change for a reaction is a measure of the thermodynamic tendencies of reaction to occur:

$G_{\mathrm{R}}{}^{\circ} = -$ (spontaneous reaction)

$G_{\mathrm{R}}{}^{\circ} = 0$ (equilibrium)

$G_{\mathrm{R}}{}^{\circ} = +$ (nonspontaneous reaction)

A negative free energy value does not guarantee that the reaction will occur, but rather it means the reaction can occur if other conditions are right. Many reactions require a catalyst to proceed toward equilibrium, or equilibrium may be hampered by kinetic factors that are discussed in Chapter 4.

Few groundwater systems have a temperature of 25 °C, so most groundwater investigations require the calculation of equilibrium constants for nonstandard temperatures. Equilibrium constants may be calculated for any temperature by using the Gibbs fee energy of the reactions. The change of free energy ($\Delta G^{\circ}_{\mathrm{R}}$) for a reaction at any temperature (T) obeys the following relationship:

$$\Delta G_{R} = \Delta G^{o}_{R} + \frac{[C]^{c}[D]^{d}}{[A]^{a}[B]^{b}} \tag{2.31}$$

where
ΔG_{R} = the Gibbs free energy of the reaction at temperature T
R = the universal gas constant (8.3143 J/mol-T or 1.98717 cal/mol-T),
T = degrees Kelvin.

The chemical activities of the reactants are those at the beginning of the reaction, and the products are those at the end of the reaction. For a reaction to proceed spontaneously there must be a net decrease in free energy, that is $\Delta G_{\mathrm{R}} < 0$. A negative value is required because the calculation is based on subtracting the reactants from the products.

At a constant temperature and pressure equilibrium occurs when $\Delta G_R^o = O$. Therefore, the equilibrium condition for Equation 2.31 is:

$$0 = \Delta G_R^o + RT \ln \frac{[C]^c [D]^d}{[A]^a [B]^b} \tag{2.32}$$

and

$$\Delta G_R^o = -RT \ln \text{Keq} \tag{2.33}$$

or in the useful form:

$$\Delta G_R^o = -RT \ln \frac{[C]^c [D]^d}{[A]^a [B]^b} \tag{2.34}$$

Equilibrium constants can be calculated for any reaction using Equation 2.34. By substituting standard conditions into Equation 2.34, the equilibrium constant can be obtained from free energy data for standard state reactions:

$$\log K = -0.175\Delta G_R^o \text{ (where } \Delta G_R^o \text{ is in kJ/mole)} \tag{2.35}$$

$$\log K = -0.733\Delta G_R^o \text{ (where } \Delta G_R^o \text{ is in kcal/mole)} \tag{2.36}$$

Defining the equilibrium constant (K) for the desired temperature as K_T, the equilibrium constant for the reference temperature (i.e., 25 °C) as K_T*, and T as the desired temperature and $T*$ as the reference temperature, Equation 2.34 may be written as:

$$\text{LogK}_T = \text{LogK}_{T^*} - \frac{\Delta H_{T^*}}{2.3R}\left(\frac{1}{T} - \frac{1}{T^*}\right) \tag{2.37}$$

As discussed later in this chapter, knowledge of Log K_T is necessary for calculating the extend of mineral saturation for each mineral of hydrogeochemical interest. Fortunately, readily available geochemical modeling codes calculate Log K_T.

Example E2.3 Calculate the Gibbs free energy

Problem: Calculate the Gibbs free energy and the equilibrium constant for water (K_w) at 25 °C and 1 atm pressure.

Solution: The dissociation of water may be represented as:

$$H_2O = H^+ + OH^- \tag{E2.3a}$$

From Appendix C, the Gibbs free energy of products and reactants are:

	ΔG_r^o (kJ/mole)
H_2O	−237.14
H^+	0
OH^-	−157.2

By substituting the appropriate $H_2O = H^+ + OH^-$ values into Equation 2.29, the Gibbs free energy of Equation E2.2a can be calculated:

$$\Delta G_r^o = -157.2 - (-237.14) \quad \text{(E2.3b)}$$

$$= 79.94 \text{ kJ/mole} \quad \text{(E2.3c)}$$

Log K_w can then be determined by substituting Equation E2.3b into Equation 2.35

$$\log K_w = (-0.1753) \times (79.94) = -14.0 \quad \text{(E2.3d)}$$

and

$$K_w = 10^{-14} \quad \text{(E2.3f)}$$

Example E2.4 Calculate Gibbs free energy and the equilibrium constant

Problem: Calculate the standard state Gibbs free energy and the equilibrium constant for dissociation of calcite at 25 °C.

Solution: The dissolution of calcite can be represented by the reaction:

$$CaCO_3 + H_2O + CO_{2(g)} = Ca^{2+} + 2HCO_3^- \quad \text{(E2.4a)}$$

From Appendix C, the Gibbs free energy of products and reactants are:

	ΔG_R^o (kJ/mole)
$CaCO_3$	−1128.79
H_2O	−237.14
CO_2	−394/37
Ca^{2+}	−552.58
HCO_3^-	−586.8

By substituting the appropriate ΔG_r^o values into Equation 29, the Gibbs free energy of reaction E2.4a can be calculated:

$$\Delta G_R^o = (-552.28 + 2(-586.8)) - (-1128.79 - 237.14 - 394.37)\text{kJ/mole} = 34.42\text{kJ/mole} \quad \text{(E2.4b)}$$

Log K can then be determined by substituting E2.4b into Equation 2.35:

$$\log K = (-0.175) \times (34.42) = -6.02 \quad \text{(E2.4c)}$$

and

$$K = 9.5 \times 10^{-7} \quad \text{(E2.4d)}$$

The calculated equilibrium constant for calcite dissolution differs from the constant in Table 2.1 because the carbon product in this example is HCO_3^- and the product in the table is CO_3^{2-}. Using Appendix C data, the calculated equilibrium constant for the system $CaCO_3 = Ca^{2+} + CO_3^{2-}$ is 3.51×10^{-9}.

Important Takeaway – The solubility of a mineral species in aqueous systems is described by a temperature dependent equilibrium constant (K or preferably K_{eq}). Knowledge of the temperature-corrected equilibrium constants are necessary for determining if a ground or surface water can dissolve or precipitate a mineral. Equilibrium constants for any water temperature can be calculated using Gibbs free energy data. Geochemical reaction programs, such as PHREEQC, calculate the equilibrium constants for most reactions of interest in hydrogeology.

2.2.3 Phase Rule

The phase rule, first formulated by J. Willard Gibbs in 1875 (Gibbs, 1948), describes the number of phases that thermodynamically may coexist in equilibrium in a heterogeneous system containing a specified number of components. The phase rule, a mathematical consequence of thermodynamic equilibrium, is used to verify the mathematical appropriateness of proposed geochemical reactions. In practice the phase rule is used to determine the number of phases (ice, water, water vapor, $CO_{(g)}$, HCO_3^-, $CaCO_{3(s)}$, etc.) that may coexist in equilibrium. The phase rule is important in hydrogeochemical investigations because the commonly used aqueous geochemical models obey the phase rule. The phase rule is frequently expressed as

$$f = (c + 2) - p \tag{2.38}$$

where

f = degrees of freedom, that is the number of independent variables which must be fixed to define the system,
c = number of components in the system,
2 = the components temperature and pressure, and
p = number of phases.

A phase is a part of a system having uniform properties that are physically and chemically distinct and mechanically separable. Each phase is separated from every other phase by a physical boundary, although a phase may exist in small portions. For example, in the water system water may exist in three phases, vapor, liquid, and solid. Mechanically separable means that each phase can be mechanically separated from other phases by processes such as filtration, sedimentation, etc. Phases may include gaseous, gaseous mixtures, homogeneous liquid solutions, uniform solids, and solid solutions. Because gases are completely miscible there can only be one gas phase. There is no theoretical upper limit on the number of possible liquid and solid phases. A system containing a single phase is homogeneous, whereas a system containing multiple phases is heterogeneous.

The number of components is the smallest number of independently variable chemical constituents, written as chemical formulas, that are needed to describe the system. The degrees of freedom may be thought of as the number of conditions, such as temperature, pressure, and concentrations, that can be arbitrarily varied. Consider a single component system consisting of pure H_2O. The phase rule predicts the following possible conditions:

Number of components	Number of phases	Degrees of freedom
1	1	2
1	2	1
1	3	0

which are graphically illustrated in Figure 2.3. The maximum number of degrees of freedom, 2, are plotted as the variables pressure and temperature. When H_2O exists as a single phase of vapor, liquid, or ice, both temperature and pressure may vary without a change in phase. However, when

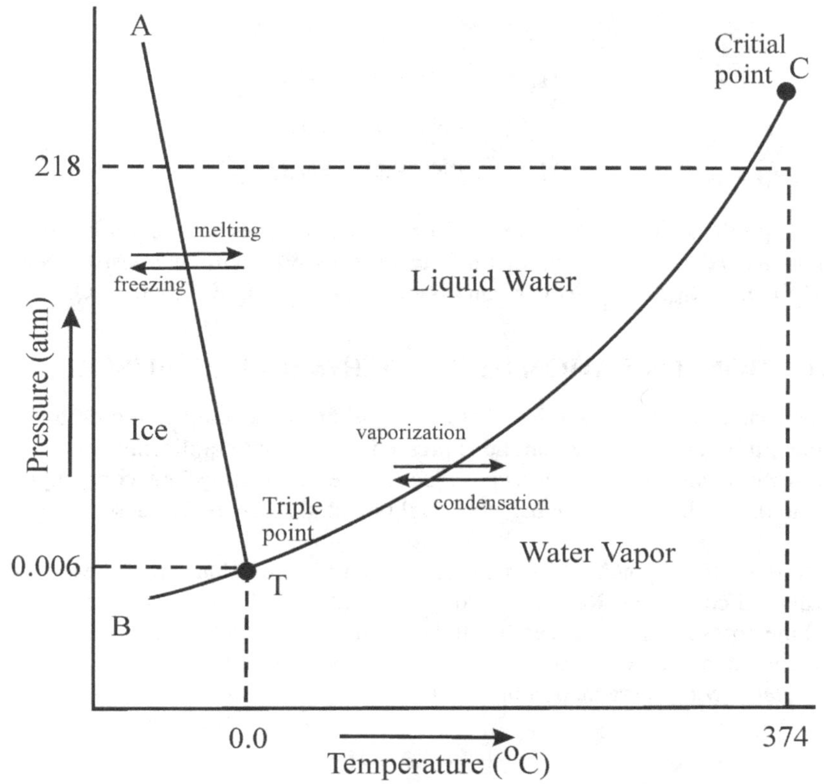

FIGURE 2.3 Graphical illustration of the phase rule for the three-phase system of water. Not to scale.

two phases coexist at equilibrium, represented by the vapor–pressure (line B–C) or freezing point (line A–T) curves, only 1 degree of freedom is possible. Using the vapor–pressure curve as an example, if we let temperature be the independent variable, an increase in temperature requires a corresponding increase in pressure to maintain equilibrium between both liquid and vapor phases. If a corresponding increase in pressure does not occur, all of the H_2O will become vapor. Three phases can only coexist at equilibrium when there are no degrees of freedom, that is, temperature and pressure cannot vary. This condition is represented by the triple point (T). The critical point C is where there is no difference between liquid and vapor H_2O.

In everyday life, the phase rule explains why the cooling water in an internal combustion engine is kept under pressure and why it is a bad idea to remove the radiator cap of an engine cooling system while the engine is still warm. Releasing the pressure causes the water to convert to vapor (i.e., flash) and potentially cause serious burns. The phenomenon of flashing also explain why thermal and CO_2 geysers erupt violently. In the case of thermal geysers, expanding warm water forces some water to flow at the surface, which reduces the pressure at depth. As the pressure is further reduced the temperatures at which the water will flash become smaller. Eventually the pressure is sufficiently reduced, some of the water becomes steam, and the geyser erupts.

Let us now consider the two component systems consisting of Na_2SO_4 and H_2O. Seven phases are needed to describe the sodium sulfate-water system

$$Na_2SO_4: \; Na_2SO_4 + xH_2O$$

$$Na_2SO_4 \bullet 7H_2O: \; Na_2SO_4 + 7H_2O$$

$$Na_2SO_4 \bullet 10H_2O: \; Na_2SO_4 + 10H_2O$$

$$Na_2SO_{4(aq)} : Na_2SO_4 + xH_2O$$

$$H_2O_{(g)} : 0Na_2SO_4 + xH_2O$$

$$H_2O_{(l)} : 0Na_2SO_4 + xH_2O$$

$$H_2O_{(s)} : 0Na_2SO_4 + xH_2O$$

The phase rule predicts that a maximum of four phases can coexist in equilibrium; however, the system would have 0 degrees of freedom. Such an aqueous system might consist of solid $Na_2SO_{4(s)}$, $Na_2SO_4 \cdot 7H_2O$, and $Na_2SO_{4(aq)}$, which can dissociate into $2Na^+$, SO_4^{2-}, and H_2O.

2.3 REDUCTION–OXIDATION (ELECTROCHEMICAL EQUILIBRIUM)

Earlier in this chapter it was shown that chemical equilibrium accompanying proton transfer reactions is temperature dependent and can be expressed in terms of equilibrium constants and Gibbs free energy. Because voltage change is a consequence of electron transfer accompanying reduction–oxidation (redox), an additional variable, electrical potential, is needed to describe electrochemical equilibrium.

The standard electrical potential for half-cell or full redox reactions are represented by the symbol E^o for standard conditions. Redox reactions are written so that they contain the portion that is oxidized and the portion that is reduced. Both the oxidized and reduced portions are known as half-cell, and the complete reaction is known as full cell. Potential is measured in volts relative to the potential associated with the reduction of H^+ ions to $H_{2(g)}$:

$$2H^+ + 2e = H_{2(g)} \tag{2.39}$$

A negative standard potential means the reaction is reducing, and a positive potential means the reaction is oxidizing. The value of the potential is an indication of the magnitude of the capacity to reduce or oxidize. Redox potential, represented by the quantities E_h or pE, cannot be directly inferred from standard potentials of $E_h{}^o$ or pE^o because the concentrations of reactants and products must be considered.

Redox potential for aqueous systems will be described in terms of E_h followed by pE. The redox potential of solutes in a solution can be calculated from knowledge of the standard potentials and effective molar concentrations using the Nerst equation:

$$E_h = E^o - \frac{RT}{nF} - \ln\frac{[\text{oxidized state or products}]}{[\text{reduced state or products}]} \tag{2.40}$$

where:

E_h = redox potential of the solution in volts,
E^o = standard potential of the redox reaction in volts,
R = gas constant 1.98717 cal/mole-°K, and 8.314×10^{-3} kJ/mole-°K,
T = temperature (°K),
n = number of electrons in the balanced half reaction.
F = Faraday constant 9.65×10^4 cal/mole. 23.06 Kcal/volt, and 96.42 kJ/volt-mole
[ ] = chemical activity of products and reactants.

At 25 °C Equation 2.40 becomes:

$$E_h = E^o + \frac{0.592}{n}\ln\frac{[\text{oxidized state or products}]}{[\text{reduced state or products}]} \tag{2.41}$$

E^o values for selected half-reactions are listed in Table 2.4. Other values can be found in many references (Latimer, 1952; Morel, 1983; Sillin and Martell, 1964, and numerous others), can be looked up on line, or can be calculated using the Gibbs free energy of the reaction ΔG_R^o:

$$-\Delta G_R^o = nFE^o \tag{2.42}$$

when rearranged becomes:

$$E^o = \frac{-\Delta G_R^o}{nF} \tag{2.43}$$

where ΔG_R^o is in kcal mole^{-1} and E^o is in volts. When performing Gibbs free energy calculations is it important to use ΔG_R^o from an internally consistent dataset.

Prior to the early 1970s, equilibrium conditions for redox reactions in aqueous systems were expressed in terms of E_h and ΔG_R^o. E_h is the energy gained by the transfer of 1 mole of electrons accompanying reaction Equation 2.39. The E in E_h symbolizes the electromotive force, and the h indicates that the redox potential is relative to the hydrogen scale. In 1967 Sillén represented electrochemical potential in terms of the negative base-10 logs of activities of electrons per liter and used the symbol pE. Since then, the use of the dimensionless pE unit has become common.

Both E_h and pE are routinely used in aqueous investigation involving redox. In practical use, E_h is the measured electrical potential of a water sample, in volts, taken using a specific ion electrode, and pE is an analytical value calculated from E_h (Figure 2.4). When possible, it is best to measure

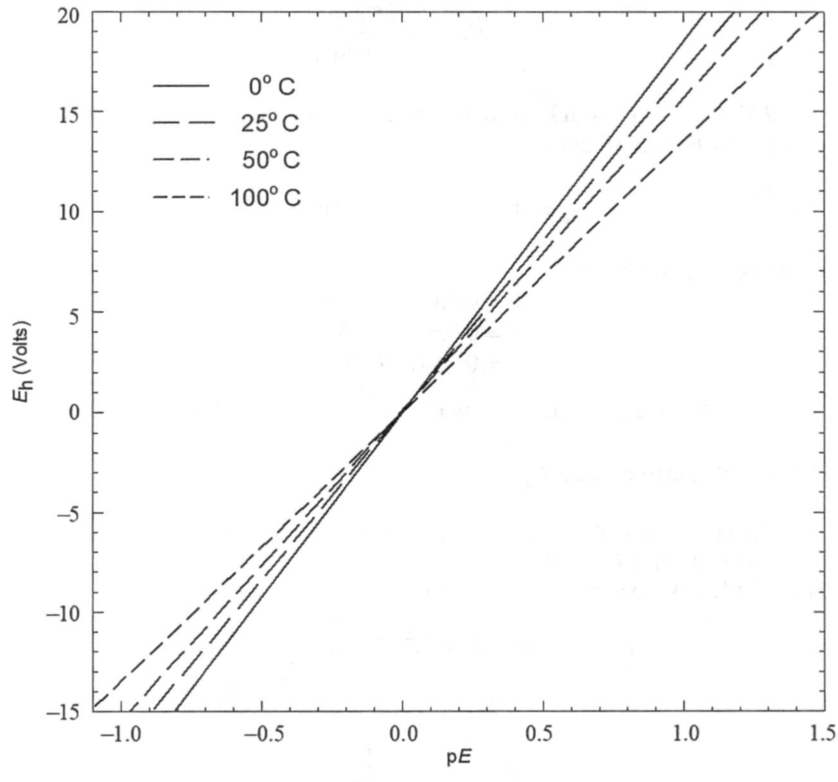

FIGURE 2.4 Comparison of E_h vs. pE at 0, 25, 50 and 100 °C. pE is a dimensionless quantity whereas E_h is in volts. The scales share a common value at 0 °C.

E_h using a flow through cell under positive pressure. Measurements need to be taken in a manner so that the water cannot come into contact with the atmosphere. When oxygen contamination occurs degassing may also occur. It is not uncommon for a reduced spring discharge to become oxidized after only flowing a meter or less from the discharge location.

E_h^o, E_h, pE^o, pE, and ΔG_R^o are related as follows:

$$pE^o = \frac{F}{2.3RT} E_h^o \tag{2.44a}$$

$$= \frac{1}{n} \log K \tag{2.44b}$$

$$K = \frac{[\text{reduced}]}{[\text{oxidized}][e]^n} \tag{2.44c}$$

$$= -\frac{1}{n} \frac{\Delta G_R^o}{2.3RT} \tag{2.44d}$$

At 25 °C

$$\text{p}E^o = 16.90 E_h^o \tag{2.45a}$$

$$= \log \frac{K}{n} \tag{2.45b}$$

$$= -\frac{0.733\ \Delta G_R^o}{\text{n}} \tag{2.45c}$$

$$pE = pE^o - \log \frac{\text{Reduced}}{\text{Oxidized}} \tag{2.46}$$

where $\Delta G_R^o = -RT \ln K$, ΔG_R^o is in kcal mole^{-1}, and E_h is in volts.

To convert between E_h and pE at any temperature:

$$E_h = 1.9858 \times 10^{-4} (T)(pE) \tag{2.47}$$

At temperatures of 0, 25, and 50 °K:

$$E_h = 0.0542\ \text{p}E\ 0\ ^\circ\text{K}$$
$$= 0.0592\ \text{p}E\ 25\ ^\circ\text{K}$$
$$= 0.0641\ \text{p}E\ 50\ ^\circ\text{K}$$

Differences between pE and E_h are graphically compared in Figure 2.4.

Example E2.5 Calculate pE and E_h

Problem: Calculate the pE and E_h of a groundwater containing $Fe^{2+} = 10^{-2.5}$ M and $Fe^{3+} = 10^{-3.4}$ M. Assume M = activity and T = 25°C.

Solution: From Table 2.4 the reduction of Fe^{2+} to Fe^{3+} is:

$$\text{Fe}^{3+} + \text{e} = \text{Fe}^{2+}; E^o = 0.771 \text{ and p}E^o = 13.0$$

Using Equation 2.46

$$\text{p}E = \text{p}E^o - \log \frac{\text{Fe}^{2+}}{\text{Fe}^{3+}} \tag{E2.5a}$$

TABLE 2.4
Standard Potential (E°) of Selected Redox Reactions Common in Groundwater Systems

Species	E^o (volts)	Source
Hydrogen		
$2H^+ + 2e \rightleftharpoons H_{2(g)}$	0	1
$H2O + 3e \rightleftharpoons \frac{1}{2}H_{2(g)} + OH^-$	–0.828	1
Oxygen		
$O_{2(g)} + 2H^+ + 2e \rightleftharpoons 2H_2O$	2.430	1
Iron		
$Fe^{3+} + e \rightleftharpoons Fe^{2+}$	0.771	1
$FeOOH_{(s)} + 3H^+ + e \rightleftharpoons Fe^{2+} + 2H_2O$	0.740	1
$Fe^{2+} + 2SO_4^{2-} + 16H^+ + 14e \rightleftharpoons FeS_{2(pyrite)} + 8H_2O$	0.362	2
$Fe(OH)_3 + 3H^+ + e = Fe^{2+} + 3H_2O$	0.975	2
Manganese		
$Mn^{3+} + e \rightleftharpoons Mn^{2+}$	1.56	1
$MnO_{2\,(pyrolusite)} + 4H^+ + 2e \rightleftharpoons Mn^{2+} + 2H_2O$	1.23	3
Nitrogen		
$N_2O_{(g)} + 2H^+ + 2e \rightleftharpoons N_{2(g)} + H_2O$	1.769	1
$NO_3^- + 4H^+ + 3e \rightleftharpoons NO_{(g)} + 2H_2O$	0.955	1
$NO_3^- + 3H^+ + 2e \rightleftharpoons HNO_2 + 2H_2O$	0.798	1
$N_{2(g)} + 8H^+ + 6e \rightleftharpoons 2NH_4^+$	0.274	1

A negative potential means the reaction is reducing, and a positive potential means the reaction is oxidizing.
Sources:
[1] www.csun.edu; Appendix H Standard Reduction Potentials
[2] Langmuir, 1997
[3] Guenther, 1975

$$= 13.0 - \log\frac{10^{-2.5}}{10^{-3.4}} = 13.0 - 0.9 = 12.1 \quad \text{(E2.5b)}$$

and from Equation 2.45a:

$$E_h = \frac{pE}{16.90} = \frac{12.1}{16.90} = 0.72 \text{ volts} \quad \text{(E2.5d)}$$

Important Takeaway – In many groundwater systems multivalent species such as iron, manganese, aluminum, and uranium are the ions of interest. Evaluating such species involves redox calculations and knowledge or an estimation of either pE or E_h *of the system is necessary. Because accurately measuring the* E_h *of a water is difficult, it may be necessary to calculate the pE of the water using the laboratory measured values of a redox pair. Many geochemical modeling programs perform the redox calculations.*

2.4 MINERAL DISSOLUTION

2.4.1 Saturation Index (SI)

In hydrogeochemistry an important use of thermodynamic calculations is to determine the thermodynamic tendency of water to dissolve or precipitate minerals using the parameter known as the saturation index (SI):

$$\mathrm{SI} = \frac{IAP}{Keq} \tag{2.48}$$

where IAP, the ion activity product, is similar to K_{eq} except that the effective molar concentrations are those of the water sampled corrected for their activity coefficient. IAP is calculated in a fashion similar to the calculation of K_{eq}:

$$IAP = \frac{[\mathrm{C}]^{\mathrm{c}}\,[\mathrm{D}]^{\mathrm{d}}}{[\mathrm{A}]^{\mathrm{a}}\,[\mathrm{B}]^{\mathrm{b}}} \tag{2.49}$$

To calculate SI, the K_{eq} must be determined for the temperature of the water sample. For convenience, SI is usually converted to log SI in groundwater investigations. By converting to log values, supersaturation is positive and undersaturation is negative. Possible saturation conditions include:

$$\log \mathrm{SI} > 0.1 \text{ (supersaturated)}$$
$$\log \mathrm{SI} = 0.0 \pm 0.1 \text{ (saturated)}$$
$$\log \mathrm{SI} < -0.1 \text{ (undersaturation)}$$

Supersaturation means the water has a thermodynamic tendency to precipitate minerals, whereas undersaturation means that the water has a thermodynamic tendency to dissolve minerals. As a first order approximation, it is generally safe to assume that a water that is saturated or supersaturated with respect to a particular mineral, such as calcite, either recharged or flowed through an aquifer that has abundant calcite. Conversely, undersaturated water suggests that the recharge zone or aquifer is deficient in that mineral. In most groundwater systems it is assumed that chemical equilibrium is attained fairly rapidly. However, the law of mass action does not account for reaction rates. In some systems it may take tens to hundreds of years to reach equilibrium, and equilibrium may never be achieved for some mineral species in groundwater systems that are subjected to frequent changes in lithology, temperature, and pressure, biological activity, or mixing with other groundwaters.

It is of great geochemical and hydrogeologic significance that K_{eq} is temperature dependent. Except for species such as HCO_3^-, which may have a gaseous phase in the normal range of groundwater temperatures and pressures, the higher the temperature the larger the K_{eq}.

Example E2.6 Calculate the SI of a groundwater

Problem: McGill hot spring, located near Ely, Nevada, discharges water at 77 °C. Using the water's solute concentrations, activity coefficients, and equilibrium constants, calculate the SI for calcite, dolomite, gypsum, and halite. The provided activity coefficients have been calculated using a calculated ionic strength of 6.41×10^{-3}.

Species	Molality	Log molality	Activity coefficient
Ca^{2+}	1.08×10^{-3}	−2.97	6.94×10^{-1}
Mg^{2+}	7.57×10^{-4}	−3.12	6.98×10^{-1}
Na^{+}	6.47×10^{-4}	−3.19	9.11×10^{-1}
K^{+}	2.55×10^{-5}	−4.59	9.01×10^{-1}
Cl^{-}	1.27×10^{-4}	−3.90	9.01×10^{-1}
SO_4^{2-}	1.55×10^{-4}	−3.82	6.92×10^{-1}
HCO_3^{-}	3.67×10^{-3}	−2.42	9.13×10^{-1}
CO_3^{2-}	4.66×10^{-5}	−4.33	6.95×10^{-1}

Solution: Use Equation 2.10 ($a_i = m_i y_i$) to calculate the activity of each species:

$$Ca^{2+}{}_{activity} = 1.08\times10^{-3}\times6.94\times10^{-1} = 7.50\times10^{-4}$$

$$Mg^{2+}{}_{activity} = 7.56\times10^{-4}\times6.98\times10^{-1} = 5.28\times10^{-4}$$

$$Na^{+}{}_{activity} = 6.47\times10^{-4}\times9.11\times10^{-1} = 5.89\times10^{-4}$$

$$K^{+}{}_{activity} = 2.55\times10^{-5}\times9.01\times10^{-1} = 2.30\times10^{-5}$$

$$Cl^{-}{}_{activity} = 1.27\times10^{-4}\times9.09\times10^{-1} = 1.15\times10^{-4}$$

$$SO_4^{2-}{}_{activity} = 1.53\times10^{-4}\times6.92\times10^{-1} = 1.06\times10^{-4}$$

$$HCO^{-}_{3\,activity} = 3.76\times10^{-3}\times9.13\times10^{-1} = 3.43\times10^{-3}$$

$$CO_3^{2-}{}_{activity} = 4.66\times10^{-5}\times6.95\times10^{-1} = 3.24\times10^{-5}$$

The ion activity product (IAP) of each species can then be calculated using Equation 2.49:

$$IAP = \frac{[C]^c\,[D]^d}{[A]^a\,[B]^b}$$

$$IAP_{Calcite} = Ca^{2+}\times CO_3^{-} = 7.50\times10^{-4}\times4.66\times10^{-5} = 2.43\times10^{-8}$$

$$IAP_{Dolomite} = Ca^{2+}\times Mg^{2+}\times(CO_3^{-})^2 = 7.50\times10^{-4}\times5.28\times10^{-4}\times\left(4.66\times10^{-5}\right)^2 = 4.16\times10^{-16}$$

$$IAP_{Gypsum} = Ca^{2+}\times SO_4^{2-} = 7.50\times10^{-4}\times1.06\times10^{-4} = 6.77\times10^{-8}$$

$$IAP_{Halite} = Na^{+}\times Cl^{-} = 5.89\times10^{-4}\times1.15\times10^{-4} = 6.74\times10^{-8}$$

The activities of the solid species $[A]^a$ and $[B]^b$ have a value of 1, so they can be ignored in the calculations. Saturation index is then calculated using Equation 2.48 and the equilibrium constants (K_{eq}) that were calculated for 77 °C:

$$SI = \frac{IAP}{Keq}$$

$$SI_{calcite} = \frac{2.43x10^{-8}}{1.12x10^{-9}} = 2.17\times10^{1} \qquad \log SI = 1.34$$

$$SI_{dolomite} = \frac{4.16x10^{-16}}{7.64x10^{-19}} = 5.44 \times 10^2 \quad \log SI = 2.74$$

$$SI_{gypsum} = \frac{7.95x10^{-8}}{1.72x10^{-5}} = 4.63 \times 10^{-3} \quad \log SI = -2.34$$

$$SI_{halite} = \frac{6.77x10^{2}}{4.81x10^{1}} = 1.41 \times 10^{-9} \quad \log SI = -8.85$$

The calculated saturation indices indicate that the thermal waters are supersaturated with respect to the carbonate minerals and are undersaturated with respect to both gypsum and halite.

Important Takeaway – SI is one of the most useful and easily calculated indices in groundwater geochemical investigation. Calculating SI requires knowledge of the K_{eq} *of mineral species of interest at the aquifer temperature. SI can be used as an indicator of which aquifer minerals a groundwater has been in contact with. Several geochemical programs readily calculated SI based on water temperate, pH, and laboratory-reported solute concentrations.*

2.4.2 Ionic Strength Effect

The inverse relationship between salinity and chemical activity at the salinity of most groundwater (Figure 2.1) may greatly affect the geochemical evolution of water as it flows through the aquifer. The groundwater may become heated and encounter additional soluble minerals. Thermal heating results in an increase in the ionic strength in most groundwater, and encountering additional soluble minerals will also increase the ionic strength. The ionic strength increases because the solubility of most minerals increases with increasing aquifer temperature. The ionic strength of warming water can decrease when soluble gases exsolve. Flow-path changes in the ionic strength may affect the IAP and hence the SI of various minerals species. This inverse relationship results in a phenomenon known as the ionic strength effect. For example, if a calcite saturated groundwater encountered and dissolved some halite, the groundwater would become more saturated with respect to halite but would become less saturated with respect to calcite. The decrease in the calcite SI is due to the increase in ionic strength of the water and the corresponding decrease in activity coefficients of both Ca^{2+} and HCO_3^-. The decrease in the activity coefficient relative to increasing ionic strength is illustrated for Mg^{2+} in Figure 2.2. It is significant to note that salinity-induced changes in mineral saturation can occur without changes in water temperature.

2.4.3 Common Ion Effect

Many aquifer minerals contain common elements that, when dissolved, contribute common ions to water. What this means is that the contribution of common ions from two or more minerals may affect the solubility of each contributing mineral by a process known as the common ion effect. The order of mineral encounter can greatly influence a groundwater's chemical composition due to the combined factors of the ionic strength and common ion effects. For example, the dissolution of both calcite ($CaCO_3$) and anhydrite ($CaSO_4$) will both release Ca^{2+}. If the groundwater encounters anhydrite after calcite has dissolved to equilibrium, the subsequent dissolution of anhydrite will affect the solubility of calcite. Some additional calcite will tend to dissolve because of the ionic strength effect. However, the excess Ca^{2+} ions provided by the more soluble anhydrite will cause some calcite to precipitate because the water will become supersaturated with respect to calcite. If the water encountered anhydrite first, the system may become flooded with Ca^{2+}, thus inhibiting the subsequent dissolution of calcite. As with the ionic strength effect, the common ion effect can occur without a change in groundwater temperature. The common ion effect is a contributing factor in the carbonate mineral cementation of sedimentary rocks.

Important Takeaway – Both the ionic strength and common ion effects can affect the SI of minerals in complex ways and need to be evaluated as part of geochemical investigations. Fortunately, many groundwater geochemical modeling programs adjust for both conditions.

2.4.4 Congruent and Incongruent Dissolution

Dissolution reactions in which the products of the reaction are all dissolved species are known as congruent reactions. Dissolution reactions in which one or more of the products of the reaction is a solid or amorphous substance are known as incongruent reactions. Except for aluminosilicate minerals, most mineral dissolution reactions of interest to hydrogeology occur congruently. The incongruent dissolution of aluminosilicate minerals is a great class of non-dynamic reactions. Except for quartz, most silicate minerals dissolve incongruently and precipitate aluminosilicate minerals, particularly clay, as the solid species. Because igneous and metamorphic rocks are composed of silicate minerals and silicate minerals are a major constituent of many sedimentary rocks, aluminosilicate and silicate clay minerals are ubiquitous soil zone weathering products, are deep sedimentary basin diagenetic products, and are fault zone hydrothermal alteration products. Silicate and aluminosilicate mineral dissolution is discussed in detail in Chapter 6 and is represented here by the weathering of the feldspar orthoclase, which is common in granite and rhyolite:

$$9H_2O + 2KalSI_3O_8 2H^+ + 4Si(OH)_4 + Al_2SI_2O_5(OH)_5 \tag{2.50}$$

The reaction in Equation 2.50 is pH dependent because hydroxyl ions are involved. The solid product, kaolinite, may further dissolve incongruently.

Chapter Problems

1. The reported average chemistry of six low temperature thermal waters in the Wasatch Range, Utah, is: 30.09 °C, pH 7.09, Ca^{2+} (13.81), Mg^{2+} (5.00), Na^+ (3.88), K^+ (0.48), HCO_3^- (8.59), Cl^- (2.32), and SO_4^{2-} (6.18). The reported ion chemistries are in meq/L. Assume there are no other ions in solution. Calculate the saturation index for calcite, dolomite, gypsum, and halite for this water. Use the K_{eq} values in Table 2.1. Assume the water temperature is 25 °C.
2. The dissolution of gypsum and halite can be represented by the reactions:

$$CaSO_4 \bullet 2H_2O = Ca^{2+} + SO_4^{2-} + H_2O$$

$$NaCl = Na^+ + Cl^-$$

Using the Gibbs free energy data in Appendix C, calculate the equilibrium constants for $CaSO_4$ and NaCl. Compare your calculated values with the values in Table 2.1. Discuss the possible reasons for any differences.

3. Using the standard potentials in Table 2.4, calculate the pE and E_h of a groundwater containing $Mn^{3+} = 10^{-2.3}$ M, and $Mn^{2+} = 10^{-3.3}$ M. Assume M = activity and T = 25°C.
4. A groundwater encounters dolomite. The water subsequently encounters gypsum and then a Na-clay. What are the processes that may affect the solute chemistry, and what are the likely processes and the qualitative effects on the major ion concentrations at each step along the flow path?
5. After reviewing Table 2.1, how much time do you think is required to form a 2 m thick soil horizon covering a carbonate vs. granitic bedrock? Assume plentiful rainfall and no mechanical erosion.

6. If you were to drive on the Bonneville Salt Flats when flooded with water, what would be the water's TDS?

Chapter notation

[ ]	effective molar concentration
( )	molar concentration (molality)
Δ	change
°C	temperature in degrees Centigrade
°K	temperature in degrees Kelvin
a, *b*, *c*, *d* superscripts	moles of species *A*, *B*, *C*, and *D*, respectively
a, *b*, *c*, *d* subscripts	moles of species *A*, *B*, *C*, and *D*, respectively
a, b, c, d	species A, B, C, and D, respectively, when used with SI
a	chemical activity or effective molar concentration []
a_{anion}	activity of anion
a_{cation}	activity of cation
a_i	constant: effective diameter of ion
cal	calorie (1 cal = 4.184 J)
cal/mole	calories per mole
deg	degree
E	total energy of system
e	electron
E_h	redox potential of a solution in volts.
E^o	standard potential of a redox reaction in volts at STP
F	Faraday constant (9.65 × 104 cal/mole; 23.06 Kcal/volt; 96.42 kJ/volt-mole)
f	degrees of freedom
G	Gibbs free energy of formation
ΔG_R	Gibbs free energy of a reaction at temperature *T*
ΔG^o_R	Gibbs free energy of formation of the reaction at (STP)
ΔG^o_p	Gibbs free energy of formation of the products at (STP)
ΔG^o_r	Gibbs free energy of formation of the reactants at (STP)
H	enthalpy
H^o	enthalpy at STP
IAP	ion activity product
J	joule (J = 0.239 cal)
K	thermodynamic equilibrium constant
K_{eq}	thermodynamic equilibrium constant
Kcal/mole	kilocalories per mole
KJ/mole	kilojoules per mole
K_{stab}	ion pair stability constant
K_w	equilibrium constant for water
M	molality; same as ( )
mg/L	milligram per liter
mole	one gram formula weight of a substance
n	number of electrons in a balanced half reaction
P	pressure or number of phases
P	number of phases
pE	base-10 log of activity of electrons/L (dimensionless)
pE^o	standard base-10 log activity of electrons L-1 at STP

pH	negative log of H^+ concentration
ppm	parts per million
R	universal gas constant (8.3143 J/mol-T; 1.98717 cal/mol-T),
S	entropy
S^o	entropy at STP
SI	saturation index
STP	standard conditions (25 °C or 298°K at 1 atm pressure)
T	temperature
y	activity coefficient
y_i	activity coefficient of species i
z_i	valance of species i

REFERENCES CITED

Bates, C.F., and R.E., Mesmer, 1976, *The hydrolysis of cations*, John Wiley and Sons.

Davies, C.W., 1962, Incomplete dissolution in aqueous solutions, in *The structure of electrolytic solutions*, W.J. Hammer (ed.), Wiley, 19–34.

Freeze, J.A., and J.A. Cherry, 1979, *Groundwater*, Prentice-Hall.

Garrels, R.M., and C.L. Christ, 1965, *Solutions, minerals, and equilibria*, Freeman, Cooper & Company.

Gibbs, J.W., 1948, *The collected works*, Yale University Press.

Guenther, W.B., 1975, *Chemical equilibrium*, Springer.

Harned, H.S., et al., 1959, The physical chemistry of electrolyte solutions, 3rd ed, *Journal of the Electrochemical Society*, 106(1).

Helgeson, H.C., 1969, Thermodynamics of hydrothermal systems at elevated temperatures and pressures, *American Journal Science*, 267, 729–804.

Helgeson, H.C., et al., 1978, Summary and critique of the thermodynamic properties of rock forming minerals, *American Journal of Science*, 278A, 1–229.

Klotz, L.M., 1950, *Chemical thermodynamics*, Prentice-Hall.

Langmuir, D., 1997, *Aqueous environmental geochemistry*, Prentice-Hall.

Latimer, W.M., 1952, *Oxidation potentials*, Prentice-Hall.

Lu, P., et al., 2022, Comparison of thermodynamic data files for PHREEQC, *Earth-Science Reviews*, 225.

Manov, G.G., et al., 1943, Values for the constants in the Debye-Hückel equation for activity coefficients, *Journal American Chemical Society*, 65, 1765–1767.

Morel, F.M.M., 1983, *Principles of aquatic chemistry*, Wiley and Sons.

Parkhurst, D.L., and C.A.J. Appelo, 2013, PHREEQC, *USGS Techniques and Methods*, 6, A43.

Pitzer, K.S., 1973, Thermodynamics of electrolytes I: Theoretical basis and general equations, *Journal Physical Chemistry*, 77, 268–277.

Pitzer, K.S., 1979, Theory: Ion interaction approach, in *Activity coefficients in electrolyte solutions*, R.M. Pytkowicz (ed.), CRC Press, 1, 157–208.

Pitzer, K.S., 1980, Electrolytes from dilute solutions to fused salts, *Journal American Chemical Society*, 102, 2902–2906.

Pitzer, K.S., and J.J. Kim, 1974, Thermodynamics of electrolytes. 4, activity and osmotic coefficients for mixed electrolytes, *Journal American Chemical Society*, 96, 5701–5707.

Robinson, R.A., and R.H. Stokes, 1970, *Electrolyte solutions*, Butterworths, London.

Sillin, L.G., and A.E. Martell, 1964, *Stability constants of metal-ion complexes*, Chemical Society, London, Special Publication 17.

Smith, R.M., and A.E. Martell, 1976, *Critical stability constants*, Springer.

Truesdell, A.H., and B.F. Jones, 1974, WATEQ, a computer program for calculating chemical equilibria of natural waters, *Journal Research, U.S., Geological Survey*, 2, 233–274.

3 Gases

3.1 NATURE OF GAS

Matter may exist in three states – solid, liquid, and gas. A given mass of a solid will occupy both a set volume and shape, and a great amount of pressure must be applied to appreciably change either. Both gases and liquids may flow. The tendency to flow is measured by the viscosity of the fluid. The viscosity of a liquid decreases with increasing temperature and the viscosity of a gas increases with increasing temperature. Low viscosity means that a fluid will readily flow, whereas high viscosity means a fluid is resistant to flow. Although liquids do not have a set shape, they will maintain a constant volume as long as the temperature does not change. Gases, on the other hand, have neither a definite shape or volume, and their shape and volume will change to fill the available space.

Shape and volume differences between solid and liquid states of matter are related to the density of each state. Density is the volume per unit mass. Except for ice, the liquid state is slightly less dense than is the solid state, because the molecules of liquids are further apart than are the molecules of solids. At sea level the density of the gaseous state of a substance is about 1/1,000 of the density of the liquid state. Gas molecules are far apart and become further apart as the confining pressure decreases. Very little pressure is required to change the shape of a gas or to compress it.

The earth's atmosphere consists of a mixture of gases that form a structureless fluid having no absolute upper boundary. Atmospheric gases are held to the earth by the gravitational force acting on the individual gas molecules. The gravitational attraction of atmospheric gases decreases with increasing elevation, causing the gas molecules to become further apart. Consequences of the elevation-density effect include the extremely high summertime temperatures in Death Valley, California (a closed basin below sea level), and the general trend of cooler temperatures and lesser amounts of O_2 at higher elevations than at sea level. This elevation reduction in O_2 is particularly noticeable when visiting high elevation areas or when climbing high mountain peaks and explains the need to pressurize aircraft. Cooler temperatures at higher elevations are also manifested in the elevation effect often observed in the stable isotopic compositions of precipitation (see Chapter 8), the general increase in precipitation amounts at higher elevations, and the general trend of wintertime snow at higher elevations.

The earth's atmosphere consists of mixture of gases – approximately 78 percent N_2 and 21 percent O_2 (Table 3.1). Most of the remaining 1 percent is Ar. The greenhouse gas CO_2 has been steadily increasing from about 0.03 to the currently 0.042 percent since 1958 (Figure 3.1). Since about 1987 the mole fraction of atmospheric CH_4 has increased from 1,625 to 1,940, and since 2001 the mole fraction of atmospheric N_2O has increased from 316 to 338 (NOAA, 2025). The atmosphere can contain up to 4 percent water vapor ($H_2O_{(g)}$). When water vapor is present, the percentage of gases is reduced proportionally.

The pressure of the atmosphere at sea level is the standard unit for gas measurements. At sea level and at 0 °C the average mass of the atmosphere pressing down on a unit surface area of the earth is considered to be one atmosphere (atm). Gas pressure is often reported in terms of the height that a column of mercury will rise under the pressure of the gas. One standard atm equals a height of 76 cm of Hg. Other useful conversions are:

$$1\ \text{atm} = 1.013\times10^5\ \frac{\text{N}}{\text{m}^2} = 14.7\frac{\text{lb}}{\text{in}^2} = 1.033\frac{\text{kg}}{\text{cm}^2} \tag{3.1}$$

DOI: 10.1201/9781003646129-4

TABLE 3.1
Gas Compositions of the 2022 Atmosphere

	Mole fraction	
Gas	**(ppm)**	**(%)**
N_2	780,840	78.084
O_2	209,460	20.946
Ar	9,340	0.934
CO_2	417	0.0417
Ne	18.18	1.82×10^{-3}
He	5.24	5.24×10^{-4}
CH_4	1.87	1.87×10^{-5}
Kr	1.14	1.14

Data do not include atmospheric moisture which locally may be as great as 3 to 4 percent. Atmospheric moisture tends to be concentrated in thin bands above the ground surface. The data do not add up to 100 percent due to measurement error and the changing CO_2 gas content.
Source: Data from NOAA, 2025. CO_2 and CH_4 contents have been adjusted to 2023 values.

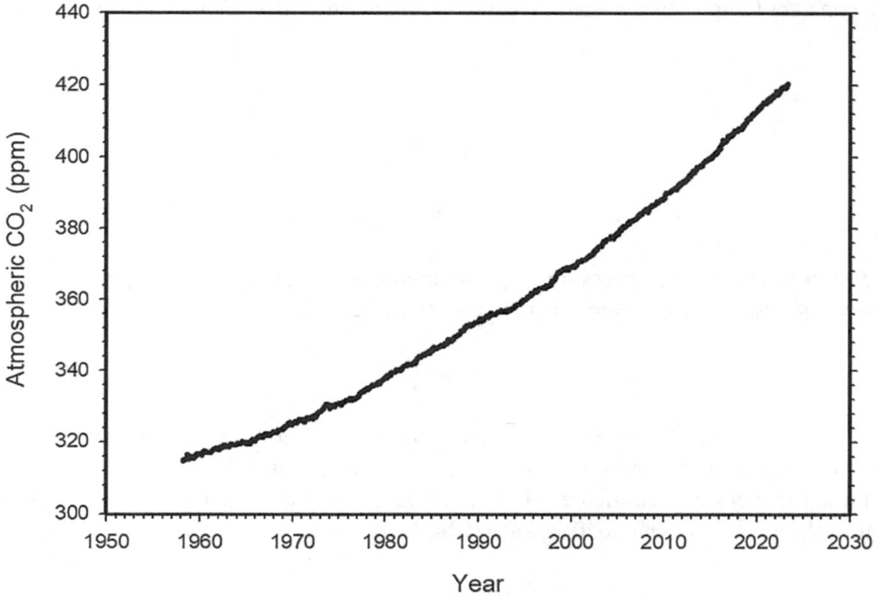

FIGURE 3.1 Temporal trend of atmospheric CO_2 at the Mauna Loa Observatory, Hawaii. The observatory is located at an elevation of 3,397 m in the north Pacific Ocean. The observatory is located about 20 degrees north latitude about 1,860 km away from the nearest mainland atmospheric pollution source. Data from NOAA, 2025.

In any closed gas–liquid system, whether it is a stoppered beaker in the laboratory or the atmosphere–ocean system, gases will attain equilibrium between the gaseous state, sometimes noted by the subscript (g), and the aqueous state, sometimes noted by the subscript (aq). When a gas and a liquid come into contact most of the gas remains as a gas, but some escapes into the liquid. Equilibrium in gas–gas and gas–liquid systems are discussed in this chapter.

Important Takeaway – Gas, unlike solids and liquids, has no set volume or shape, and the density of gas decreases with elevation. The pressure of the atmosphere at sea level is the standard unit of gas measurement, which is 1 atm at sea level and at 0 °C.

3.2 GAS LAWS

The dissolved gas content has not been a focus in most groundwater investigations despite the fact that dissolved gas, particularly O_2 and CO_2, is critically important in the chemical evolution of most groundwater, the biologic health of surface water, acid mine drainage, and bioremediation. The dissolved CO_2 is largely responsible for regulating water's pH and the dissolution of carbonate minerals, and dissolved O_2 is critical in redox reactions. In recent years the importance of dissolved gas in groundwater systems has been brought to the forefront in some investigations. For example, the escape of coal bed methane to shallow aquifers and to the surface environment due to poor methane extraction well design and construction, and well bore failure can result in serious public health concerns. The evaluation of gas–water interactions involve the use of the well-established gas laws.

Gas molecules are far apart and are relatively free from mutual physical and chemical interactions. The gas laws are a series of mathematical relationships that describe how gases physically behave under different conditions. In 1662, Robert Boyle found that the volume of a fixed amount of air at a constant temperature varied inversely with pressure. Boyle's law may be expressed as:

$$V = \frac{K_T}{P} \tag{3.2a}$$

and

$$PV = K_T \tag{3.2b}$$

where V, P, and K_T are volume, pressure, and a temperature-dependent constant, respectively. A useful derivative of Boyles' law, where T is held constant, is:

$$P_1V_1 = P_2V_2 \tag{3.3}$$

where the subscripts 1 and 2 represent the same gas at two different pressures and volumes.

In 1787, J. Charles determined that the volume of a gas expands by the same fraction above its initial volume for every degree increase in temperature above 0 °C. This fraction is approximately 1/273 of its volume at 0 °C. In equation form, Charles's law is:

$$\frac{V_1}{V_2} = \frac{T_1}{T_2} \tag{3.4}$$

Only absolute temperatures are used in gas law calculations, so Equation 3.4 may be represented as:

$$V_T = V_o\left(1 + \frac{T}{273.16}\right) \tag{3.5}$$

where V_T is volume at temperature T (°K) and V_o is volume at 0 °C. It is apparent that at 0 °K the volume would theoretically become zero. It can be shown that:

$$V = K_p T \tag{3.6}$$

where mass and pressure are held constant, and K_p is a proportionality constant. It can also be shown that:

$$P_T = \left(\frac{P_o}{273.16}\right)(273.16 + T) \tag{3.7}$$

where P_T and P_o are pressures at temperatures T and 0 °K, respectively.

Equations 3.2b, 3.6, and 3.7 describe a gas as a function of pressure, volume, and temperature when one of the variables remains constant. The three critical variables are linked to each other by a single equation known as the ideal gas law:

$$PV = nRT \tag{3.8}$$

where n is the number of moles of a gas and R is the universal gas constant per mole. R is:

$$R = 1.987 \frac{\text{cal}}{\text{mole}^\circ\text{K}} = \frac{8.3143\ \text{J}}{\text{mole}^\circ\text{K}} \tag{3.9}$$

where cal is calories. Implicit in the concept of an ideal gas is the idea that each gas component in a gas mixture behaves independently of the other gas components and that there are no physical or chemical interactions between different gases in the mixture. Dalton's law of partial pressure, which is the basis of this assumption, states that the individual pressure exerted by each gas in a mixture of gases is the partial pressure of that gas, and that the sum of these partial pressures is the total pressure of the gas. Dalton's law may be written as:

$$P_T = P_1 + P_2 + P_3 + \ldots P_n\,(\text{V, T constant}) \tag{3.10}$$

Where P_T is the total pressure of all gases and the numerical subscripts are the pressures exerted by each individual gas. Because the sum of all of the gases in a mixture is 100 percent, each gas in the mixture maybe represents a percentage of the total mixture. Equation 3.10 is valid when both volume and temperature are held constant.

Dalton's law may also be expressed in terms of partial volumes:

$$V_T = V_1 + V_2 + V_3 + \ldots V_n\,(\text{P, T constant}) \tag{3.11}$$

where the V_T is the total volume and the numerical subscripts are the volumes of each individual gas.

Important Takeaways – An ideal gas behaves independently of the other gas components, which means there are no physical or chemical interactions between different gases in a mixture of gases. A gas does not have a defined shape or volume and expands or contracts to fill the available space. The gas laws are a series of mathematical relationships that describe how gases physically behave under different conditions. The relationship between the volume, temperature and pressure of a gas is described by the ideal gas law (PV = nRT). Although there is no mathematical justification, the gas laws help explain why your garage, household cupboards, and closets become full. Accumulation will occur to fill the available space.

3.3 GASES AND SOLUTIONS – GAS EXCHANGE PROCESSES

The gaseous components of a gas–liquid solution may come from the vaporization of the solvent or from the introduction of gases into the solvent. Most gases in non-contaminated aqueous systems are from external sources because water is the only solvent of significance. Water is not very volatile at typical surface and groundwater temperatures. Surface and groundwaters may acquire gas from the atmosphere, the soil zone, chemical reactions with soil zone and aquifer minerals, biologic processes, magmatic fluids, thermocatalytic processes, and anthropogenic sources. These gases are usually in low concentrations and form dilute solutions. More concentrated solutions may exist in high temperature geothermal systems where considerable water vapor may exist near the surface, in geologic regimes where abundant CH_4, CO_2, or other gases evolve, and in regions where there is soil and groundwater contamination.

Gases perform critical roles in chemical weathering, the geochemical evolution of groundwater, and in modulating the extent of groundwater contamination. As examples, the availability of CO_2 is a primary factor controlling the dissolution of carbonate rocks, the presence or absence of O_2 helps regulate acid mine drainage as well as methane production from landfills, and the escape of N_2 is the principal natural inorganic mechanism for removing nitrogen compounds from waters. Additionally, many hydrocarbon compounds are volatile at common aquifer temperatures.

3.3.1 IDEAL SOLUTIONS (GAS–LIQUID SYSTEMS)

At a fixed temperature the gaseous component of a gas–liquid solution will exert a definite vapor pressure. If the gas mixture is composed of ideal gases, each gas will exert a partial pressure as described by Dalton's law. Raoults' law and Henry's law both describe equilibrium behavior between gas and the liquid phases of ideal gases in aqueous solutions. Raoult's law describes equilibrium between a liquid and its vapor (e.g., water and water vapor). It is also used to describe the volatilization of organic contaminants, such as benzene, which may have accumulated in the soil zone. Raoult's law states that, for ideal gases, the vapor pressure of the gas of each liquid component in the solution is proportional to the mole fraction of the liquid in the solution:

$$P_i = X_i P_i^o \tag{3.12}$$

where P_i and X_i are the vapor pressure (i.e., partial pressure) and the mole fraction of liquid of component i, respectively. It is often helpful to think of partial pressure as concentration. P_i^o is the partial pressure of the pure component i at a given temperature. In real solutions, Raoult's law describes solutions where $X_i \rightarrow 1$. In other words, Raoult's law is valid for single component liquids. Most surface and groundwaters only consist of the single solvent water, so Raoult's law predicts that the partial pressure of $H_2O_{(g)}$ is the same as the theoretical partial pressure of water (i.e., $P_i = P_i^o$). In an ideal solution, mixing of various liquid components occurs without a change in the molecular energies of the components or in a change in the total volume of the mixture. The chemical activities of the components are equal to their concentrations:

$$a_i = X_i \tag{3.13}$$

where a_i is the chemical activity of component i.

The mole fraction of a gas is equal to the ratio of the partial pressure of the gas relative to the total pressure of the gas phase:

$$X_i = \frac{P_i}{P_T} \tag{3.14}$$

The total vapor pressure of an ideal multi-solvent solution may therefore be described by the equation:

$$P_T = P_1 + P_2 + P_3 + \ldots P_n = X_1P_1^o + X_2P_2^o + X_3P_3^o + \ldots X_nP_n^o \tag{3.15}$$

Equation 3.15 is useful for predicting the behavior of various volatile components in contaminated soils. From Equations 3.13 and 3.14 it is evident that in ideal solutions:

$$a_i = P_i \tag{3.16}$$

and from Equation 3.11 it is evident that Raoult's law predicts that a plot of a_i vs. X_i is a straight line (Figure 3.2).

Henry's law describes the equilibrium between dissolved gases and liquids. Henry's law states that at a constant temperature the amount of a dissolved gas in a liquid is proportional to the partial pressure of the gas above the liquid. The law is strictly valid for ideal dilute gases in solution, which do not chemically react with either the liquid or with other gases in the system.

In the simplest of terms Henry's law may be expressed as:

$$K_i = \frac{X_i}{P_I} \tag{3.17}$$

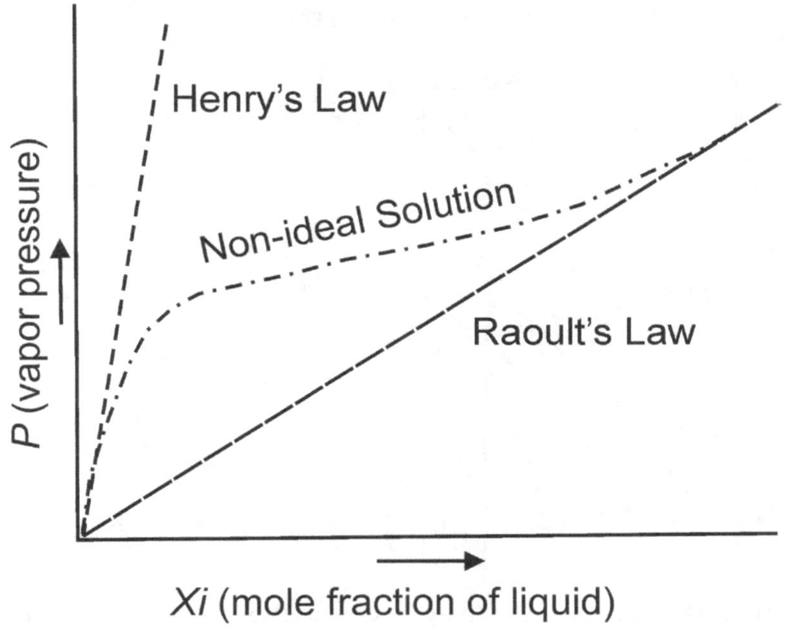

FIGURE 3.2 Partial pressure vs. mole fraction for solutions obeying Raoult's and Henry's laws. Raoult's law is a good approximation for nonideal solutions where X_i is great and Henry's law is applicable to very dilute nonideal solutions. Both Raoult's law and Henry's law assume that the gas fraction is nonreactive.

where K_i, which is also known as H_s in the literature, is the temperature-dependent Henry's constant. X_i is the mole fraction of substance i in the liquid phase, and P_i is the partial pressure of substance i in the gas mixture.

In the literature there are eight variants of Henry's law constants (Figure 3.3). These variants either express Henry's law in terms of the solubility of the gas in the liquid (a.k.a., Henry solubility or H_s), or in terms of volatility (a.k.a., Henry volatility or H_v). The most commonly used variant for H_s in groundwater investigations is H_s^{cp}, which has values of $\frac{mol}{m^3 - Pa}$ and is adopted herein. For simplicity, values of H_s^{cp} are used herein as K_i values. Approximate values of H_s^{cp} and C for gases of hydrologic interest at 25 °C and 1 atm are listed in Table 3.2. The conversion factor for converting H_s^{cp} $(\frac{mol}{m^3 - Pa})$ values in Table 3.3 to $\frac{mol}{L - atm}$ is 1.1033×10^2.

Henry's law values decrease with increasing temperature (Table 3.3). What this means is that gas solubility is unlike the solubility of most mineral species in that gases become less soluble with increasing temperature (Figure 3.4). This inverse relationship with temperature occurs because the dissolution of most gases is exothermic, whereas the dissolution of most minerals is an endothermic reaction. This inverse temperature relationship between temperature and gas solubility has many practical consequences. For example, this relationship describes why warm carbonated drinks rapidly lose CO_{2g}, whereas cold drinks retain their fizz. The inverse relationship also means that cold groundwater

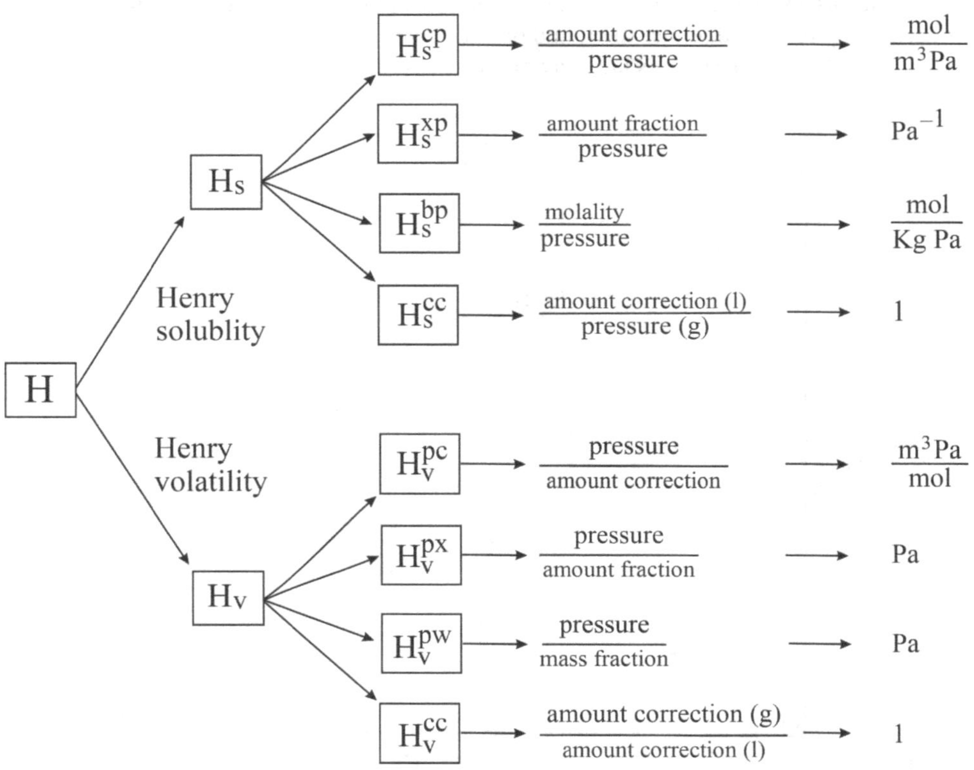

FIGURE 3.3 The eight possible variants of Henry's law constants (used with permission from Rolf Sander). The H_s^{cp} variant is most commonly used in groundwater investigations. Pascal (Pa) is the standard SI unit of pressure measurement. One Pa equals 10^{-5} bar or atm. Sander (2023) compiled a comprehensive list of constants for about 4,600 organic and inorganic species. C is a constant used in van't Hoff equation discussed later. The Sander tables are particularly impressive for the compilation of inorganic species.

TABLE 3.2
Selected Henry's Law Constants at 25 °C and 1 atm

Species	H_s^{cp}	C
O_2	1.26×10^{-3}	1,600
CH_4	1.4×10^{-5}	1,600
CO_2	3.9×10^{-2}	2,400
H_2S	1.0×10^{-3}	2,100
N_2	6.4×10^{-6}	1,300
NH_3	5.9×10^{-1}	4,200
SO_2	1.3×10^{-2}	2,900

The H_s^{cp} values are in $\frac{\text{mol}}{\text{m}^3 - \text{Pa}}$ and can be converted to $\frac{\text{mol}}{\text{L} - \text{atm}}$ using the conversion factor of 1.0133×10^2. See Sander (2023) for a comprehensive compilation of constants.

TABLE 3.3
Temperature Dependent Henry's Law Constants $\left(\frac{\text{mol}}{\text{L} - \text{atm}}\right)$ for Selected Gases at 1 atm

°C	O_2	N_2	CO_2	H_2S	SO_2
0	2.18×10^{-3}	1.05×10^{-1}	7.64×10^{-2}	2.08×10^{-1}	3.56
5	1.91×10^{-3}	9.31×10^{-4}	6.35×10^{-2}	1.77×10^{-1}	3.01
10	1.70×10^{-3}	8.30×10^{-4}	5.33×10^{-2}	0.52×10^{-1}	2.53
15	1.52×10^{-3}	7.52×10^{-4}	4.55×10^{-2}	1.31×10^{-1}	2.11
20	1.38×10^{-3}	6.89×10^{-4}	3.92×10^{-2}	1.15×10^{-1}	1.76
25	1.26×10^{-3}	6.40×10^{-4}	3.39×10^{-2}	1.02×10^{-1}	1.46
30	1.16×10^{-3}	5.99×10^{-4}	2.97×10^{-2}	9.09×10^{-2}	1.21
35	1.09×10^{-3}	5.60×10^{-4}	2.64×10^{-2}	8.17×10^{-2}	1.00
40	1.03×10^{-3}	5.28×10^{-4}	2.36×10^{-2}	7.41×10^{-2}	0.84
50	9.32×10^{-3}	4.85×10^{-4}	1.95×10^{-2}	6.21×10^{-2}	--

Source: Data from Langmur, 1997, and Sander, 2015.

in chemical equilibrium with limestone will become calcite supersaturated and may precipitate calcite cement in the aquifer matrix when the groundwater warms due to deep circulation.

The general relationship between a_i and X_i as predicted by Henry's law also plots as a straight line (Figure 3.2). On the other hand, the behavior of a nonideal gas does not plot as a straight line. Raoult's law may be used as a good approximation for nonideal gases where the mole fraction of the component (X_i) is large. Henry's law may be used as a good approximation for a nonideal gas where the mole fraction of the gas is small. A small mole fraction means a dilute solution.

The temperature dependence of Henry's law constants can be calculated using the van't Hoff equation:

$$k_H = H_s^{cp} \exp\left[-C\left(\frac{1}{T} - \frac{1}{T^o}\right)\right] \tag{3.18}$$

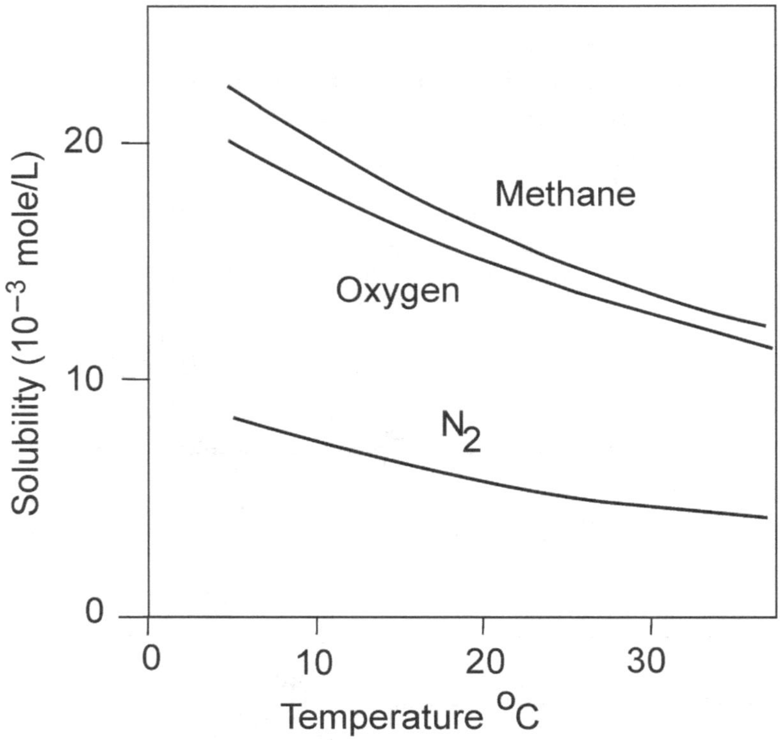

FIGURE 3.4 The temperature-dependent solubility of nitrogen, oxygen, and methane. As with other gases, solubilities decrease with increasing temperature.

where
k_H = Henry's law constant at the desired temperature,
H_s^{cp} = Henry's law constant at the standard temperature (T^o),
T = desired temperature in kelvin,
T^o = standard state temperature (298 °K),
exp = the exponential function, and
C = a constant for the desired species with the dimension of kelvin (see Table 3.2).

Henry's law can also be expressed as:

$$a_i = K_i X_i \tag{3.19}$$

Example E3.1 Henry's law

Problem: Calculate the concentration of O_2 in water (in mg/L) that is in equilibrium with the atmosphere at 25 °C. Assume a liter of water has a mass of 1,000 g.

Solution: The concentration of dissolved oxygen in equilibrium with the atmosphere can be calculated using Henry's law (Equation 3.14), the Henry's law constant for O_2, and the partial pressure of O_2 in the atmosphere from Table 3.1. Henry's law is:

$$K_i = \frac{X_i}{P_i} \tag{E3.1a}$$

and

$$X_i - K_i \times P_i \qquad \text{(E2.1b)}$$

The O_2 equilibrium constant at 25 °C is 1.26 × 10^{-3} mole/L-atm, the atmospheric O_2 partial pressure is ~ 0.21 atm, and the molecular weight of 1 mole of O_2 is ~32 g. Substituting these values into E3.1b:

$$O_{2aq} = \frac{1.26 \times 10^{-3}\ \text{mole}}{\text{L} - \text{atm}} \times 0.21\,\text{atm}$$

and solving for $O_{2(aq)}$:

$$O_{2(aq)} = \frac{1.26 \times 10^{-3}\ \text{mole}}{\text{L} - \text{atm}} \times \frac{32\ \text{g}}{\text{mole } O_2} \times \frac{1000\ \text{mg}}{\text{g}} \times 0.21\,\text{atm} = \frac{8.45\ \text{mg}}{\text{L}}$$

Example E3.2 Methane gas-methane dissolved concentrations

Problem: There is no drinking water standard for methane. The U.S. Department of Interior has established a warning level of 10 milligrams of methane per liter of water (mg/L). Levels of dissolved methane greater than 28 mg/L could release potentially flammable amounts of methane. What are the pressures of methane gas escaping from a gas extraction well that will result in aqueous concentrations of 10 and 28 mg/L?

Solution: From Table 3.2, Henry's law constant at 25 °C is:

$$1.4 \times 10^{-5}\ \frac{\text{mole}}{\text{m}^3 - \text{Pa}} = 1.4 \times 10^{-3}\ \frac{\text{mole}}{\text{L} - \text{atm}}$$

The solubility of $CH_4\left(10\frac{mg}{L}\right)$:

$$K_{CH_4} \times P_{CH_4} = 100\frac{\text{mg}}{\text{L}} \times \frac{\text{g}}{10^3\ \text{mg}} \times \frac{\text{mole CH}_4}{16.04\ \text{g CH}_4} = \frac{6.2344 \times 10^{-4}\ \text{mole}}{\text{L}}$$

$$P_{CH_4} = \frac{\text{Solubility CH}_4}{K_{CH_4}} = \frac{\dfrac{6.2344 \times 10^{-4}\ \text{mole}}{\text{L}}}{1.4 \times 10^{-3}\ \dfrac{\text{mole}}{\text{L} - \text{atm}}} = 0.45\ \text{atm}$$

and the solubility of $CH_4(28\frac{mg}{L})$:

$$K_{CH_4} x P_{CH_4} = 28.0\frac{\text{mg}}{\text{L}} = \frac{\text{g}}{10^3\ \text{mg}} x \frac{\text{mole CH}_4}{16.04\ \text{g CH}_4} = \frac{1.746 \times 10^{-3}\ \text{mole}}{\text{L}}$$

$$P_{CH_4} = \frac{\text{Solubility CH}_4}{K_{CH_4}} = \frac{\dfrac{1.746 \times 10^{-3}\ \text{mole}}{\text{L}}}{1.4 \times 10^{-3}\ \dfrac{\text{mole}}{\text{L} - \text{atm}}} = 1.25\ \text{atm}$$

Important Takeaways – Unlike solids, the solubility of a gas decreases with increasing temperature (drink your carbonated soda cold). Raoult's law describes the behavior between an ideal gas and its liquid phase. The vapor pressure of the gas of each liquid component in a solution is proportional to the mole fraction of the liquid in the solution. Henry's law describes the equilibrium between dissolved gases and liquids. At a constant temperature, the amount of a dissolved gas in a liquid is proportional to the partial pressure of the gas above the liquid.

3.3.2 Nonideal Solutions

Strict use of the gas laws assumes that the gases are ideal and behave as predicted by Boyle's law. Simple gases such a H_2 obey Boyle's law over a wide range of temperatures and pressures and will have a constant pressure–volume product (*PV*) at all pressures.

However, most gases, including many of significance to groundwater, such as H_2O, CO_2, and CH_4, do not strictly follow Boyle's law and deviate greatly from the pressure–volume relationship at elevated pressures (Figure 3.2). In the atmosphere, deviations from ideal behavior are relatively unimportant because N_2 and O_2, which together constitute approximately 99 percent of atmospheric gases (Table 3.1), behave in a near-ideal fashion at pressures up to 1 atm. CO_2 behaves less ideally at lower pressures. In the atmosphere, where log P_{CO2} is only about $10^{-3.2}$ atmospheres, this deviation is of minor importance. In most groundwater the partial pressure of CO_2 gas is considerably higher (log $P_{CO2} \approx 10^{-2}$ atm; Rightmire and Hanshaw, 1973), and nonideal behavior becomes more significant.

G. N. Lewis introduced the thermodynamic quantity fugacity (*f*) to account for nonideal gas behavior. Fugacity may be thought of as the idealized pressure of a gas in a manner similar to considering chemical activity as the idealized concentration of a solute. In a mixture of gases, the fugacity of a gas approaches the partial pressure of the gas as the total gas pressure of the mixture approaches zero:

$$\frac{f_{\mathrm{i}}}{P_{\mathrm{i}}} \rightarrow 1 \text{ as } P_{\mathrm{T}} \rightarrow 0 \tag{3.20}$$

where f_{i} and P_{i} are the fugacity and partial pressure of component i, respectively. Fugacity and partial pressure have common units and are related to each other by a dimensionless gas activity coefficient (y):

$$f_{\mathrm{i}} = y_i P_{\mathrm{i}} \tag{3.21}$$

Gas activity coefficients for most gases approach 1 as total pressure approaches 0. What this means is that at relatively low pressures, partial pressure may be used for fugacity in most thermodynamic calculations.

3.4 CO_2 SYSTEM

CO_2 is the most important and, in most cases, the only gas of significance in groundwater systems. The reactions between CO_2 water and the dissolved species are:

$$\mathrm{CO}_{2(\mathrm{aq})} + \mathrm{H_2O} = \mathrm{CO}_{2(\mathrm{aq})} + \mathrm{H_2O} \tag{3.22}$$

$$\mathrm{CO}_{2(\mathrm{aq})} + \mathrm{H_2O} = \mathrm{H_2CO_3} \tag{2.23}$$

It is common to consider all $CO_{2(aq)}$ as H_2CO_3. Using the law of mass action, Equation 2.23 can be expressed as:

$$\mathrm{K}_{\mathrm{CO_2}} = \frac{[\mathrm{H_2CO_3}]}{[\mathrm{H_2O}]\left[\mathrm{CO}_{2(\mathrm{g})}\right]} \tag{2.24}$$

Because the activity of water is 1, Equation 2.24 can be written as the partial pressure of CO_2:

$$K_{CO_2} = \frac{[H_2CO_3]}{[P_{CO_2}][\gamma CO_{2(g)}]} \tag{2.25}$$

where γ is the CO_2 activity coefficient. Equation 2.25 can be used to calculate the P_{CO_2} and H_2CO_3 contents of waters when the other variable is known. The activity coefficient for uncharged dissolved gases such as CO_2 and O_2 is often greater than 1, but the coefficients decrease with increasing ionic strength. The activity coefficients of gases are also affected by the types of other electrolytes in solution. For example, the activity coefficient of CO_2 is greater in a KCl solution than in a NaCl solution. The effect of decreasing the activity coefficient according to the composition of the electrolyte solution is known as the salting-out-effect. Most groundwater systems have an ionic strength that is less than 0.2, and the activity of dissolved gases are usually represented as unity. In such systems Equation 2.2.5 can be expressed as:

$$K_{CO_2} = \frac{[H_2CO_3]}{[P_{CO_2}]} \tag{2.26}$$

Example E3.3 Calculate dissolved CO_2 content in groundwater

Problem: Compare the solubility of CO_2 in groundwater in equilibrium with the atmosphere and in the soil zone of a typical aquifer at 25 °C. Assume no loss of CO_2 due to chemical reactions.

Solution: The partial pressures of CO_2 in the atmosphere and in typical soil zones can be approximated as about $10^{-3.5}$ and 10^{-2} atm, respectively. From Table 3.3 $K_{CO_2} = 3.39\times10^{-2}\ \frac{mol}{L-atm}$. The solubility is calculated by rearranging Henry's law (Equation 3.17).

Solubility CO_2 at $10^{-3.5}$ atm is:

$$\frac{3.39\times10^{-2}\text{ mole}}{\text{L}-\text{atm}}\times10^{-3.5}\text{ atm}=\frac{1.07\times10^{-5}\text{ mole}}{\text{L}}$$
$$=\frac{1.07\times10^{-5}\text{ mole}}{\text{L}}\times\frac{44.01\text{ g}}{\text{mole}}\times\frac{1000\text{ mg}}{\text{g}}=\frac{0.47\text{ mg}}{\text{L}}$$

and the solubility CO_2 at 10^{-2} atm:

$$\frac{3.39\times10^{-2}\text{ mole}}{\text{L}-\text{atm}}\text{x }10^{-2}\text{atm}=\frac{3.39\times10^{-4}\text{ mole}}{\text{L}}$$
$$=\frac{3.39\times10^{-4}\text{ mole}}{\text{L}}\times\frac{44.01\text{ g}}{\text{mole}}\times\frac{1000\text{ mg}}{\text{g}}=\frac{14.92\text{ mg}}{\text{L}}$$

The two orders of magnitude difference between atmospheric and soil zone CO_2 concentrations explain why soil zone CO_2 is so important in chemical weathering.

Problems

1. Calculate the concentration of CO_2 (in mg/L) in rain and in the typical soil zone water at 10 and 20 °C.

2. Using the answers from question one, calculate the concentration of Ca^{2+} of water that has been in contact with $CaCO_3$. Assume calcite equilibrium.
3. A 1L container contains 0.5 mole of a gas at 25 °C. The measured pressure of the gas is 1.02 atm. What is the activity coefficient for the gas?
4. Groundwater from the thermal spring at Pah Tempe, Utah, exsolves $CO_{2(g)}$ that escapes to the atmosphere. The solute concentrations are listed here. The spring water temperature is 40.2 °C. The total CO_2 discharge is 4.4 L/sec.

Ca^{2+}		Mg^{2+}		Na^+		K^+		HCO_3^-		Cl^-		SO_4^{2-}	
mg/L	meq/L	mg/L	meq/L	mg/L	meq/L	mg/L	meq/L	mg/L	meq/L	mg/L	meq/L	mg/L	meq/L
735.4	36.7	143.3	11.8	2520	109.6	219.6	75.6	1277	20.9	3595	101.4	1783	37.1

a. What is the concentration (in mg/L) of the dissolved gas?
b. Assuming the $CO_2(g)$ is derived at depth from the thermal decomposition of $CaCO_3$ and that $CaCO_3$ + heat = CaO + CO_2, what is the annual mass of calcite metamorphosed (in kg)?

Chapter Notation

Subscripts

1,2,3 etc.	components 1, 2, and 3
aq	aqueous
g	gaseous
i	component i
n	component n = last component in a series of components
°C	temperature in degrees Centigrade
°K	temperature in degrees Kelvin
a_i	chemical activity of component i
atm	atmosphere (unit of pressure)
cal	calories
cm	centimeter
f	fugacity
f_i	fugacity of component i
J	joule (J = 0.239 calories)
kg	kilogram
K_i	temperature dependent Henry's constant listed as H_s^{cp} in the tables. In the literature K_i may also be shown as K_h.
K_T	temperature dependent constant
K_p	proportionality constant
L	liter
lb	pound
log	logarithm to base 10
m	meter
n	number of moles of a gas
N	Newton
P	pressure of a gas
P_{CO2}	partial pressure of $CO_{2(g)}$
P_i	partial pressure of component i
P_i^o	partial pressure of pure component I at a set temperature

P_o	pressure at °K
P_t	total pressure
P_T	pressure at temperature T
R	universal gas constant
T	temperature
V	volume
V_o	volume at °C
V_t	total volume
V_T	volume at temperature T
X_i	mole fraction of component i
y_i	dimensionless gas activity coefficient of component i

REFERENCES CITED

Langmur, D., 1997, *Aqueous environmental geochemistry*, Prentice Hall.

NOAA, 2025, noaa.gov/ccgg/trends/

Rightmire, C., and B. Hanshaw, 1973, Relationship between the carbon isotopic composition of soil CO_2 and dissolved carbonate species in ground water, *Water Resources Research*, 9(4), 958–967.

Sander, R., 2015, Compilation of Henry's law constants (version 4.0) for water as a solvent, *Atmospheric Chemistry Physics*, 15, 4399–4981.

Sander, R., 2023, Compilation of Henry's law constants (version 5.0.0) for water as solvent, *Atmospheric Chemistry Physics*, 23.

4 Kinetics

4.1 INTRODUCTION

It is tempting to assume that all rock–water interactions of hydrogeologic interest proceed rapidly to thermodynamic equilibrium when either excess reactants or products are available. Thermodynamic modeling is comforting because measured and fixed parameters such as temperature, solute and dissolved gas concentrations, equilibrium constants, ΔH, ΔS, and ΔG, and computational results such as SI and moles of dissolved and precipitated substances are readily quantifiable.

Although the results of thermodynamic modeling provide valuable insights into hydrogeochemical processes, such modeling does not always yield results that are anticipated in light of known geochemical, geologic, hydrogeologic, hydrodynamic, and biologic conditions. It is not uncommon, for example, to find water in carbonate aquifers to be undersaturated with respect to carbonate minerals and to find water in cool temperature, shallowly circulating aquifers to be supersaturated with respect to quartz.

Thermodynamics is time independent and is only concerned with the energy relationships between reactants and products. Thermodynamics answers the critical questions: Can a reaction occur, and in which direction will the reaction proceed? Thermodynamics does not answer the time- and mechanism-dependent questions: How fast will a reaction occur, what additional processes or mechanisms are required to initiate a reaction, and what are the mechanisms or steps of the reaction? The answers to such questions, which may be critical to resolving a hydrogeochemical problem, belong to the field of investigation known as chemical kinetics.

There are two important differences between kinetic and thermodynamic treatments of hydrogeochemical problems. First, kinetics depends on knowledge of the specific reaction pathways (i.e., which specific species, including intermediate products, are reacting), whereas thermodynamics only requires knowledge of the initial state of the reactants and final state of the products to determine the ΔG, ΔH, and ΔS of the overall reaction. And second, kinetics is concerned with all parallel reactions that lead to the same products, whereas thermodynamics only requires knowledge of the most rapid reaction pathway. The study and application of kinetics is more difficult than thermodynamics in that it is both time and pathway dependent. There is no general relationship between the thermodynamics of reactions and the rates at which reactions occur.

Chemical kinetics is studied at three levels of inquiry: phenomenological, mechanistic, and statistical mechanical. Phenomenological kinetics attempts to develop rate constants in terms of macroscopically observable quantities such as composition, temperature, pressure, volume, and time. Mechanical kinetics attempts to describe the individual steps involved in overall reactions. Statistical mechanical kinetics attempts to evaluate the energies of the individual reaction steps, the details of breaking and forming chemical bonds, and the transition states in chemical reactions.

Although considerable progress has been made in the field of chemical kinetics, the application of kinetics to real world hydrogeologic problems remains limited when compared to the application of thermodynamics. In light of the difficulty of quantitatively applying kinetics to most

DOI: 10.1201/9781003646129-5

hydrogeochemical investigations, the purpose here is to describe the qualitative application of phenomenological and mechanistic kinetics to hydrogeochemistry. Having a general knowledge of qualitative kinetics can provide insight into geochemical reactions that have not behaved as thermodynamically anticipated and to help explain some observed phenomena.

4.2 EQUILIBRIUM THERMODYNAMICS AND KINETICS

Although there is no universal quantitative link between chemical equilibria (thermodynamics) and the rate of chemical reactions (kinetics), these two concepts can be reconciled by the ideas embodied in transition-state theory. Transition-state theory says that reaction rates are controlled by the difference between the energy of the reactants and the energy required for the transition state, which is also known as the activation energy. The investigation of activation energies is part of the field of inquiry known as linear free energy relationships or LFER. The principle of activation energy is illustrated in Figure 4.1 where: 1) a ball is stuck in a depression and 2) there is a mixture of $O_{2(g)}$ and $H_{2(g)}$. From a thermodynamic perspective it is obvious that the ball should spontaneously roll down the hill and the gas mixtures should become water. The energy of the reactants, as measured by the high elevation on the hill and the ΔH of the ½$O_{2(g)}$ + $H_{2(g)}$ reaction, are greater than the energy of the products, as measured by the elevation of the bottom of the hill and the ΔG of the gas reaction. However, a child waiting at the bottom of the hill for the ball would have a long wait, and the owner of the gas mixture might die of thirst waiting for water to form because the reactions are kinetically hindered. In order for the ball to roll down the hill or some water to form, some additional energy must be supplied to elevate the ball above the low ridge and to make the gas reaction proceed. This additional energy requirement is the activation energy.

Fortunately, the activation energy needed for the chemical combination of ½$O_{2(g)}$ and $H_{2(g)}$ is great, and these gases will coexist indefinitely in the atmosphere. However, when sufficient energy is applied to a mixture of the two gases the reaction explosively creates H_2O. Less kinetically hindered reactions include methane explosions that can occur in coal mines and in houses with a gas line leak.

Important Takeaways – Thermodynamics is only concerned with the state of the most rapid initial reactants and the final product, and assumes the reactions occur without delay. Kinetics is concerned with all intermediate reactions and the time required for all reactions to occur.

Many reactions that are thermodynamically possible cannot proceed until an external energy source or other mechanism provides the necessary activation energy.

4.3 CLASSIFICATION OF KINETIC REACTIONS

Several classification schemes are used to describe and understand chemical kinetics. The schemes include classifying reactions according to the

1. number of reaction steps,
2. interaction between phases,
3. molecularity and reaction order, and
4. sequence of reactions.

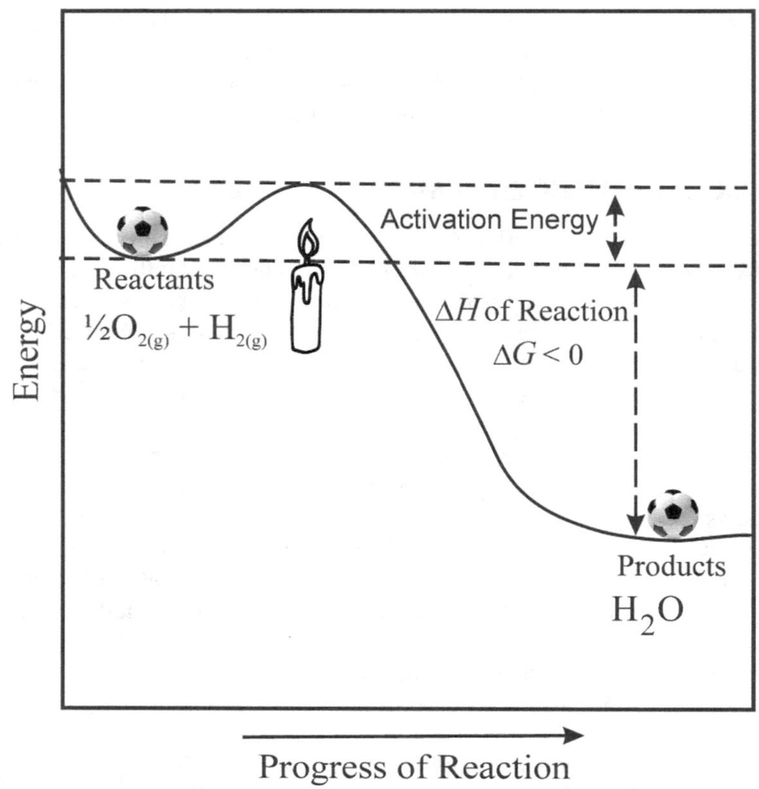

FIGURE 4.1 Thermodynamics vs. kinetics. According to thermodynamics the ball should roll down the hill and oxygen should spontaneously form water. However, the ball is in a small depression and needs a little push to complete its roll down the hill, and the reaction $O_{2(g)}$ + $2H_{(g)}$ needs an energy source to proceed to completion.

Elementary reactions are chemical reactions that occur as a single step. For example, the dissolution of NaCl into Na^+ and Cl^- involves a single reaction step and proceeds as written at the molecular level:

$$NaCl = Na^+ + Cl^- \tag{4.1}$$

Overall reactions may be written as a single step or multiple steps; however, they are the net result of a series of elementary reactions. For example, sulfate reduction by the bacterial decomposition of organic matter may be written as the single reaction:

$$SO_4^{2-} + 2CH_2O = H_2S + 2CO_3^- \tag{4.2}$$

but the reaction actually involves several intermediate reactions that include the breakdown of complex organic molecules and several reactions between SO_4^{2-} and bacterial enzymes.

Reactions are either homogeneous or heterogeneous. Homogenous reactions are reactions that occur between molecules or ions in the same phase (i.e., gas, liquid, or solid). Heterogenous reactions are chemical reactions that occur between various chemical species at the interface of two different phases and include the dissolution of minerals and reactions between gas and water. Although both reaction types are important to hydrogeochemistry, the kinetics of homogeneous reactions are better understood than are the kinetics of heterogeneous reactions.

Molecularity refers to the number of reactant molecules in an elementary reaction. Molecular conditions and their reactant concentration are shown in Table 4.1. Reaction order is the sum of the exponents of the reactant species concentration. As a general rule, Equation 4.3 is a first order reaction, Equation 4.4 and Equation 4.5 are second order reactions, and Equation 4.6 and Equation 4.7 are third order reactions. It is important to note that molecularity and reaction order are not a priori identical and that stoichiometric relationships do not necessarily indicate reaction order. Most reactions of hydrologic interest are either first or second order because it is difficult for three reactant species to arrive at the same location, with the proper orientation, at the same time.

Multi-step reactions consist of a series of elementary reactions that occur as sequential (consecutive), parallel (concurrent), or as a combination of sequential and parallel reactions. Each elementary reaction in an overall reaction may also be reversible or nonreversible. Consider the overall reaction:

$$A \rightarrow D \tag{4.8}$$

where A represents the initial reactant and D represents the final product. If Equation 4.8 is sequential it actually represents three elementary reactions: $A \rightarrow B$, $B \rightarrow C$, and $C \rightarrow D$, which may be depicted as:

$$A \rightarrow B \rightarrow C \rightarrow D \tag{4.9}$$

where A is only a reactant, B and C are first products and then reactants, and D is the final product. For example, the conversion of soil zone $CO_{2(g)}$ to carbonic acid may be written as:

$$CO_{2(g)} + H_2O \rightarrow H_2CO_3$$

TABLE 4.1
Molecular Conditions and Reaction Concentration of Various Molecular Combinations

Molecularity	Reaction	Concentration	
Unimolecular	$A \rightarrow$ Products	C_A	(4.3)
Unimolecular	$2A \rightarrow$ Products	$C_A{}^2$	(4.4)
	$A + B \rightarrow$ Products	C_AC_B	(4.5)
Bimolecular	$3A \rightarrow$ Products	$C_A{}^3$	(4.6)
	$A + B + D \rightarrow$ Products	$C_AC_BC_D$	(4.7)

At time (t) = 0 the initial concentration (C^o) = concentration (C). A, B, and D are reactants, and C is concentrations of the reactant species A, B, and D. Concentration is usually expressed in moles, and time in second, minute, hour, day, or year.

but it actually involves two reactions:

$$CO_{2(g)} + H_2O \rightarrow CO_{2(aq)} + H_2O$$

$$CO_{2(aq)} + H_2O \rightarrow H_2CO_3$$

If reaction 4.8 can occur from two parallel, but not necessarily related, reactions A → B → D, and A → C → D, the overall reaction would be represented as:

$$\begin{array}{ccc} & \rightarrow B \rightarrow & \\ A \rightarrow\mid & & \mid\!\rightarrow D \\ & \rightarrow C \rightarrow & \end{array} \quad (4.10)$$

The parallel reaction $CO_{2(aq)} \rightarrow HCO_3^-$ is described in Equations 4.22 to 4.25.

Important Takeaway – Chemical reactions may involve single or multiple steps. Multi-step reactions consist of a series of single steps that occur sequentially (A → B → C → D). *In some systems parallel but unrelated reactions can yield products from the same reactants*

$$\begin{array}{ccc} & \rightarrow B \rightarrow & \\ A \rightarrow\mid & & \mid\!\rightarrow D \\ & \rightarrow C \rightarrow & \end{array}$$

4.4 KINETIC REACTION RATES

The reaction rate of the slowest elementary reactions in a sequence, known as the rate-limiting step, controls the reaction rate of the overall reaction. In sequential reactions the rate of the slowest elementary reaction is the rate-limiting step. In parallel reactions the rate of the fastest reaction pathway is the rate-limiting step because concentrations of the final product are not dependent upon the bottleneck caused by the slow reaction pathway. Reaction rates are quantified in terms of the parameter known as the reaction rate constant (k), whose units are the reciprocals of time $\left(\frac{1}{t}\right)$. A few reactions, such as some radioactive decay processes, are nonreversible elementary reactions. The rate constant for unimolecular and bimolecular reactions may be represented as:

$$A \xrightarrow{k} \text{Products} \quad (4.11)$$

and

$$A + B \xrightarrow{k} \text{Products} \quad (4.12)$$

where A and B are reactants.

The rate constant of reversible unimolecular and bimolecular elementary reactions is the sum of the rate constants of the forward and reverse reactions:

$$A \xrightarrow[K^{-1}]{K_1} \text{Products} \quad k = k_1 + k^{-1} \quad (4.13)$$

and

$$A + B \underset{K^{-1}}{\overset{K_1}{\longleftrightarrow}} \text{Products} \qquad k = k_1 + k^{-1} \tag{4.14}$$

where k is the overall reaction rate constant and k_1 and k^{-1} are the forward and reverse reaction rate constants, respectively. Some chemical transformations are considerably more complex. Nonreversible sequential reactions may be represented as:

$$A \xrightarrow{K_1} B \xrightarrow{K_2} \qquad k = k_1 + k_2 \tag{4.15}$$

where B is an intermediate product, or as:

$$A \xrightarrow{K_1} B \xrightarrow{K_2} C \xrightarrow{K_2} \qquad k = k_1 + k_2 + k_3 \tag{4.16}$$

where B and C are intermediate products. Nonreversible parallel reactions may be represented as:

$$\begin{aligned} &A \xrightarrow{k_1} B \\ &\qquad\qquad k = k_1 + k_2 \\ &A \xrightarrow{k_2} C \end{aligned} \tag{4.17}$$

where A is the reactant and B and C are products.

Example 4.1 Ozone kinetics

The ozone layer, located 10 to 50 km above the earth's surface, forms a protective shield against harmful ultraviolet radiation (UV). Stratospheric ozone is naturally produced by the combination of oxygen atoms (O) with an oxygen molecule (O_2):

$$O_2 + UV \rightarrow O + O \qquad \text{E4.1a)}$$

$$2O + O_2 \rightarrow 2O_3 \qquad \text{(E4.1b)}$$

Ultraviolet radiation (UV) also plays a role in O_3 destruction:

$$O_3 + UV \rightarrow O + O_2 \qquad \text{(E4.1c)}$$

$$O + O_3 \rightarrow 2O_2 \qquad \text{(E4.1d)}$$

Anthropogenic gas contaminants such as chlorine-rich CFCs (chlorofluorocarbons) and nitric oxide (NO) also destroy O_3 molecules. CFCs were used as refrigerants and as a propellant in aerosol products. Due to concerns about the deteriorating ozone layer, the Montreal Protocol banned the global use of CFCs in 1988. In 1994 CFCs were banned for use in the United States. The concentration of CFCs in groundwater systems have been used to date the recharge of piston flow systems (Chapter 11). Nitrous oxides remain a byproduct of fossil fuel combustion.

The $O_2 - O_3$ relationship can be represented by the reaction:

$$2O_3 \rightarrow 3O_2 \qquad \text{(E4.1f)}$$

The reaction in Equation E4.1f can be kinetically represented as:

$$O + O_3 \underset{k^{-1}}{\overset{k_2}{\rightleftarrows}} 2O_2 \quad \text{(fast reverse reaction)} \qquad \text{(E4.1g)}$$

and

$$O + O_3 \xrightarrow{k_2} 2O_2 \quad \text{(slow reverse reaction)} \tag{E4.1h}$$

The reverse reaction in E4.1h is extremely slow and can be ignored. The change rate of oxygen can be determined by adding the contributions from E4.1g and E4.1h:

$$\frac{d[O]}{dt} = k_1[O_3] - k^{-1}[O_2][O] - k_2[O][O_3] \tag{E4.1i}$$

The rates at which chemical reactions take place depend upon several factors including the

1. type of reactants,
2. temperature of the reaction,
3. concentration of reactants, and
4. presence or absence of catalytic substances.

Chemical reactions rates proceed differently from one substance to another. Fundamental differences between substances may be attributed to the structures of the atoms, molecules, and ions that make up the reacting substances and the type of chemical bonds involved. For example, noble gases tend to remain chemically inactive regardless of changes in temperature and pressure, whereas H^+ and OH^- are chemically very reactive.

Temperature has a profound influence on reaction rates. Many reaction rates are doubled by a 10 °C increase in temperature, and the reaction rate, as measured by the reaction rate constant, can vary by several orders of magnitude over a 100 °C temperature change (Lasaga, 1981). In 1889, Arrhenius suggested that the rate constant is exponentially related to temperature according to the relationship:

$$\frac{d\ln k}{dT} = \frac{E_a}{RT^2} \tag{4.18}$$

where E_a is the activation energy, R is the universal gas constant, and T is the absolute temperature. The Arrhenius equation, which predicts a linear relation between ln$k$ and T, verifies our intuitive understanding that reaction rates increase with increased temperature. E_a is temperature independent, and Equation 4.18 can be integrated to:

$$k = Ae^{-Ea/RT} \tag{4.19}$$

where A is a pre-exponential or frequency factor. Activation energy is primarily a function of the reactants. However, some evidence suggests that the frequency factor may be weakly temperature related. Typical units of A are $\frac{1}{\text{sec}}$ for first order reactions and $\frac{1}{\text{moles}-\text{sec}}$ for second order reactions. Arrhenius's model incorporates two principles. First, reactant molecules must attain the necessary activation energy before spontaneous reactions can occur. Second, the frequency of molecular interactions, and hence the rate of chemical reactions, increases with increased temperatures. The usefulness of the Arrhenius equation is based on the fact that experimentally derived log k vs. $\frac{1}{T}$ data plot along a straight line. Rate constants for a wide range of temperatures may therefore be extrapolated from a plot of experimental values (Figure 4.2). Changes in the slope of log k vs $\frac{1}{T}$ plot can be useful for diagnosing changes in reaction mechanisms as a function of reaction temperature variations.

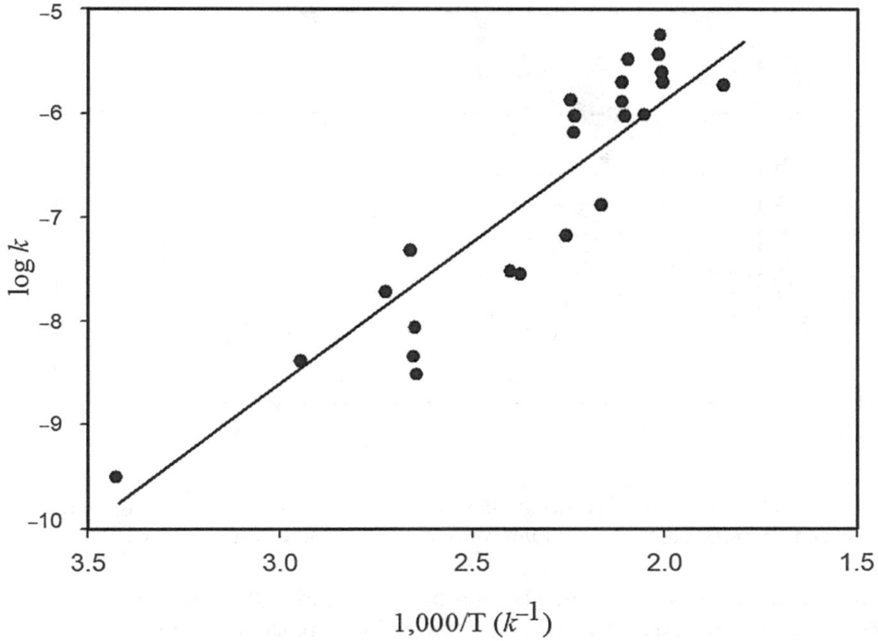

FIGURE 4.2 Arrhenius plot of experimental determinations of various quartz mineral–water reactions. The line is the least squares fit of the data and represents the reaction rate constant. Modified after Lasaga (1981).

TABLE 4.2
Summary of the Differential Equations and Analytical Solutions That Represent the Rate Laws

Reaction order	Differential reaction equation	Analytical solution	Units of k
Zero	$A \xrightarrow{k} P \quad k = \frac{-dC_A}{dt}$	$C_A = C_A^o - kt$	t^{-1}
First	$A \xrightarrow{k} P \quad kC_A = \frac{-dC_A}{dt}$	$C_A = C_A^o \, e^{-kt}$	t^{-1}
Second	$A + B \xrightarrow{k} P \quad kC_AC_B = \frac{-dC_A}{dt}$	$C_A = \frac{C_A^o\left(C_A^o - C_B^o\right)}{C_A^0 - C_B^0 - e^{-kt\{(C_A^o - C_B^o)\}t}}$	$C^{-1}t^{-1}$

At time (t) = 0 the initial concentration (C^o) = concentration C. A and B are species. Concentration is usually express in moles, and time is expressed in second, minute, hour, day or year. The minus sign indicates that the concentration decreases with time.

The order of chemical reactions is an important factor in kinetic reaction rates. Differential equations and analytical solutions relating reaction rates, concentrations of reactants, and time to zero, first, and second order reactions are listed in Table 4.2. For zero order reactions, such as the reduction of SO_4^{2-} (concentration > 2 mmol/L) by the bacteria *Desulfovibrio*, a plot of concentration vs. time yields a straight line (Berner, 1971). What this means is that quantity of reactant converted to product per unit time increment remains constant. The relationship between concentration vs. time

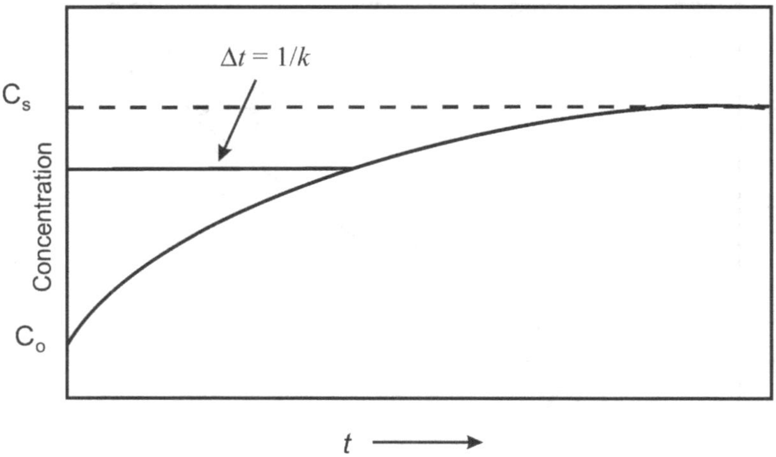

FIGURE 4.3 Sketch of the change in concentration of products as a function of time. C_o represents the initial concentration and C_s represents the equilibrium concentration. Modified after Lasaga (1981).

for first order reactions is exponential. The significance of this is that the time required to achieve equilibrium is dependent upon $1/k$ and that the conversion of reactants to products proceeds rapidly at first and then slows down as equilibrium is approached (Figure 4.3).

The analytical solution for a first order reaction (Table 4.2) can be written as:

$$\ln C_A = \ln C_A^o - kt \tag{4.20}$$

Equation 4.20 means that a plot of $\ln C_A$ vs. time is a straight line and that k is the slope of the line.

First order reactions may be either true or pseudo first order. True first order reactions are unimolecular reactions and have only one reactant (A), whereas pseudo first order reactions include bimolecular and termolecular reactions that have two or three reactants (i.e., A, B, and C). Pseudo first order behavior occurs when the change in concentrations of reactants B and C are small relative to changes in the concentration of reactant A. In other words, reactants B and C are in such large concentrations that they are reaction buffers, and their concentrations remain essentially constant.

Radioactive decay is a true first order reaction of great significance. Rate laws for radioactive decay are usually presented in terms of half-life ($t_{½}$), which is the time required to consume one-half of the remaining reactants. By substituting $C_A = ½C_A^o$ into Equation 4.20, the equation can be expressed as the radioactive decay half-life equation:

$$t_{\frac{1}{2}} = \frac{0.693}{k} \tag{4.21}$$

In the case of ^{14}C decaying to ^{14}N ($k = 1.29 \times 10^{-4}$/yr), the half-life predicted by Equation 4.21 is 5,370 years.

Pseudo first order reactions include sulfate reduction by bacteria when the sulfate concentrations are less than 2 mmol/L and the hydration of carbon dioxide, forming carbonic acid. The hydration of carbon dioxide is extremely important in hydrogeochemistry because it is the first in a series of reactions that pH buffer most groundwater and which are largely responsible for the dissolution of carbonate rocks.

The hydration of CO_2 and related chemical reactions are particularly interesting because they can be used to illustrate fundamental differences between thermodynamic and kinetic approaches to evaluating hydrochemical reactions. From a thermodynamic perspective, the conversion of $CO_{2(aq)}$ to HCO_3^- involves the following sequential reactions:

$$CO_{2(aq)} + H_2O \rightarrow H_2CO_{2(aq)} \rightarrow HCO_3^- + H^+ \tag{4.22}$$

Because $CO_{2(aq)}$ and $H_2CO_{3(aq)}$ may be thermodynamically treated as the same and are usually expressed as $H_2CO_{3(aq)}$, the reaction in Equation 4.22 is typically abbreviated to:

$$CO_{2aq} + H_2O = HCO_3^- + H^+ \tag{4.23}$$

The thermodynamic treatment of reaction in Equation 4.23 is kinetically deficient for two reasons. First, the reaction:

$$CO_{2(aq)} \rightarrow H_2CO_{3(aq)} \tag{4.24a}$$

is the rate limiting step because it occurs slowly relative to the overall reaction:

$$H_2CO_{3(aq)} \rightarrow HCO_3^- \tag{4.24b}$$

At 0 °C the half-life of reaction 4.24a is about 6 minutes (Jones et al., 1964), whereas reaction 4.24b occurs rapidly. Second, there are two parallel and competing pathways for the conversion of $CO_{2(aq)}$ to HCO_3^- . One pathway is represented by Equation 4.23, and the other pathway is:

$$CO_{2(aq)} + HO^- \rightarrow HCO_{3(aq)}^- \tag{4.25}$$

The reaction rate of Equation 4.25 is pH dependent, whereas the reaction rate of Equation 4.23 is not.

Catalysts are important in many chemical reactions. They allow a reaction to follow a reaction pathway with a lower activation energy than does the pathway that is followed by non-catalyzed reactions. Common examples of catalysts include metals in contact with solutions where the metals provide electrons that facilitate redox reactions, H^+ ions that catalyze hydrolysis reactions accompanying weathering of aluminosilicate minerals, and hydrothermal alteration. Additionally, microbial activity catalyzes many inorganic reactions.

Important Takeaway – Numerous factors affect reaction rates: reaction type, temperature, concentration, and catalyst. The rate of the overall reaction is controlled by the rate of the slowest reaction in the sequence. This is known as the rate-limiting step, which can be affected by the reacting substances, temperature of the reaction, concentration of the reactants, and other factors.

4.5 KINETIC PRECIPITATION AND DISSOLUTION MECHANISMS

Hydrogeochemical reactions can be broadly classified as either mineral precipitation or weathering–dissolution. Reaction rates for these processes are dependent upon the rates of both chemical transport and chemical transformation. First, the reactants must be transported to each other or to the phase interface. In the case of mineral precipitation transportation is usually by diffusion.

Second, the chemical reaction or transformation occurs. Third, the products of the reaction are transported away from each other.

4.5.1 Mineral Precipitation

From a thermodynamic perspective the precipitation of a mineral in an aqueous solution occurs when the condition of mineral supersaturation is attained. Mineral precipitation involves two processes: nucleation and crystal growth. Nucleation is the formation of embryo crystals that precede the growth of true crystals. During the early stages of nucleation, the size of embryo crystals increases until an activation energy barrier is encountered that inhibits additional precipitation. This activation energy barrier is the result of the interface between the growing crystal body and the solution. If the energy barrier can be overcome, true crystal growth will spontaneously proceed and continue until the condition of supersaturation no longer exists. However, the formation of stable nuclei does not always proceed spontaneously in supersaturated solutions because the minimum activation energy (ΔG_a) has not been reached. It is important to remember that ΔG_a is not accounted for in thermodynamic expressions.

Nucleation may be described in terms of the free energy of formation of a nucleus (ΔG_n) that consists of the energy released by the formation of chemical bonds (ΔG_b) plus the energy required to create the surface of the nuclei (ΔG_s):

$$\Delta G_n = \Delta G_b + \Delta G_s \tag{4.26}$$

For nucleation to proceed and crystal growth to occur, ΔG_b must equal or exceed the energy requirement for surface creation. ΔG_b is always negative. Crystal growth spontaneously proceeds when $\Delta G_n < 0$. Thermodynamic models assume that in a supersaturated solution $G_n \geq G_a$; however, this condition is seldom true in homogeneous solutions.

Three factors influence the activation energy required to sustain nucleation and crystal growth:

1. the degree of mineral supersaturation of the solution, described by the saturation index (SI),
2. the interfacial energy (μ) required at the solid and the liquid interface, and
3. the radii of the nuclei or molecules.

For spherical nuclei, the energy released by chemical bond formation increases logarithmically with increases in SI according to the relationship:

$$\Delta G_b = -jKT \ln \frac{IAP}{Keq} \tag{4.27}$$

where j is the number of ions or molecules in the nucleus, k is the rate constant, T is the absolute temperature, and IAP/K_{eq} is the saturation index. Refer to Chapter 2 for a review of saturation index. Because $j = 4\pi r^3/3V$, where V is the molecular volume, the energy released by bond formation decreases exponentially with increased nuclei radii. For spherical nuclei, the energy required to create surface bonds is also exponentially related to the nuclei radii according the relationship:

$$\Delta G_r = 4\pi r^2 \mu \tag{4.28}$$

where μ is the interfacial energy. Interfacial energy may be thought of as the energy per unit area of the interface between the precipitate and the matrix. Equation 4.28 can be rewritten as:

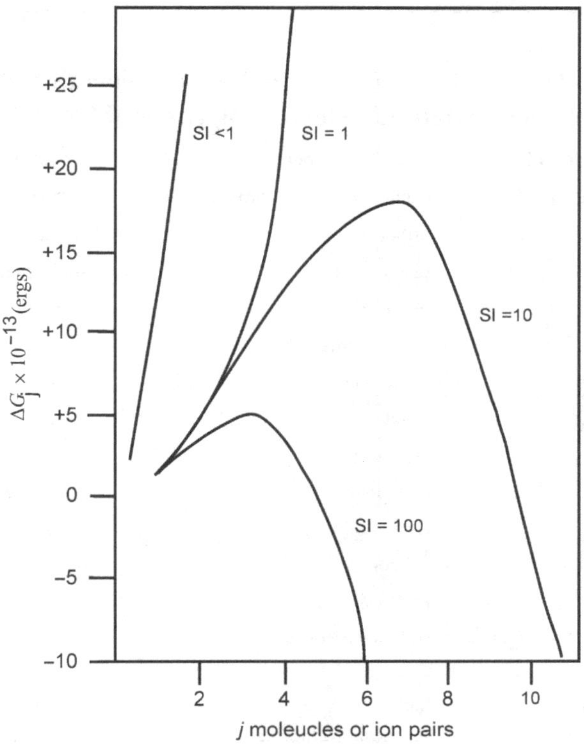

FIGURE 4.4 The nucleation rate depends on the degree of mineral saturation, particularly supersaturaton. Free energy of nucleation formation as a function of nucleus size calculated from saturation index SI $\left(\frac{\text{IAP}}{\text{Keq}}\right)$. The nucleus size is regulated by the number of molecules or ion pairs. Modified after Stumm and Morgan (1981).

$$\Delta G_n = \frac{-4\pi r^3}{3V} kT \ln \frac{IAP}{Keq} + 4\pi r^2 \mu \tag{4.29}$$

The dependence of nucleation on the combined factors of the degree of supersaturation, interfacial energy requirements, and particle radius is illustrated in Figure 4.4. In the figure, ΔG_n is plotted as a function of r and the mineral saturation $\left(\frac{\text{IAP}}{\text{Keq}}\right)$ for a constant set of V and μ values.

Homogeneous nucleation is unlikely, and for all degrees of mineral supersaturation nucleation can only begin when the interfacial energy requirement has been minimized. The ease of nucleation increases with increased supersaturation. Once a critical nucleus size has formed, additional precipitation will proceed spontaneously, with a net decrease in free energy. In aqueous systems the minimization of the activation energy requirement is commonly attained by heterogeneous precipitation, where precipitates form on existing solid phases such as inorganic substances having similar atomic structures, organic skeletal particles, clays and sand in the aquifer skeleton, and bio colloids (Stumm and Morgan, 1981). With high degrees of supersaturation, nucleation rates exceed crystal growth rates, and fine-grained precipitates and poorly formed crystals predominate. At low levels of supersaturation, crystal growth rates exceed nucleation rates, and most of the excess dissolved material is precipitated on a limited number of nuclei (Berner, 1981), resulting in large individual crystals.

TABLE 4.3
Dissolution Rate Mechanisms of Various Minerals and Compounds Arranged in Increasing Order of Solubility

Compound	Species	Reaction type
$Ca_5(PO_4)OH$	calcium apatite (hydroxyapatite)	surface reaction
KAl_3SiO_8	orthoclase	surface reaction
$NaAlSi_3O_8$	plagioclase	surface reaction
$BaSO_4$	barite	surface reaction
AgCl	chlorargyrite (silver chloride)	transport
$SrCO_3$	strontianite	surface reaction
$CaCO_3$	calcite	surface reaction
$PbSO_4$	anglesite	mixed
$SrSO_4$	celestine	surface reaction
SiO_2	Opaline	surface reaction
$CaSO_4 \cdot H_2O$	gypsum	transport
$MgSO_4 \cdot 2H_2O$	magnesium sulfate	transport
$Na_2CO_3 \cdot 10H_2O$	sodium carbonate	transport
KCl	sylvite	transport
NaCl	halite	transport
$MgCl \cdot 6H_2O$	magnesium chloride	transport

Source: Data from Berner (1981).

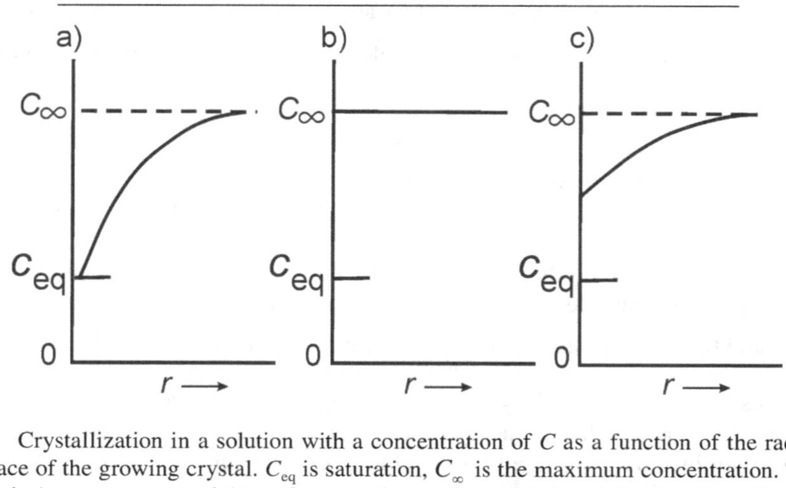

FIGURE 4.5 Crystallization in a solution with a concentration of C as a function of the radial distance r from the surface of the growing crystal. C_{eq} is saturation, C_{∞} is the maximum concentration. Three systems are represented: a) transport control, b) surface reaction control, and c) mixed transport and surface reaction control. Modified after Berner (1981).

Precipitation reactions may be transport-controlled, surface-reaction controlled, or controlled by a combination of these processes (Table 4.3, Figure 4.5). The rate of crystal growth may be strongly influenced by the control mechanism. Concentrations of reactants at the crystal surface in transport-controlled systems are largely regulated by the hydrodynamic condition of the system. During pure transport control the rate of mineral precipitation may be faster than the rate at which new reactants can be transported to the crystal face (Figure 4.5a). When this occurs the reactant concentrations

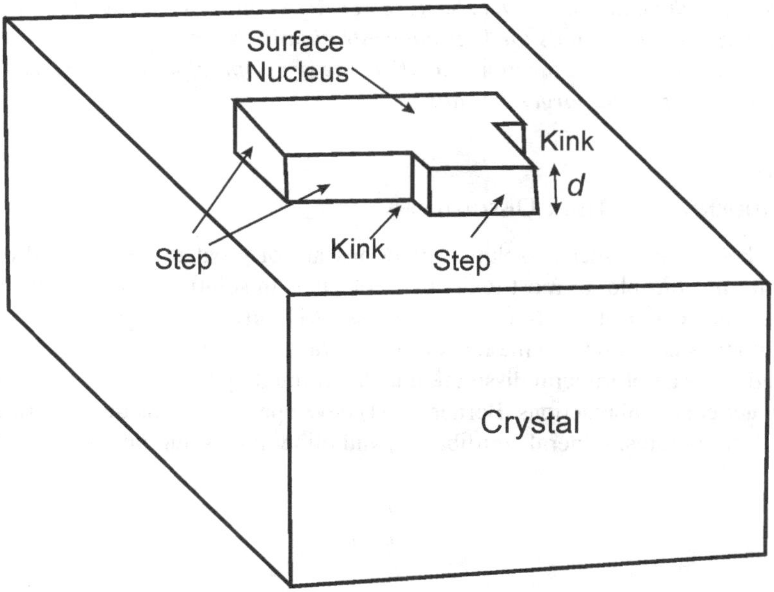

FIGURE 4.6 Idealized representation of steps and kinks accompany the growth of a crystal. The dimension *d* represents one atom. Modified after Berner (1981).

may become limited due to the slow rate of molecular diffusion. The concentration of reactants at the crystal surface may decline to equilibrium levels, and mineral precipitation may be retarded.

During pure surface-reaction controlled precipitation the rate of precipitation is so slow that the movement of ions to the mineral surface, by molecular diffusion or other processes, can keep the concentration of reactants at the crystal surface equal to the concentration in the bulk of the solution (Figure 4.5b). What this means is that the precipitation continues regardless of the hydrodynamic condition of the aquifer.

Two general mechanisms have been proposed for surface-reaction controlled precipitation: 1) surface nucleation, and 2) dislocation. During surface nucleation a nucleus attaches to an otherwise flat mineral surface (Figure 4.6). Such attachments are difficult to initiate in solutions having a low degree of supersaturation because one-dimensional surfaces have a higher interfacial-energy requirement than do two- or three-dimensional surfaces. Once attached, the initial nucleus creates steps and kinks on the mineral surface that are energetically favorable locations for additional mineral attachment. Steps are two-dimensional surfaces, and kinks are three-dimensional surfaces. As precipitation proceeds, steps and kinks are created and are filled in stepwise along the surface of the crystal. Eventually, the last kink will reach the edge of the crystal and nucleation will have to once again be initiated on a completely flat surface.

The idea of dislocation-controlled precipitation is that mineral crystallization rarely forms perfect crystals. Instead, crystallization may proceed in a spiral fashion, known as screw dislocations. After nucleation is initiated, there will always be steps and kinks on the crystal surface for additional mineral precipitation.

Important Takeaways – Mineral precipitation can only occur when the minimum activation energy is achieved. This energy requirement may be greater than the thermodynamic energy necessary for the reaction.

Mineral precipitation can be transportation- or surface-reaction controlled. Transportation controlled means the rate of crystallization is faster than the rate at which reactants can be delivered to the mineral surface. Surface-reaction controlled means the concentration at the mineral surface equals the bulk concentration in the solution.

The degree of mineral supersaturation can affect the size of precipitated crystals. Low supersaturation tends to yield larger crystals.

4.5.2 Weathering and Other Dissolutions

Chemical weathering and other dissolution reactions are of great interest in hydrogeochemistry because they are responsible for most ionic and molecules in solution. Most weathering and other dissolution reactions occur at solid–liquid interfaces. It is only natural to assume that the rate of mineral dissolution is controlled by the rate of groundwater flushing. However, this is not necessarily true. Instead, the rate of mineral dissolution is determined by the rate of flushing relative to the rate of rock–water chemical reactions. Berner (1971) developed equations that describe the relationship between flushing rates, mineral equilibrium, and mineral dissolution:

$$C = \frac{k\,C_{eq}}{k + k_f} \tag{4.30a}$$

$$R = \frac{kk_f C_{eq}}{k + k_f} \tag{4.30b}$$

where:

C = concentration of dissolved species i (mass per unit volume),
R = rate of dissolution (mass per unit volume of solution per unit time (t); $R = k(C_{eq} - c)$, C_{eq} = saturation concentration,
k = dissolution rate constant,
k_f = flushing frequency (F/V),
F = inflow − outflow = flushing rate (volume per unit time), and
V = fixed volume of water with concentration c.

From Equations 4.30a and 4.30b it is apparent that an increase in the flushing rate (F) will cause a decrease in the concentration (c) and an increase in the rate of dissolution (R). What this means is that the dissolution rate is accelerated by flushing due to a decreased concentration. When the flushing rate is great $k_f >> k$, Equations 4.30a and 4.30b may be simplified to:

$$C \approx \frac{k}{k_f} C_{eq} \ll C_{eq} \tag{4.31a}$$

and

$$R \approx kC_{eq} \tag{4.31b}$$

Because high flow rates inhibit saturation, the maximum rate of dissolution kC_{eq} is independent of the flushing rate. During sluggish groundwater flow the opposite is true, as $K_f << k$, and Equations 4.30a and 4.30b may be simplified to:

$$C \approx C_{eq} \tag{4.32a}$$

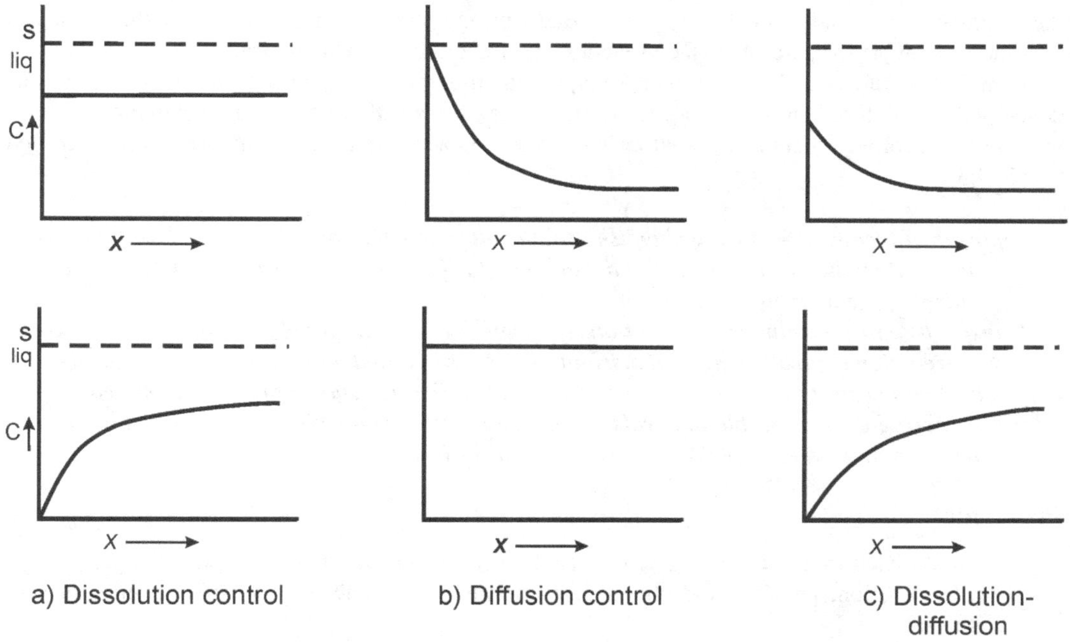

FIGURE 4.7 The chemical weathering distribution of a solution with a concentration of C that has been released by the dissolution of a solid mineral phase: a) reaction/dissolution controlled, b) diffusion controlled, and c) mixed controlled. The distance from the mineral surface is represented by x, and time of the reaction is represented by t. Modified after Jimenez-Mena et al., 2019.

and

$$R \approx K_f C_{eq} \tag{4.32b}$$

During sluggish groundwater flow, saturation of the solution can occur, and the rate of mineral dissolution is controlled by the rate of flushing. Equations 4.30 to 4.32 assume that the pore waters are near saturation so that the reactivity of the minerals can be neglected and that other rate limiting factors are not important. Such assumptions may be reasonable for highly soluble minerals in many groundwater systems; however, these assumptions are seldom valid for minerals with low solubility.

In addition to affecting mineral precipitation, surface-reaction controlled and chemical transport-controlled mechanisms also have a profound effect on mineral dissolution. The process of mineral dissolution differs significantly from mineral precipitation in that three-dimensional nucleation does not have to occur. Surface reactions are the actual detachments of ions or molecules from crystal surfaces. During surface-reaction controlled dissolution the rate of ion detachment is slow relative to the rate of ion build-up in the solution at the crystal surface. The ions are rapidly moved away from the crystal surface by advection and diffusion, which means that the concentration of ions at the crystal surface remains constant with time (Figure 4.7). Because surface-reaction controlled dissolution originates at points of excess energy, such as dislocation features, mineral surfaces are partially dissolved. Such partial dissolution often results in well-developed crystallographic controlled dissolution features such as etch pits. The rough weathered texture of carbonate rocks in arid environments is an example of surface-reaction controlled dissolution. Here, water accumulating on the mineral surface is not continuously flushed away. Transport-controlled dissolution, also known as diffusion-controlled dissolution, involves the formation of a thin layer of liquid film or solid phase about the mineral surface. This film is usually depleted in some cations. Diffusion throughout the liquid film is usually too fast to be the rate-limiting step; however, molecular diffusion can become as slow as the rate of surface reactions (Schnoor, 1990). During transport-controlled dissolution the rate of mineral dissolution is so rapid that mineral etching occurs everywhere, and well-developed dissolution features do not form. In general, the weathering of more soluble minerals is transport-controlled, whereas the weathering of less soluble minerals is surface-reaction controlled (Table 4.3).

Important Takeaways – During sluggish groundwater flow the rate of mineral dissolution is controlled by the flushing rate, but during high flushing rates mineral dissolution is independent of the flushing rate.

Transport-controlled mineral dissolution results in smooth-weathered mineral surfaces. Surface-reaction controlled mineral dissolution results in pitted and etched mineral surfaces. The weathering of more soluble minerals tends to be transport-controlled, whereas the weathering of less soluble minerals is surface-reaction controlled. Limestone in an arid environment is an example of surface reaction control.

Problems

1. How do each of the following effect reaction rates: reactant concentration, temperature of the reaction, physical and chemical properties of the solvent, and the presence of a catalyst?
2. Explain the general differences in reaction rates for zero, first, and second order reactions.
3. The reaction rate for $A + B \rightarrow C$ can be expressed as rate $= \frac{[A]}{t} = \frac{1}{2}\left(\frac{[C]}{t}\right)$. Write an equation for a mineral that is consistent with the rate.
4. A mineral precipitates from a saturated solution. Select three minerals and discuss the differences when considering the precipitation from a thermodynamic and kinetic perspective.
5. How can a change in the seasonal groundwater flow rate affect the SI of common aquifer minerals, and how might this change affect your sampling plan and your interpretation of what the water chemistry means when evaluating the groundwater system?

Chapter notation

Subscripts

aq	aqueous
g	gas
°C	temperature in degrees Centigrade
π	pi (= 3.14159)
μ	interfacial energy
$\approx$	approximately equal to
<	less than
>	greater than
>	greater than or equal to
A, B, C, D	reactants or products in a chemical reaction
$A^{e^{-Ea/RT}}$	pre-exponential or frequency factor (A) raised to $e^{-Ea/RT}$
C	concentration of dissolved species i (mass per unit volume)
C_{eq}	saturation concentration
C_i	concentration of reactant or product i, where i may be A, B, C, or D
C_{io}	initial concentration of reactant or product i, where i may be A, B, C, or Δ change
$d\ln k$	derivative of the natural log of k
dT	derivative of temperature

E_a	activation energy
F	flushing rate; inflow – outflow = F (volume per unit time)
ΔG	Gibbs free energy of formation
ΔG_a	minimum activation energy – minimum Gibbs free energy required for spontaneous reaction
ΔG_b	Gibbs free energy released by formation of a chemical bond
ΔG_n	Gibbs free energy required to form nucleus
ΔG_s	Gibbs free energy required to create surface of nuclei
ΔH	enthalpy
IAP	ion activity product
j	number of ions or molecules in the nucleus
k	reaction rate constant
kf	flushing frequency (F/V)
ky	reaction rate constant of forward reaction where y = reaction 1, 2, 3, etc.
$k\text{-}y$	reaction rate constant of reverse reaction where y = reaction 1, 3, etc.
K_{eq}	thermodynamic equilibrium constant
L	liter
Log	logarithm to the base 10
Ln	natural logarithm (logarithm to the base e); ln a = 2.303 log a
mM	millimole
P	product
r	molecular radius
R	universal gas constant (8.3143 J mole^{-1} T^{-1}; 1.98717 calorie mol^{-1} T^{-1}),
R	rate of dissolution (mass per unit volume of solution per unit time)
sec	second
SI	saturation index
ΔS	entropy
t	time
$t_{½}$	half-life (time)
T	temperature
V	molecular volume
V	fixed volume of water with concentration c
yr	year

REFERENCES CITED

Berner, R.A., 1971, *Principles of chemical sedimentology*, McGraw Hill.

Berner, R.A., 1981, Kinetics of weathering and diagenesis, in *Kinetics of geochemical processes*, A.C. Lasaga and R.J. Kirkpatrick (eds.), *Mineralogical Society of America, Reviews in Mineralogy*, 8, 111–134.

Jimenez-Mena, N., et al., 2019, On the interplay between intermetallic controlled growth and hot tearing susceptibility in Al-to-steel welding with additional interlayers, *Materials & Design*, 180.

Jones, P., et al., 1964, The hydration of carbon dioxide, *Journal of Chemical Education*, 41, 610–612.

Lasaga, A.C., 1981, Rate laws of chemical reaction, in *Kinetics of geochemical processes*, A.C. Lasaga and R.J. Kirkpatrick (eds.), *Mineralogical Society of America, Reviews in Mineralogy*, 8, 1–68.

Schnoor, J.L., 1990, *Kinetics of chemical weathering: A comparison of laboratory and field weathering rates in Aquatic chemical kinetics*, Wiley and Sons, 475–504.

Stumm, W., and J.J. Morgan, 1981, *Aquatic chemistry,* John Wiley and Sons.

5 Organic Chemistry

5.1 INTRODUCTION

Organic chemistry is the study of the organization, structure, properties, and reactions of organic carbon compounds. The modern world is dependent on the manipulation of organic compounds. Examples of this manipulation include hydrocarbon-based fuels, pharmaceuticals, plastics, clothing, agriculture fertilizers, pesticides, and herbicides, your phone, and all synthetic compounds. A biproduct of organic carbon manipulation has been air and water pollution. Our purpose here is to gain an understanding of the structure and basic properties of organic compounds of interest to hydrogeology.

Most carbon compounds contain hydrogen and may also include the halogens, oxygen, nitrogen, phosphorous, and other elements. Organic compounds may be naturally occurring or anthropogenic and can occur in water as pure compounds, mixtures of compounds, and as dissolved molecules. Organic carbon compounds are import to hydrogeology in three ways: 1) as a source of $CO_{2(g)}$, 2) as natural organic contributions to aqueous geochemistry, and 3) as a source of groundwater contamination.

A common measure of the total dissolved organic matter in water is the dissolved organic carbon (DOC). DOC is determined in the laboratory by converting the organic matter to CO_2 and then measuring the amount of the gas. Because DOC only includes the carbon portion of organic matter, DOC is an incomplete measure of the total dissolved organic matter. In addition to the total carbon, the dissolved organic matter (DOM) includes nitrogen, hydrogen, phosphorous, sulfur, and other fractions of dissolved elements and molecules. In groundwater the concentration of DOM is typically about twice as great as DOC.

Organic matter, like inorganic matter, is considered to be dissolved if it will pass through a 0.45 μm filter. The remaining fraction of carbon is known as particulate organic carbon (POC). DOC and DOM, both natural and anthropogenic, are common in marine and freshwater systems. The typical ranges of DOC concentrations in aqueous environments are shown in Figure 5.1. When the DOC is 10 mg/l or more, the water can have a noticeable brown color. Locally, surface and groundwater systems can have DOC concentrations appreciably greater than shown in Figure 5.1. For example, in the confined aquifer in San Luis Valley, Colorado, the DOC is a great as 300 mg/L (Figure 5.2).

5.2 PROPERTIES OF ORGANIC COMPOUNDS

Organic compounds, like other compounds, can occur in three temperature dependent phases:

solid < melting point,
liquid is between melting and boiling point, and
gas > boiling point (measured at 1 atm – 760 mm of mercury).

Boiling temperatures generally increase with increasing molecular weight.

The specific gravity of an organic liquid is the ratio of the liquid weight to its volume relative to the same ratio of water. The specific gravity of water is arbitrary assigned a value of 1 at 4 °C. Pure water has a maximum density at ~4 °C (Figure 1.5). The specific gravity of organic liquids is typically measured at 20 °C. In most low temperature hydrogeologic investigation it is not necessary

DOI: 10.1201/9781003646129-6

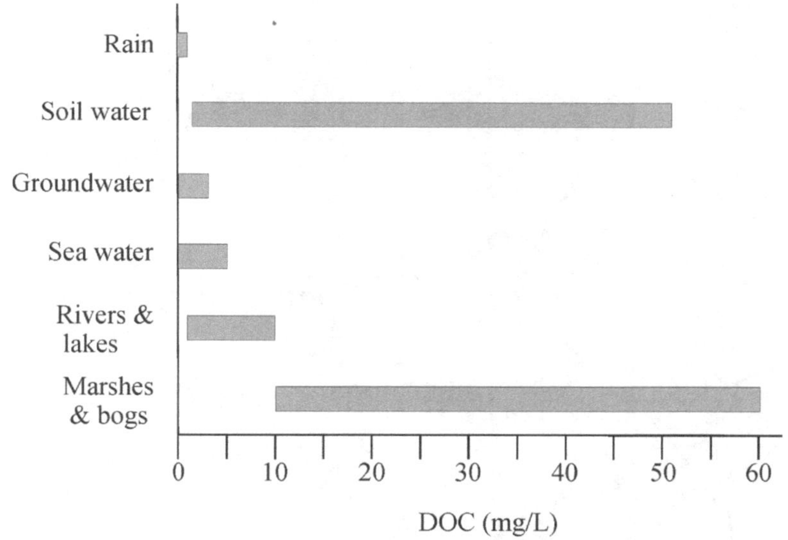

FIGURE 5.1 Typical ranges of DOC concentrations in fresh and saline water environments (data from Drever, 1982; Chaudhary, 2021; Whithead, 2024).

to compensate for the fact that the specific gravities of water and organic liquids are measured at different temperatures because at 20 °C water has a density of 0.9982.

When the specific gravity of an immiscible organic liquid is less than 1, the liquid will float, and when it is greater than 1 it will sink. The idea of floaters and sinkers is important in many groundwater contamination investigations, particularly when hydrocarbons are involved. Non-miscible hydrocarbons with a density less than 1, such as the largest fraction of gasoline, are known as a LNAPL (light nonaqueous phase liquid). Liquid hydrocarbons with a density greater than one are known as a DNAPL (dense nonaqueous phase liquid). Refer to texts such as Fetter et al. (2018), and Drever (1982) for a through discussion of the fate of LNAPs and DNAPLs.

The water solubility of an organic substance is extremely important in that the water solubility of organic liquids describes the equilibrium condition between liquid organic substance and the amount of the liquid that will dissolve in water. Recall that the water solubility of an inorganic solid is measured by the equilibrium constant K. The solubility of an organic liquid in either water or in an organic solvent is measured by the octanol-water partition coefficient (K_{ow}). Octanol is an 8-carbon chain alcohol. K_{ow}, also known as $P_{octanol}$ is calculated as:

$$K_{ow} = P_{octanol} = \frac{C_{octonol}}{C_{water}} \tag{5.1}$$

where
K_{ow} = the octanol-water partition coefficient,
$C_{octanol}$ = concentration of the organic liquid in octanol, and
C_{water} = concentration of organic liquid in water.

The organic liquid is mixed with equal amounts of octanol and water. The unit less calculated K_{ow} value is usually presented as a logarithm. In environmental investigation we commonly use log K_{ow}, whereas in the pharmaceutical industry log P is commonly used. Log K_{ow} provides a measure

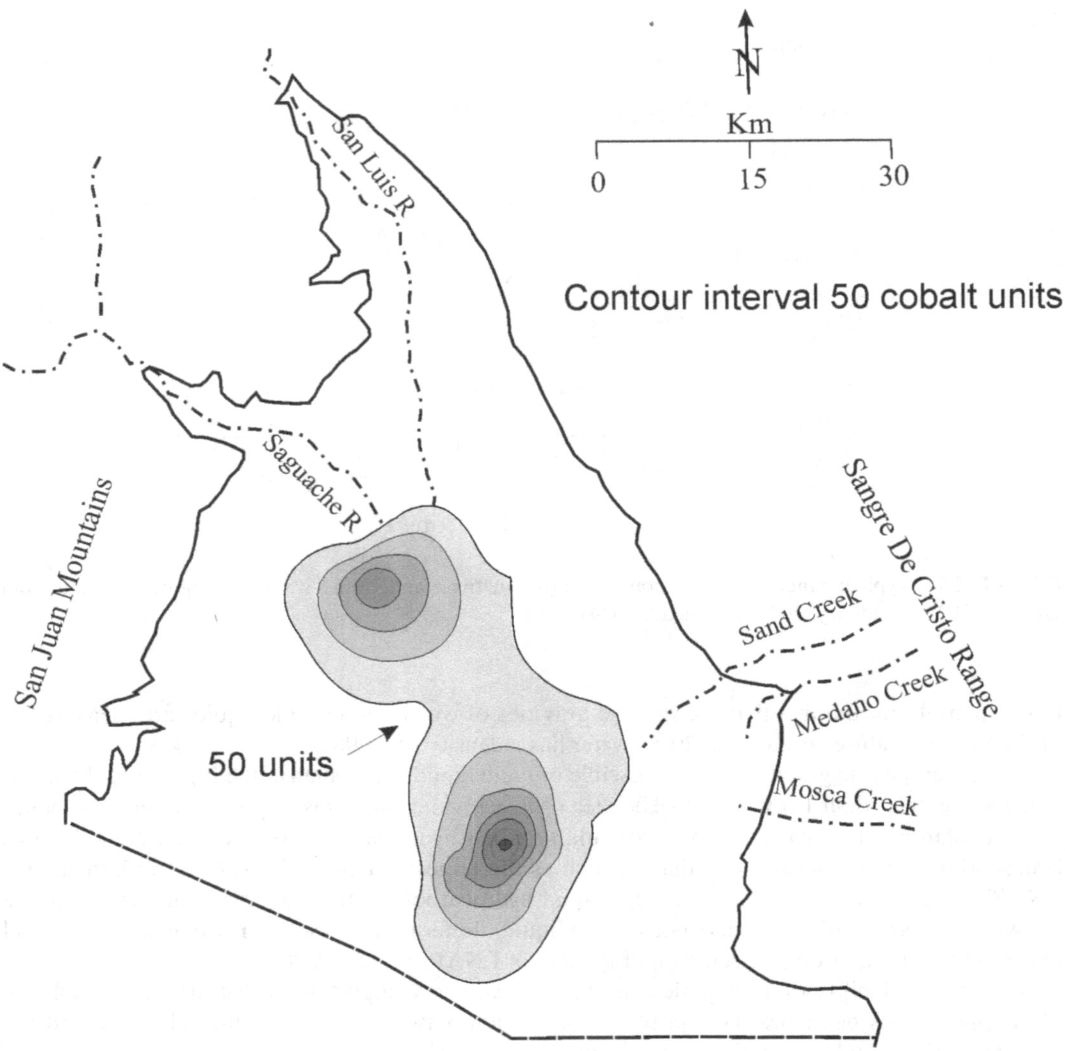

FIGURE 5.2 San Luis Valley, Colorado, confined aquifer cobalt color units. The groundwater is extremely colored due to the anaerobic decay of buried carbon-rich swamp sediments. The decay is accompanied by the release of $CH_{4(g)}$, much of which has converted to $CO_{2(g)}$. Although there is no standard conversion between cobalt color units and DOC, one cobalt unit is approximately 1 mg/L (after Mayo, 2020).

of the lipophilic vs. hydrophilic nature of a compound. In environmental and water investigations, log K_{ow} is useful for determining the solubility of organic compounds. The concept of log P is used to help understand how well a drug will meet the two requirements: dissolving in the digestive system and accessing the biological target of the drug. Log K_{ow} values typically range between −3 and +10. Large values mean that the organic liquid has a tendency to dissolve in an organic solvent, and smaller values mean that the organic liquid has a tendency to dissolve in water. The specific gravity and octanol-water partition coefficients for some common organic pollutants are listed in Table 5.1, and values for hundreds of organic liquids are available online.

TABLE 5.1
Specific Gravity and Octanol-Water Partition Coefficients (K_{ow}) for Some Common Organic Pollutants

Compound	Formula	Specific gravity	Water solubility at 25 °C (mg/L)	log K_{ow}
Aliphatic hydrocarbon				
n-heptane	C_7H_{16}	0.68	3.0	4.50
n-octane	C_8H_{18}	0.70	0.7	5.15
Aromatic hydrocarbon				
Benzene	C_6H_6	0.88	1,750	2.11
Ethylbenzene	C_8H_{10}	0.87	150	3.15
Toluene	$C_6H_5CH_3$	0.88	530	2.73
Chlorinated Hydrocarbon				
Bromoform	$CHBrCl_2$	2.90	3,010	2.40
Carbon tetrachloride	CCl_4	1.59	8,000	1.97
Chlorobenzene	C_6H_5Cl	1.11	465	2.84
Methylene chloride	CH_2Cl_2	1.37	20,000	2.56
1,1-Dichlorethane	C_2H_2Cl	1.18	5,500	1.79
1,2-Dichlorethane	$C_2H_4Cl_2$	1.15	8,500	1.48
2-Chlorophenol	C_8H_4ClOH	1.24	29,000	1.18
Vinyl chloride	CH_2CHCl	gas	2,670	2.42
Polynuclear aromatic hydrocarbon (PAH)				
Acenaphthene	$C_{12}H_{10}$	1.07	3.42	4.0
Anthracene	$C_{14}H_{10}$	1.30	0.07	4.5
Methyl naphthalene	$C_{10}H_7CH_3$	1.03	25.4	4.11
Naphthalene	$C_{10}H_8$	1.15	32.0	3.0

Source: Data from Bedient et al., 1999; Drever, 1982; Wasik et al., 1981.

Example 5.1 Calculate the DDT octanol-water partition coefficient

DDT, a synthetic organic halogen insecticide, is a persistent organic pollutant that has had great environmental impacts. DDT accumulates in the soil after repeated spraying and ultimately into waterways. Waterway contamination was a major factor in the precipitous decline in the fish-eating bald eagle population. The use of DDT was banned in the United States in 1972, and the eagle population subsequently recovered.

Calculate the DDT octanol-water partition coefficient assuming the following DDT equilibrium concentrations:

$$C_{octano} = 15.0 \text{ mg/L}$$

$$C_{water} = 0.05 \text{ mg/L}$$

The $P_{octanol}$ can be calculated using Equation 5.1:

$$\frac{15.0\,\text{mg}/\text{L}}{0.5\,\text{mg}/\text{L}} = 300$$

and

$$\text{Log10}(300) = 2.48 \quad \text{(E5.1)}$$

The positive value of E5.1 explains why DDT accumulates in the fatty tissues of organisms.

Vapor pressure measures the tendency of a solid or liquid organic substance to undergo a phase change to the vapor state. The vapor pressure of a substance varies by temperature. A substance with a high vapor pressure has a low boiling point. Higher temperatures result in greater vapor pressures and hence greater volatility. Compounds that have a high vapor pressure at room temperature are known as a volatile organic compound (VOC). Temperature–vapor pressure relationships are particularly important in air pollution studies involving VOCs. The energy needed to vaporize a given amount of volatile solvent is the enthalpy of vaporization.

Most VOCs in the earth's atmosphere are emitted by plants and are responsible for the distinctive smell of scents, perfumes, rose gardens, pine forests, the outdoor environment including the sea shore, and air pollutants. The use of many VOCs of commercial interest, including fossil fuels, dry cleaning, paints and coatings, and solvents are regulated in the United States at both the state and federal levels.

The tendency of gases to rise in the atmosphere or to sink is controlled by the vapor density (V_d) which can be calculated as:

$$V_d = \frac{PM}{RT} \quad (5.2)$$

where
V_d = vapor density,
P = equilibrium vapor pressure (atm),
M = gram molecular weight,
R = gas constant (0.082 L-atm/mol/K), and
T = temperature in K.

The V_d of the atmosphere at sea level and at 25 °C is arbitrarily defined as 1 atm. Most VOCs disperse rapidly into the atmosphere. However, a gas with V_d greater than 1 will sink and can flow along the land surface with potentially catastrophic consequences. For example, in 1986 a limnic eruption of $CO_{2(g)}$ escaping from deep waters in Lake Nyos, Cameroon, suffocated 1,700 people and killed 3,000 livestock by asphyxiation (Kling et al., 1987).

In Chapter 3 Henry's law was described in terms of both Henry solubility (H_s; Equation 3.17) and Henry volatility (H_v). Because we are interested in how much of an organic gas or liquid will dissolve in water (H_s) and how much of the organic liquid will escape into the atmosphere (H_v), it is often useful to describe Henry's law in terms of both Henry solubility and Henry volatility. A common form of the equation describing H_s is Equation 3.17. A common form of H_v is:

$$H_v^{pc} = \frac{p}{C_a} \quad (5.3)$$

where

H_v^{pc} = Henry's volatility constant $\left(\frac{m^3 - Pa}{mol}\right)$,

p = partial pressure of gas (atm at given temperature), and
C_a = equilibrium concentration of gas in solution $C = C$.

The larger the H_v^{pc}, the greater the volatility.

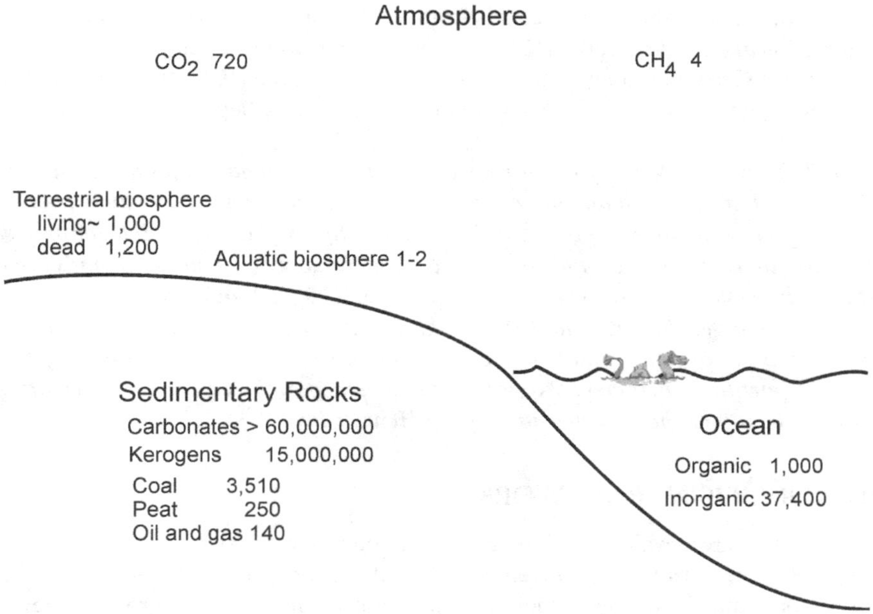

FIGURE 5.3 An approximation of the amount of carbon stored in the earth's carbon reservoirs (values are in gigatons of carbon: GtC). The data is from various sources.

5.3 THE CARBON CYCLE

Almost all of the earth's carbon is tied up in carbonate and organic rich sedimentary rocks (Figure 5.3). Inorganic marine carbon is a distant second reservoir. The earth's atmosphere-terrestrial-lithosphere carbon reservoir accounts for less than 0.01 percent of the total sequestered carbon. If the earth's carbonate rocks were to release their sequestered carbon, the earth would have an atmosphere similar to Venus.

Carbon in the form of CO_2 is treated as an inorganic molecule in most groundwater investigations. However, organic CO_2 makes up a significant proportion of the atmosphere-terrestrial-lithosphere carbon reservoir and is important in climate change investigations and for carbon-14 dating of groundwater age. Most naturally produced atmospheric CO_2 comes from ocean atmospheric exchange (42.84%), plant and animal respiration (28.56%), soil zone respiration and organic decomposition (28.56%), and volcanic eruptions (0.03%; IPCC, 2007). The oxidation of organically derived $CH_{4(g)}$ is a contributor to soil zone and atmospheric CO_2.

Atmospheric methane, which has a residence time of about 12 years compared to centuries for CO_2, is released to the atmosphere by natural (36%) and anthropogenic (64%) sources. About 78, 13, and 10 percent of natural methane emissions are from wetlands, termites, and ocean methane-producing microbes, respectively. Anthropogenic emission includes fossil fuels – production and use (33%), livestock farming (27%), landfills and waste treatment (16%), biomass burning (11%), rice agriculture (9%), and other (4%; Bousquet et al., 2006).

The atmospheric conversion of CH_4 to CO_2 may be represented as:

$$2CH_4 + 3O_2 = 2CO + 4H_2O \tag{5.4}$$

$$2CO + O_2 = 2CO_2 \tag{5.5}$$

Knowledge of the organic CO_2 contribution to groundwater is critical when estimating ^{14}C groundwater ages because about 50 percent of soil zone $CO_{2(aq)}$ is from root zone respiration and the decay of dead organic matter and about 50 percent is from the dissolution of inorganic carbonate minerals. What this means is that the $\delta^{13}C$ of the HCO_3^- is from a mixture of sources. Each of the sources have different $\delta^{13}C$ and ^{14}C isotopic compositions. About one-half of the HCO_3^- is from both organic and inorganic carbon, and all of the ^{14}C is from organic sources (see Chapters 9 and 11).

Important Takeaways – Most of the earth's carbon is permanently sequestered in carbonate rocks. The relatively small amount of carbon in the earth's atmosphere-lithosphere-terrestrial-biosphere reservoirs has a relatively short half-life, and is responsible for almost all of the natural inorganic and organic aqueous carbon. Organic carbon compounds are important to hydrogeology in three ways: 1) as a source of $CO_{2(g)}$*, 2) as natural organic contributions to aqueous geochemistry, and 3) as a significant source of groundwater contamination.*

The tendency of an organic liquid compound to dissolve in water is measured by the octanol-water partition coefficient (K_{ow}*), and the tendency of the compound to volatilize and escape to the atmosphere is measured using Henry's law.*

5.4 HYDROCARBON FRAMEWORK

Organic compounds are covalently bonded and may be polar or nonpolar. The basic framework for most organic molecules is an arrangement of bonded hydrogen and carbon atoms. Hydrogen–carbon frameworks may have one or more attached molecular functional groups. They are called functional groups because they function and behave as a molecule. Molecules that only consist of the hydrogen–carbon framework are called hydrocarbons. Basic hydrocarbon frameworks include aliphatic (straight or branched) and aromatic benzene ring (Figure 5.4).

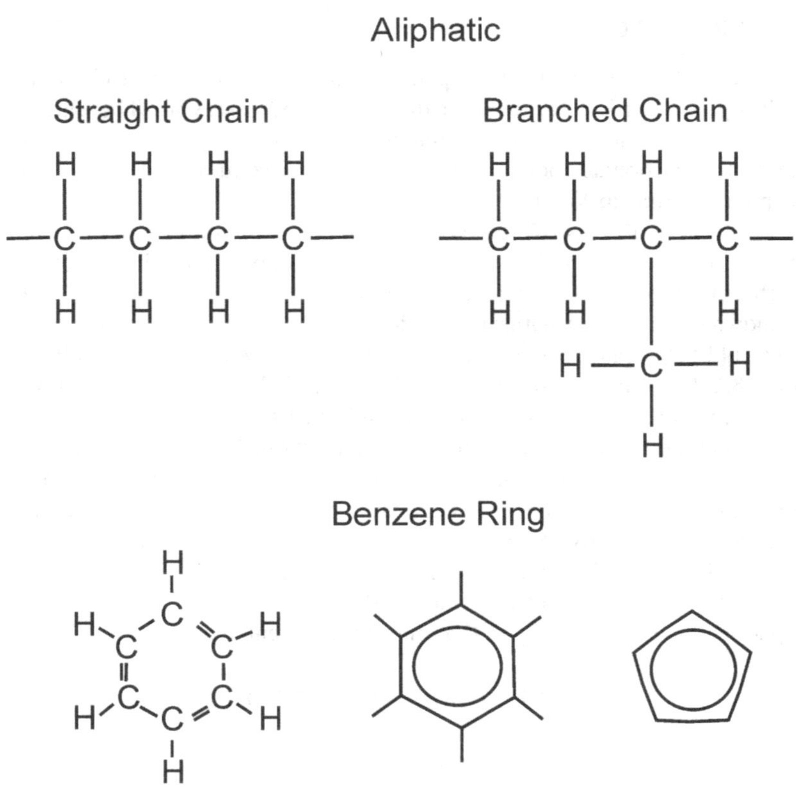

FIGURE 5.4 Structural representations of the three primary hydrocarbon frameworks. The benzene ring may be graphically represented in numerous forms.

Aliphatic hydrocarbons with more than one carbon atom have carbon atoms that are joined by a carbon bond. There are three types of aliphatic hydrocarbons: alkanes, alkenes, and alkynes (Figure 5.5). Alkanes have single covalent carbon bonds (—C or C—C) and are the least reactive of the hydrocarbons. Alkanes are often call saturated hydrocarbons or paraffins because they have the maximum number of hydrogen atoms relative to the number of carbon atoms. Alkenes are unsaturated hydrocarbons (or olefins) that contain at least one C = C double bond. Alkynes are unsaturated hydrocarbons that contain at least one C ≡ C triple bond. The single —, double = or triple ≡ markers that proceed or join atoms or molecules represent single, double, or triple covalent bonds, respectively.

Aromatic hydrocarbons contain a benzene ring, which consists of six carbon atoms joined in a ring structure with alternating single and double covalent bounds (Figure 5.5). More complex molecules form when functional groups become attached to a benzene ring such as chloride ions forming polychlorinated biphenyl (PCB). When two or more rings join together, they are known as polycyclic aromatic hydrocarbon (PAH). Naphthalene ($C_{10}H_8$) is an example of a PAH.

Organic structural isomers are compounds that have the same number of atoms of each element, but the atoms have different structural arrangements or different chemical bonds. For example, there are three molecular forms of $C_4H_{10}O$: butanol (H_3C—$(CH_2)_3$—OH), methyl propyl ether (H_3C—$(CH_2)_2$—O—OH_3), and diethyl ether ((H_3CCH_2—$)_2O$). Each have the same molecular formula but different structural arrangements of the atoms. Structural isomers are also common in

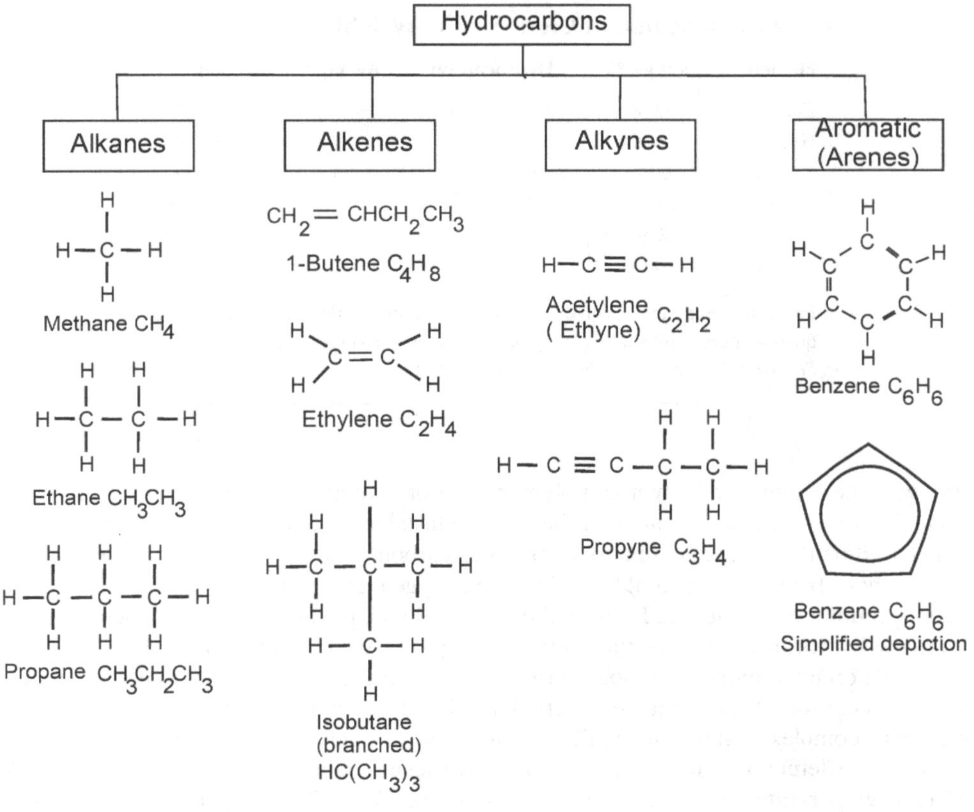

FIGURE 5.5 Examples of hydrocarbon sub-frameworks. The benzene ring is often represented in the simplified form. Alkanes (saturated hydrocarbons) only have a single bond between carbon atoms, alkenes have at least one carbon–carbon bond, and alkynes have at least one carbon–carbon triple bond. Alkenes and alkynes are known as unsaturated hydrocarbons.

TABLE 5.2
Carbon Chain Terminology and Examples

Continuous carbon atoms	Name	Example
1	Methyl or meth	Methane (CH_4)
2	Ethyl or eth	Ethane (C_2H_6)
3	Propyl or prop	Propane (C_3H_8)
4	Butyl or but	Butane (C_4H_{10})
5	Pentyl or pen	Pentane (C_5H_{12})
6	Hexyl or hex	Hexane (C_6H_{14})
7	Hepyl or hep	Heptane (C_7H_{16})
8	Octyl or oct	Octane (C_8H_{18})
9	Nonyl or non	Nonane (C_9H_{20})
10	Decyl or dec	Decane ($C_{10}H_{22}$)
11	Undecyl or undec	Undecane ($C_{11}H_{24}$)
12	Dodecyl or dodec	Dodecane ($C_{12}H_{26}$)

TABLE 5.3
The Composition of Crude Oil by Weight

Element	Range %	Hydrocarbon	Average %	Range %
Carbon	83–85	Alkane (paraffin)	30	5–60
Hydrogen	10–14	Naphthene	49	30–60
Nitrogen	0.1–2	Aromatic	15	3–30
Oxygen	0.05–1.5	Asphaltic	6	remainder
Sulfur	0.005–6.0			
Metals	<0.1			

From the wide range of hydrocarbon concentrations, it is apparent that the quality of crude oil varies greatly between hydrocarbon reservoirs.
Source: Data from Speight, 2014; Hyne, 2001.

mineralogy, where they are known as polymorphs. For example, $CaCO_3$ has three polymorphs: stable calcite, metastable aragonite, and the very metastable vaterite. The name of a carbon compound is based on the longest number of continuous carbon atoms in the carbon chain (Table 5.2).

Hydrocarbons from coal tar (coal heated without oxygen) and petroleum products are significant sources of groundwater pollution. Hydrocarbons in crude oil are a mixture of alkanes, cycloalkane, and aromatic hydrocarbons that contain nitrogen, oxygen, sulfur and metals (Table 5.3). The solubility of crude oil hydrocarbon fractions varies greatly. Crude oil is converted to useful products by a heating process (distillation) known as cracking. Cracking, which breaks carbon–carbon bonds in long-chain complex organic molecules, is based on the fact that the different components of crude oil have different boiling temperatures and will separate into more useful alkane and alkenes at different temperatures in the cracking process (Table 5.4). The compounds benzene, toluene, ethylbenzene, and xylene, called BTEX, are associated with gasoline and to a lesser extent with coal tar and diesel fuel. Organic additives and potential contaminants such as methyl tert-butyl (MTBE: $(CH_3)_3COCH_3$), and ethanal are often added to gasoline to increase the octane rating.

TABLE 5.4
Common Hydrocarbon Distillation Fractions and Uses

Fraction	Carbon atoms per molecule	Boiling point °C	Uses
Liquefied petroleum gas (LPG)		−40	Cooking, heating
Butane	1 to 4	−12 to −1	Cooking, heating
Gasoline	5 to 10	−1 to 110	Fuel, benzene, fuel stock
Kerosene	11 to 13	105 to 205	Jet fuel, fuel
Fuel oil/diesel	14 to 18	205 to 290	Diesel fuel, heating oil
Lubrication oils	19 to 40	360 to 530	Lubricating oil, grease
Residuum	>40	>560	Asphalt

Source: Data from Speight, 2014; Fetter et al., 2018, and other sources.

Important Takeaway – Most organic molecules are based on one of three hydrocarbon structures: straight chain aliphatic, branched chain aliphatic, and the benzene ring. When a functional group is not attached to a framework, the molecule is simply known as a hydrocarbon. Crude oil contains a complex mixture of hydrocarbons that can be separated for commercial use by heating.

5.5 ORGANIC CARBON FUNCTIONAL GROUPS

A functional group is an atom or a group of atoms in a molecule that have similar chemical properties and chemically behave in a similar fashion when bonded to either a carbon or hydrogen atom in a hydrocarbon framework. The functional group may replace either a carbon or hydrogen atom in the framework. Functional groups allow us to organize and better understand the more than 20 million known organic compounds in a meaningful fashion. A functional group consists of two or more atoms of O, S, N, and/or P bonded together. Functional groups containing either oxygen or nitrogen are typically polar and tend to be quite soluble in water. Nonpolar functional groups tend to be relatively insoluble. Some functional groups bond with a carbon or hydrogen atom in the absence of a hydrocarbon framework. There are dozens of functional groups, and only those most important to hydrogeologic investigations are discussed here.

In organic chemistry it is common to use the designation R in both chemical formulas and in the structural representations of functional groups. R usually represents the attaching hydrogen or carbon of the hydrocarbon framework and may represent the remainder of the functional group molecule.

The carboxyl group constitutes about 90 percent of all dissolved organic carbon in most surface and groundwater systems. Carboxylic acids are unusual in that they are composed of two bonded functional groups: 1) the carboxyl group (C—O) and 2) the hydroxyl group (—OH). In the simplest form, the carboxyl group can be written in condensed form as —CO_2H or —COOH (Figure 5.6). In the figure, R represents the attachment to hydrocarbon framework.

Dissociation of carboxyl acid can be described by its dissociation constant:

$$\text{RCOOH} = \text{RCOO}^- + \text{H}^+ \tag{5.6}$$

and

$$K_a = \frac{a_{\text{RCOO}^- + \text{H}^+}}{a_{\text{RCOOH}}} \tag{5.7}$$

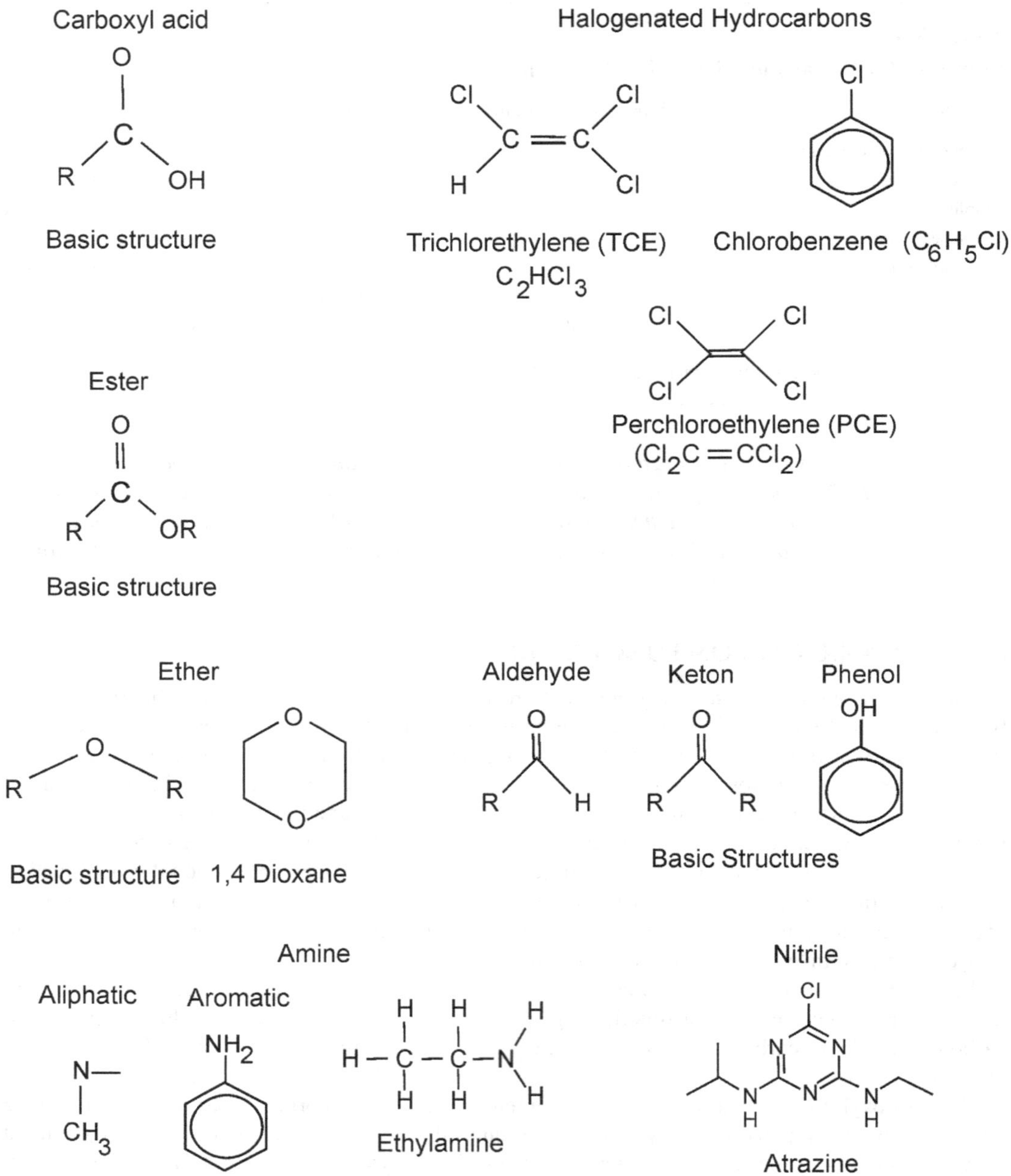

FIGURE 5.6 Structures of selected functional groups and organic molecules containing a functional group. R represents the attached hydrocarbon framework.

The pK_a of most carboxylic acids is in the range of 3.5 to 5.0, which means that most completely dissolve in typical surface and groundwater. A second important feature of carboxyl group molecules is their ability to complex with transition metals and aluminum. What this means is carboxylic acids promote the dissolution and transport of metals in surface and groundwater. Example of carboxylic acids are acetic acid (vinegar), citric acid (citrus fruit), and oxalic acid (spinach).

Halogenated hydrocarbons, containing the organic halides Cl (chloro), Br (bromo), and F (fluoro), are among the most important sources of groundwater contamination. They include both aliphatic and aromatic subclasses. Some hydrogenated hydrocarbons are widely used as commercial solvents due to their low flammability and high vapor pressure, and others are quite dense, causing them to sink in groundwater systems. As early as 1949, the notorious polluter trichloroethene $\left(\text{TCE; Cl}_2C = \text{CHCl}\right)$ was recognized as a surface and groundwater contaminant in the United Kingdom (Lyne and McLachlan, 1949). Between 1970 and the mid-1980s it was recognized that widespread use of chlorinated solvents such as TCE, perchloroethylene or tetrachloroethene (PCE), and others (Figure 5.6) has resulted in significant groundwater contamination worldwide (Rivett, 2010).

Alcohols have one or more hydroxyl groups (—OH) that substitute for hydrogen ions. Alcohols are very soluble in water and are only found in trace quantities because they readily degrade. The basic structure of an alcohol is:

$$\text{R—OH} \tag{5.8}$$

Esters include an organic acid and at least one acidic hydroxyl. They have the general structure:

$$\text{ROCOOR} \tag{5.9}$$

Organic esters are a component of animal fats and vegetable oils. Esters of carboxylic acids are used as high-grade solvents in the manufacture of plastic, resins, lacquers, and synthetic lubricants. Esters in groundwater are almost always a contaminant and include pesticides such as dimethrin ($C_{19}H_{26}O_2$) and several industrial chemicals.

Ethers occur naturally and as contaminants in surface and groundwater. They have an oxygen atom bonded between the two carbon atoms (Figure 5.6). The general structure of ethers is:

$$\text{R—O—R} \tag{5.10}$$

Two ethers, tetrahydrofuran $(CH_2)_4O$, and dioxane $(O(C_2H_4)_2O$, are common groundwater contaminants. Tetrahydrofuran is miscible in water and resistant to biodegradation and is used as solvent for resins such as PVC and adhesives. 1,4-dioxane is miscible in water and very resistant to biodegradation, has been used as a solvent in lacquers, paints, and inks, and has been added as a stabilizer in chlorinated solvents to prevent degradation.

Both aldehydes and ketones are functional groups that contain a carbonyl group. Aldehydes have the structure $\text{R—CH} = \text{O}$ and ketones have the structure $\text{R—C}(=)\text{—R}$ (Figure 5.6). The most commonly known ketone is propanone $(CH_3)_2CO$, which is known as the widely used solvent acetone.

Phenols, sometime known as phenolics, contain one or more hydroxyl group (–OH) molecule bounded to a benzene ring (Figure 5.6). In the simplest form, phenol is aromatic with the formula C_6H_5OH. Phenols are produced by plants and are a synthetic parent compound used in many commercial products, including disinfectants. Phenols are a groundwater contaminant due to their industrial and agricultural uses. Phenol is an active component in the DNAPL contaminant creosote, which has a long history of use as a wood preservative for railroad ties, telephone poles, and piers.

Nitrogen-bearing organic compounds found in surface and groundwater include amino acids, amines, and nitriles. Amines are derivatives of ammonia (NH_3) with one or more of the hydrogen atoms replaced (Figure 5.6). This substitution (—NH_2) is called the amino group. Aliphatic amines contain the H and alkyl substitute (CH_3), and aromatic amines have the nitrogen atom attached to an aromatic ring. Amines make up a large number of industrial contaminants including wastes from herbicides. The wastes include ethylamine ($CH_3CH_2NH_2$), methylamine (CH_2NH_2), and naphthylamine ($C_{10}H_7N_7HN_2$).

Amino acids contain both amino and carboxyl functional groups. Proteins are composed of alpha-amino acids. Although there are more than 500 natural amino acids, only 22 alpha amino acids appear in the DNA of all lifeforms (Rother and Krzycki, 2010). The concentration of amino acids in groundwater is relatively insignificant.

Nitriles are organic compounds that contain the —C ≡ N functional group, with a triple bond between the carbon and nitrogen atom. The prefix cyano is commonly used instead of nitril. Inorganic compounds containing the —C ≡ N functional group are called cyanides. The common nitrile contaminant atrazine is a complex chlorinated herbicide (Figure 5.6).

Important Takeaway – A functional group is the bonding of two or more atoms of O, S, N, and/or P to a hydrocarbon framework or to a carbon atom. The bonding of functional groups to hydrocarbon frameworks are responsible for most of the 20 million-plus known organic compounds and for the industrial syntheses and manufacture of organic compounds that are used in commercial and defense industries, agriculture, paints and coatings, and the cleaning of non-water-soluble organic compounds. Worldwide, some of these organic compounds have been involved in uncontrolled releases that have resulted in significant organic surface and groundwater contamination.

5.6 HUMIC SUBSTANCES

Humic substances are colored organic compounds that form during biomass biodegradation. Humic substances range in color from yellow to brown to black and represent most of the organic matter in near-surface sediments and soil. Humic substance is a catchall term that includes the biomass residual as well as humic acid, fluvic acid, and humin and hymatomelanic acid. Both humic and fluvic acid are complex mixtures of carboxyl and phenolate hydrocarbons. Humic acid is soluble in neutral and alkaline water, and fluvic acid is soluble at any pH including very acidic water. Humic acids are very large molecules, ranging in size from 10,000 to 100,000 atoms. Fluvic acids molecules are considerably smaller. Humic and fluvic acid are general terms that include a large number of molecules, and the structures of most of these molecules have not been described. Almost all dissolved organic carbon (DOC) in surface and groundwater is fluvic acid, and most DOC in soil solutions is humic acid. DOC is responsible for the dark color of many European river waters.

Humins are dark-colored, insoluble organic compounds that make up about 50 percent of soil matter. Their chemical structures are complex long-chained carbons that are difficult to characterize. Hymatomelanic acid is the part of humic acid that is soluble in ethanol.

Chapter notation

C_a = equilibrium concentration of gas in solution $\left(\frac{\text{mol}}{\text{m}^3 \text{ water}}\right)$
$C_{octanol}$ = concentration of the organic liquid in octanol, and
C_{water} = concentration of organic liquid in water
DOC – dissolved organic carbon
DOM – dissolved organic matter
H_s – Henry solubility
H_v – Henry volatility
H_v^{pc} = Henry's volatility constant $\left(\frac{\text{m}^3 - \text{Pa}}{\text{mol}}\right)$
K_a – dissociation constant
$K_{air\text{-}water}$ – air water partition coefficient
K_{ow} = the octanol-water partition coefficient
M = gram molecular weight
P = equilibrium vapor pressure (atm)

p = partial pressure of gas (atm at given temperature
POC – particulate organic carbon
R = gas constant (0.082 L-atm/mol/K)
R = functional group
T = temperature in K
μmole – micromole
V_d – vapor density
VOC – volatile organic

Problems

1. A pipeline that transports gasoline from Texas to Arizona develops a leak and spills 3,500 gallons of gasoline. Much of the gas remains near the surface, but some encounters a shallow unconfined aquifer.
   a. What is the fate of the gasoline that remains above the water table and of the gasoline that encounters the groundwater system?
   b. How would you determine the concentrations of the dissolved phase and of the vapor phase of a liquid spill?
   c. Assume the spill was tar sand crude oil being transported from Canada to Texas. What would be the fate of the crude oil that encountered the groundwater system?
2. Groundwater DOC is typically reported as mg/L. Develop a conversion factor to convert mg/L to micromoles/L (μmole/L).
3. A chemical manufacturing facility has experienced a major leak. A large volume of each of the chemicals listed has escaped from a different location in the facility and encountered an unconfined groundwater system. The following table lists various parameters of selected organic chemical at 25 °C. The table also includes the parameter $K_{air\text{-}water}$, which describes the volatility of an aqueous compound into the air.

$$K_{air-water} = \frac{H_v^{pc}}{C_{aq}}$$

	Vapor				
Pressure		H_v^{pc}		Solubility	
Chemical	**(P_a)**	**(m^3- P_a)/mol**	**(mol/m^3)**	**log K_{ow}**	**$K_{air\text{-}water}$**
Phenol	5.5×10^{-1}	6.23×10^{-2}	8.83×10^{2}	1.48	2.52×10^{-5}
Benzene	1.27×10^{4}	5.57×10^{2}	2.28×10^{1}	2.13	2.25×10^{-1}
Pyrene	6.0×10^{-4}	9.18×10^{1}	6.53×10^{-4}	5.11	$3/71 \times 10^{-4}$
DDT	2.0×10^{-5}	2.82×10^{-6}	2.82×10^{-6}	6.53	2.86×10^{-3}

   a. Rank each chemical according to its tendency to dissolve in the water and the tendency of the aqueous phase to volatilize and escape to the atmosphere.
   b. Assuming only one spill can be remediated at a time, in what order should the remediation proceed, and why? Do not consider the danger each chemical represents to the environment in your answer.

REFERENCES CITED

Bedient, P.B., et al., 1999, *Ground water contamination transport and remediation*, 2nd ed, Prentice-Hall.
Bousquet, P., et al., 2006, Contribution of anthropogenic and natural sources to atmospheric methane variability, *Nature*, 443(7110), 439–443.
Chaudhary, A. 2021, Qualitative and quantitative analysis of water, *Applied Water Science*, 1.

Drever, J.I., 1982, *The geochemistry of natural waters – surface and groundwater environments*, 3rd ed, Prentice Hall.

Fetter, C.W., et al., 2018, *Contaminant hydrology*, 3rd ed, Waveland Press.

Hyne, N., 2001, *Nontechnical guide to petroleum geology, exploration, drilling, and production*, Serbiula Librum 2.0.

IPCC, 2007, Fourth assessment report: Intergovernmental panel on climate change, *World Meteorological Organization*, 52.

Kling, G.W., et al., 1987, The 1986 Lake Nyos gas disaster in Cameroon, West Africa, *Science*, 236(4798), 169–175.

Lyne, F.A., and T. McLachlan, 1949, Contamination of water by trichloroethylene, *The Analyst*, 74(882), 513.

Mayo, A.L., 2020, Mapping paleo-lacustrine environments and long-term climatic trends using groundwater chemistry, *Hydrogeology Journal*, 28, 987–1002.

Rivett, M.O., 2010. The legacy of chlorinated solvents in groundwater: An overlooked 1940s contribution, *Environmental Chemistry Group Bulletin (Royal Society of Chemistry)*, 1, 5–14.

Rother, M., and J.A. Krzycki, 2010, Selenocysteine, pyrrolysine, and the unique energy metabolism of methanogenic archaea, *Archaea*, 1–14.

Speight, G, J., 2014, *The chemistry and technology of petroleum*, 4th ed, CRC Press.

Wasik, S.P., 1981, *Octanol/water partition coefficients and aqueous solubilities of organic compounds*, U.S. Department of Commerce, NBSIR 81-2406.

Whithead, P., 2024, *Total organic carbon (TOC) and its measurement*, www.elgalabwater.com/blog/total-organic-carbon-toc

6 Major and Trace Aqueous Species

6.1 INTRODUCTION

Groundwater acquires almost all of its dissolved content by interactions with soil zone gases and minerals encountered in the soil zone and in the aquifer along the flow path. Major species are primarily from natural sources, and trace species may or may not occur in a groundwater system. When trace species are present, anthropogenic processes are often a significant factor. Most carbonate, sulfate, silicate, and aluminosilicate mineral dissolution occurs in the soil zone and the shallow subsurface, where free hydrogen and oxygen ions are abundant and the recharge waters are undersaturated with respect to most minerals. Despite the fact that more than 75 percent of the upper continental crust is composed of igneous and siliceous sedimentary rocks, which primarily consist of aluminosilicate minerals (Table 6.1, Figure 6.1), the solute contributions from aluminosilicate mineral weathering are small compared to the contributions from the weathering of carbonate rocks and carbonate cemented sediments. This disparity is evident when the major ion concentrations in surface and groundwater are considered. In most fresh water, calcium and magnesium, and sometimes sodium, are the major cations, and bicarbonate and sometimes sulfate are the major anions (Table 6.2). Both potassium and chloride are considered major ions, but their concentrations are usually quite small when compared to other major ions. The disproportionally large contribution from carbonate and sulfate minerals is due to the large solubility differences of carbonate and sulfate minerals relative to aluminosilicate minerals (Table 2.1). At 25 °C, calcite and gypsum have solubilities of 4.67×10^{-9} and 2.62×10^{-5}, respectively, whereas aluminosilicate minerals have solubilities of 10^{-19} to 10^{-20}.

Iron, nitrogen, silica, and noble gases may be locally important in groundwater investigations. Iron is often a concern in volcanic terrain and near mining sites. Aqueous nitrogen, which may be anthropogenic, can be a serious health concern, and silicate mineral weathering is important in aluminosilicate terrains and in many geothermal systems. Noble gas concentrations can provide valuable insight into groundwater recharge temperatures and, by extension, recharge elevations in mountainous terrain.

6.2 CARBONATE SYSTEM

More than 99 percent of the earth's crustal carbon is stored in carbonate and other sedimentary rocks. Most of this carbon is in calcite and dolomite. The abundance of carbonate rocks and the relatively high solubility of calcite make the weathering of Ca-limestone and Ca-Mg-limestone the most important contributors to the salinity in almost all fresh groundwater. Dissolution of small amounts of soil zone and aquifer calcite is a major contributor in many non-carbonate sedimentary environments, igneous rock terrains, and some arid region soils that contain only a small percentage of calcite, carbonate detritus, carbonate cement, and fossil shells. The neutralizing potential of carbonate minerals is extremely important in acid-producing environments such as mine drainage, industrial wastes, and acid rain. For example, coal mines in the western United States typically have mine drainage that is acid free due to the abundance of carbonate cement in the non-coal-bearing bedrock (Mayo et al., 2000).

The complex carbonate-CO_2 system is regulated by the interaction of three primary variables – pH, temperature, and soil zone CO_2 – and by a host of secondary variables including open vs. closed groundwater conditions, the common ion and ionic strength effects, groundwater mixing, external CO_2 inputs, and soil zone evaporation.

DOI: 10.1201/9781003646129-7

TABLE 6.1
Average Composition of Crustal Rocks in mg/kg

Element	Igneous rocks	Sandstone	Argillaceous rocks	Carbonate rocks
Si	285.0	260.0	33.9	–
Al	359.0	32.1	80.1	9.0
Fe	79.5	18.6	38.8	8.2
Ca	42.2	22.4	22.5	272.0
Na	36.2	3.9	4.9	393.0
Mg	28.1	8.1	16.4	45.3
K	17.6	13.2	24.9	2.4
Mn	25.7	0.4	0.6	0.8
S	0.9	0.9	1.9	4.6
Cl	0.4	13.8	15.3	113.0

Igneous, sedimentary, and argillaceous rocks primarily consist of silicate minerals that contain abundant oxygen. Oxygen is the most abundant element in the earth's crust and is not represented in the table.
Source: Data from Hem, 1985; Cocker, 1998.

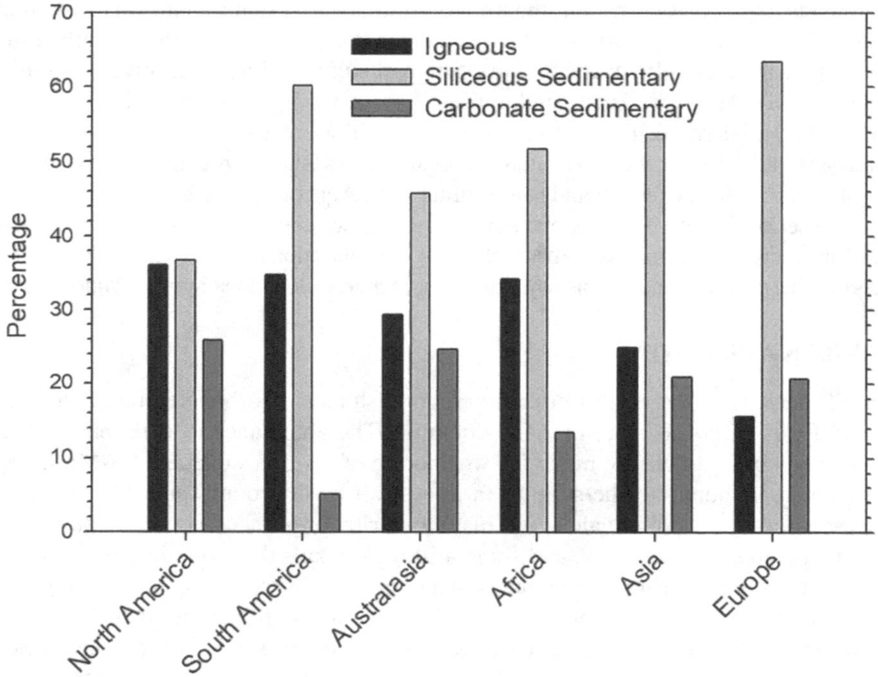

FIGURE 6.1 Average lithologies of shallow crustal land masses. Igneous includes plutonic and volcanic terrains, siliceous sediments include alluvial, semi-consolidated, loess, and dunes, and carbonate includes carbonate and mixed sediments (data from Dürr et al., 2005).

TABLE 6.2
Common Abundances of Major and Minor Dissolved Inorganic Elements in Groundwater

Major constituents (>5 mg/L)	
Calcium	Bicarbonate
Sodium	Chloride
Magnesium	Sulfate
Potassium	
Silicon	
Minor Constituents (0.1–10 mg/L)	
Carbonate	Boron
Strontian	Fluoride
Nitrate	

All other elements are considered trace in most groundwater.
Source: Data from Freeze and Cherry, 1979.

6.2.1 CO_2 AND PH CONTROL

Weak carbonic acid is the most ubiquitous and important acid in almost all surface and groundwater systems. The interaction between dissolved CO_2 and carbonate minerals regulates the water's pH. Surface and groundwater can acquire CO_2 from a variety of sources, the most common of which are atmospheric CO_2 in precipitation, soil zone root respiration, and organic material decay. Although precipitation is the primary source of groundwater recharge water in most groundwater systems, CO_2 in precipitation is of minor importance to groundwater chemistry where the soil zone contains organic material. In bare bedrock, terrain precipitation may be the only CO_2 source. The contribution of atmospheric $CO_{2(g)}$ in precipitation is relatively small. As recently as 1960 the atmospheric CO_2 was in the range of $10^{-3.5}$ atm, whereas in 2023 the content increased to $10^{-3.38}$ atm (i.e., from 320 to 420 ppm). This increase is great in terms of global climate change and near-surface ocean chemistry but is not consequential in most groundwater systems. Water in the soil zone acquires most of its $CO_{2(g)}$, typically in the range of 10^{-2} atm, from the decay of organic material and root zone respiration. Root zone respiration, sometimes known as soil respiration, is the process where plant roots absorb O_2 and release CO_2. Aerobic mediated bacterial reduction of organic matter in the soil zone and the subsequent release of CO_2 can be represented as:

$$\underset{\text{(organic matter)}}{CH_2O} + O_2 \rightarrow CO_2 + H_2O \tag{6.1}$$

When in contact with water, CO_2 readily dissolves to equilibrium. By convention, the dissolved CO_2 is referred to as carbonic acid (H_2CO_3):

$$CO_{2(g)} + H_2O = H_2CO_3 \tag{6.2}$$

The equilibrium relationship between CO_2 and carbonic acid is proportional to the partial pressure (i.e., fugacity) of the CO_2. In most groundwater systems the partial pressure and fugacity are nearly equal. The equilibrium relationship is typically represented as:

$$K_{CO_2} = \frac{a_{H_2CO_3}}{P_{CO_2} a_{H_2O}} \tag{6.3}$$

In most groundwater systems the activity of water is ~1, so Equation 6.3 can be expressed as:

$$a_{H_2CO_3} = K_{CO_2} P_{CO_2} \tag{6.4}$$

Carbonic acid activity is commonly represented as the P_{CO_2} of the water and is one of the outputs of most geochemical modeling programs. Carbonic acid dissociates into H^+ and bicarbonate ions:

$$H_2CO_3 = H^+ + HCO_3^- \tag{6.5}$$

which in turn can dissociate into carbonate ions:

$$HCO_3^- = H^+ + CO_3^{2-} \tag{6.6}$$

Equilibrium constants for Equations 6.5 and 6.6, usually represented as K_1 and K_2, respectively, are commonly represented as:

$$K_1 = \frac{a_{H^+} a_{HCO_3^-}}{a_{H_2CO_3}} \tag{6.7}$$

$$K_2 = \frac{a_{H^+} a_{CO_3^{2-}}}{a_{HCO_3^-}} \tag{6.8}$$

The $CO_2 \rightarrow CO_3^{2-}$ system (Equations 6.2, 6.5, and 6.6) is both temperature (Table 6.3) and pH dependent. At pH 6.4, both H_2CO_3 and HCO_3^- are in equal proportions (Figure 6.2). Below about pH 6 almost all of the dissolved carbon is H_2CO_3. Above about pH 7 carbon is HCO_3^-, and above pH 10.33 CO_3^{2-} is the dominate species. Figure 6.2 illustrates that in most groundwater systems, HCO_3^- is the primary anion and CO_3^{2-} seldom occurs.

The pH of precipitation in the western United States is typically in the range of 5 to 5.6, and in the eastern United States it is usually in the range of 4.3 to 4.8 (Brenner and Brenner, 1987; Langmuir, 1997). These pH values mean that most precipitation is a weak acid. The problem of acid rain is more severe in the eastern United States due to the burning of fossil fuels, particularly coal. In pure water the pH cannot exceed 7 at 25 °C because pK_w = 14 (Table 6.3). Raising the pH above 7 requires the dissolution of other species such as carbonate minerals that are pH dependent. The dissolution of carbonate minerals consumes H^+ ions, raises the pH, and increases the concentration of HCO_3^-.

The solubility of most minerals in groundwater-mineral systems increases with increasing temperature because the reactions are endothermic (i.e., consume heat). What this means is that as the depth of groundwater circulation increases the applicable equilibrium constants become larger, and additional mineral dissolution can occur. In the near surface environment groundwater temperature is approximately the average annual air temperature. At greater depths the temperature increases due to the geothermal gradient. Gas solubility has an inverse relationship with temperature because gas-H_2O reactions are exothermic (see Chapter 3). Unlike solid mineral species, the solubility of

TABLE 6.3
Selected Carbonate System Equilibrium Constants at 0 to 45 °C

T (°C)	pK_W	pK_{CO_2}	pK_1	pK_2	pK_{cal}	pK_{arag}	pK_{dol}
0	14.94	1.11	6.58	10.63	8.38	8.22	26.56
5	14.73	1.19	6.52	10.55	8.39	8.24	16.63
10	14.53	1.27	6.46	10.49	8.41	8.26	16.71
15	14.35	1.34	6.42	10.43	8.43	8.28	16.79
20	14.17	1.41	6.38	10.38	8.45	8.31	16.89
25	14.00	1.47	6.35	10.33	8.48	8.34	17.00
30	13.83	1.52	6.33	10.29	8.51	8.37	17.12
35	13.68	1.48	6.31	10.25	8.54	8.41	17.25
40	13.53	1.68	6.30	10.22	8.58	8.45	17.39
45	13.40	1.67	6.29	10.20	8.62	8.49	17.39

pK = −logK

$\log K_{a1} = -356.3094 - 0.06091964T + 21834.37/T + 126.8339 \times \log T - 1684915/T^2$

$\log K_{a2} = -107.8871 - 0.03252849T + 5151.79/T + 38.92561 \times \log T - 563713.9/T^2$

Sources: pK_{CO_2}, pK_{cal}, and pk_{arag} are from Plummer and Busenberg (1982); pK_{dol} is from Langmuir (1971)

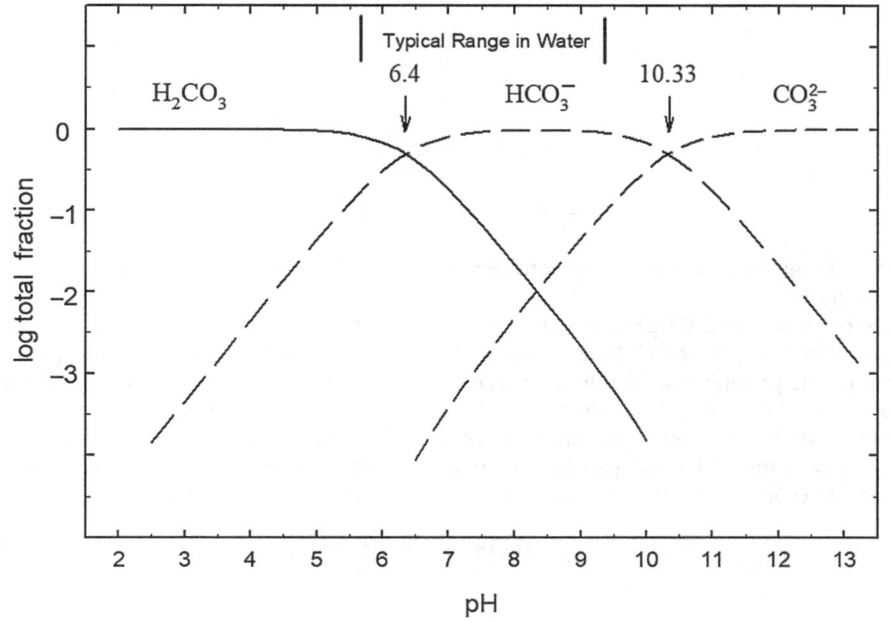

FIGURE 6.2 The fraction of carbon species in the $CO_2 \rightarrow CO_3^{2-}$ system as a function of pH at 25 °C.

CO_2, like other gases, decreases with increasing temperature. The consequence of this inverse relationship is that as groundwater temperature increases, the dissolved gas can thermodynamically change from the aqueous to gaseous phase, and mineral precipitation can occur. In open systems the gases exsolve and escape to the vadose zone. In closed systems only the partial pressure of the gas can keep the gas in the aqueous phase. In the case of the CO_2-carbonate mineral system, thermal heating can result in the precipitation of carbonate mineral cement in porous sedimentary rocks and as veins of carbonate minerals in fractured rock. It is this carbonate mineral precipitation process that is responsible for most carbonate mineral cement in sedimentary rocks.

Important Takeaway – Groundwater acquires most of its carbon from soil zone $CO_{2(g)}$, which is derived from root zone respiration and organic decay. The P_{CO_2} in the soil zone is typically about 10^{-2} atm. Most dissolved groundwater carbon is bicarbonate (HCO_3^-). As the groundwater temperature increases, $CO_{2(g)}$ exsolves and carbonate minerals can precipitate.

Example 6.1 Calculate the pH of precipitation

Problem: Calculate the pH difference between pure rain water in 1960 and 2021 at 25 °C.
Solution: First calculate the pH of atmospheric precipitation in 1960 (i.e., CO_2 = 320 ppm); 320 ppm CO_2 is approximately $10^{-3.5}$ atm. The activity of carbonic acid can be calculated using Equation 6.4 and the K_{CO_2} from Table 6.3 ($10^{-6.35}$):

$$a_{H_2CO_3} = K_{CO_2} P_{CO_2} \tag{E6.1a}$$

$$= 10^{-1.47} x 10^{-3.5} = 10^{-4.97}$$

From the K_1 equilibrium expression:

$$K_1 = \frac{a_{H^+} a_{HCO_3^-}}{a_{H_2CO_3}} \tag{E6.1b}$$

$$a_{H_2CO_3} x\, K_1 = a_{H^+} x\, a_{HCO_3^-} \tag{E6.1c}$$

$$= 10^{-6.35} \times 10^{-4.97} = 10^{-11.32}$$

From the charge balance in E6.1c the activities of H^+ and HCO_3^- are equivalent, so $H^+ = 10^{-5.66}$, and the pH is 5.66.

Then calculate the 2021 pH where the atmospheric CO_2 content was 420 ppm.

After substituting P_{CO_2} of $10^{-3.38}$ into equation E6.1a and recalculating, the resulting pH is 5.60. Comparing the precipitation in pure rain water between 1960 and 2021 the difference is trivial. Because the pH of most precipitation in the United States is less than 5.6, and the average is 4.6 to 4.9 in the eastern United States, precipitation pH is influence by atmospheric gases in addition to CO_2. Most of the additional gas of consequence is anthropogenic $SO_{2(g)}$, which combines with water and oxygen to contribute H^+ and SO_4^{2-}, creating acid rain according to:

$$SO_{2(g)} + H_2O + ½O_2 = 2H^+ + SO_4^{2-} \tag{E6.1d}$$

6.2.2 Calcite and Dolomite

The two carbonate minerals of primary hydrogeologic interest are calcite ($CaCO_3$) and dolomite ($CaMg(CO_3)_2$), which in rock form can collectively be called limestone. The dolomite Ca–Mg ratio varies from 1:1 to 1:0.5. Dolomite is sometimes called dolostone. Locally, other metal carbonates such as copper (azurite), iron (siderite), lead (cerussite), manganese (rhodochrosite), and strontium (strontianite) may be important (Table 2.1). The dissolution of Ca–Mg-limestone is the primary source of dissolved Mg^{2+} in most groundwater. In pure water, calcite and dolomite are relatively stable but become soluble in the presence of CO_2 due to fact that in water $CO_{2(g)}$ forms carbonic acid, which dissociates into bicarbonate and H^+ ions (Equation 6.5). The H^+ ions in turn are responsible for the dissolution of calcite and dolomite. This reaction is familiar to every beginning geology student when they placed a drop of acid on a mineral to see if it will fizz. The dissociation of calcite and dolomite into Ca^{2+}, Mg^{2+}, and HCO_3^- can be expressed as:

$$CO_2 + H_2O = H_2CO_3 \tag{6.9}$$

$$H_2CO_3 = H^+ + HCO_3^- \tag{6.10}$$

$$CaCO_3 + H+ = Ca^{2+} + HCO_3^- \tag{6.11}$$

and

$$CaMg(CO_3)_2 + 2H^+ = Ca^{2+} + Mg^{2+} + 2HCO_3^- \tag{6.12}$$

From these equations it is evident that the dissolution of calcite or dolomite doubles the concentration of bicarbonate ions. This doubling, half from modern carbon sources and half from usually ancient carbon sources, has a profound effect on the stable carbon isotopic composition ($\delta^{13}C$; Chapter 9) and on ^{14}C carbon dating (Chapter 11).

On bare carbonate bedrock and in arid environments carbonate mineral weathering is relatively slow due to several factors including the low partial pressure of $CO_{2(g)}$ in bare bedrock water, a limited amount of vadose zone organic matter, and the absence of deep seasonal flushing of recharge water in arid regions. Despite the fact that carbonate mineral weathering is relatively slow in arid regions, the presence of extensive solution weathering in many thick Paleozoic carbonate rock sequences attests the earlier wetter climatic conditions. Extensive solution weathering is evident in many arid region karst terrains and by the presence of limestone caves and solution channels. For example, when floating the Colorado River in the arid Grand Canyon, the sights of Vaseys Paradise, Cheyava Falls, Roaring Falls, and Thunder and Tapeats springs issuing from the steep carbonate canyon walls are visual and audio evidence of ancient wetter climatic times.

At 25 °C calcite equilibrium is expressed as:

$$K_{cal} = a_{Ca^{2+}} + a_{CO_3^{2-}} = 4.67 \times 10^{-9} \tag{6.13}$$

Calcite has a polymorph aragonite ($K_{arag} = 6.03 \times 10^{-9}$) that, despite its lower solubility, often preferentially precipitates. Metastable aragonite alters to the more stable calcite form. The weathering of dolomite is considerably less vigorous than pure calcite limestone due to its lower solubility. Dolomite equilibrium is expressed as:

$$K_{Dol} = a_{Ca^{2+}} + a_{Mg^{2+}} + (a_{CO_3^{2-}})^2 = 10^{-17} \tag{6.14}$$

6.2.3 Open and Closed Groundwater Systems

The first step in evaluating groundwater CO_2-carbonate interactions is to determine if the groundwater system is open or closed. Garrels and Christ (1965) describe in detail the dissolution of carbonate minerals for several open and closed system CO_2 conditions using a series of approximations. Open systems means that the aquifer is open to the atmosphere and there is a regular, but not necessarily steady, supply of recharge water, atmospheric and soil CO_2, and other gases. Unconfined aquifers are open systems. In these systems there can be a continuous release of reaction products in response to the addition of CO_2. Assuming the supply of carbonate minerals are not a limiting factor, the buildup of reaction products continues until mineral saturation occurs. Quasi-open conditions also occur in confined aquifers where there is an external CO_2 source such as the thermal decomposition of carbonate rocks and volcanic emanations. In closed systems there is no new supply of recharge water, and the supply of atmospheric and soil gases, such as CO_2, become exhausted by reactions in the up-gradient portion of the aquifer. Confined aquifers are closed systems. It is important to keep in mind that groundwater flow in all closed systems originated as an open system. Once the initial open system $CO_{2(g)}$ is exhausted, additional CO_2-driven carbonate dissolution is not possible in the absence of reactions driven by increases in TDS, the common ion effects, methanogenesis, etc.

The calcite system includes Ca^{2+}, H^+, CO_3^{2-}, HCO_3^-, and OH^-. At 25 °C the equilibrium governing equations are:

$$K_{CaCO_3} = \frac{\left[Ca^{2+}\right]\left[CO_3^{2-}\right]}{\left[CaCO_{3(s)}\right]} = 10^{-8.48} \tag{6.15}$$

$$K_{H_2CO_3} = \frac{\left[H^+\right]\left[HCO_3^-\right]}{\left[H_2CO_3\right]} = 10^{-6.4} \tag{6.16}$$

$$(K_2) = K_{HCO_3^-} = \frac{\left[H^+\right]\left[CO_3^{2-}\right]}{\left[HCO_3^-\right]} = 10^{-10.3} \tag{6.17}$$

$$K_{H_2O} = \frac{\left[H^+\right]\left[OH^-\right]}{\left[H_2O\right]} = 10^{-14} \tag{6.18}$$

$$K_{CO_2} = \frac{\left[H_2CO_3\right]}{P_{CO_2}} = 10^{-1.47} \tag{6.19}$$

The electrical charge balance for calcite may be expressed as:

$$2m_{Ca^{2+}} + m_{H^+} = 2m_{CO_3^{2-}} + m_{HCO_3^-} + m_{OH^-} \tag{6.20}$$

where m is the moles of the species. In most groundwater systems the concentration of H^+, OH^-, and CO_3^{2-} are small relative to Ca^{2+} and HCO_3^-, so Equation 6.20 usually may be simplified to:

$$2m_{Ca^{2+}} + m_{H^+} = m_{HCO_3^-} \tag{6.21}$$

When the P_{CO_2} is known or estimated, it is possible to calculate the equilibrium concentrations of Ca^{2+} and HCO_3^- as a function of pH. From Equation 6.19:

$$\left[H_2CO_3\right] = P_{CO_2} K_{CO_2} \tag{6.22}$$

At 25 °C and a soil gas P_{CO_2} of $10^{-3.5}$ atm:

$$\log\left[H_2CO_3\right] = \log P_{CO_2} + \log K_{CO_2} \tag{6.23}$$

$$= -3.5 + (-1.47) = -4.97$$

From Equation 6.16 $\left[HCO_3^-\right]$ can be calculated as a function of pH:

$$\left[HCO_3^-\right] = \frac{\left[H_2CO_3\right]K_1}{\left[H^+\right]} \tag{6.24}$$

and

$$\log HCO_3^- = \log K_1 + \log\lfloor H_2CO_3 \rfloor + pH \tag{6.25}$$
$$= -6.35 - 4.97 + pH = -11.32 + pH$$

In a similar fashion:

$$\log CO_3^{2-} = \log K_2 + \log\left[HCO_3^-\right] + pH \tag{6.26}$$
$$= -10.3 + (-11.32 + pH) + pH$$
$$\text{Log } H^+ = K_{H_2O} + \log OH^- \tag{6.27}$$
$$= -14 + OH^- \text{ Log } OH^- = K_{H_2O} + \text{Log } OH^- = K_{H_2O} + \log H^+ \tag{6.28}$$
$$= -14 + pH \text{ Log } Ca^{2+} = \log K_{CaCO_3} + \log\left[CO_3^{2-}\right] \tag{6.29}$$
$$= -8.48 + \left(-10.3 - \left(\left(-11.32 + pH\right) + pH\right)\right)$$

In the pH range of most groundwater, the solubility of carbonate minerals is approximately the cubed root of the CO_2 pressure (Langmuir, 1997). For example, at 25 °C the concentration of Ca^{2+} from the dissolution of calcite can be expressed as:

$$Ca^{2+}\left(\text{mmol/kg}\right) = 9.057(P_{CO_2})^{0.3165} \tag{6.30}$$

Equation 6.22 is pH independent, and the log of H_2CO_3 plots as a straight line on a pH-log concentration on the diagram (Figure 6.3). Equations 6.23 to 6.29 are also plotted on the figure. Similar plots can be made for P_{CO_2} values other than $10^{-3.5}$ by substituting the P_{CO_2} value into the equations. At pH 10.33, HCO_3^- and CO_3^{2-} exist in equal concentrations. Figure 6.3 also shows that the concentration of carbonic acid is pH independent and the Ca^{2+} concentration is inversely related to pH, with the highest concentrations occurring in the more acidic water. This relationship holds regardless of the P_{CO_2}. In closed systems carbonate mineral dissolution may result in lower concentrations and non-linear concentration-pH relationships may occur (Langmuir, 1997). In addition to being pH dependent, the concentration of carbonate mineral dissolution is P_{CO_2} dependent (Figure 6.4). The total carbonic acid concentration increases linearly in the CO_2 range of most groundwater (i.e., $>10^{-4}$ bar). The concentrations of Ca^{2+} and HCO_3^- increase and track parallel. With increasing CO_2, the concentration of CO_3^{2-} becomes insignificant. If the CO_2-charged groundwater encounters carbonate minerals in a closed system, the CO_2 becomes depleted and the concentration of dissolved carbonate products will be less than if the dissolution occurred in an open system where additional CO_2 is available.

Important Takeaway – The dissolution of carbonate minerals are temperature and pH dependent. pH is largely controlled by the groundwater P_{CO_2} and the interaction with carbonate minerals. The solubility of CO_2 and carbonate minerals and the concentration of reaction products can be calculated using temperature-dependent equilibrium constants.

Unlike most minerals, H_2O-CO_2 reactions are exothermic, which means the solubility of CO_2 decreases with increasing temperature. The significance of this is that with increasing aquifer temperature, carbonate minerals tend to precipitate from groundwater.

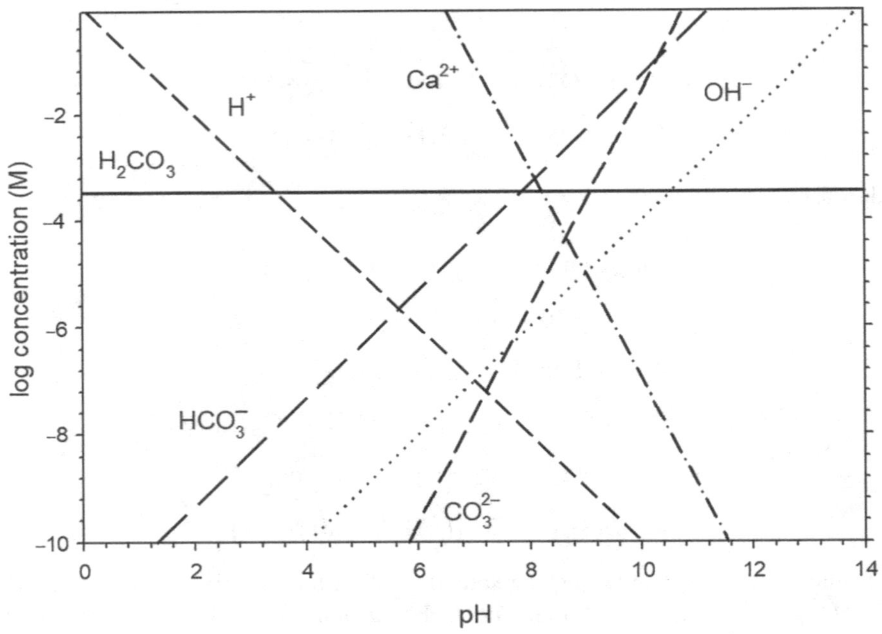

FIGURE 6.3 Plot of the concentrations of species in equilibrium with calcite in an open system at various values of pH, 25 °C, and P_{CO_2} = −3.5 atm. Ion pairs, which are in very low concentrations, are not included.

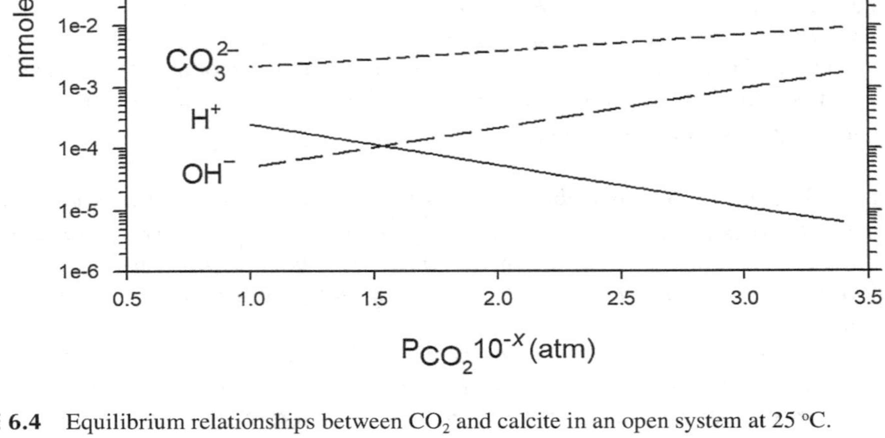

FIGURE 6.4 Equilibrium relationships between CO_2 and calcite in an open system at 25 °C.

6.2.4 Other CO_2–Carbonate Considerations

Several additional factors such as groundwater mixing, the common ion and ionic strength effects, contributions of CO_2 from non-soil zone sources, and evaporation can influence groundwater carbonate concentrations.

Root zone respiration and aerobic decay of organic material are the most common soil zone processes responsible for groundwater CO_2. Root zone respiration consumes CO_2 and drives the pH up, making groundwater basic, and aerobic decay drives the pH down, making groundwater acidic. Soil zone water typically has a CO_2 content ~ 10^{-2} atm, whereas the atmosphere has a CO_2 content of ~$10^{-3.5}$ atm. Root zone respiration, regulated by O_2 flux into the soils zone, can be represented as:

$$6CO_2 + 6H_2O \rightarrow C_6H_{12}O_6 + 6O_2 \text{ (photosynthesis)} \tag{6.31}$$

The aerobic decay of dead plant matter in the soil zone, which is responsible for most soil zone and groundwater CO_2, can be represented as:

$$C_6H_{12}O_6 + 6O_2 \rightarrow 6CO_2 + 6H_2O \text{ (aerobic decay)} \tag{6.32}$$

These two redox reactions are critical to the evolution of most groundwater.

The common ion effect, such as the dissolution of gypsum in a groundwater that previously encountered and became saturated with respect to calcite, drives the pH down by the liberation of CO_2. Additional carbonate mineral dissolution is thermodynamically possible because the increased salinity due to gypsum dissolution reduces the carbonate mineral activity coefficients by the ionic strength effect. If the water subsequently encounters small amounts of halite or other soluble non-calcium-bearing minerals, the ionic strength effect may result in additional calcite dissolution. The common ion effect depends on the order of encounter. For example, Back and Hanshaw (1970) found that when calcite-saturated groundwater in carbonate terrain encounters more soluble gypsum, the common ion effect results in calcite supersaturation and calcite precipitation. The reaction of a calcite-saturated water encountering gypsum along the flow path can be represented as:

$$\underset{\text{(gypsum)}}{CaSO_4 \bullet H_2O} + 2HCO_3^- \rightarrow CaCO_3 + SO_4^{2-} + 2H_2O + CO_2 \tag{6.33}$$

Precipitation of $CaCO_3$ and liberation of CO_2 occurs because the high solubility of gypsum (K_{gyp} = 4.93 × 10^{-5} at 25 °C; Table 2.1) is appreciably greater than the solubility of calcite (K_{cal} = 4.67 × 10^{-9}). What this means is that even though the groundwater is Ca^{2+} saturated prior to encountering gypsum, gypsum dissolution will flood the systems with additional Ca^{2+}, causing calcite to become supersaturated and $CaCO_3$ to precipitate. If the order of encounter is gypsum first and calcite second, the groundwater would be calcite-Ca^{2+} saturated prior to encountering calcite, and calcite dissolution would be inhibited.

If a carbonate mineral saturated groundwater encounters a soluble non-Ca bearing minerals, such as small amounts of halite, the dissolution of halite will increase the TDS and reduce the activity coefficient of the carbonate mineral due to the ionic strength effect. When the resulting activity coefficient is large, the water will become carbonate mineral undersaturated, and additional carbonate mineral dissolution can occur. Many groundwaters flowing in carbonate-clastic rock sequences have Na^+ concentrations that are significantly in disequilibrium with halite and other Na-bearing minerals. These waters are often in carbonate mineral disequilibrium where the Ca^{2+} and/or Mg^{2+} reacting equivalents are significantly less than the HCO_3^- reacting equivalents. A common source

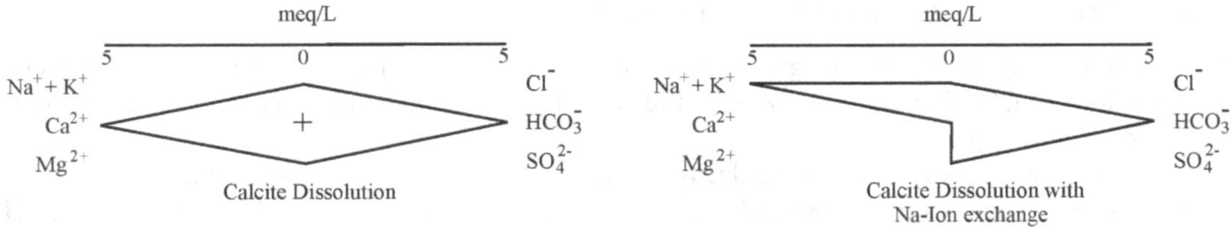

FIGURE 6.5 Stiff diagrams showing a) an initial water that has only dissolved calcite and b) the effect of complete Na-ion exchange. The figures do not represent calcite saturation.

of the Na^+ surplus and the Ca^{2+}-Mg^{2+} deficit is ion exchange, were the Ca^{2+} and Mg^{2+} ions are exchanged for Na^+ on Na-clay. This process can be represented as:

$$Ca^{2+} + \text{Na-clay} = 2Na^+ + \text{Ca-clay} \tag{6.34}$$

and

$$Mg^{2+} + \text{Na-clay} = 2Na^+ + \text{Mg-clay} \tag{6.35}$$

Ion exchange is readily apparent when the ion chemistries have been plotted as Stiff diagrams (Figure 6.5). See Chapter 12 for a discussion of Stiff diagrams.

Anaerobic sulfate reduction is a microbial-mediated process that is common in organic-rich sediments, particularly when the groundwater has been thermally heated by the geothermal gradient, proximity to cooling magma, or other thermal sources. Sulfate-reducing microorganisms (sulfate reducing bacteria and SBA and sulfate reducing archaea – SRA) utilize sulfate as the terminal electron acceptor and expel hydrogen sulfide. Sulfate reduction is obvious by the rotten egg smell accompanying the liberated hydrogen sulfide gas (HS^-). Sulfate reduction drives the pH down by consuming H+ ions. The liberated CO_2 can either raise or lower the pH depending on the fate of the resulting carbonic acid. Sulfate reduction of organic matter can be represented as:

$$\underset{\text{(organic)}}{2CH_2O} + SO_4^{2-} + H^+ \rightarrow 2CO_2 + HS^- + 2H_2O \tag{6.36}$$

Some sulfate-reducing microorganisms play a role in the anaerobic oxidation of methane and drive the pH up due to the increase in bicarbonate concentration:

$$CH_4 + SO_4^{2-} \rightarrow HCO_3^- + H_2O \tag{6.37}$$

Common nitrogen processes that can affect groundwater CO_2 include nitrate reduction from ammonia and denitrification. Nitrate, common in surface and groundwater, is considered to be a health hazard in concentrations greater than 10 mg/L in the United States. and in concentrations greater than 50 mg/L in the European Union. Nitrogen loading can be from natural organic soil zone processes, agricultural waste, wastewater treatment facilities, commercial fertilizer, and in rare instances the oxidization of nitrogen-bearing clay minerals. Both of these processes drive the pH up. Nitrate reduction can be represented as:

$$NO_3^- + 2H^+ + \underset{\text{(organic)}}{2CH_2O} \rightarrow NH_4^+ + 2CO_2 + H_2O \tag{6.38}$$

And denitrification can be represented as:

$$5CH_2O + 4NO_3^- + 4H^+ \rightarrow N_2 + 5CO_2 + 7H_2O \tag{6.39}$$

Aluminosilicate mineral weathering consumes CO_2 and drives the pH up. These processes are commonly incongruent with residual clay minerals as a solid phase. Because the solubility is small, the concentrations of products are quite low. The bicarbonate product in the shallow soil zone may precipitate as calcite when the soil zone dries out. Despite the small solubility, over the course of million years the accumulation of weathered granite can become a few to tens of meters thick. Exposures of thick aluminosilicate mineral weathering product can be seen in road cuts in the Idaho Batholith and the Sierra Nevada Range in California. Such thickness attests to stable tectonism, extremely long weathering episodes, and possibly elevated air temperatures and higher levels of atmospheric CO_2. The weathering of orthoclase is an example of aluminosilicate mineral dissolution:

$$\underset{(K-\text{feldspar})}{2KAlSi_3O_8} + 2CO_2 + 11H_2O \rightarrow \underset{(\text{kaolinite})}{Al_2SI_2O_4(OH)_4} + 2K^+ + 2HCO_3^- \tag{6.40}$$

In calcium-rich igneous rock terraines, soil zone calcite precipitation is often an end product of slow aluminosilicate mineral dissociation. Calcite often precipitates in the residual water when Ca^{2+} and HCO_3^- seeps into the soil zone and evaporates. Soil zone precipitation of calcite due to evaporation is also a common phenomenon in arid region soils. CO_2 gas is released to the atmosphere by CO_2 exsolution accompanying soil zone evaporation:

$$Ca^{2+} + 2HCO_3^- \rightarrow CaCO_3 + CO_{2(g)} + H_2O \tag{6.41}$$

In arid regions, where soil zone-bedrock groundwater flushing is not a frequent event, $CaCO_3$ accumulated in the B horizon is known as caliche. In wetter silica-rich environments, where flushing is frequent, clay minerals and iron oxides tend to accumulate. In wet silica-rich tropical regions, the aluminum ore mineral (bauxite) accumulates in the B horizon.

Groundwater mixing can affect the concentration of carbonate ions. In Figure 6.4 the concentrations of Ca^{2+} and HCO_3^- are not linear with changes in CO_2. This fact becomes significant when two waters with different solute concentrations mix. For example, in Figure 6.6, if a water in equilibrium with calcite with a Ca^{2+} concentration of *A* encounters a similar water with a concentration of *B*, the mixed Ca^{2+} concentration will fall along the mixing line *A-B*. Because line *A-B* is in the undersaturated region of the figure, the resulting water will be calcite undersaturated, and additional calcite dissolution can occur, resulting in a Ca^{2+} concentration of *S*. Groundwater mixing is common in coastal regions where freshwater encounters brackish and saline water and at many spring discharge locations where shallow and deep groundwaters mix.

Locally, CO_2 can be acquired from other sources and mechanisms such as methanogenesis, natural leakage from oil and natural gas reservoirs, anthropogenic leakage from petroleum-producing wells, thermal decomposition of carbonate rocks (i.e., metamorphism), mantle and volcanic emanations, and other anthropogenic releases. Non-soil zone sources can increase the groundwater P_{CO_2} to 1 atm or more and can be the driving force in the chemical evolution of groundwater. Some of these processes can often be identified by their HCO_3^- and other ion concentrations and their $\delta^{13}C$ composition. Two of these processes are methanogenesis and thermal decomposition of carbonate minerals.

Methanogenesis – methane fermentation by anaerobic decay – is a microbial anaerobic respiration process that is common in organic-rich environments such as coal-bearing sediments, buried swamp deposits, warming permafrost regions, and organic-rich lake bed environments. Methanogenesis is the anoxic biogenic degradation of organic matter yielding CH_4 and CO_2 (Hunt 1979,

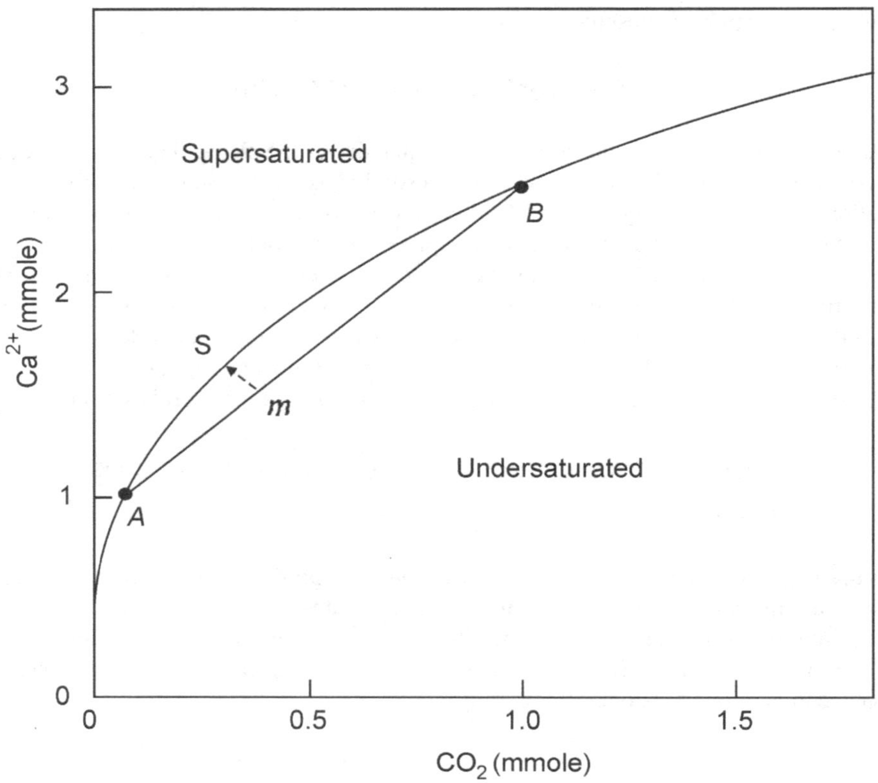

FIGURE 6.6 Equilibrium concentration of Ca^{2+} in calcite saturated water with varying CO_2 concentrations. Line *A-B* represent the mixing of waters *A* and *B*, *m* represents the mixed Ca^{2+} concentration which become undersaturated. S represents the mixed saturation concentration.

Whiticar et al., 1986, Wolin and Miller, 1987). The terminal electron acceptor is carbon instead of oxygen. This process drives the pH down, but the resulting carbonic acid from the liberated CO_2 can increase the pH by promoting the dissolution of carbonate minerals. The methanogenic release of CO_2 and methane can be represented as:

$$\underset{(\text{organic})}{C_6H_{12}O_6} + 2H_2O \rightarrow 2H_3CO_2^- + 2H^+ + 2CO_2 + 4H_2O \tag{6.42}$$

and

$$2H_3CO_2^- + 2H^+ \rightarrow 2CH_4 + 2CO_2 \tag{6.43}$$

In confined aquifer methanogenic environments, it is not uncommon for both CO_2 and CH_4 to become over-pressured. For example, in the San Luis Valley of Colorado, both CH_4 and CO_2 evolve from free-flowing artesian wells. In some instances, well water will briefly burn when ignited. The in situ CO_2 results from a cascade of reactions (Equations 6.2 to 6.5) culminating in elevated TDS and HCO_3^- concentrations. The HCO_3^- $\delta^{13}C$ has been used to map the extent of the methanogenic CO_2 and hence the extent of the ancient swamp deposits now buried beneath non-organic-rich sediments (Figure 6.7). Methane over-pressurization has also been attributed to explosive crater

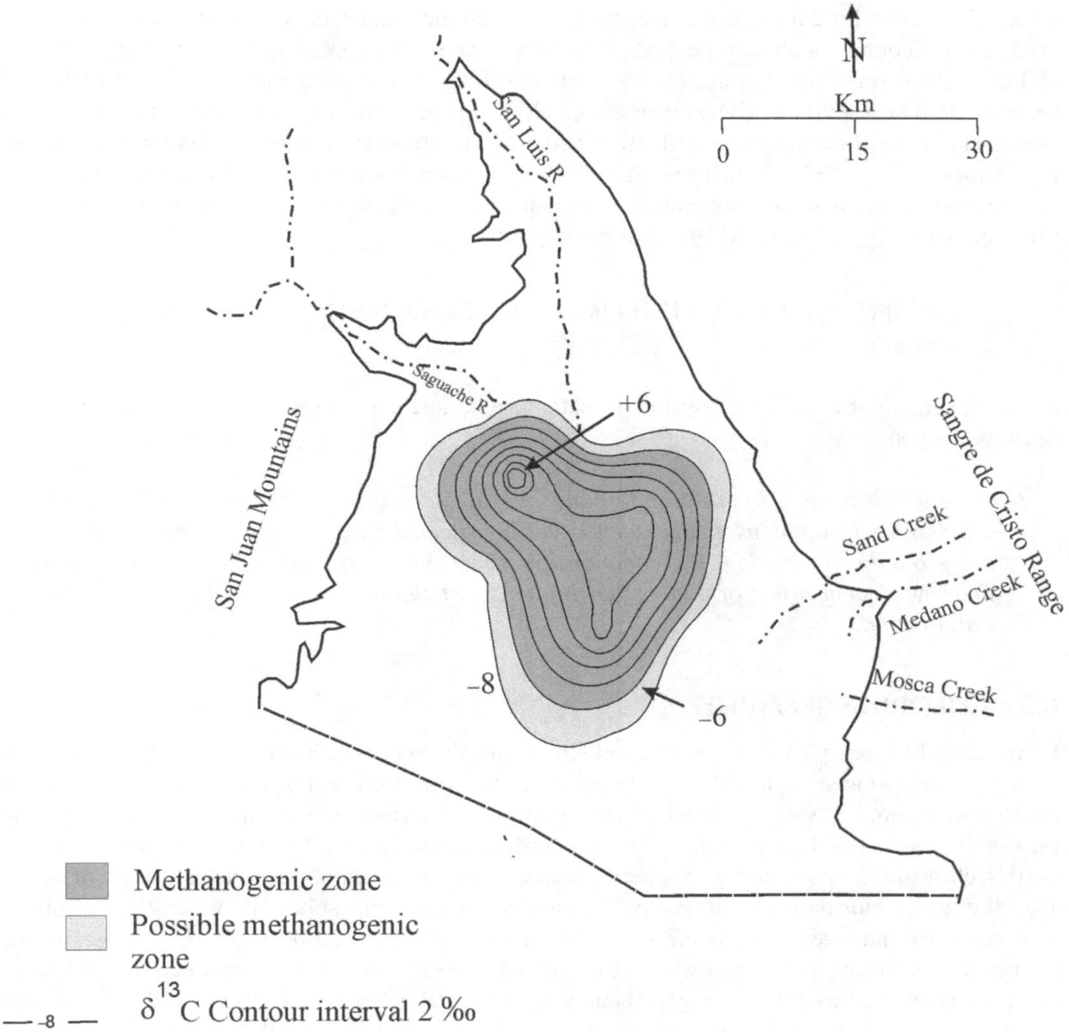

FIGURE 6.7 Contour map depicting the extent of buried organic-rich swamp deposits in the upper confined aquifer in the San Luis Valley, Colorado. The extent of methanogenic CO_2 was mapped using the HCO_3^- $\delta^{13}C$ as a surrogate. $\delta^{13}C > -4‰$ was considered to be derived from methanogenic-sourced CO_2 (after Mayo, 2020).

formation in Russian Arctic permafrost regions and to the ever-increasing release of methane from warming arctic permafrost in North America (Chuvilin et al., 2020; Walter et al., 2007).

Carbonate rock metamorphism is an important source of CO_2 along oceanic ridges, in some volcanic fields, and from deeply buried continental carbonate sequences. Thermal decomposition of carbonate rocks decreases the pH in water by the liberation of CO_2. The thermal decomposition of carbonate rocks, particularly calcite limestone, has been a common industrial process since ancient Egyptian times and locally can result in CO_2 over-pressured groundwater systems. The ancient Egyptians used lime-gypsum as cement for the pyramids, and the Colosseum in Rome (80 AD) is a concrete structure. The thermal decomposition of limestone can be represented as:

$$CaCO_3 + \text{heat} \rightarrow CaO + CO_2 \tag{6.44}$$

where CaO is the lime used to make cement. This commercial process is so ubiquitous that lime production accounts for about 4 percent of the atmosphere's anthropogenic CO_2 (Andrew, 2019).

Pure calcite limestone is rare, but the metamorphism of dolomite and calc-silicate rocks has been found to be a source of CO_2 (Evans et al., 2008; Mayo and Muller, 1997; Kerrick and Caldeira, 1998). The thermal decomposition of siliceous-carbonate rocks (Equation 6.45) is common during high temperature (>450 °C) metamorphic reactions. At lower temperatures the reaction pathways are reduced, but even at low temperatures CO_2 can be released. For example, a dolomite-silica reaction can yield a P_{CO_2} of 1 atm at 190 °C according to:

$$\underset{\text{(dolomite)}}{3\text{CaMg}(\text{CO}_3)_2} + 4\text{SiO}_2 + \text{H}_2\text{O} + \text{heat} = \underset{\text{(talc)}}{\text{Mg}_3(\text{Si}_4)\text{O}_{10}(\text{OH})_2} + 3\text{CaCO}_3 + 3\text{CO}_2 \tag{6.45}$$

A good example of massive exposures of metamorphic talc can be seen in Death Valley's Warm Springs Canyon.

Important Takeaway – In addition to atmospheric and soil zone CO_2, carbonate ion concentrations in aqueous environments can be affected by a variety of process including groundwater mixing, the common ion effect, ionic strength effect, ion exchange, the flow system order of mineral encounter, contributions of non-soil zone CO_2, and evaporation.

6.3 CHLORIDE-BROMIDE

Elemental chlorine (Cl_2) is a gas that seldom naturally occurs. Chloride (Cl^-), the most common aqueous halogen, is found in large concentrations in marine, playa lake, and closed basin environments, and in small concentrations in almost all surface and groundwater. Other locally important aqueous halogens include fluorine, bromine, and iodine. More than 75 percent of the earth's chlorine is sequestered in ocean waters, which have chloride concentrations of about 19,400 mg/L. Chloride is a conservative species in most aqueous solutions. When the salinity of surface or groundwater exceeds the solubility of gypsum, about 2,000 mg/L, Cl^- is usually the major anion. Surface and groundwater in playa and closed basin environments can have Cl^- concentrations as great as 150,000 mg/L (Hamdani et al., 2018; Mayo and Tingey, 2021). Such high concentrations can occur because halite has a weak ionic bond ($K_{eq} = 10^{1.54}$; Table 2.1). Aqueous chloride only forms chloride complexes when the salinity is extremely high. Chlorine is not readily involved in redox reactions.

The chloride content of precipitation may sometimes be used to estimate the percentage contribution of marine aerosol sulfate in coastal precipitation and, by extension, the contribution of marine aerosol sulfate to groundwater recharge. For coastal groundwater this relationship has been defined as (Cortecci and Longinelli, 1970):

$$A = \frac{\text{SO}_4^{2-}\ \text{seawater}}{\text{SO}_4^{2-}\ \text{observed}} \times 100 = \frac{(\text{Cl}^-)\text{observed}}{7.1(\text{SO}_4^{2-})\text{observed}} \times 100 \tag{6.46}$$

where A is the percent of sea spray sulfate. This relationship assumes that all chloride comes from marine aerosols and that the value of the Cl^-/SO_4^{2-} ratio in the marine aerosol component in precipitation is 7:1. Groundwater can acquire appreciable amounts of chloride and sulfate from the soil zone and other aquifer sources, and caution should be employed when attempting to apply this relationship to groundwater systems. Sulfate in non-coastal-zone continental precipitation and in

TABLE 6.4
The Mean Chlorine Abundance in Terrestrial Rocks

Source	Cl (ppm)	Source	Cl (ppm)
ultramafic rocks	100	shale and clay	100
mafic rocks	160	sandstone	20
intermediate rocks	180	sandstone	20
acidic rocks	200	limestone	130
syenites	430	schist	350
nepheline syenite	2,170	gneiss	200
amphibolite	300		

Source: Data from Kuroda and Sandell, 1953; Johns and Huang, 1967.

dry fall tends to have a terrestrial origin rather than a saline water origin (Berner and Berner, 1987; Munger, 1982), and Equation 10.2 should be used with caution.

Most dissolved chloride in surface and groundwater is from natural sources, but locally anthropogenic factors such as road deicing salts, wastewater disposal, industrial leaks, and oil field brine disposal can be important chloride sources. Natural sources include precipitation, dissolution of minerals in igneous and sedimentary rocks, sea water intrusion, and evaporation. The chloride content of most precipitation is from oceanic Na^+–Cl^--bearing aerosols. Inland chloride concentrations in precipitation are greatest along the coast and vary greatly by location, season, and storm. Much of the oceanic Na^+ is lost during inland atmospheric transport (Moller, 1990). At the coast concentrations can range from 0.16 to ~ 4 mg/L, and inland concentrations are typically less than 0.5 mg/L (Gordon, 2013; Holden, 1961; Hutton, 1976).

Except for halite (NaCl) and sylvite (KCl), other chlorine-bearing minerals do not occur in great abundances and are generally not important in groundwater investigations. The average content in various crustal rocks is summarized in Table 6.4. Chloride can substitute for hydroxide ions in aluminosilicate minerals such as biotite and amphibole and is found in fluid inclusions. The slow dissolution of aluminosilicate minerals and the dissolution of carbonate minerals can release small amounts of chloride. In laboratory reports chloride ion concentrations are typically listed as a major ion, and Cl^- is one of the anions routinely plotted on Stiff diagrams and trilinear plots (Chapter 12). When the laboratory analysis is converted to meq/L, the results are often compared to the Na^+ concentration to help sort out the origins of both the Cl^-- and Na^+.

Chloride/bromide ratios can be useful for investigating groundwater systems (Alcala and Custodio, 2008; Aliewi et al., 2022; Davis et al., 1998; Naily, 2018). The chloride-bromine ratio (R) can be calculated using either molar concentration or mass concentration:

$$R = \frac{r\mathrm{Cl}}{r\mathrm{Br}} = 2.25\frac{p\mathrm{Cl}}{p\mathrm{Br}} \tag{6.47}$$

where (r) is in moles and (p) is in mg/L. In most aqueous studies mg/L are used for calculating R. Typical R values based in mg/L are listed in Table 6.5. Although both chloride and bromide are monovalent anions with a negative charge, the notation Cl/Br without a charge is frequently used when discussing chloride/bromide ratios. In groundwater investigations is it common to plot the Cl/Br ratio of a water sample relative to the total chloride concentration of known endmembers to help sort out the origin of the target water. Chloride concentrations and Cl/Br rations may be useful indicators of sea water intrusion and other Cl sources (Figure 6.8). Saline water intrusion is a natural

TABLE 6.5
Typical Cl/Br ratios in Aqueous Solutions

Environment	Cl/Br (mg/L)	Source
Precipitation	50–150	1
Agriculture pollution	90–200	2, 3
Urban runoff	10–100	1
Oil field water	200–400	1
Sea water	292	
Sewage	300–600	1
Septic tanks	400–1,100	4
Halite dissolution	10^3–10^4	1

Sources:
[1] Davis et al., 1998
[2] Naily, 2018
[3] Alcala and Custodio, 2008
[4] Katz et al., 2011

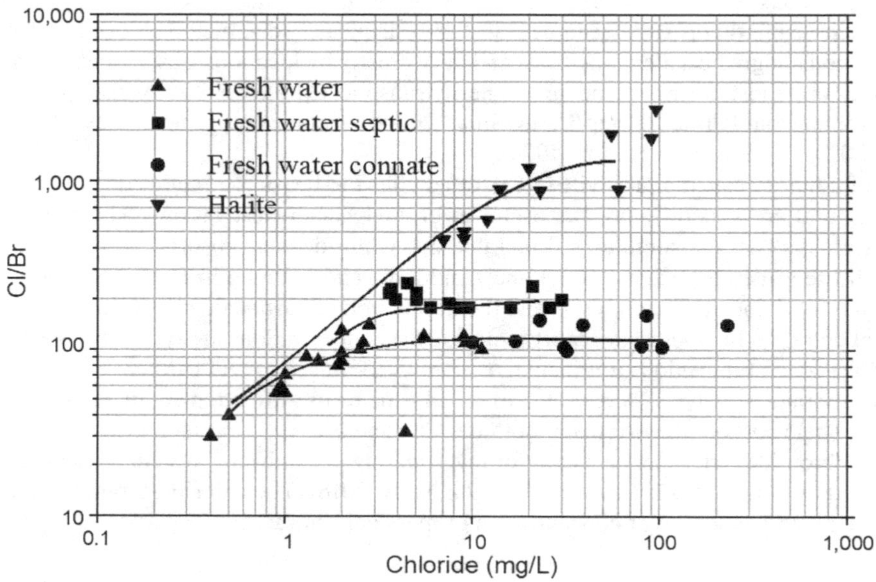

FIGURE 6.8 The Cl/Br mass ratio vs. Cl of groundwater is often a useful indicator of the Cl source and sometimes can be used to identify Cl sources (modified after Naily, 2018).

occurrence that can be enhanced by groundwater extraction in coastal areas and on the fringes of closed basins. Near the coast, simple contouring of Cl^- or Cl^- Na^+ concentrations may be used to determine the extent of sea water intrusion. Further inland, where chlorine concentrations are lower and where agricultural and other anthropogenic activities can introduce Cl^- into groundwater systems, Cl/Br, Na/Cl, and Cl/HCO_3 ratios have been used to identify sea water intrusion several kilometers from the ocean (Lakshmanan, 2013; Panjaitan et al., 2018). The underlying principle of using Cl/Br ratios in sea water intrusion investigations is the fact that the sea water Cl/Br ratio of 292:1 remains stable even at low chlorine concentrations.

Chloride-36 has been used to estimate the mean residence time (age) of groundwater that recharged 10^5 and 10^6 years ago. Cl^{36} groundwater dating is discussed in Chapter 11. Chloride has also been used to estimate groundwater recharge in arid and semi-arid environments using the chloride-mass balance method (CMB). The CMB is an inexpensive method for estimating the aerial mass flux of water crossing the plane of the water table. The underlying assumptions of the CMB are: 1) the chloride in the groundwater only originates from precipitation that falls directly on the aquifer, 2) the chloride is conserved in the system, 3) the atmospheric chloride-mass flux has not changed over time, and 4) there is no recycling, loss, or concentration of chloride in the aquifer (Wood and Sanford, 1995). If these conditions are met, the recharge can be calculated using the relationship:

$$Q = \frac{P(\mathrm{Cl_p})}{\mathrm{Cl_{gw}}} \tag{6.48}$$

where Q is the groundwater recharge flux (length/time), p is the average annual precipitation in units of length per time, Cl_p is the average precipitation-weighted chloride concentration in units of mass/volume, and Cl_{gw} is the average concentration of chloride in the groundwater in units of mass per volume. All units must be consistent. Equation 6.48 is only applicable to local recharge. Wood (1999) suggested a modification for regional applications:

$$Q = \frac{P(\mathrm{Cl_p})}{\mathrm{Cl_{uz}}} \tag{6.49}$$

where Cl_{uz} is the chloride concentration of the unsaturated zone. The unsaturated zone concentration is determined from a core sample. Other modifications to Equation 6.48 have been proposed to account for account for runoff into local depressions, surface runoff, and other conditions (Eriksson and Khunakasem, 1969; Maurer et al., 1996; Somarantne and Smettem, 2014). The CMB method has been used to estimate recharge as low as 0.05 to 0.1 mm/y. CMB should be limited to areas with annual recharge of less than 300 mm/yr (Scanlon, 2004). Because the chloride concentration in precipitation varies greatly between storm events and has seasonal and interannual cycles, satisfying assumption 3 and eliminating other factors can be difficult.

Important Takeaway – Chloride concentrations in groundwater can exceed 150,000 mg/L. Most groundwater Cl^- is from natural sources including precipitation, dissolution of igneous and sedimentary rock minerals, sea water intrusion, and evaporation. Locally anthropogenic sources such as road deicing salts, wastewater disposal, industrial leaks, and oil field brine disposal can be significant. Cl/Br ratios can be useful in sorting out the chloride source, particularly sea water intrusion. The chloride-mass balance (CMB) method is sometimes useful for estimating groundwater recharge rates in arid and semi-arid environments. 36Cl (Chapter 11) can be used as a groundwater age dating tool for very old groundwater.

6.4 SULFUR

Sulfur is locally important in groundwater systems, and like carbon it cycles through the atmosphere, hydrosphere, biosphere, and lithosphere (Figure 6.9). Sulfur is present as solid, gas, and aqueous species. Most of the earth's accessible sulfur is sequestered in lithospheric rocks, although the oceans contain appreciable amount of mobile sulfur (Table 6.6). Sulfur, the 10th most abundant

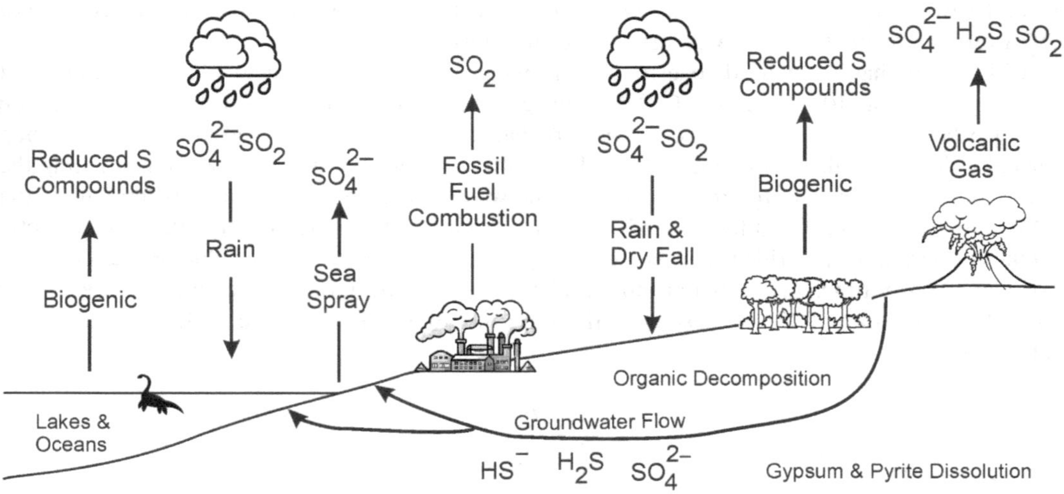

FIGURE 6.9 The sulfur cycle. Atmospheric sulfur is from sea spray, volcanic gas, biogenic sources, and fossil fuel combustion. Although precipitation contains anthropogenic sulfur, most groundwater sulfur is from the dissolution of gypsum, with lesser amounts from pyrite.

lithospheric element, is found in sulfate and sulfide minerals, organic and biomass materials, and coal and petroleum. Elemental sulfur occurs in evaporite sedimentary rocks, at hot springs, fumaroles, volcanic vents, and as massive deposits in carbonate cap rocks overlying salt domes. Common solid species include sulfide minerals such as pyrite (FeS_2) and galena (PbS) and sulfate minerals such as gypsum ($CaSO_4 \bullet H_2O$) and anhydrite ($CaSO_4$). Gases include the common atmospheric sulfur compounds sulfur dioxide (SO_2) and carbonyl sulfide (OCS), and H_2S, which has the rotten egg smell associated with the microbial breakdown of organic matter, volcanic emanations, and natural gas deposits. Sulfur is a critical component in biologic matter including amino acids, B vitamins, and the substrate coenzyme A (an organic molecule). In surface and groundwater, as well as in ocean water, sulfate $\left(SO_4^{2-}\right)$ is the most common form of sulfur.

6.4.1 Atmospheric Sulfur

Sulfur is not a major atmospheric gas (Table 6.6). The partial pressure of the most common atmospheric species, SO_2, is only 10^{-6} atmospheres. Prior to the industrial revolution, the global cycling of sulfur was primarily regulated by volcanic SO_2 and H_2S, which oxidize to sulfate aerosols, the oceanic dispersion of atmospheric sulfur aerosols including dimethylsulfide (DMS), and anaerobic respiration by sulfate-reducing bacteria (Bates et al., 1992; Andres and Kasgnoc, 1998; Martin, 2018). DMS ($(CH_3)_2S$) is a product of the enzymatic breakdown of dead phytoplankton. DMS is responsible for the distinct smell of the sea along the coast. During periods of high activity, volcanoes can account for more than 80 percent of natural atmospheric sulfur (Lucas and Akimoto, 2007).

In the atmosphere, most sulfur dioxide, DMS, and volcanic and anthropogenic sulfur are oxidized to carbonyl sulfide (COS), which has an atmospheric half-life of about 2 years but may be as long as 47 years in the lowermost stratosphere (Karu et al., 2023). Other sulfur gases have a very short atmospheric half-life. In the literature, carbonyl sulfide is sometime referred to as OCS. Oxidized atmospheric sulfur dioxide (SO_2) returns to the earth as weak sulfuric acid (H_2SO_4) and as dry fall. Locally, atmospheric dry fall and precipitation may acquire sulfate from wind-blown soil and playa sulfate dust. Since the industrial revolution, the burning of fossil fuels, biomass burning, agricultural activities, and the smelting of sulfide ore has more than doubled the atmospheric sulfur (Kooijman et al., 2010). Over the oceans most atmospheric SO_2 is from oceanic sources, and over land most is from anthropogenic activities (Lucas and Akimoto, 2007).

TABLE 6.6
Major Worldwide Sulfur Reservoirs

Reservoir	Forms of Sulfur	(kg)
Atmosphere	Gas: DMS, H_2S, SO_2, OCS Aerosols: H_2SO_4, $(NH_4)_2SO_4$	4.8×10^9
Pedosphere and biosphere	Sulfates, organic sulfur, pyrite	2.9×10^{14}
Rivers and lakes	SO_4^{2-}	3×10^{12}
Fresh groundwater	SO_4^{2-}, H_2S, HS^-	3.5×10^3
All groundwater	SO_4^{2-}, H_2S, HS^-	7.6×10^3
Oceans	SO_4^{2-}	1.3×10^{18}
Marine biota		2.4×10^{11}
Marine sediments	Pyrite, gypsum, anhydrite	3×10^8
Lithosphere	Gypsum, anhydrite, sulfides, sulfur	2.4×10^{19}

Source: Data from Smith, 2001; Palya et al., 2008; Schoonen, 2016; Kooijman et al., 2010.

Most of the excess atmospheric sulfur is recycled to the ocean. The pre-industrial revolution sulfur landfall deposition rate of about 43 Tg/y has increased to about 65–90 Tg/y today, and the sulfur flux from rivers to the oceans has increased from 104 to 213 Tg/y. A Tg is 10^{12} grams. The increase in atmospheric sulfur deposition and sulfur in streamflows means that the sulfur content of groundwater recharge has also increased. Weathering of lithospheric rocks releases about 72 Tg/y, most of which is carried to the oceans by streams. Surface and groundwater contain a small but mobile portion of the earth's sulfur (Table 6.6).

In unpolluted areas, sulfate concentrations in precipitation are often less than 1 mg/L (0.02 meq/L), but it may be 10 mg/L (0.21 meq/L) or more in areas subjected to appreciable natural or anthropogenic air pollution (Berner and Berner, 1987). For example, in 1964 the mean SO_4^{2-} in Sierra Nevada and Salt Lake City snow was reported as 0.93 mg/L (0.019 meq/L) and 2.25 mg/L (0.047 meq/L), respectively (Feth et al., 1964). Salt Lake City is located on the eastern edge of the Basin and Range physiographic province some 500 miles farther inland from the Pacific Ocean than is the Sierra Nevada Range. Some of the Salt Lake City SO_4^{2-} is the result of desert dust blown into the atmosphere. In 1984 the mean SO_4^{2-} in Salt Lake City precipitation increased to 5.4 mg/L (0.112 meq/L; Christensen et al., 1984). In more recent years there has been an overall decrease in atmospheric sulfate in North America and Europe and an increase in Asia (Manktelow et al., 2007). These changes are believed to be the result of long-term air pollution controls in the United States and Europe and an increase in fossil fuel burning, particularly coal, in Asia.

6.4.2 Biogenic Sulfur

Microbiological conversions of sulfur compounds are important sources of sulfur in many groundwater systems, particularly in organic-rich reducing environments. The biologic sulfur cycle is illustrated in Figures 6.10 and 6.11. Microorganisms abound in the freshly deposited substrate of marine and terrestrial aqueous environments and in other aqueous environments where sulfur compounds are readily available. Biologic mediation involves intermediate sulfur oxidation states that are used as electron acceptors and donors. Sulfate-reducing microorganisms (SRM) and sulfur-oxidizing microorganisms (SOM) substitute sulfur compounds and sulfur-bearing ions for free oxygen as part of their metabolic processes.

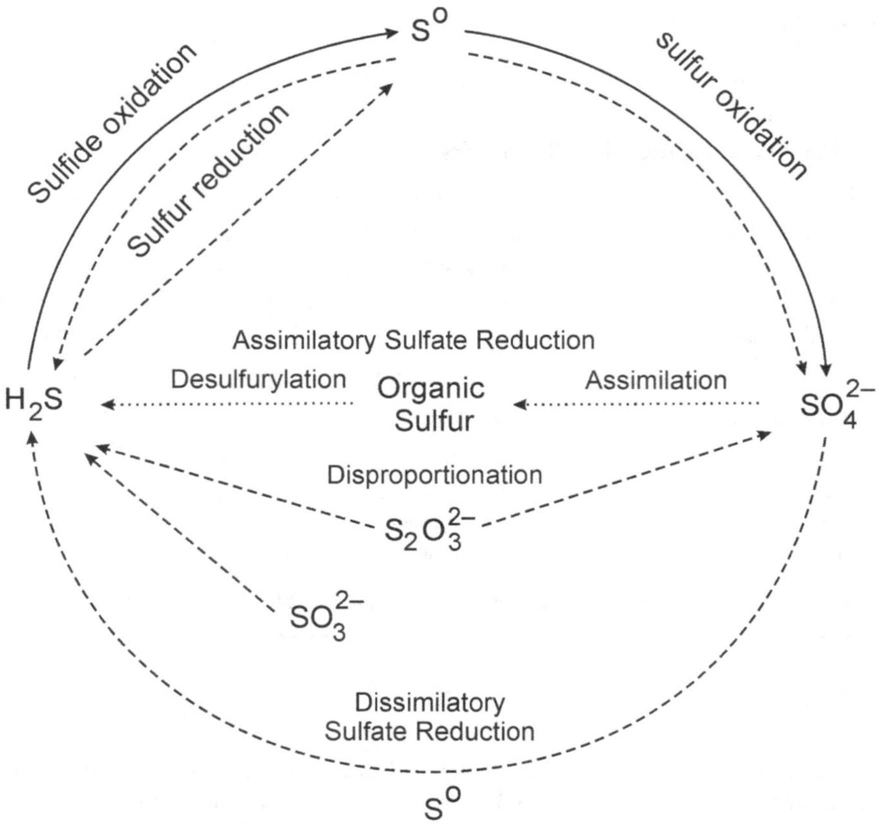

FIGURE 6.10 The biologic sulfur cycle – solid lines are aerobic, dash lines are anaerobic, and dotted lines are aerobic. Details of assimilatory and dissimilatory reduction are shown in Figure 6.11.

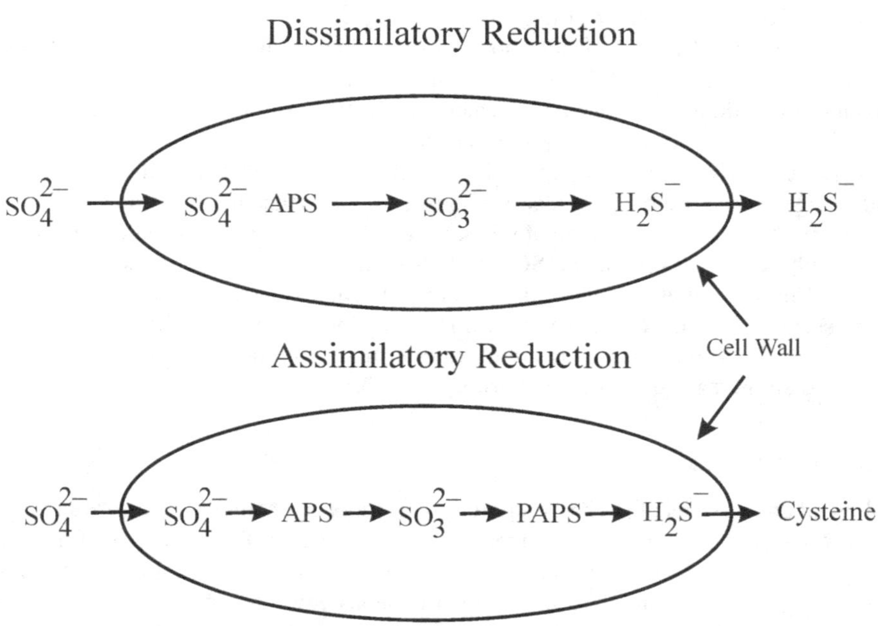

FIGURE 6.11 Details of anaerobic biologic sulfate reduction. APS and PAPS are synthesized organic compounds. Cysteine is a protein-building amino acid.

Sulfur compounds are metabolized along two primary reduction pathways: assimilatory and dissimilatory (respiratory), although some sulfur is metabolically oxidized (Kushkevych et al., 2021). In dissimilatory reduction SO_4^{2-} is converted to H_2S or HS^-. During assimilatory reduction the resultant sulfide is incorporated into the organism's metabolites, which are directed toward the organism's nutritional needs. The end product of assimilatory sulfate reduction is cysteine (an amino acid), whereas the end product of dissimilatory sulfate reduction is sulfide (Figure 6.11). Biological oxidation converts H_2S to SO_4^{2-} with an intervening step of S°. Sulfur-oxidizing bacteria, particularly thiobacilli, are particularly harmful to cultural stonework because they oxidize sulfides to sulfate, resulting in sulfuric acid, which then dissolves the carbonate minerals that bind the quartz and other minerals in the stonework (Ranalli et al., 2019).

6.4.3 Sulfur in Groundwater

Sulfur is the second most common groundwater anion. Groundwater acquires sulfur from several sources including atmospheric dry fall and precipitation, the dissolution of calcium-sulfate minerals in evaporite and non-evaporite sedimentary rocks, the oxidation of sulfide minerals, and the biologic oxidation and reduction of organic material. Most groundwater sulfur occurs in the form of sulfate (SO_4^{2-}) which has a +6 valance (Figure 6.12). Sulfur also occurs in the −2, +2, and +4 valance states. Sulfide either as hydrogen sulfide ($H_2S_{(aq)}$, pH ~ <7) or bisulfate (HS^-, pH ~ >7) prevails

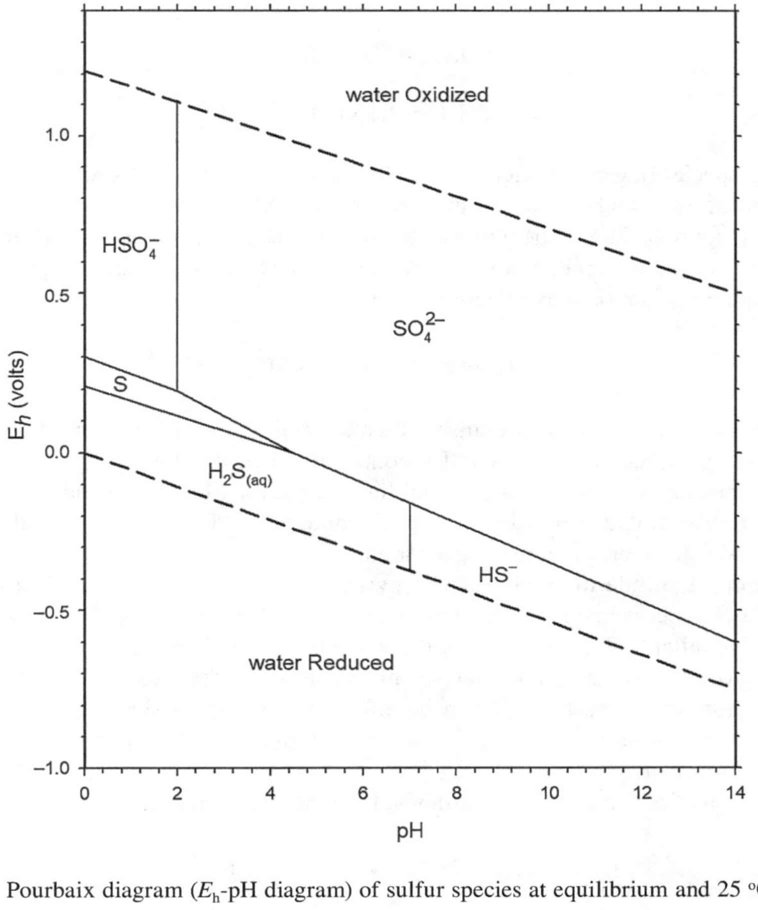

FIGURE 6.12 Pourbaix diagram (E_h-pH diagram) of sulfur species at equilibrium and 25 °C and 96 mg/L SO_4^{2-}.

under reducing conditions. The presence of H_2S, common in oil field brines and some geothermal waters, can be detected by the rotten egg smell.

Some groundwater sulfate is due to atmospheric dry fall and precipitation, particularly when the sulfate in terrestrial windblown aerosols has been concentrated by evapotranspiration (Bao and Reheis, 2003; Mayo and Klauk, 1991). Terrestrial windblown aerosols are particularly important in arid regions containing playa lakes; however, elevated sulfur concentrations in groundwater generally indicates sources other than the atmosphere. Most groundwater sulfate is from the dissolution of the highly soluble evaporite minerals gypsum ($CaSO_4{\bullet}nH_2O$) and anhydrite ($CaSO_4$). Gypsum, the most common sulfate mineral, is deposited from sulfate-saturated lake and sea water. It occurs as evaporite beds and as an accessory mineral in carbonate rocks, and it is associated with halite, hot springs, and volcanic vapors. Massive amounts of both gypsum and anhydrite occur in salt dome cap rocks. Gypsum has great economic significance as it is the primary ingredient in dry wall (gypsum board), which is ubiquitous in the house building industry. Anhydrite is most frequently found in evaporite deposits associated with gypsum. When exposed to water, anhydrite readily transforms into gypsum. Locally, magnesium sulfate, which is usually in the hydrate form ($MgSO_4 \bullet nH_2O$), can precipitate in evaporite and supergene environments.

The dissolution of ionically bonded calcium-sulfate minerals is not pH or redox dependent, and in the typical range of groundwater temperatures it may contribute as much as 2,000 mg/L to the TDS of water due to the relatively high equilibrium constants (gypsum $K_{eq} = 2.62 \times 10^{-5}$; anhydrite $K_{eq} = 4.93 \times 10^{-5}$; Table 2.1). Gypsum and anhydrite dissolution may be respectively represented as:

$$CaSO_4 = Ca^{2+} + SO_4^{2-} \tag{6.50}$$

$$CaSO_4 \bullet 2H_2O = 2H_2O + Ca^{2+} + SO_4^{2-} \tag{6.51}$$

Sulfur-bearing species in groundwater may be chemically and isotopically altered during oxidation and reduction processes, and the final form of the liberated SO_4^{2-} is controlled by both the pH and E_h of the solution (Figure 6.12). Sulfide liberated by the dissolution of sulfide minerals may be oxidized to elemental sulfur and to sulfate with or without biological participation. Inorganic conversion of sulfide to sulfate by oxidation may be represented as:

$$\text{Oxidation sulfide} \rightarrow SO_4^{2-} \tag{6.52}$$

In organic-rich sedimentary and in aluminosilicate terrains, the oxidization of sulfide minerals, particularly pyrite, can be an aqueous sulfur contributor. Locally, thermal groundwater, reducing buried organic matter, volcanic emissions, and fugitive hydrocarbon brine and vapor can also contribute sulfur. Sulfur dioxide (SO_2) is a deadly volcanic gas and, when detected at a non-erupting volcano, may indicate an eruption may occur soon.

The oxidation of sulfide minerals such as pyrite (FeS_2), galena (PbS), chalcopyrite ($CuFeS_2$), and realgar (As_2S_2) occurs in open groundwater systems and elsewhere, such as mine environments where free O_2 is available. The oxidation of these and other sulfide minerals are locally responsible for metal groundwater contamination and for most acid mine drainage (AMD). In coal and hard rock mine environments, most AMD can be attributed to pyrite oxidation. Sulfide oxidation is exothermic, and in coal mines with high sulfur content, pyrite oxidation can result in spontaneous combustion and mine fires.

The oxidization of pyrite and the release of sulfate may be represented as:

$$2FeS_2 + 7O_2 + 2H_2O \rightarrow 2Fe^{2+} + 2\,SO_4^{2-} + 4H^+ \tag{6.53}$$

Initially the reaction yields ferrous ions (Fe^{2+}) and sulfuric acid (H_2SO_4). The sulfuric acid quickly dissociates into SO_4^{2-} and H^+ (Equation 6.52). The ferrous ions can further be oxidized by O_2 into ferric ions (Fe^{3+}) that hydrolyzes into $Fe(OH)_3$ (rust) and releases additional H^+ ions:

$$4Fe^{2+} + O_2 + 10H_2O \rightarrow 4Fe(OH)_3 + 8H^+ \quad (6.54)$$

This sequence of processes began about 2 billion years ago during the Great Oxidation Event (GOE). The GOE was initiated by the rise of oceanic photosynthesis, resulting in the earth's atmosphere acquiring its free molecular O_2 and a following cascade of geochemical processes. After the GOE, atmospheric O_2 oxidized the abundant ferrous iron that had accumulated in some sediments and sedimentary rocks. Some of this oxidized iron was deposition as the iron ores in North America and elsewhere. Prior to the GOE the limited oxidation of iron-bearing aluminosilicate and sulfide minerals resulted in the accumulation of ferrous iron. Without the vast accumulation of banded iron ores around Lake Superior and the nearby coal deposits in the Appalachian Mountains, the North American industrial revolution would not have been possible, and the United States may have remained a largely agrarian society.

Most groundwater sulfide is thought to form in situ near the discharge area by the bacterial reduction of sulfate ions that accompany the decomposition of organic material. Groundwater sulfide, H_2S or HS^-, is stable under reducing conditions and has concentrations of only a few milligrams per liter because of the very low solubility of sulfide minerals (Table 2.1). Sulfate reduction is mediated by archaea (single-celled organisms) and sulfate-reducing bacteria that are known by the acronym SRB (Aravena and Mayer, 2009; Baker and Banfield, 2003):

$$SO_4^{2-} \xrightarrow{\text{(reduction)}} \text{sulfide} \quad (6.55)$$

SRB are anaerobes that acquire energy from the oxidation of organic matter by using sulfate as the electron acceptor. In the biologic terminal electron accepting process (TEAP), sulfate reduction occurs after iron reduction and before methanogenesis (see Table 1.12; Christensen et al., 2000). The terminal electron acceptor is the last compound to receive an electron in the complex electron transport chain. Common electron acceptors in groundwater systems include oxygen, nitrate, iron (III), manganese (IV), sulfate and carbon dioxide. In organically contaminated groundwater tetrachloroethylene (PCE), trichloroethylene (TCE), dichloroethene (DCE), and vinyl chloride (VC) are terminal electron acceptors. The general form of sulfate reduction is:

$$SO_4^{2-} + 9H^+ + 8e \rightarrow HS^- + 4H_2O \quad (6.56)$$

and dissimilatory sulfate reduction may involve organic carbon and $H_{2(g)}$:

$$SO_4^{2-} + \underset{\text{(organic)}}{2CH_2O} \rightarrow H_2S + 2HCO_3^- \quad (6.57)$$

and

$$SO_4^{2-} + 2H^+ + 4H_2 \rightarrow H_2S + 4H_2O \quad (6.58)$$

The amount of available organic carbon and other nutrients can be limiting factors in sulfate reduction. Groundwater systems with petroleum hydrocarbons, lake sediments, and marine environments have sulfate reduction coefficients that are typically orders of magnitude greater than most groundwater (Miao et al., 2012).

Important Takeaway – Sulfur, the second most abundant anion in groundwater, is largely derived from the dissolution of calcium-sulfate minerals in sedimentary rocks. The oxidization of pyrite and other sulfide minerals contributes smaller amounts of sulfur to groundwater systems and are responsible for AMD. Biological mediation is important in many sulfide mineral oxidization and sulfate reduction processes.

6.5 IRON AND MANGANESE

Iron in combination with other elements is found in more than 1,000 minerals. It occurs in oxide, hydroxide, halide, carbonate, chloride, sulfate, sulfide, and silicate minerals (Table 2.1). Iron, typically in the ferrous form (Fe^{2+}), is an important element in igneous and metamorphic rocks and is a key element in the dark-colored minerals pyroxene, amphibole, biotite, magnetite and olivine. Iron readily substitutes for magnesium in olivine ($(Mg,Fe)_2SiO_4$), where magnesium and iron range from forsterite (Mg_2SiO_4) to fayalite (Fe_2SiO_4). In sedimentary environments iron is typically in the ferric form in minerals such as pyrite or marcasite (FeS_2), siderite ($FeCO_3$), and magnetite $\left(Fe^{2+}Fe_2^{3+}O_4\right)$. Pyrite is associated with other sulfide minerals in quartz veins and in sedimentary and metamorphic rocks and is common in coal beds and other organic-rich reducing environments. Siderite occurs in hydrothermal veins and is a common diagenetic mineral in shale and sandstone. Magnetite is one of the common iron ores, and iron oxides and hydroxides, such as hematite (Fe_2O_3) and ferrous hydroxide ($Fe(OH)_2$), are common iron-bearing minerals.

Despite iron's lithospheric abundance (Table 6.1), in aqueous solutions iron is only a trace element due to the extremely small solubilities of aluminosilicate minerals and the fact that, once released, iron's solubility is redox controlled. When ferrous iron minerals are weathered, the free iron can be in either the ferrous or ferric state. In dissolved form Fe is typically reduced (Fe^{2+}). When oxidized, Fe^{3+} readily forms an iron oxide and either remains suspended in solution or precipitates. In solution, iron is reduced when the oxygen content is low and the water is slightly acidic (Figures 6.13 and 6.14). When reduced groundwater is re-oxygenated, it precipitates slightly soluble hydrous ferric oxide (Fe^{3+}-oxide) and oxyhydroxides, which are deposited in soils and sedimentary rocks. These oxides are responsible for the red and yellow color of sedimentary rocks and soils. Classic examples of this staining are the colorful red beds in the western United States and elsewhere and the Redwall Limestone in the Grand Canyon. The dark desert varnish on many exposed rocks in arid regions consists of clay particles mixed with manganese and iron oxides. Iron and manganese ions stain laundry and plumbing features, clog plumbing pipes, create a metallic taste and add color to drinking water, and can cause health issues.

In most natural waters, Fe^{2+} tends to be non-complexed and Fe^{3+} forms ion complexes, many of which are solid species. Common aqueous forms of iron in water containing sulfur and carbon are E_h and pH dependent. In Figures 6.13 and 6.14 it is evident that small changes in pH or E_h will results in great changes in iron solubility. What this means is that iron solubility can vary greatly along groundwater flow paths, and it is dependent on the flow path mineralogy and dissolved oxygen content. In highly acidic unconfined and confined aquifers iron is Fe^{2+}. In the typical pH range, aqueous iron occurs as Fe^{2+} in confined aquifers and precipitates as ferric hydroxide when the system becomes unconfined (i.e., oxygenated). This process is common at mine drainage sites and is observed as iron staining in toilet bowls and sinks in regions where the groundwater is derived from iron rich bedrock, such as volcanic terrain.

The steep slopes of the pH dependent boundary between Fe^{2+} and $Fe(OH)_3$ or $FeCO_3$ in Figures 6.13 and 6.14 explains why acid mine drainage (AMD) containing iron and other trace metallic elements is common in some eastern U.S. coal mines and coal handling facilities, and why it rarely occurs in western U.S. coal mines. The western U.S. coal deposits are commonly associated with sediments containing carbonate minerals. Dissolution of the carbonates consumes the H^+ liberated by iron oxidation and buffers the mine drainage causing the iron to precipitate. Such pH buffering occurs less frequently in eastern U.S. coal facilities. A similar pH buffering phenomenon occurs in some metallic mines. In hard rock mines the abundance of pH buffering minerals is

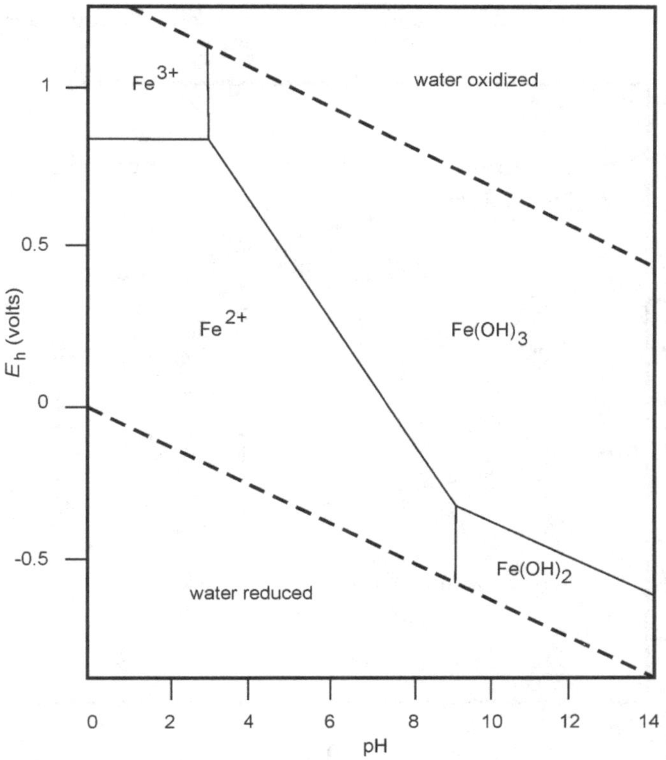

FIGURE 6.13 Pourbaix diagram of iron species in equilibrium relative to E_h and pH conditions at 25 °C.

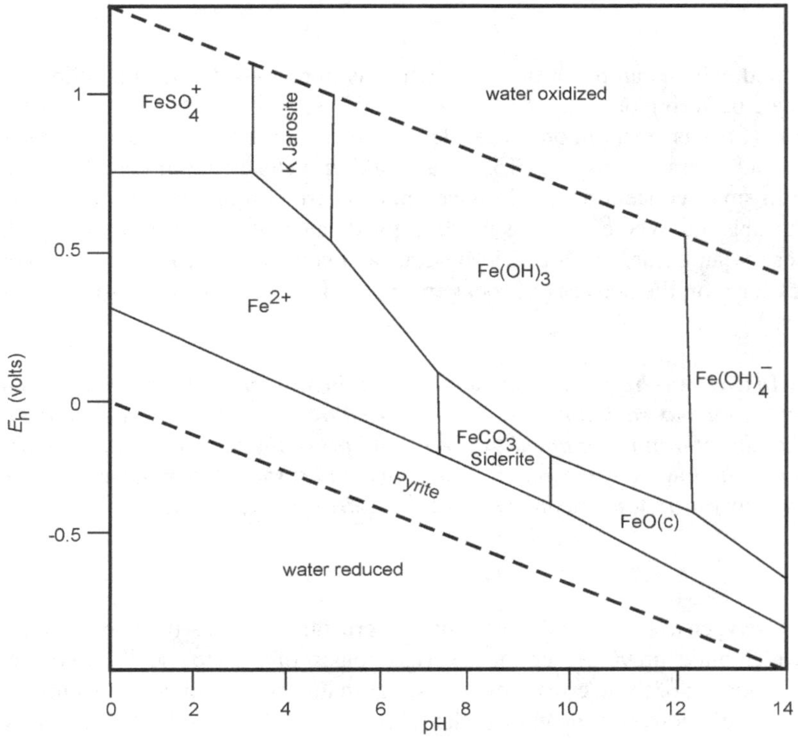

FIGURE 6.14 Pourbaix diagram of iron species in a solution containing sulfur and CO_2 in equilibrium relative to E_h and pH at 25 °C, dissolved CO_2=1,000 mg/L HCO_3^-, and SO_4^{2-} = 96 mg/L.

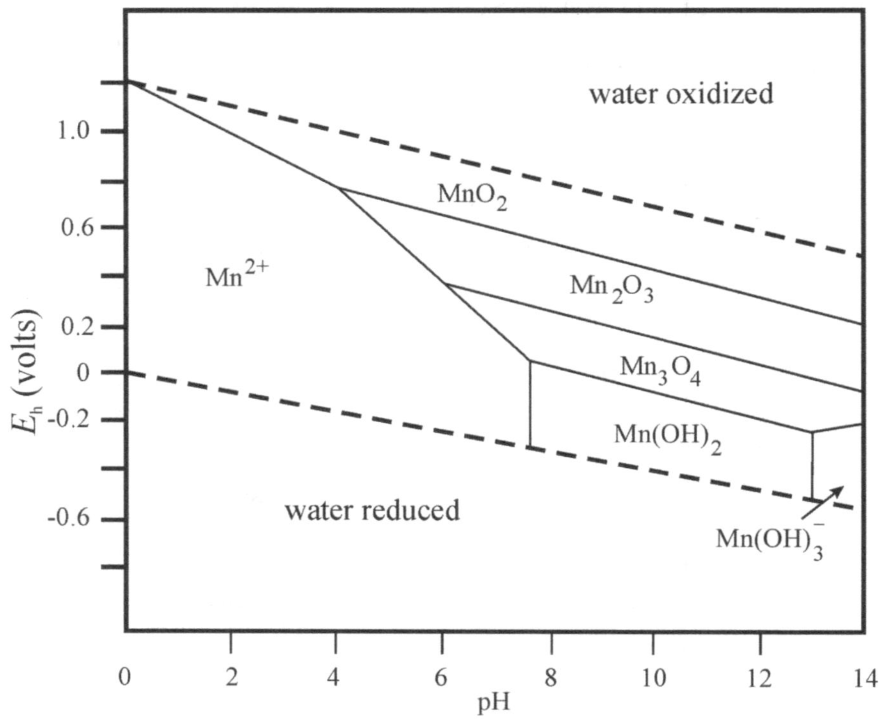

FIGURE 6.15 Pourbaix diagram of the behavior or manganese in aqueous solutions.

generally limited, which can result in AMD, whereas in many soft-rock metallic mining environments carbonate buffering occurs.

Manganese (Mn) is a common trace element that has a similar aquatic behavior as iron (Figure 6.15) and is associated with iron as a drinking water contaminant. Manganese is a neurotoxin even in small concentrations. Reduced manganese is found in igneous rocks, particularly biotite and hornblende, where it may substitute for divalent ions of similar size. Like iron, most soil manganese is an oxide, but it can also occur as manganese carbonate (rhodochrosite). Manganese oxides form on the surfaces of rocks in arid regions as well as deposits in fresh and ocean environments.

Important Takeaway – Iron is an abundant element in most crustal rocks, but it is only a trace element in aqueous environments. Once released from aluminosilicate and other minerals, the solubility of iron is regulated by the E_h and pH of the aqueous system. In acidic water and water with low oxygen contents, iron is in the Fe^{2+} state. When the water is oxygenated, Fe2+ readily precipitates as an iron oxide or hydroxide, carbonate, or sulfide.

6.6 SILICA

Silicon follows oxygen as the second most abundant crustal element. All igneous and most metamorphic rocks, which make up 90 percent of the crust, consist of silicate and aluminosilicate minerals. Despite the abundance of these crustal minerals, the shallow crust contains a significant proportion of carbonate and siliceous sedimentary rocks (Figure 6.1). The basic building block of all silicate and aluminosilicate minerals is the silica tetrahedron (Figure 6.16). The tetrahedron, consisting of a Si^{4+} atom that is surrounded by four closely packed oxygen atoms (O^{2-}), is chemically expressed

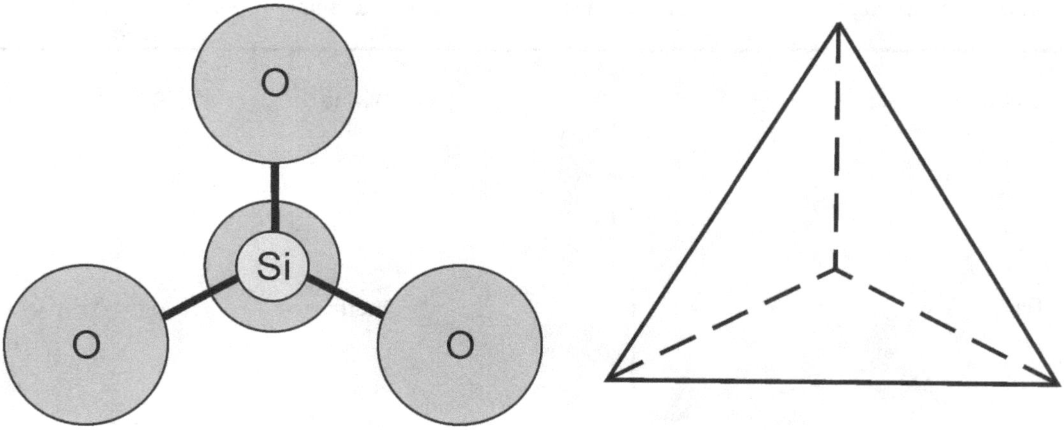

FIGURE 6.16 The silica tetrahedron structure consists of a silicon atom in the center surrounded by four oxygen atoms. Ions of Ca, Na, Mg, Fe, and Al can bond to the oxygen atoms in a variety of structural configurations (see Figure 6.17).

as SiO_4^{2-} and is commonly known as SiO_4. The silica–oxygen bond is strong, resulting in minerals that have very small equilibrium constants (Table 2.1), making them highly resistant to weathering. A detailed discussion of the silica tetrahedron is beyond the scope of this book, but a brief summary is helpful for understating silicate mineral dissolution. Each oxygen atom in the tetrahedron has the potential to bond with another Si, Ca, Mg, Na, K, Fe, or Al atom. Oxygen bonds joining one or more tetrahedrons to form complex molecules give rise to six different tetrahedron structural classes and groups of silicate minerals (Figure 6.17). Most igneous and metamorphic minerals have a single chain, double chain, sheet, or three dimensional tetrahedron linkage.

In his landmark paper, Goldich (1938) described the resistance of silicate and aluminosilicate minerals to weathering (Figure 6.18). The weathering scheme is generally the inverse of Bowen's reaction series for silicate melts and is known as the Goldich weathering series. If you recall from Geology 101, Bowen's reaction series describes the order of silicate and aluminosilicate mineral crystallization from a silicate melt. The weathering series makes intuitive sense in that the first minerals to crystalize from a melt are the most chemically stable at high temperatures and pressures (i.e., olivine, pyroxene, amphibole, and Ca-Na-plagioclase and will be least stable at the earth's surface. The weathering series explains why quartz is the dominant mineral in mature sandstone and in beach sand and why mafic silicate and aluminosilicate minerals rarely occur in soil and sedimentary rock. The least resistant minerals to weathering have a relatively small SiO_2 content and a large proportion of ionic bonds. The more resistant minerals have a large SiO_2 content and a large proportion of covalent bonds.

Most silicate minerals weather congruently, and most aluminosilicate minerals weather incongruently. Congruent weathering results in a net loss of solid material by releasing cations, HCO_3^-, and silicon in the form of H_4SiO_4 to aqueous solutions. Incongruent weathering not only releases ions but results in the buildup of clay in soil and the shallow subsurface environment. Recall that incongruent reactions are nonreversible. Because incongruent weathering products form at or near the earth's surface temperature and pressure, they are commonly as resistant to erosion as is quartz (Table 6.7). Thick B and C soil horizons often contain significant percentages of residual clay in addition to the parent bedrock minerals. Three processes, often operating simultaneously, contribute to the chemical weathering of silicate and aluminosilicate minerals. The processes are: 1) attack by weak acidic water (H^+), 2) hydrolysis resulting in clay mineral formation, and 3) oxidation of iron. Weak acidic water is largely the result of atmospheric and soil zone CO_2 (Equations 6.2 to 6.6).

Tetrahedron Linkage	Configuration	Mineral Example	Possible Elemental Constituents
Single		Olivine	Mg-Fe-SiO_4
Rings	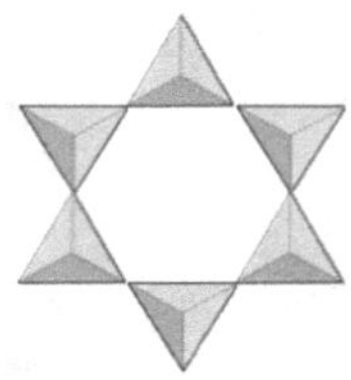	Tourmaline	Mg-Fe-Al-SiO_4
Single chain	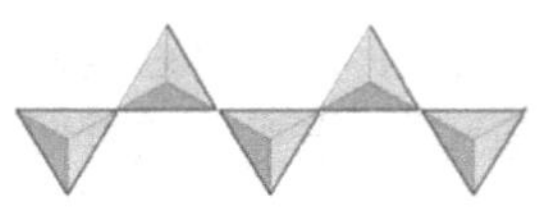	Pyroxene (augite)	Ca-Na-Mg-Fe-Al-SiO_4
Double chain	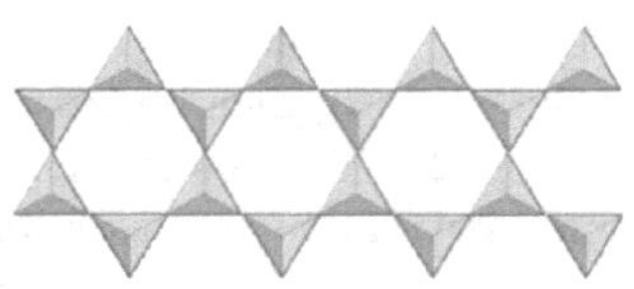	Amphibole (hornblende)	Ca-Na-Mg-Fe-Al-SiO_4
Sheet		Kaolinite	$AlSiO_4$
		Mica (muscovite)	K-Al-SiO_4
		Mica (biotite)	K-Mg-Fe-Al-SiO_4
3-dimensional	Framework (tectosilicate)	Feldspar (orthoclase)	K-Al-SiO_4
		Feldspar (plagioclase)	Na-Ca-Al SiO_4
		Quartz	SiO_4

FIGURE 6.17 Structural configurations of silica tetrahedron linkages and their associated mineral groups and elements that attach to the oxygen atoms.

Hydrolysis is the process where a water molecule, specifically the attached H^+ or OH^- ions, replace an element in the mineral, break the mineral's chemical bond, and form a new mineral, usually clay. Iron oxidation is an important factor in the dissolution of iron-bearing silicates. The congruent dissolution of forsterite (Mg-olivine) by a weak acid is an example of silicate mineral dissolution:

$$\underset{\text{(forsterite)}}{Mg_2SiO_4} + 4H^+ = 2Mg^{2+} + \underset{\text{(orthosilicic acid)}}{H_4SiO_4} \tag{6.59}$$

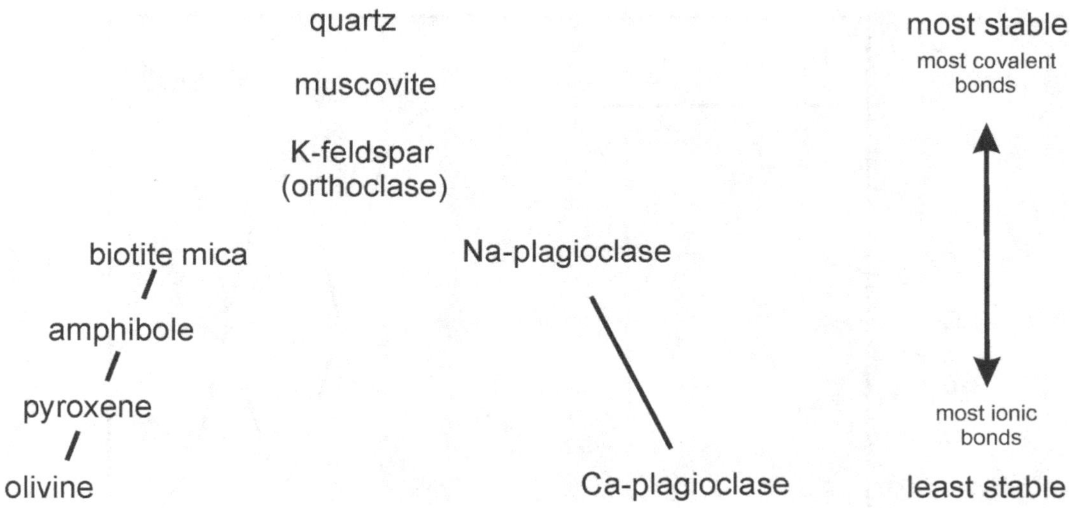

FIGURE 6.18 Goldrich weathering series (Goldich, 1938) is the reverse of Bowen's reaction series. Bowen's reaction series describes the order crystallization of silica and aluminosilicate minerals in a silicate melt. In addition to quartz, clay minerals and iron and aluminum oxides are also very resistant to weathering.

TABLE 6.7
Resistance of minerals to chemical weathering arranged from the most stable to the least stable. Minerals marked with an asterisk (*) usually crystalize from a silicate melt.

Iron and aluminum oxides (e.g., hematite Fe_2O_3 and gibbsite ($Al(OH)_3$; bauxite)
Quartz*
Clay minerals – products of incongruent dissolution
Muscovite*
K- and Na-feldspar*
Biotite*
Amphiboles*
Pyroxenes*
Ca-feldspar*
Carbonates
Sulfates
Olivine*
Chlorides

Forsterite can also dissolve incongruently by a weak acid and hydrolysis:

$$\underset{\text{(forsterite)}}{5MgSiO_4} + 8H^+ + 2H_2O \rightarrow \underset{\text{(serpentine)}}{Mg_6(OH)_8Si_4O_{10}} + H_4SiO_4 + 4Mg^{2+} \tag{6.60}$$

During the reactions in Equations 6.59 and 6.60, magnesium and orthosilicic acid are released to the solution. Orthosilicic acid, also known as silicic acid, is a polyprotic acid that can dissociate into $H_3SiO_4^-$ and $H_3SiO_4^{2-}$:

$$H_4SiO_4 = H^+ + H_3SiO_4^{2-} \tag{6.61}$$

$$H_3SiO_4 = H^+ + H_3SiO_4^{2-} \tag{6.62}$$

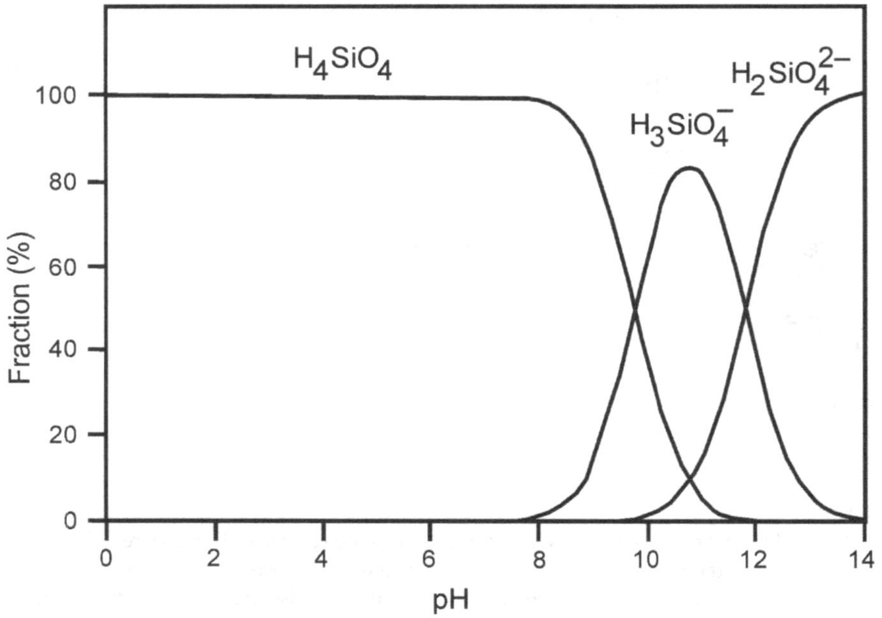

FIGURE 6.19 The pH dependence of polyprotic silicic acid at 25 °C. In the pH range of most groundwater, H_4SiO_4 is the dominant species. At pH ~ 9, H_4SiO_4 and $H_3SiO_4^-$ are in equal concentrations.

In the normal pH range of groundwater, H_4SiO_4 is the dominant species (Figure 6.19). However, at higher silica concentrations, silicic acid can precipitate as quartz and amorphous silica (Figure 6.20). The dissolution and precipitation of both quartz and amorphous silica are pH independent;

$$\underset{\text{(quartz)}}{SiO_2} + 2H_2O = H_4SiO_4 \quad (6.63)$$

At 25 °C and pH 7 the solubility of quartz and amorphous silica are 12 and 120 mg/L, respectively (Freeze and Cherry, 1979). Shallow groundwater is commonly supersaturated with respect to quartz, and deeper groundwater systems are commonly saturated. Supersaturation occurs because the rate of the forward reaction in Equation 6.63 is kinetically faster than the rate of the reverse reaction. Achieving equilibrium is dependent on the downstream flow distance and the flow velocity (De Caritat, 1990). Fast-flowing shallow aquifers and hydrothermal systems are most likely to have quartz disequilibrium. There is no clear kinetic theory that describes the dissolution of quartz (Crundwell, 2017). Although quartz and amorphous silica dissolution are pH independent, the rate of dissolution is both pH and temperature dependent (Figure 6.20). In acidic water the dissolution rate is constant but increases several orders of magnitude in basic solutions.

Common incongruent weathering processes of iron-bearing pyroxenes and amphiboles and feldspars are promoted by weak acids combined with hydrolysis and iron oxidation. For example, the weathering of iron-containing hornblende involves several intermediate steps. Weathering starts with acid attack and hydrolysis, which breaks the pyroxene molecule. This step is followed by the oxidization of iron, forming the solid phase ferric hydroxide. These processes can be summarized as a single reaction:

$$\underset{\text{(pyroxene)}}{4CaFeSi_2O_6} + 8H^+ + O_2 + 6H_2O \rightarrow 4Ca^{2+} + 4FeOOH + 8H_2SiO_3 \quad (6.64)$$

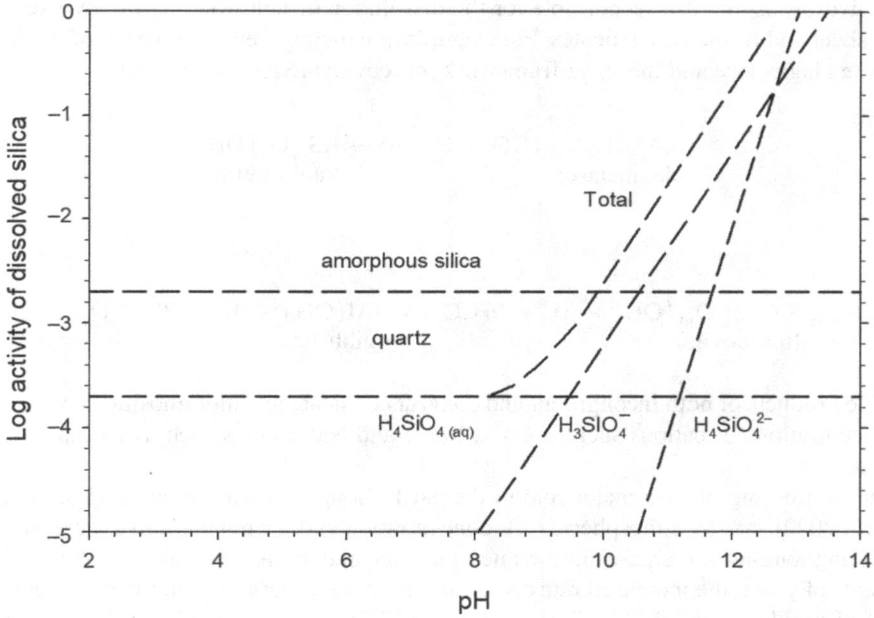

FIGURE 6.20 The equilibrium of silicic acid and the solid phases quartz and amorphous silica.

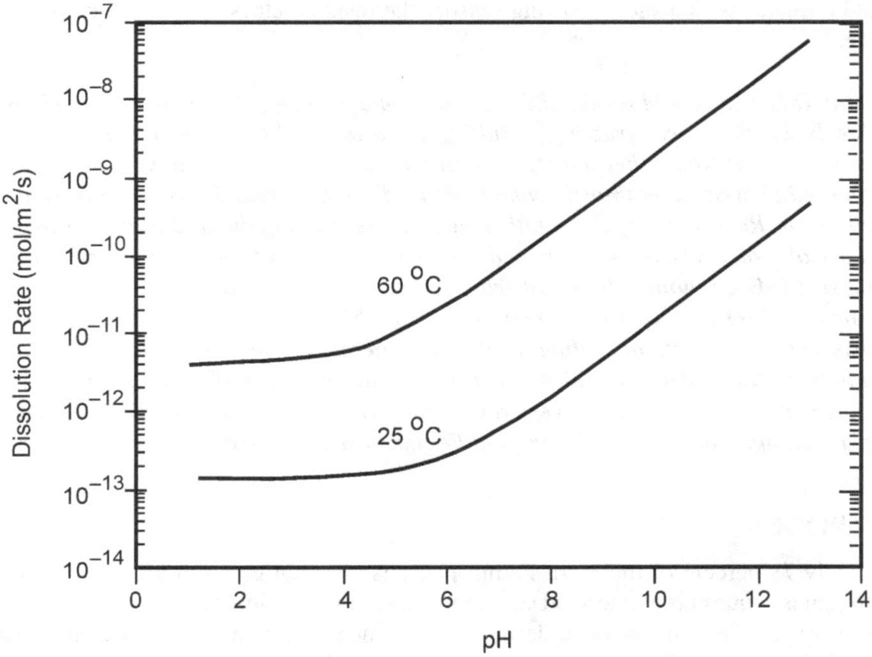

FIGURE 6.21 The rate of quartz dissolution at 25 °C and 60 °C as a function of pH (after Crundwell, 2017; https://pubs.acs.org/doi/10.1021/acsomega.7b00019).

Acid and hydrolysis are also responsible for the dissolution of non-iron-bearing pyroxene, amphibole, and sheet and framework silicates. For example, the incongruent weathering of the framework tectosilicate plagioclase and the sheet framework muscovite are represented as:

$$\underset{\text{(plagioclase)}}{CaAl_2Si_2O_8} + H_2O + H^+ \rightarrow + \underset{\text{(kaolinite)}}{Al_2Si_2O_5(OH)_4} \tag{6.65}$$

and

$$\underset{\text{(muscovite)}}{KAl_3Si_3O_{10}(OH)_2} + H^+ + 9H_2O \rightarrow \underset{\text{(gibbsite)}}{3Al(OH)_3} + K^+ + 3H_4SiO_4 \tag{6.66}$$

The soluble products of both incongruent and congruent silicate and aluminosilicate weathering are small concentrations of cations such as Ca^{2+}, Mg^{2+}, and Na^+, anions such as Cl^-, and dissociated silicic acid.

Silicate weathering plays a major role in the earth's long-term carbon cycle (Bufe et al., 2021; Deng et al., 2022). As the atmospheric CO_2 concentration and air temperatures increase, there are corresponding long-term increases in the rate of silicate and aluminosilicate dissolution. Over tens of thousands of years, the increased rates of aluminosilicate mineral weathering will slightly buffer the effect of rapid climate change. Temperature and CO_2 weathering relationships are especially sensitive to climatic zones. Deep weathering, often tens to hundreds of feet thick, in some granitoid bodies may be a testament to warmer and CO_2-rich paleoclimatic conditions. In warm tropical climates where organic soil zone CO_2 is high and groundwater flushing is great, incongruent weathering yields shallow soils that are depleted in Ca, Na, Fe, and Mg and enriched in Al and silica. Locally, this enrichment is great enough that the soil can be mined for aluminum ore (bauxite). In cold climates with a short growing season, the residual clays are often enriched in Ca, Na, Fe, and Mg.

Important Takeaways – Most silicate minerals weather congruently. Most aluminosilicate minerals weather incongruently, resulting in residual clay and/or an iron oxide solid. Except for quartz and other pure SiO_2 *minerals, silicate and aluminosilicate minerals are weathered by weak acid (mostly commonly* H^+ *from* CO_2 *reactions), hydrolysis, and iron oxidization. Regardless of the weathering process, the silicate and aluminosilicate mineral dissolution is extremely slow, and the concentrations of product ions in solutions are quite small. Even though the weathering is slow, over the course of millions of years the accumulated residual clay can be tens of meters thick.*

The resistance of silicate and aluminosilicate minerals to weathering is described by the Goldich weathering series, which essentially is the reverse of Bowen's reaction series for silicate melts – where quartz is very resistant to weathering and low silica-bearing pyroxenes and amphiboles and feldspars and Fe-mica are less resistant.

6.7 NITROGEN

Approximately 78 percent of the earth's atmosphere is nitrogen gas, and 97.8 percent of the total global nitrogen is sequestered in terrestrial rocks that are inaccessible to most groundwater systems. Except for saltpeter deposits in some desert regions, most sedimentary rocks contain such small quantities of nitrogen that their chemical forms and isotopic compositions are virtually unknown. The nitrogen cycle is largely biochemical, involving physical and anthropogenic processes in a complex series of interactions in the atmosphere, biosphere, lithosphere, and hydrosphere (Figure 6.22). Mobile nitrogen occurs as nitrogen gas (N_2), ammonia (NH_3), ammonium (NH_4^+), nitrite (NO_2^-),

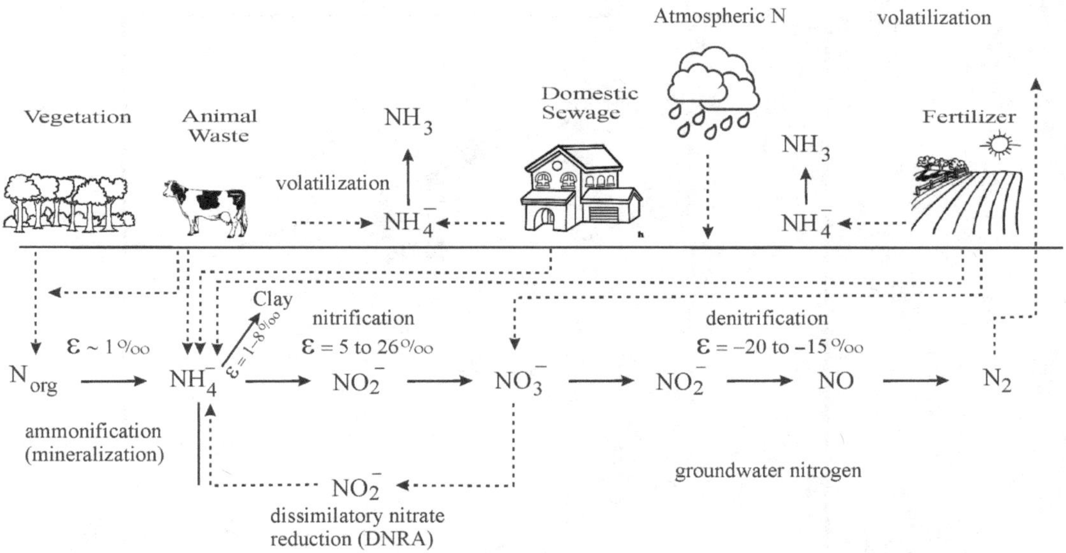

FIGURE 6.22 The terrestrial nitrogen cycle. Groundwater acquires nitrogen from both natural and anthropogenic sources. Aqueous nitrogen undergoes a series of transformations with N_2 gas as the final product.

nitrate $\left(NO_3^-\right)$, nitrous oxide $\left(N_2O\right)$, and nitric oxide (NO). These species are transformed during redox processes (Figure 6.23) including nitrogen fixation, ammonification (a.k.a. mineralization), nitrification, and denitrification. Nitrogen transformation may be either inorganic (equilibrium based) or biologic (kinetic based).

Nitrogen fixation is the reduction of relatively inert $N_{2(g)}$ into reactive species such as ammonia, nitrite, and nitrate. Natural fixation during lighting and by ultraviolet rays produce atmospheric nitric oxide, but most natural fixation is by free-living (non-symbiotic) bacteria such as cyanobacteria (blue green algae) and soil zone symbiotic bacteria. Anthropogenic fixation accompanying fossil fuel emissions releases NO_x to the atmosphere. Modern agriculture relies heavily on nitrate and ammonia fertilizers produced from atmospheric gas. Nitrogen fixation by lightning, industry, and biological nitrogen fixation (BNF) can be represented as:

$$N + O_2 \rightarrow 2NO \text{ (lightning)} \tag{6.67}$$

$$N_2 + 3H_2 \rightarrow 2NO_3 \text{ (anthropogenic)} \tag{6.68}$$

$$N_2 + 8H^+ + 8e \rightarrow 2NH_3 + H_2 \left(\text{BNF}\right) \tag{6.69}$$

Nitrification is the oxidization of reduced nitrogen, mostly ammonia $\left(NH_3\right)$ and ammonium $\left(NH_4^+\right)$, into nitrite and nitrate:

$$NH_3 + O_2 \rightarrow NO_2^- + 3H^+ + 2e \tag{6.70}$$

$$NO_2^- + H_2O \rightarrow NO_3^- + 2H^+ + 2e \tag{6.71}$$

$$2NH_4^+ + 3O_2 \rightarrow 2NO_2^- + 2H_2O + 4H^+ \tag{6.72}$$

$$2NO_2^- + O_2 \rightarrow 2NO_3^- \tag{6.73}$$

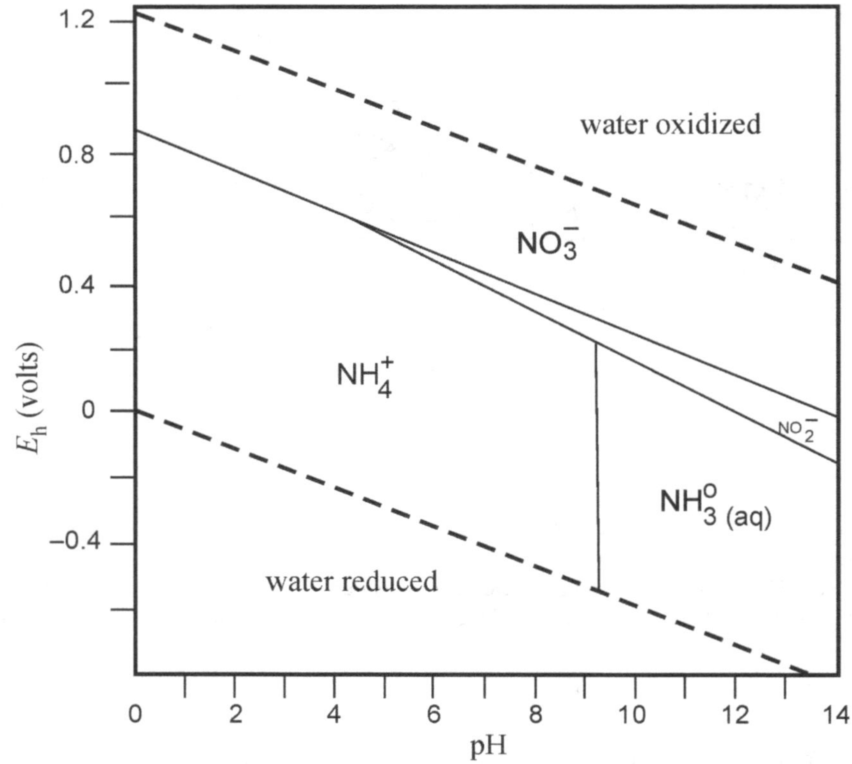

FIGURE 6.23 Pourbaix diagram of the dominant nitrogen aqueous species as a function of E_h and pH at nitrogen activity of 10^{-2} molar and 25 °C.

Ammonification, sometimes known as mineralization, produces ammonia and ammonium during the biologic breakdown of organic matter. Both aerobic and anaerobic decomposers such as bacteria and fungi participate in ammonification. Ammonification can be represented as a two-step process:

$$R-NH_2 + H_2O \rightarrow NH_3 + R-OH \quad (6.74)$$

$$NH_3 + H_2O \rightarrow NH_4^+ + OH^- \quad (6.75)$$

which is sometime shortened to:

$$RNH_2 + 2H_2O \rightarrow NH_4^+ + ROH \quad (6.76)$$

where R is an alkyl group. Once ammonium is converted to nitrite and then to nitrate by nitrification, nitrogen is available for uptake by plants and may become dissolved and transported by surface and groundwater. Soil zone and aqueous nitrate can be converted to $N_{2(g)}$ by denitrification. Most denitrification occurs in anoxic environments where the nitrate or nitrite is used as a terminal electron acceptor instead of free oxygen. Denitrification involves a series of steps:

$$NO_3^- + 2e + 2H^+ \rightarrow NO_2^- + H_2O \quad (6.77)$$

$$NO_2^- + e + 2H^+ \rightarrow NO + H_2O \quad (6.78)$$

$$2NO + 2e + 2H^+ \rightarrow N_2O + H_2O \tag{6.79}$$

$$N_2O + 2e + 2H^+ \rightarrow N_2 + H_2O \tag{6.80}$$

which can be summarized as:

$$2NO_3^- + 7e + 8H^+ \rightarrow N_2 + 4H_2O \tag{6.81}$$

Nitrogen concentrations in surface and groundwater are reported as nitrate-N, combined nitrate/nitrite, and total N. Nitrate, the most common form of freshwater nitrogen, typically has a natural groundwater concentration that is <2 mg/L and a natural stream and lake concentration that is <4 mg/L (Mueller et al., 1995). The natural concentration of aqueous NH_3 is generally <0.1 mg/L. Investigations of aqueous nitrogen often involve the use of the nitrogen isotope ^{15}N (see Chapter 10).

Nitrogen is the most widespread groundwater contaminant. The U.S. Environmental Protection Agency has set the U.S. nitrate drinking water standard as 10 mg/L, whereas the European Union standard is 50 mg/L for nitrate and 0.5 mg/L for nitrite. The World Health Organization (WHO) has set 50 mg/L as a recommended maximum. Nitrogen toxicity results from the body's natural breakdown of nitrate into nitrite results in blue baby syndrome (methemoglobinemia), which threatens the oxygen-carrying capacity of the blood. The primary sources of nitrogen pollution are agriculture, industry, sewage, septic tanks, and landfills. For example, the average annual nitrogen flux to the Gulf of Mexico from the heavily agricultural Mississippi River Basin (MRB) increased from about 243 Gt/yr in the 1900s to about 378.5 Gt/yr in the early 2000s. The basin drains about 41 percent of the total contiguous United States. About 62 percent is nitrate-N and 38 percent is organic nitrogen. The nitrogen is primarily agricultural sourced, with 41, 26, 10, 9, 7, and 7 percent from farm fertilizer, atmospheric deposition, animal manure, legume crops, urban runoff, and wastewater treatment, respectively (Goolsby and Battaglin, 2001; Tian et al., 2020).

Important Takeaway – The nitrogen cycle is largely biochemical, involving physical and anthropogenic processes in a complex series of interactions in the atmosphere, biosphere, lithosphere, and hydrosphere. Most groundwater nitrogen is in the form of nitrate. Anthropogenic nitrogen from agricultural sources and the burning of fossil fuels is a significant groundwater contaminant in many worldwide locations.

6.8 NOBLE GAS

The atmosphere is the primary terrestrial reservoir of the noble gases helium (He), neon (Ne), argon (Ar), krypton (Kr), and xenon (Xe). Except for argon, their atmospheric abundances are quite small (Table 6.8). The table also includes the isotopes of potential use in groundwater studies. Noble gases have been of hydrogeologic interest since the 1970s (Herzberg and Mazor, 1979; Mazor, 1972). Since then, considerable progress has been made in using noble gases as aqueous geochemical tracers for dating groundwater based on the accumulation of ^3He (Chapter 11) and for calculating groundwater recharge temperatures using their temperature-dependent solubility. Knowledge of recharge temperature can be correlated with recharge elevation in mountainous terrain, seasonal recharge, and paleoclimate recharge (Aeschbach-Hertig et al., 1999; Manning and Solomon, 2003). Noble gas concentrations in groundwater are derived from three sources: 1) dissolved atmospheric gas in equilibrium with the recharge water, 2) isotopes from radiogenic decay, particularly ^4He, and 3) excess air, which may be soil gas air bubbles trapped in soil pores (Aeschbach-Hertig et al., 2000; Stute and Schlosser, 1993).

The concentration of dissolved noble gas in fresh water can be described as (Manning and Solomon, 2003):

$$C_i = C_i^E + C_i^A \tag{6.82}$$

TABLE 6.8
Atmospheric Noble Gas Concentrations and a Summary of Their Isotopic Compositions

	Atmospheric mole fraction				
Noble gas	ppm	%	Stable isotopes	Radiogenic isotopes	Isotopes of potential hydrologic interest
Ar	9,340	0.934	3	23	^{36}Ar, ^{38}Ar, ^{40}Ar
Ne	18.18	1.12×10^{-3}	3	17	–
He	5.24	5.24×10^{-4}	2	7	^{3}He, ^{4}He
Kr	1.14	1.14×10^{-4}	5	29	^{81}Kr
Xe	0.0087	8.7×10^{-6}	7	2	^{129}Xe, ^{136}Xe

Source: Data from Ozima and Podosek, 2002; Stute and Schlosser, 1993.

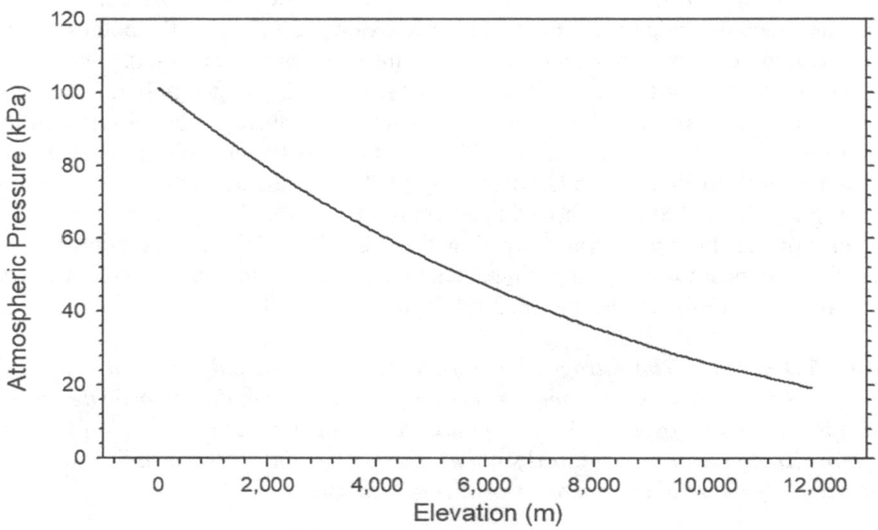

FIGURE 6.24 Changes in atmospheric pressure as a function of elevation.

Where C_i is the concentration of noble gas i in fresh water, C_i^E is the noble gas equilibrium component, and C_i^A is the excess air component. The equilibrium component of dissolved noble gases in groundwater can be represented by a modified form of Henry's law (Stute and Schlosser, 1993):

$$C_x = \beta(T,S)\frac{P_x}{P_o} \tag{6.83}$$

Where C_x is the concentration of the gas, β is the Bunsen coefficient, which is a function of temperature (T) and water salinity (S), P_x is the partial pressure of the gas at the water table, and P_o is the atmospheric pressure at sea level (101.25 kPa). The decrease in atmospheric pressure with increasing elevation is illustrated in Figure 6.24 and can be calculated by:

$$P = P_o\left(\frac{288.16}{288.16 - 0.0065H}\right)^{-5.255877} \tag{6.84}$$

where H is the elevation in meters, P_o is the standard atmospheric sea level pressure, 288.16 is standard sea level temperate in degrees kelvin, and −5.258877 is a constant that includes the gas law constant and other factors. There are numerous online elevation-atmospheric pressure calculators.

The effects of temperature and salinity on gas solubility are species dependent. Xe, Kr, and Ar are considerably more sensitive to temperature than are He and Ne (Figure 6.25). What this means is that the altitude effect on He and Ne is minimal and is quite large on Xe and Kr. In most groundwater systems the recharge zone temperature can be approximated by the mean annual air temperature. All of the gases have a decreasing solubility with increasing salinity (Smith and Kennedy, 1983). The salting out effect is the opposite of the temperature effect, with He being most affected and the effect on Xe being almost negligible.

The key to using noble gas concentration as a surrogate for recharge elevation or in climate studies is the fact that in groundwater systems the partial pressure of noble gases are established at the unsaturated–saturated interface, where the relative humidity is about 100 percent. Unlike atmospheric gases, such as CO_2, the partial pressure of noble gases remains nearly constant in closed systems such as confined aquifers and in ice, provided the flow path does not leak to other aquifers and does not mix with other groundwaters. Such mixing is common at many springs, where mixed local and regional groundwater systems discharge.

In deep groundwater systems the concentration of ^4He from the radiogenic decay of U and Th is often considerably greater than the concentration of atmospheric ^3He, which makes ^4He not useful for paleo-temperature studies. The $^3He/^4He$ ratio can be used for evaluating mantle gas contributions to groundwater (Ballentine and Burnard, 2002; Craig et al., 1978; Crossey et al., 2009). The concentration of ^{40}Ar from the decay of ^{40}K is usually considerably less than the contribution from the atmosphere and does not affect the use of Ar.

Excess air is most easily detected by analyzing the Ne concentration and is often reported as the Ne excess relative to equilibrium Ne. Several diffusion and equilibrium based mathematical models have been developed to account for excess air. The models include: 1) the PR-model, which is based on diffusion-controlled fractionation of the excess gas (Aeschbach-Hertig et al., 2000; Peeters et al., 2003), 2) the PD-model, which is a diffusion-controlled model that assumes only partial degassing (Lippmann et al., 2003), 3) the UA-model, which assumes the trapped air completely dissolves and the resulting dissolve gas is unfractionated (Stute and Schlosser, 1993), and 4) the CE-model, which is a closed system equilibrium model that assumes that trapped air bubbles in the soil and upper part of the aquifer will equilibrate with the groundwater (Aeschbach-Hertig et al., 2000). The most applicable model for groundwater noble gas analysis is the CE-model (Aeschbach-Hertig et al., 2008). In the simplest form, the CE-model can be expressed as:

$$C_{\text{iw}}^{\text{eq}} = \frac{1+H_i}{1+BH_i} \tag{6.85}$$

Where $C_{\text{iw}}^{\text{eq}}$ is the equilibrium concentration of the noble gas in the water phase, H is the Henry constant (dimensionless), and B is the ratio of final trapped bubbles to water volume (dimensionless). Because of the complexity of the analysis, particularly when several gases are involved, computer codes such as PANGA (Jung and Aeschbach, 2018) are commonly used to analyze noble gas data.

Important Takeaway – Although the concentrations of noble gases in groundwater are small, they can be used in unmixed groundwater systems to estimate the mean annual air temperature at the recharge area, which can be correlated with the recharge elevation in mountainous terrain.

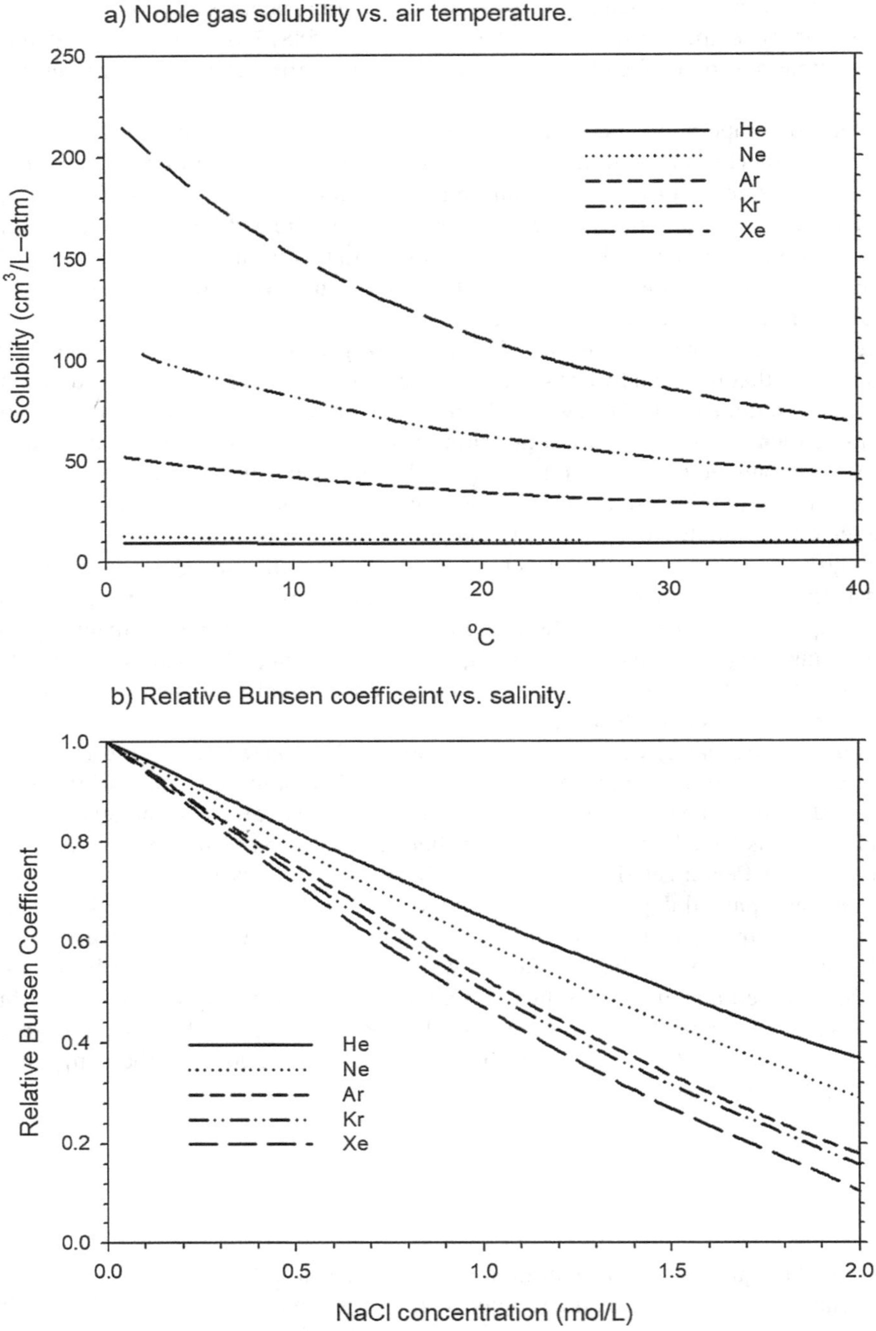

FIGURE 6.25 The solubility of noble gases in water as a function of a) air temperature, and b) salinity (data from IAPWS, 2004).

6.9 MINOR CONSTITUENTS

Minor constituents in aqueous systems include metals, nonmetallic elements, and radioactive elements. Trace constituents are uncommon in most water and when present usually occur in low concentrations. Low concentrations are the result of the dissolution of relatively rare minerals, the low solubility of many trace and aluminosilicate minerals (Table 6.9), and the interactions of trace elements with other aqueous constituents. Aqueous trace constituents are from both natural reactions and anthropogenic activities. The metals may be organized into sub-groups representing their location on the periodic chart of the elements: alkali metals, alkaline earth metals, and transition

TABLE 6.9
Dissolution Reactions and Equilibrium Constants of Selected Trace Elements in Aqueous Solutions

Mineral		Reaction	log K	Source
Alkali metal				
Li	$Li_2CO_{3(s)}$	$Li_2CO_{3(s)} = 2Li+ + CO_3^{2-}$	−2.71	1
	Spodumene	$LiAl(Si_2O)_{6(s)} + 2H_2O + 4H^+ = Li^+ + Al^{3+} + 2H_4SiO_4$	10.2	1
	$LiF_{(S)}$	$LiF_{(S)} = Li^+ + F^-$	−2.76	1
Sr	Strontianite	$SrCO_{3(s)} = Sr^{2+} + CO_3^{2-}$	−9.03 to −9.28	1
Alkaline earth metals				
Ba	Witherite	$BaCO_{3(s)} = Ba^{2+} + CO_3^{2-}$	−8.29 to −8.74	1, 3
	Barite	$BaSO_{4(s)} = Ba^{2+} + SO_4^{2-}$	−8.84 to −10.3	1, 3
Be	Bromellite	$BeO_{(s)} + H_2O = Be^{2+} + 2OH^-$	−19.7 to −25.9	1
	$BeCO3_{(s)}$	$BeCO_{3(s)} = Be^{2+} + CO_3^{2-}$	−6.55 to −6.88	1
	Chrysoberyl	$Be(AlO_2)_{2(s)} + 8H^+ = Be^{2+} + 2Al^{3+} + 4H_2O$	21.1 to 27.0	1
		$Be^{2+} + H_2O = BeOH^+ + H^+$	−5.4	4
Sr	Strontianite	$SrCO_{3(s)} = Sr^{2+} + CO_3^{2-}$	−9.03 to −9.28	1
Transition metals				
Ag	Chlorargyrite	$AgCl_{(s)} = Ag^+ + Cl^-$	−9.75	1
Co	Transvaalite	$Co(OH)_{2(s)} = CO^{2+} + 2OH^-$	−14.9 to −15.4	1
	Sphaerocobaltite	$CoCO_3$	10^{-10}	
	Cattierite	CoS_2		
Cr	Eskolaite	$Cr_2O_{3(s)} + 2H^+ = 2Cr^{3+} + 6OH^-$	6.72 to 7.35	1
	$Cr(OH)_{3(s)}$	$Cr(OH)_{3(s)} = Cr^{3+} + 2OH^-$	20 to 37.4	2
	$CrO_{3(s)}$	$CrO_{3(s)} + H_2O = Cr_2O_7^{2-} + 2H^+$	7.13	2
Cu	Chalcocite	$Cu_2S_{(s)} + H^+ = 2Cu^+ + HS^-$	−33.8 to −35.6	1, 3
	Cuprite	$Cu_2O_{(s)} + H_2O = 2Cu^+ + 2H^+$	−1.55.	3
	Chalcopyrite	$CuFeS_{2(s)} + 2H^+ = CU^{2+} + Fe^{2+} + 2HS^-$	−30.8 to −35.3	1
Mn	Rhodochrosite	$MnCO_3 = Mn^{2+} + CO_3^{2-}$	−9.3 to −11.4	1, 3
	Pyrolusite	$MnO_{2(s)} + 4H^+ + 2e = Mn^{2+} + H_2O$	41.3	3
Mo	Molybdite	$MoO_{3(s)} + H_2O = MoO_4^{2-} + 2H^+$	−10.7 to −12.1	1
Pb	Cerussite	$PbCO_{3(s)} + Pb^{2+} + CO_3^{2-}$	−12.8 to −13.5	1, 3
	Galena	$PbS(s) + H^+ = Pb^{2+} + HS^-$	−12.782	3
U	Uraninite	$UO_{2(s)} + 2H_2O = U^{4+} + 4OH^-$	−52.2 to −60.7	3
Zn	Sphalerite	$ZnS_{(s)} + H^+ = Zn^{2+} + HS^-$	−11.2 to −12	1, 3

(Continued)

TABLE 6.9 *(Continued)*
Dissolution Reactions and Equilibrium Constants of Selected Trace Elements in Aqueous Solutions

Metals				
Al	Gibbsite	$AlOH_{3(s)} + 3H^+ = Al^{3+} + 3H_2O$	7.54	3
	Boehmite	$ALOOH + 3H^+ Al^{3+} + 2H_2O$	8.58	3
	Kaolinite	$Al_2SI_2O_5(OH)_4 + 6H^+ = 2Al^{3+} + 2H_4SiO_4 + H_2O$	7.44	3
Bi	Bismite	$Bi_2O_{3(s)} + 3H_2O = 2Bi^{3+} = 6OH^-$	−68.1 to −78	1
Cd	Otavite	$CdCO_{3(s)} = Cd^{2+} + CO_3^{2-}$	−13.77	3
Hg	Cinnabar	$HgS_{(s)} + H^+ = Hg^{2+} + HS^-$	−39.1 to −39.8	1
Pb	Cerussite	$PbCO_{3(s)} + Pb^{2+} + CO_3^{2-}$	−12.8 to −13.5	1, 3
	Galena	$PbS(s) + H^+ = Pb^{2+} + HS^-$	−12.782	3
Sb	Stibnite	$Sb_2S_{3(s)} + 6H_2O = 2Sb\ OH_3^- + 3HS^- + 3H^+$	−54.4 to −60.2	1
Sn	Cassiterite	$SnO_{2(s)} + 2H_2O = Sn^{4+} + 4OH^-$	−61 to −64.4	
Non-metalic				
As	Orpiment	$As_2S_{3(s)} + 3H_2O = 2AsO_3^{2-} + 6H^+$	−60.97	3
	Realgar	$AsS = 3H_2O = H_3AsO_3^{0-} + HS^- + 2H^+ + e$	−19.75	3
	Arsenopyrite	$FeAsS + 7H_2O = Fe^{2+} + H_3AsO_{3(aq)} + 11e + SO_4^{2-}$		
B	$BF_{3(s)}$	$BF_{3(s)} + 3H_2O = B(OH)_3 + 3H^+ + 3F^-$	−3.78	1
	Borax	$Na_2(B_4O_5(OH)_48H_2O = 2Na^+ + 4B(OH)_4^- + 2H^+ + H_2O$	−25.48	2
	Colemanite	$Ca_2(B_3O_4(OH)_3H_2O + 4H_2O = Ca^{2+} + 3B(OH)_4^- + H^+$	−17.54	2
Se	Selenite	$SeO_{2(s)} + H_2O = SeO_3^{2-} + 2H^+$	−6.49 to −7.87	1

Some aqueous trace elements, such as aluminum, are abundant in crustal rocks. Equilibrium constants are for 25 °C.

Sources:

1 Van't Dack et al., 1983
2 Parks, 2005
3 Drever, 1997
4 Brown and Ekberg, 2016

metals. Alkali metals are in the first row of the periodic table, alkaline earth metals are in the second row, and transition metals are in rows 3–12. Alkali metals are shiny, soft, and highly reactive. They readily lose an outermost electron and form cations with a +1 electrical charge. Alkali earth metals are shiny and reactive and form cations with a +2 oxidization state. Transition metals are lustrous and hard and are good conductors of electricity. They form compounds with two or more oxidization states. They bind to ligands and form complex ions.

6.9.1 Alkali Metals

The common alkali metals Na and K are found in almost every aqueous system. Less common alkali metals include lithium (Li), rubidium (Rb), and cesium (Cs). The demand for lithium batteries has made lithium-bearing minerals one of the most sought-after minerals and has a lead to the worldwide search for and development of lithium ore deposits. Native lithium occurs in both liquid and solid forms and is found in pegmatite minerals, brine deposits, saline playas and salt flats, geothermal brines, and lithium-rich clay. Lithium pegmatite minerals include spodumene ($LiAlSi_2O_6$), lepidolite, and petalite ($LiAlSi_4O_{10}$). In aqueous solutions lithium primarily exists as Li^+. Li^+ reacts with water, forming lithium hydroxide ($LiOH_{(aq)}$) and lithium carbonate (Li_2CO_3):

$$Li^+ + O^- = LiOH_{(aq)} \tag{6.86}$$

$$2Li^+ + CO_3^{2-} = Li_2CO_3 \tag{6.87}$$

Other lithium salts include LiF, LiCl, and LiBr.

Although the crustal abundance of rubidium is greater than the more common metals such as copper, lead, and zinc, it is not a primary element in any mineral (Butterman and Reese, 2003). It is often a trace constituent in minerals such as lepidolite [$(K(Li,Al)_3(Al,Si,Rb)_4O_{10}(F,OH)_2$], pollucite ($Cs_2Al_2Si_4O_{12}$), carnallite, and leucite, where it substitutes for potassium. Pollucite, a zeolite mineral with rubidium as a substituting element, is locally a significant Rb ore mineral. In the aqueous systems, rubidium is usually in the form Rb^+ and as rubidium hydroxide (RbOH). Carnallite is a K-Mg-Cl evaporite mineral. Leucite is a K-feldspar where rubidium is a substituting element.

Cesium is a rare element that is 30 times less abundant than rubidium (Butterman et al., 2004). Cesium occurs only in a few minerals, such as avogadrite [$(KCs)BF_4$], some beryls, rhodizite, and pollucite. Two-thirds of the known reserves of the economically available cesium occurs in pollucite and is found in Canada. About 20 percent is in Zimbabwe. In water, cesium is usually in the forms Cs^+ and is readily transported. Non-radioactive cesium is of little concern in most groundwater systems. Locally, radioactive ^{137}Cs is a persistent groundwater contaminant due to past nuclear weapons testing and nuclear reactor accidents. Cesium is preferentially sorbed onto clay because it has small hydration energy.

6.9.2 Alkali Earth Metals

The alkali earth metals Mg and Ca are ubiquitous in aqueous solutions. The less common aqueous elements include beryllium (Be), strontium (Sr), barium (Ba), and the radiogenic element radium (Ra). Radium is constantly being replenished by the decay of uranium and thorium. Beryllium has a crustal abundance similar to cesium. There are about 40 minerals that contain beryllium, but the most common beryllium mineral is the silicate beryl ($Be_3Al_2Si_6O_{16}$), which is found in granites, pegmatites, and in mica schist. Other beryllium-bearing minerals include bertrandite, bromellite, chrysoberyl, and beryllonite. Trace amounts are also found in coal. Beryllium-containing minerals are very insoluble, and most beryllium is released to the environment by atmospheric emissions from burning fossil fuel and industrial waste streams such as smelting and mining ores. Leaching of plagioclase, some micas, and beryllium hydroxide also releases free Be^{2+} to the aqueous environment. Atmospheric beryllium accumulates in the soil via wet and dry fall. Acidic water supports free Be^{2+}, but in the pH range of most water, beryllium is in the form beryllium hydroxide. At elevated pH beryllium mobility increases due to the formation of Be-hydroxide complexes (Boland et al., 2023). Hydrolysis of beryllium with water molecules can results in numerous Be-hydroxides, where the simplest form can be represented as:

$$Be^{2+} + H_2O = BeOH^+ + H^+ \tag{6.88}$$

Strontium has chemical properties similar to calcium and potassium and frequently substitutes in small quantities for these elements in Ca- and K-aluminosilicate minerals. In sedimentary environments, dissolution of the soluble strontium carbonate and sulfate minerals strontianite ($SrCO_3$) and celestine ($SrSO_4$) can release appreciable amounts of Sr^{2+} (Table 2.1). The Sr to Ca ratio in most limestone is about 1:1,000, which means that most water flowing through carbonate rocks contains some Sr^{2+}.

Barium (Ba) occurs in more than 200 minerals including sulfides, halide, oxides, carbonates, sulfates, phosphates, and silicates. The concentration of Ba^{2+} in most natural waters is controlled by the dissolution of the relatively common sulfate mineral barite ($BaSO_4$):

$$BaSO_4 = Ba^{2+} + SO_4^{2-} \tag{6.89}$$

The median concentration of Ba^{2+} in U.S. public water supplies and rivers is about 0.45 mg/L (Hem, 1985). Barite forms when Ba-rich fluids flow through bedded sedimentary rocks. It forms in evaporite deposits and hydrothermal veins, as hot spring deposits, and as a replacement mineral.

6.9.3 Transition Metals

The transition metals include elements 21 to 28, 39 to 46, and 72 to 78 on the periodic table. The numbers refer to the number of protons in each element. Most of the transition metals are rare in surface and groundwater and will not be discussed here. Transition metals have more than one oxidation state and are stable in water. In addition to the common transition metals iron and manganese, aqueous solutions can contain chromium (Cr), copper (Cu), cobalt (Co), mercury (Hg), lead (Pb), and zinc (Zn).

Chromium is a lithophile element, meaning it has an affinity for silicate minerals rather than as metallic chromium. It is found in ultramafic and mafic rocks, particularly as the minerals peridotite and chromite ($FeCr_2O_4$). Chromium is also transported in hydrothermal fluids. Ultramafic and mafic rocks have elevated concentrations of Fe and Mn. Chromium can exist in +2, +3, +5, and +6 oxidization states. In most geologic environments Cr^{3+} is the stable form, but in aqueous environments it readily oxidizes to the toxic Cr^{6+} (Figure 6.26). The most common chromium mineral is chromite,

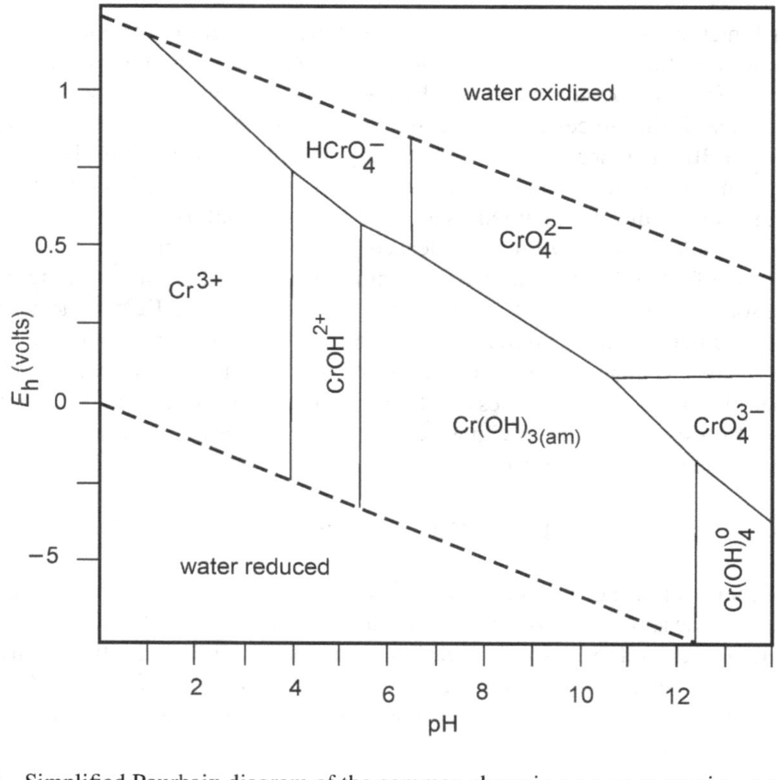

FIGURE 6.26 Simplified Pourbaix diagram of the common chromium aqueous species as a function of E_h and pH at total chromium of 10×10^{-5} molar and 25 °C.

which is highly resistant to weathering. Most chromium in aqueous solutions has an anthropogenic source. Mining and industrial uses of chromium result in an annual release of 75,000 tons of Cr (Hua et al., 2005). Perhaps the best-known example of chromium groundwater contamination is the class action lawsuit *Anderson, et al. v. Pacific Gas & Electric* (PG&E) and several subsequent class action filings. The case is better known as the Erin Brockovich case after the movie of the same title. Between 1952 and 1966, PG&E disposed of 370 million gallons of wastewater at its Hinkley, California, natural gas compressor station. The water, which contained chromium to help prevent scaling in the evaporation cooling towers, was released into unlined evaporation ponds when the TDS in the cooling tower water became too high. The wastewater in the ponds percolated to the underlying shallow alluvial aquifer and contaminated numerous down-gradient wells (Figure 6.27).

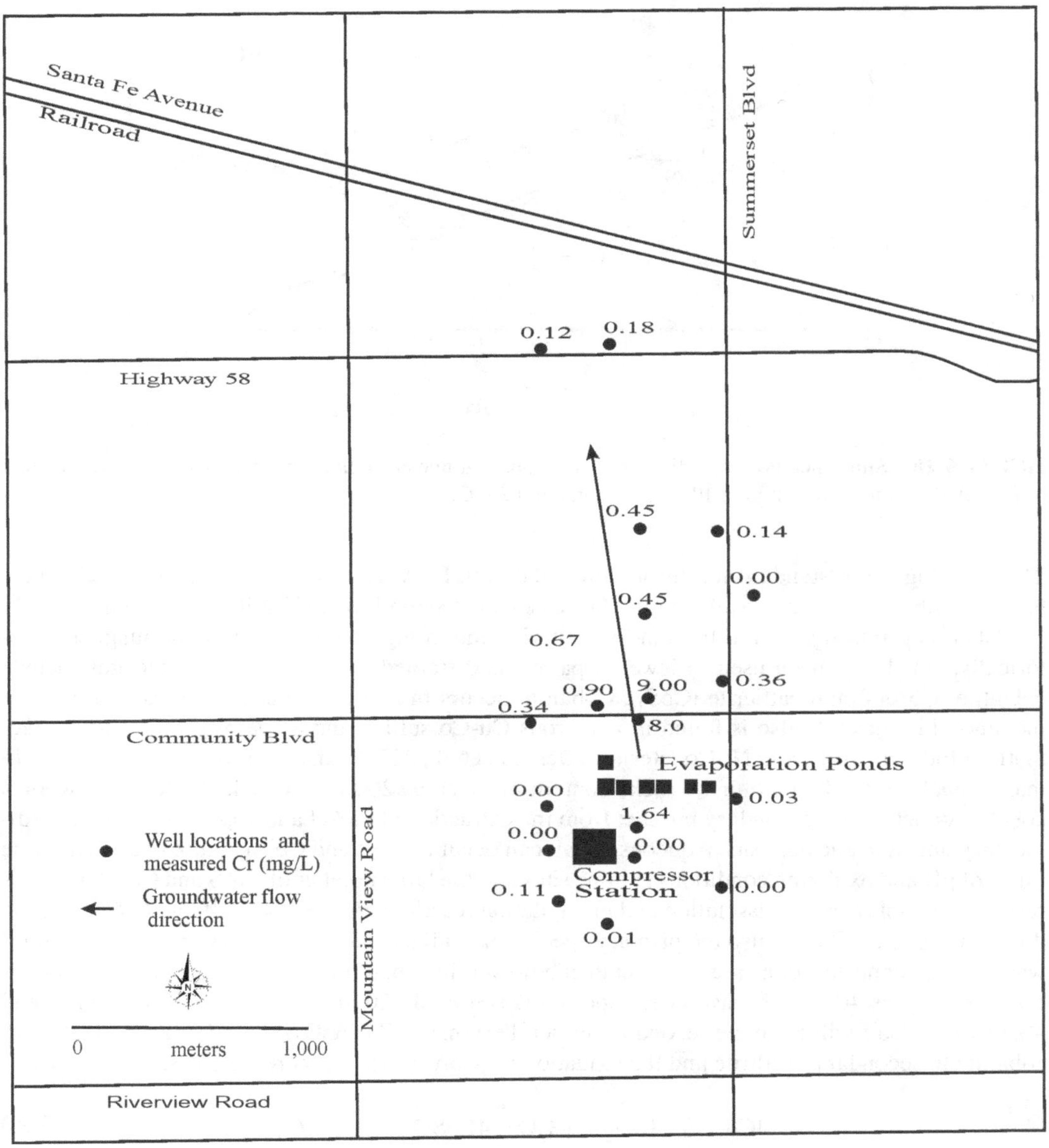

FIGURE 6.27 Maximum chromium concentrations in shallow alluvial well water in the vicinity of the PG&E natural gas compressor station at Hinkley, California. The wells were sampled between 1965 and 1972.

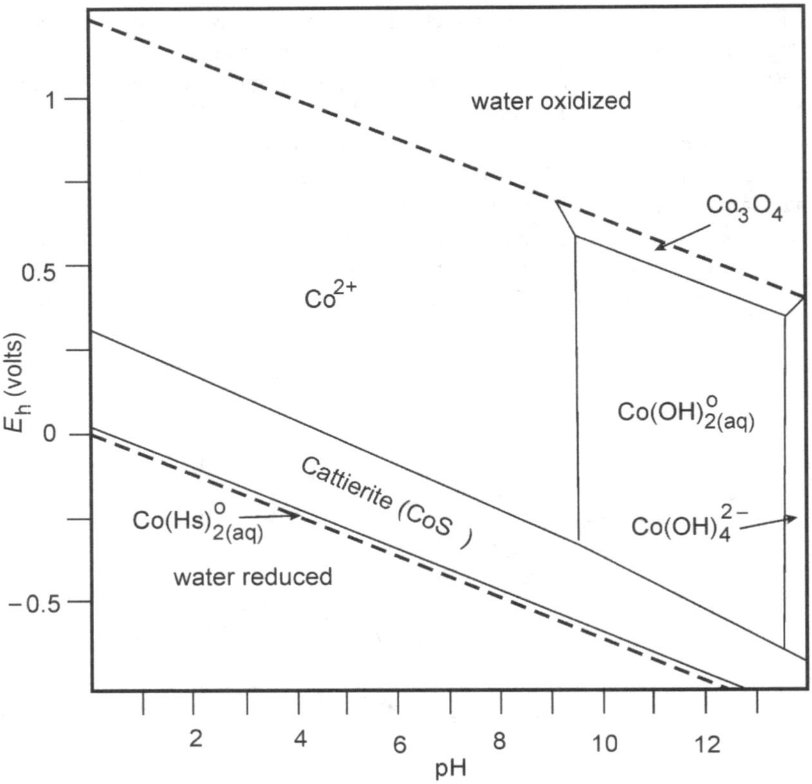

FIGURE 6.28 Simplified Pourbaix diagram of the cobalt sulfide and oxides in aqueous species as a function of E_h and pH at total chromium of 10×10^{-12} molar and 25 °C.

The drinking water standard for chromium is 0.05 mg/L. At Hinkley, down-gradient well waters had concentrations as great as 9.0 mg/L. The cases were settled for \$333 million in damages.

Cobalt is primarily used in lithium-ion batteries and in high-strength alloys and magnets. Historically, cobalt has been used in jewelry, paints, and stained glass. Primary cobalt ores include cobaltite (CoAsS) and cattierite (CoS_2). Cobaltite occurs in hydrothermal deposits and in contact metamorphic rocks. It also is found in numerous Cu-Co sulfide and oxide minerals, Ni-Cu magmatic sulfide minerals, and Ni-Co laterites (Dehaine et al., 2021). Arsenic is a common element in many cobalt minerals. Ultramafic rocks contain as much as 200 ppm cobalt. Most of the world's cobalt production is a secondary product from the extraction of nickel and copper minerals in sedimentary and volcanic deposits. Aqueous cobalt can occur as Co^{2+} and Co^{3+} (Figure 6.28). The wide range of pH and oxidizing conditions at mine sites and the large number of Co-S and Co-O minerals results in complex cobalt dissolution and precipitation reactions. At pH < 4, Co^{2+} and $CoSO_4^o$ are the dominant species. Co^{2+} is also the primary species in oxidized water with a pH <9. In most natural water, cobalt concentrations are undetectable; however, in cobalt mine and tailing water concentrations as great as 400 mg/L have been reported (Ziwa et al., 2020). In oxidizing near-neutral and alkali water, Co^{2+} will form the secondary mineral erythrite ($Co_3(AsO_4)_2 \bullet 8H_2O$). The oxidation of cobaltite to secondary erythrite and the oxidation of (CoAs2) can be represented as:

$$4CoAsS + 13O_2 + 6H_2O = 4CoSO_4 + 4H_3AsO_4 \tag{6.90}$$

$$3CoSO_4 + 2H_3AsO_4 + 8H_2O = Co_3\left(AsO_4\right)_2 \bullet 8H_2O(\text{erythrite}) + 3H_2SO_4 \tag{6.91}$$

$$CoAs_2 + 2H_2O + 3O_2 = Co^{2+} + 2H_2AsO_4^- \tag{6.92}$$

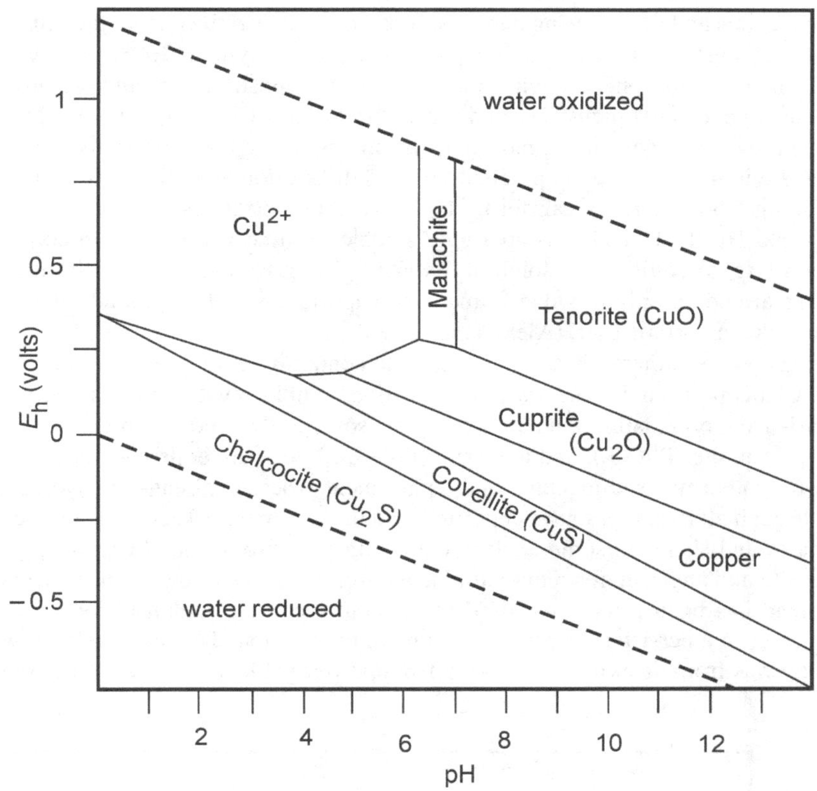

FIGURE 6.29 Pourbaix diagram of the copper in aqueous solutions at 25 °C with carbon = 10^{-2} molar, sulfur = 10^{-3} molar, and copper = 10^{-5} molar.

The subsequent dissolution of erythrite in acidic water mine wastewater can be represented as:

$$Co_3\left(AsO_4\right)_2 \bullet 8H_2O + 2H^+ = 3Co^{2+} + 2H_2AsO_4^- + 8H_2O \tag{6.93}$$

Copper is a critical element that has been used by civilization for thousands of years. Copper ore minerals include chalcopyrite ($CuFeS_2$), bornite (Cu_5FeS_4), malachite ($Cu_2CO_3(OH)_2$), chalcocite (Cu_2S), azurite ($Cu_3(CO_3)_2(OH)_2$), cuprite (Cu_2O), and enargite (Cu_2AsS_4). Copper minerals occur in igneous rocks from magmatic-hydrothermal mineralization, in volcanic and sedimentary rocks from hydrothermal fluids, and from chemical precipitation in sedentary and evaporation basins. Near-surface weathering of copper minerals, particularly in the oxidized zone above copper ore deposits, results in secondary copper minerals such as malachite, azurite, and cuprite. Most copper minerals are relatively insoluble in natural waters, where concentrations may be a few *u*g/L. In solution, copper occurs as Cu^+ and Cu^{2+} (Figure 6.29). In oxygenated water, most dissolved copper is Cu^+. Cu^{2+} can form stable complexes with NH_2, SH, and OH groups, and in typical natural waters as much as 98 percent of the dissolved copper may be complexed with dissolved organic matter (Rader et al., 2019). Dissolution of copper from plumbing pipes by acidic water can raise the copper concentrations, and acid mine drainage can result in copper concentrations in the mg/L range.

6.9.4 Other Metals

Mercury is a persistent, toxic, bio-accumulating pollutant. The primary source of mercury in aqueous solution is cinnabar (HgS), which is associated with hydrothermal activity in silica-carbonate and volcanic terrains and in hot spring deposits. Most mercury is refined from the mineral cinnabar. It is also a trace constitute in some gold and silver deposits (Rytuba, 2003). Mercury is released into the aqueous and atmospheric environments from ore deposit mining and ore processing, iron and steel production, cement industries, and organic-rich shale. Cinnabar is refined by roasting and collecting the mercury vapor. In groundwater systems, mercury concentrations are redox and pH regulated, where it undergoes redox precipitation and dissolution as well as adsorption–desorption reactions (Figure 6.30). Mercury can form pH-dependent chloride and hydroxide complexes such as $Hg(OH)_2$, and $HgCl_2$. Cinnabar is sparingly soluble in mine wastewater. In confined aquifers the free metal ($Hg_{(aq)}$) equilibrium solubility is about 25 *u*g/L, and in open systems equilibrium concentrations are commonly less due to mercury vaporization. Most groundwater systems have concentrations that are <0.01 ug/L (Aleku et al., 2024).

Lead, a chalcophile mineral, is a persistent, toxic contaminant in many surface and groundwater systems. Chalcophiles are minerals where a cation combines with sulfur. Lead in surface and groundwater is from both natural and anthropogenic sources. The most common lead minerals are galena (PbS), cerussite ($PbCO_3$), and anglesite ($PbSO_4$). These minerals and lead ore deposits are the result of hydrothermal, sedimentary, and replacement processes. Lead-rich hydrothermal fluids circulating through all rock types have deposited lead in veins and pockets. Lead has been deposited with carbonates and shales in some ancient sedimentary environments. Secondary lead minerals such as cerussite and anglesite form in oxidizing environments overlying primary lead deposits.

Because lead has been mined and used throughout recorded human history, toxic lead contamination has been a persistent byproduct of human civilization. The most well-known historical lead poisoning was from drinking acidic wine from pewter goblets. In the modern era the primary

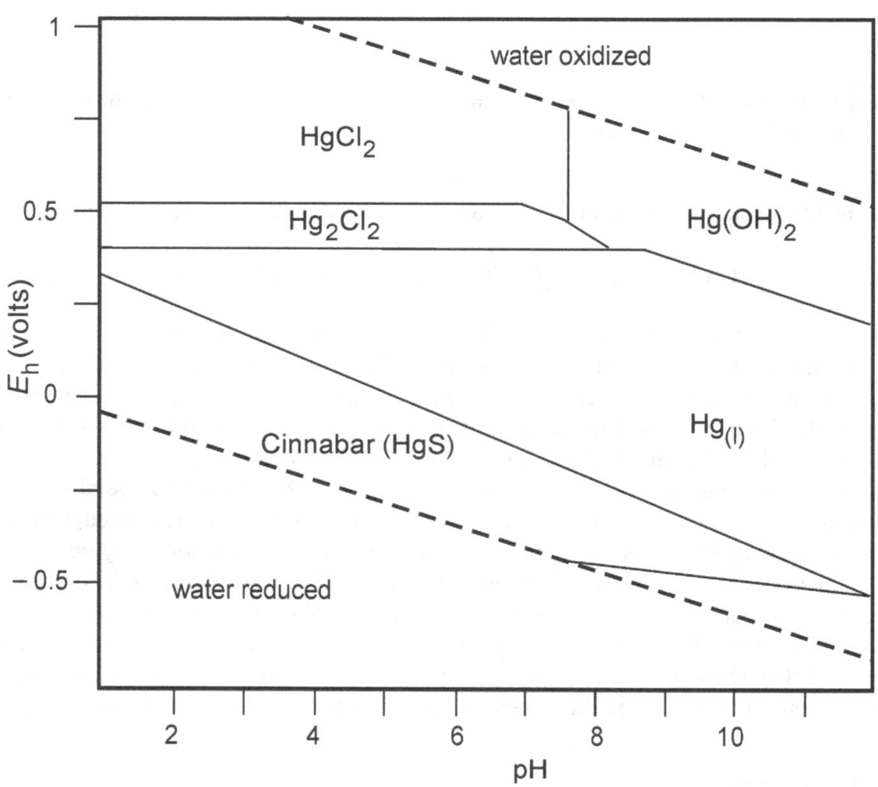

FIGURE 6.30 Pourbaix diagram of mercury in a solution containing Hg at 25 °C.

sources of aqueous lead have been from water in pipes fused together with a lead amalgam, lead bullets, atmospheric lead, and the subsequent lead dry and wet fall. Atmospheric lead is from the earlier use of leaded gasoline in automobile engines, the smelting of sulfide ores, and the burning of coal. Tetraethyl lead allowed engine combustion to have higher fuel-air combustion ratios, which produced more power. Leaded gasoline was banned in the United States. in 1966. In 1991, lead shot for shooting water fowl was banned in the United States. Lead in domestic drinking water from lead-containing pipes remains a major environmental issue.

In aqueous systems, lead occurs as Pb^{2+}, hydroxide complexes, and carbonate and sulfate ion pairs (Figure 6.31). pH is the primary factor controlling the dissolution of metallic sulfide minerals such as lead. In natural groundwaters, elevated lead concentrations are usually associated with oxygenated (DO ≥ 2 mg/L), acidic water (pH ≤ 6.5). In neutral to alkaline water, soluble lead will remain in solution unless Pb-carbonates (cerussite and hydrocerussite) and Pb-orthophosphates precipitate out of the solution (Jurgens et al., 2019; Brown et al., 2019).

Common zinc ores include sphalerite (ZnS), smithsonite ($ZnCO_3$), hemimorphite ($ZnSiO_3(OH)_2 \cdot H_2O$), and zincite (ZnO). Sphalerite is one of the most common sulfide minerals and is found in skarns, hydrothermal deposits, sedimentary exhalative deposits, massive volcanic sulfide deposits, and Mississippi-Valley Type deposits. (MVT). Sedimentary exhalatives are deposits of zinc-lead that form by the discharge of fluids on the seafloor, and MVT include the replacement of sphalerite, galena, pyrite, pyrrhotite, and marcasite in carbonate host rocks. Smithsonite and hemimorphite are found in the oxidized zone of zinc deposits. Zinc is a widely dispersed element in natural and contaminated waters because zinc is widely used in brass and bronze fixtures, as a galvanizing coating to inhibit corrosion in steel, and as a white pigment in paint (ZnO).

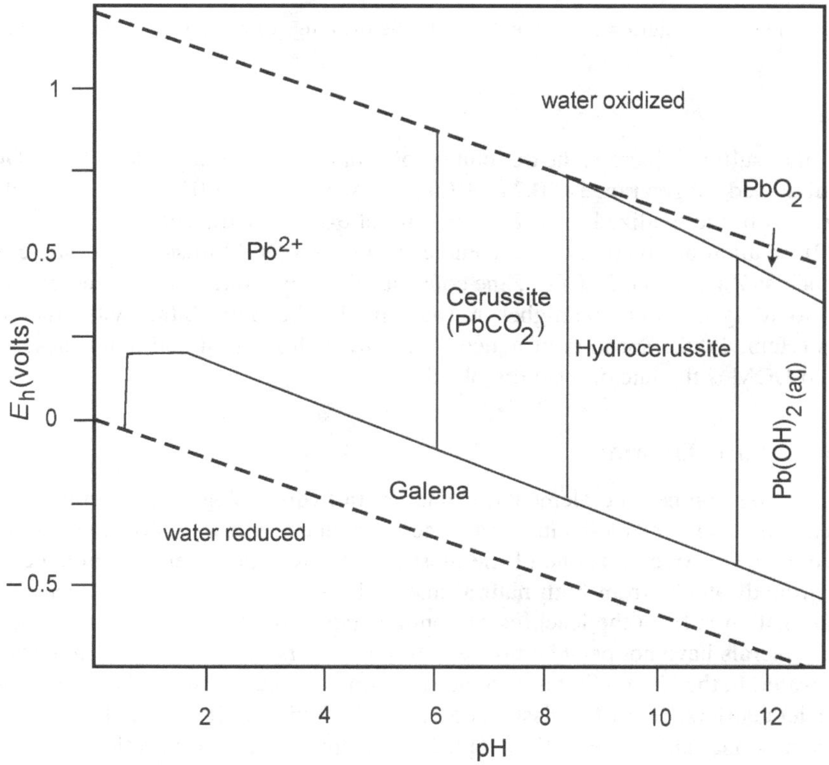

FIGURE 6.31 Pourbaix diagram of lead at 25 °C.

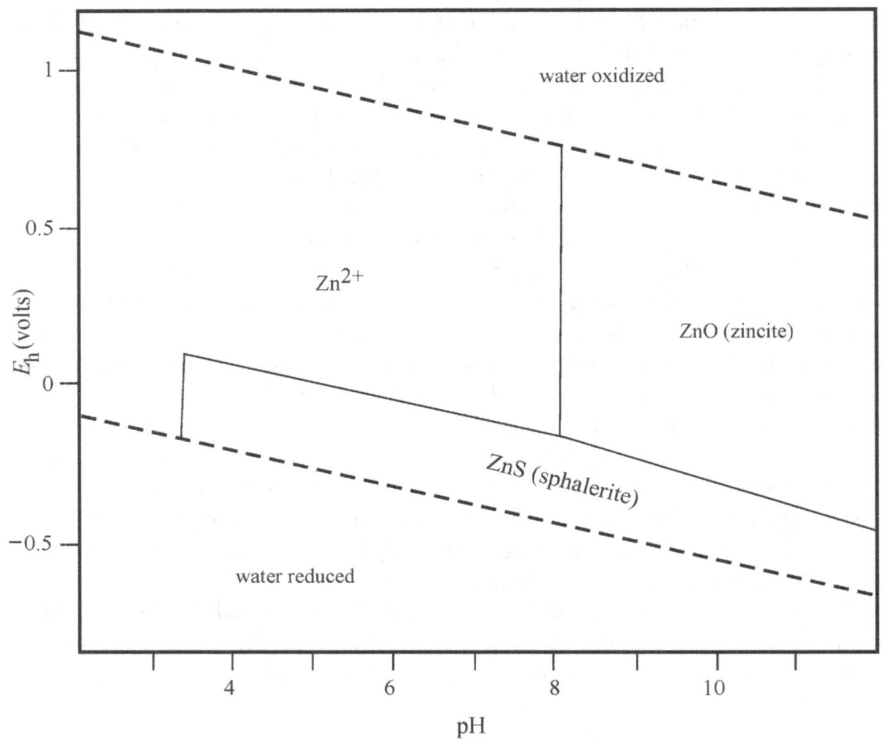

FIGURE 6.32 Pourbaix diagram of zinc at 25 °C. In the pH range of most groundwater, zinc is in the very mobile form Zn^{2+}.

As with other sulfide minerals, the dissolution of sphalerite is pH dependent, but is DO independent in the dissolved oxygen range of 0.2 to 8.7 mg/L (Acero et al., 2007). In aqueous solutions with a pH ≤ 8, zinc is in the oxidized form Zn^{2+}, making it extremely transportable with the fluid flow (Figure 6.32). In alkaline solutions, zinc complexes with Zn-DOM (dissolved organic matter) and molecules such as $ZnOH^+$ and $ZnCO_3^-$. Zinc concentrations in natural surface and groundwater are typically 10 to 40 *u*g/L but can be higher in water that has been in contact with zinc-coated pipes and fixtures (Hem, 1985). Sorption of aqueous zinc by hydrous iron and manganese oxides, clay minerals, and DOM is the fate of most dissolved zinc.

6.9.5 Nonmetallic Elements

The most common nonmetallic elements of interest in hydrogeology are arsenic (As), antimony (Sb), and selenium (Se). Arsenic-containing minerals include realgar (AsS), arsenopyrite (FeAsS), and orpiment (As_2S_3). Arsenic is one of the most investigated trace elements in aqueous systems. Arsenic in groundwater is from both natural and anthropogenic sources; however, most groundwater contamination is from the leaching of young deposits of alluvial sediments where the arsenic-bearing minerals have not been completely dissolved. Arsenic also occurs in volcanic gas and geothermal water. In the United States, arsenic is common in the Southwest basin fill aquifers, New England glacier aquifers, Texas-Mississippi coastal and alluvial aquifers, and the upper Midwest. In the United States, one-half of the wells sampled for arsenic had a concentration of 1 *u*g/L or more, and about 7 percent had concentrations exceeding the 10 ug/L maximum Contaminant level (MCL; USGS, 2025). In the Southwest United States, about 16 percent of the wells exceed the MCL.

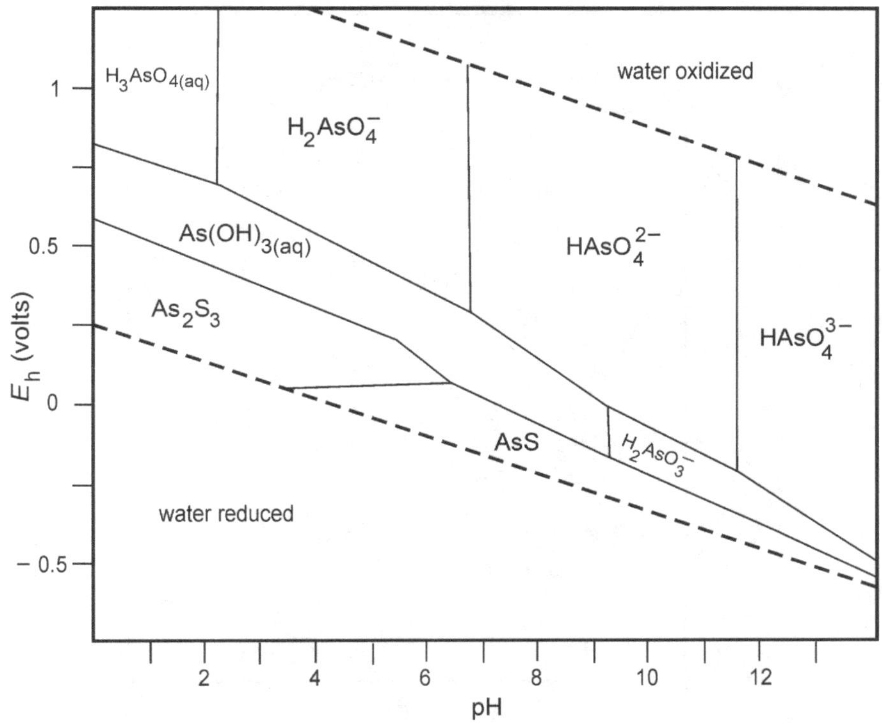

FIGURE 6.33 Pourbaix diagram of arsenic at 25 °C.

Aqueous arsenic occurs in oxidation states –2, 0, +3, and +5. In water, arsenic transformations are regulated by pH, redox, concentrations of organic matter, and other elements, particularly Fe and Mn. In natural waters, arsenic occurs as arsenite As^{3+} and arsenate As^{+5} (Figure 6.33). The oxidant state in aerobic water is +5, and in anoxic water it is +3 (Kanel et al., 2023). In oxic water at pH 2, 2 to 7, 7 to 11, and >11, arsenic is usually in the forms H_2ASO_4, $H_2AsO_4^-$, $HAsO_4^{2-}$, and HsO_4^{3-}, respectively. In reducing water in the pH range 2 to 9, arsenic is H_3AsO_4.

Antimony is introduced into aqueous systems by mineral weathering, microbial activity, and anthropogenic activates including mining, wastewater effluent disposal, and pesticide applications. Most aqueous antinomy is from mining and industrial sources. Stibnite, the most common antimony mineral, is deposited by hydrothermal systems and is often associated with other sulfide minerals. In most natural water, the concentration of Sb is <1 *u*g/L. Mining area antimony concentrations as great as 9,300 *u*g/L have been reported (Sun et al., 2022). Dissolved antimony in reduced water is Sb^{+3} and Sb^{+5} in oxidized water. Sb^{+3} is adsorbed on iron-manganese oxide/hydroxide and can be oxidized to Sb^{+5}. The dominant species in acidic environments are $Sb(OH)_3$, $Sb(OH)_2^+$, and SbO^+. $Sb(OH)_4^-$ and $Sb(O)_2^-$ occur in alkaline water (Figure 6.34).

Selenium minerals are rare, and most selenium occurs as a replacement element in sulfide minerals. Se minerals include selenides that form in reducing hydrothermal solutions, secondary selenites and selenates that form in oxidized near-surface waters, and anhydrous selenites in volcanic fumaroles (Krivovichev et al., 2019). In the presence of iron under reducing conditions, selenium can precipitate as ferroselite ($FeSe_2$). Ferroselite is common in upper Cretaceous Rocky Mountain shales, making it a common groundwater contaminant associated with phosphate and other mining activities. Selenium can occur in several oxidation states, including selenide (–2), selenite ($+4; SeO_3^{2-}$), and selenate $\left(+6; SeO_4^{2-}\right)$. Selenium also occurs in some organic compounds in biologic

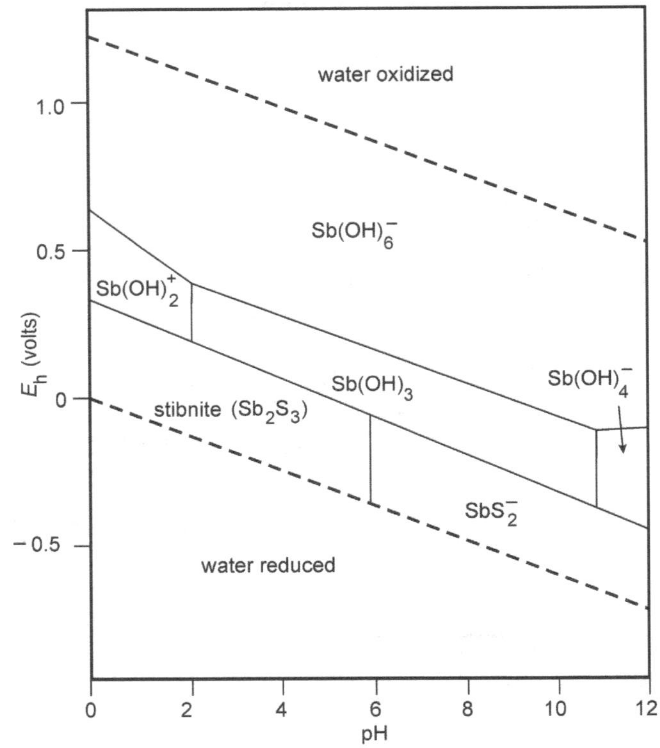

FIGURE 6.34 Pourbaix diagram of antimony at 25 °C.

communities (Sharma and Sharma, 2021), most notably as dimethyl selenide ($(CH_3)_2Se$), dimethyl selenone ($(CH_3)_2SeO_2$), and dimethyl diselenide ($(CH_3)_2Se_2$). Most natural aqueous selenium is from the weathering of selenium-bearing minerals. Anthropogenic selenium is from coal burning, mining and smelting of sulfide ores, oxidized mine drainage, agricultural drainage, and sewage disposal. The Se concentration on most groundwater is less than 1 *u*g/L, but concentration as great as a few mg/L have been reported in mine drainage and agricultural irrigation drainage where oxidizing water encounters selenium-bearing minerals.

Important Takeaway – In aqueous systems, trace constituents include metals, nonmetallic elements, and radioactive elements. The dissolution of alkali and alkali earth elements, such as Li, are not pH or redox dependent. The transition metals and metals are commonly affected by pH and redox conditions. Although aqueous trace constituents can be the result of weathering, anthropogenic activities including metal and non-metal mining and refining, wastewater disposal, and energy production are major contributors of aqueous trace elements. Unlike the major ions, very low concentrations of many trace elements are harmful to human and other animal health.

Problems

1. A groundwater acquires $CO_{2(g)}$ by the thermal decomposing of calcite. The gas then migrates to a shallow carbonate aquifer with a temperature of 25 °C. Assume the P_{CO_2} is 1 atm. Calculate the equilibrium concentrations of Ca^{2+} using the equilibrium constants K_1, K_2, and pK_{cal}. Then calculate the Ca^{2+} using Equation 6.30. Compare the results.

2. A groundwater recharges in a carbonate aquifer containing Ca-Mg-limestone. The water then becomes confined before encountering gypsum in a Na-clay. The water ultimately discharges at a spring that is also the discharge location of an unconfined carbonate aquifer. Both the confined and unconfined aquifers discharge equal amounts of water.
   a. What are the likely chemical compositions of both aquifer systems as the waters flow along their respective flow paths?
   b. What is the likely composition of the stream water that is supported by the spring?
3. Phosphate is mined from the Phosphoria Formation in southeast Idaho. The Phosphoria Formation is a phosphate-bearing shale that overlies a thick sequence of carbonate rocks. Selenium concentrations as great as 1 mg/L have been measured in mine waste drainage and in springs located downgradient of open pit mines and mine waste piles.
   a. What is the likely oxidation state of the selenium in the Phosphoria Formation, mine waste piles, mine wastewaters, and spring discharge waters?
   b. What is the likely source of the selenium?
4. In the West Bank, Palestine, the average chloride concentration in rainwater is 9.18 mg/L, the average annual precipitation is 550 mm/y, and groundwater Cl^- concentrations range from 19 to 59 mg/L (Marei et al., 2010).
   Using the chloride-mass balance method, calculate the range of groundwater annual recharge rates.
5. List the possible sources of groundwater sulfur. What criteria would you use to determine if groundwater sulfur is from each of these sources?
6. A mine sedimentation pond was filled with fine-grained sediments. The sediments contain As, Cd, Cr, Cu, Hg, Pb, Se, and Zn. After the pond was no longer used for mining purposes, slightly acidic surface water continues to flow through the pond sediment and subsequently encounters an oxidized shallow aquifer.
   a. Estimate the concentrations of each of the trace constituents in the groundwater.
   b. What will be the fate of each constituent as the aquifer water becomes confined?

REFERENCES CITED

Acero, P., et al., 2007, Sphalerite dissolution kinetics in acidic environment, *Applied Geochemistry*, 22(9), 1872–1883.

Aeschbach-Hertig, W., et al., 1999, Interpretation of dissolved atmospheric noble gases in natural waters, *Water Resources Research*, 35(9), 2779–2792.

Aeschbach-Hertig, W., et al., 2000, Paleotemperature reconstruction from noble gases in ground water taking into account equilibrium with trapped air, *Nature*, 405, 1040–1044.

Aeschbach-Hertig, W., et al., 2008, Modeling excess air and degassing in groundwater by equilibrium partitioning with a gas phase, *Water Resources Research*, 44(8).

Alcala, F.J., and E. Custodio, 2008, Using the Cl/Br ratio as a tracer to identify the origin of salinity in aquifers in Spain and Portugal, *Journal of Hydrology*, 359(1–2), 189–207.

Aleku, D.L., et al., 2024, Mercury in groundwater – source, transport and remediation, *Applied Geochemistry*, 170, 2024.

Aliewi, A., et al., 2022, A new modified chloride mass balance approach based on aquifer hydraulic properties and other sources of chloride to assess rainfall recharge in brackish aquifers, *Hydrological Processes*, 36(3).

Andres, R.J., and A.D. Kasgnoc, 1998, A time-averaged inventory of subaerial volcanic sulfur emissions, *Journal Geophysical Research*, 103(D19), 25251–25261.

Andrew, R.M., 2019, Global CO_2 emissions from cement production, 1928–2018, *Earth System Science Data*, 11, 1675–1710.

Aravena, R., and B. Mayer, 2009, Isotopes and processes in the nitrogen and sulfur cycles, in *Environmental isotopes in biodegradation and bioremediation*, C.M. Aelion, P. Höhener, D. Hunkeler and R. Aravena (eds.), CRC Press, 203–246.

Back, W., and B.B. Hanshaw, 1970, Comparison of chemical hydrogeology of the carbonate peninsulas of Florida and Yucatan, *Journal of Hydrology*, 10(4), 330–368.

Baker, B.J., and J.F. Banfield, 2003, Microbial communities in acid mine drainage, *FEMS Microbiology Ecology*, 44(2), 139–152.

Ballentine, C.J., and P.G. Burnard, 2002, Production, release and transport of noble gases in the continental crust, in *Reviews in mineralogy and geochemistry-Noble gases in geochemistry and cosmochemistry*, D. Porcelli, C.J. Ballentine and R. Weiler (eds.), *Mineralogical Society of America*, 4, 481–538.

Bao, H., and M.C. Reheis, 2003, Multiple oxygen and sulfur isotopic analyses on water-soluble sulfate in bulk atmospheric deposition from the southwestern United States, *Journal of Geophysical Research Atmospheres*, 108(D14).

Bates, T.S., et al., 1992, Sulfur emissions to the atmosphere from natural sources, *Journal Atmospheric Chemistry*, 14, 315–337.

Berner, R.E., and R.A. Berner, 1987, *The global water cycle: Geochemistry and environment*, Prentice Hall, 397.

Boland, S., et al., 2023, Beryllium contamination and its risk management in terrestrial and aquatic environmental settings, *Environmental Pollution*, 320.

Brown, C.J., et al., 2019, Factors affecting the occurrence of lead and manganese in untreated drinking water from Atlantic and Gulf coastal plain aquifers, eastern United States –dissolved oxygen and pH framework for evaluating risk of elevated concentrations, *Applied Geochemistry*, 101, 88–102.

Brown, L., and C. Ekberg, 2016, *Hydrolysis of metal ions*, Wiley, 155–178.

Bufe, A., et al., 2021, Co-variation of silicate, carbonate and sulfide weathering drives CO_2 release with erosion, *Nature Geoscience*, 14, 211–216.

Butterman, W.C., and R.G. Reese, 2003, *Mineral commodity profiles rubidium*, U.S. Geological Survey, Open-File Report 03-045.

Butterman, W.C., et al., 2004, *Mineral commodity profiles cesium*, U.S. Geological Survey, Open-File Report 2004-1432.

Christensen, R.C., et al., 1984, *Quality and quantity of runoff and atmospheric deposition in urban areas of Salt Lake County, Utah*, U.S. Geol. Survey, Water-Resources Investigations Report 84-4011.

Christensen, T.H., et al., 2000, Characterization of redox conditions in groundwater contaminant plumes, *Journal of Contaminant Hydrology*, 45(3–4), 165–241.

Chuvilin, E., et al., 2020, Evidence of gas emissions from permafrost in the Russian Arctic, *Geosciences*, 10(10).

Cocker, M., 1998. Distribution of selected elements in stream sediments, stream hydrogeochemistry, and geology of the Flint River Basin, Georgia, *Geogia Geological Society*, 129.

Cortecci, G., and Longinelli, A., 1970, Isotopic composition of sulfate in rain water, Pisa, Italy, *Earth and Planetary Science Letters*, 8(1), 36–40.

Craig, H., et al., 1978, Helium isotopic ratios in Yellowstone and Lassen Park volcanic gases, *Geophysical Research Letters*, 5, 897–900.

Crossey, L., et al., 2009, Degassing of mantle-derived CO_2 and He from springs in the southern Colorado Plateau region – neotectonic connections and implications for groundwater systems, *Geological Society of America Bulletin*, 121(7), 1034–1053.

Crundwell, F.K., 2017, On the mechanism of the dissolution of quartz and silica in aqueous solutions, *ACS Omega*, 2(3), 1116–1127.

Davis, S.N., and R.J.M. Dewiest, 1966, *Hydrology*, Wiley and Sons, 463.

Davis, S.N., et al., 1998, Uses of chloride bromide ratios in studies of potable water, *Groundwater*, 36(2), 338–350.

De Caritat, P., 1990, Persistence of quartz disequilibrium in groundwater flow, *Terra Nova*, 2(1), 53–59.

Dehaine, Q., et al., 2021. Geometallurgy of cobalt ores: A review. *Minerals Engineering*, 160.

Deng, K., et al., 2022, A global temperature control of silicate weathering intensity, *Nature Communications*, 13.

Drever, J.I., 1997, *The geochemistry of natural waters*, Prentice Hall.

Dürr, H.H., et al., 2005, Lithologic composition of the Earth's continental surfaces derived from a new digital map emphasizing riverine material transfer, *Global Biogeochemical Cycles*, 19(4).

Eriksson, E., and V. Khunakasem, 1969, Chloride concentration in groundwater, recharge rate and rate of deposition of chloride in the Israel Coastal Plain, *Journal of Hydrology*, 7(2), 178–197.

Evans, M.J., et al., 2008, Degassing of metamorphic carbon dioxide from the Nepal Himalaya, *Geochemistry Geophysics Geosystems*, 9(4).

Feth, J.H., et al., 1964, *Chemical composition of snow in the northern Sierra Nevada and other areas*, U.S. Geological Survey Water-Supply Paper 1535-J.

Freeze, R.A., and J.A. Cherry, 1979, *Groundwater*, Prentice-Hall Inc.

Garrels, R.M., and C.L. Christ, 1965, *Solutions, minerals and equilibria*, Freeman Cooper & Company.

Goldich, S.S., 1938, A study in rock-weathering, *Journal of Geology*, 46(1), 17–58.

Goolsby, E.A., and W.A. Battaglin, 2001, Long-term changes in concentrations and flux of nitrogen in the Mississippi River Basin, USA, *Hydrological Processes*, 15(7), 1209–1226.

Gordon, B., 2013, *Outside diameter stress corrosion cracking of stainless steel in light water reactors*, NACE Corrosion Conference 2013, Paper 2539.

Hamdani, I., et al., 2018, Seasonal and diurnal evaporation from a deep hypersaline lake: The Dead Sea as a case study, *Journal of Hydrology*, 562, 155–167.

Hem, J.D., 1985, *Study and interpretation of the chemical characteristics of natural water*, 3rd ed, U.S. Geological Survey, Water-Supply Paper 2245.

Herzberg, O., and E. Mazor, 1979, Hydrological applications of noble gases and temperature measurements in underground water systems: Examples from Israel, *Journal of Hydrology*, 41(3–4), 217–231.

Holden, A., 1961, Concentration of chloride in fresh waters and rain water, *Nature*, 192, 961–962.

Hua, T., et al., 2025, Back to chromite as a mineralogical strategy for long-term chromium pollution control, *Nature Communications*, 16.

Hunt, J.M., 1979, *Petroleum geochemistry and geology*, Freeman, New York.

Hutton, J.T., 1976, Chloride in rainwater in relation to distance from ocean, *Search*, 7(5), 207–208.

IAPWS, 2004, Guideline on the Henry's constant and vapor-liquid distribution constant for gases in H_2 O and D_2O at high temperatures, *International Association for the Properties of Water and Steam*, 2–5.

Johns, D., and W.H. Huang, 1967, Distribution of chlorine in terrestrial rocks, *Geochimica et Cosmochimica Acta*, 31(1), 35–49.

Jung, M., and W. Aeschbach, 2018, A new software tool for the analysis of noble gas data sets from (ground) water, *Environmental Modelling and Software*, 103, 120–130.

Jurgens, B., et al., 2019, Assessing the lead solubility potential of untreated groundwater of the United States, *Environmental Science & Technology*, 53(6).

Kanel, S.R., et al., 2023, Arsenic contamination in groundwater: Geochemical basis of treatment technologies, *ACS Environmental Au*, 3(3), 135–152.

Karu, E., et al., 2023, Carbonyl sulfide (OCS) in the upper troposphere/lowermost stratosphere (UT/LMS) region: Estimates of lifetimes and fluxes, *Geophysical Research Letters*, 50(19).

Katz, B.G., et al., 2011, Using Cl/Br ratios and other indicators to assess potential impacts on groundwater quality from septic systems: A review and examples from principal aquifers in the United States, *Journal of Hydrology*, 397(3–2), 151–166.

Kerrick, D.M., and K. Caldeira, 1998, Metamorphic CO_2 degassing from orogenic belts, *Chemical Geology*, 145(3–4), 213–232.

Kooijman, A.M., et al., 2010, Nutrient cycling in *Inland drift sand landscapes*, KNNV Publishing, 139–156.

Krivovichev, V., et al., 2019, Selenium minerals: Structural and chemical diversity and complexity, *Minerals*, 9, 455.

Kuroda, P.K., and E.B. Sandell, 1953, Chlorine in igneous rocks; some aspects of the geochemistry of chlorine, *Geological Society America Bulletin*, 64(8), 879–896.

Kushkevych, I., et al., 2021, Microscopic methods for identification of sulfate-reducing bacteria from various habitats, *International Journal of Molecular Sciences*, 22(8).

Lakshmanan, E., 2013, Identification of seawater intrusion by Cl/Br ratio and mitigation through managed aquifer recharge in aquifers North of Chennai, India, *Journal of Groundwater Research*, 2, 155–162.

Langmuir, D., 1971, The geochemistry of some carbonate ground waters in central Pennsylvania, *Geochimica et Cosmochimica Acta*, 35(10), 1023–1045.

Langmuir, D., 1997, *Aqueous environmental chemistry*, Prentice Hall.

Lippmann, J., et al., 2003, Dating ultra-deep mine waters with noble gases and ^{36}Cl, Witwatersrand Basin, South Africa, *Geochimica et Cosmochimica Acta*, 67(23), 4597–4619.

Lucas, D.D., and H. Akimoto, 2007, Contributions of anthropogenic and natural sources of sulfur to SO_2, $H_2SO_{4(g)}$ and nanoparticle formation, *Atmospheric Chemistry and Physics Discussions*, 7, 7679–7721.

Manktelow, P.T., et al., 2007, Regional and global trends in sulfate aerosol since the 1980s, *Geophysical Research Letters*, 34(14).

Manning, A.H., and D.K. Solomon, 2003, Using noble gases to investigate mountain-front recharge, *Journal of Hydrology*, 275(3–4), 194–207.

Marei, A., et al., 2010, Estimating groundwater recharge using the chloride mass-balance method in the West Bank, Palestine, *Hydrological Sciences Journal*, 55(5), 780–791.

Martin E., 2018, Volcanic plume impact on the atmosphere and climate, O- and S-isotope insight into sulfate aerosol formation, *Geosciences*, 8(6).

Maurer, D.K., et al., 1996, *Subsurface flow to Eagle Valley from Vicee, Ash, and Kings Canyons, Carson City, Nevada, estimated from Darcy's law and the chloride-balance method*, USGS Water Resources Investigation Report 96-4088.

Mayo, A.L., 2020, Mapping paleo-lacustrine environments and long-term climate trends using groundwater chemistry, *Hydrogeology Journal*, 28, 987–1002.

Mayo, A.L., and R.H. Klauk, 1991, Contributions to the solute and isotopic groundwater geochemistry, Antelope Island, Great Salt Lake, Utah, *Journal of Hydrology*, 127, 307–335.

Mayo, A.L., and A. Muller, 1997, Low temperature diagenetic – metamorphic and magmatic contributions of external CO_2 gas to a shallow ground water system, *Journal of Hydrology*, 194(1–4), 286–304.

Mayo, A.L., E.C. Petersen, and C. Kravits, 2000, Chemical evolution of coal mine drainage in a non-acid producing environment, Wasatch Plateau, UT, *Journal of Hydrology*, 236(1–2), 1–16.

Mayo, A.L., and D.G. Tingey, 2021, Shallow groundwater chemical evolution, isotopic hyperfiltration, and salt pan formation in a hypersaline endorheic basin: Pilot Valley, Great Basin, USA, *Hydrogeology Journal*, 29, 2219–2243.

Mazor, E., 1972, Paleotemperatures and other hydrological parameters deduced from noble gases dissolved in groundwaters; Jordan Rift Valley, Israel, *Geochimica et Cosmochimica Acta*, 36(12), 321–1336.

Miao, Z., et al., 2012, Sulfate reduction in groundwater: Characterization and applications for remediation, *Environ Geochem Health*, 34(4), 539–550.

Moller, D., 1990, The Na/Cl ratio in rainwater and the sea salt chloride cycle, *Tellus*, 42b, 254–262.

Mueller, D.K., et al., 1995, *Nutrients in ground water and surface water of the United States – an analysis of data through 1992*, U.S. Geological Survey Water Resources Investigations Report 95-4031.

Munger, J.W., 1982, Chemistry of atmospheric precipitation in the north-central united states: Influence of sulfate, nitrate, ammonia and calcareous soil particulates, *Atmospheric Environment*, 16, 1633–1645.

Naily, W., 2018, Cl/Br ratio to determine groundwater quality, in *IOP conference series, Earth and Environmental Science*, 118(1).

Ozima, M., and F.A. Podosek, 2002, *Noble gas geochemistry*, Cambridge Press.

Palya, A., et al., 2008, Storage and mobility of organic nitrogen in the deep continental crust: Evidence from partially melted metasedimentary rocks, *Geochimica et Cosmochimica Act*, 72(12), A721.

Panjaitan, D., et al., 2018, Determining sea water intrusion in shallow aquifer using chloride bicarbonate ratio method, in *IOP conference series, Earth and Environmental Science*, 205.

Parks, J., 2005, Sorption of boron and chromium onto solids of environmental significance: Implications for sampling and removal in water treatment, *Critical Reviews in Environmental Science and Technology*, 5, 81–114.

Peeters, F., et al., 2003, improving noble gas-based paleoclimate reconstruction and groundwater dating using $^{20}Ne/^{22}Ne$ ratios, *Geochimica et Cosmochimica Acta*, 67(4), 587–600.

Plummer, L.N., and E. Busenberg, 1982, The solubilities of calcite, aragonite and vaterite in CO_2-H_2O solutions between 0 and 90°C, and an evaluation of the aqueous model for the system $CaCO_3$-CO_2-H_2O, *Geochimica et Cosmochimica Acta*, 46(6), 1011–1040.

Rader, K.J, et al., 2019, The fate of copper added to surface water: Field, laboratory, and modeling studies, *Environmental Toxicology and Chemistry*, 38(7), 1386–1399.

Ranalli, G., et al., 2019, *Biodeterioration – including cultural heritage in encyclopedia of microbiology*, 4th ed, Academic Press, 491–509.

Rytuba, J.J., 2003, Mercury from mineral deposits and potential environmental impact, *Environmental Geology*, 43, 326–338.

Scanlon, B.R., 2004, Evaluation of methods of estimating recharge in semiarid and air regions in the southwestern U.S., in *Groundwater recharge in a desert environment*, J.F. Hogan, F.M. Phillips and B.R. Scanlon (eds.), *American Geophysical Union Water Science and Application*, 9, 235–249.

Schnoonen, M.A., 2016, Sulfur cycle, in *Encyclopedia of geochemistry*, W.M. White (ed.), Springer International Publishing.

Sharma, S., and S.K. Sharma, 2021, Chapter 2 – Selenium distribution and chemistry in water, in *Selenium contamination in water*, Wiley.

Smith, S.P., and B.M. Kennedy, 1983, The solubility of noble gases in water and in NaCl brine, *Geochimica et Cosmochimica Acta*, 47(3), 503–515.

Smith, W.H., 2001, *Encyclopedia of biodiversity*, 2nd ed, Elsevier, 147–156.

Somarantne, N., and K. Smettem, 2014, Theory of the generalized chloride mass balance method for recharge estimation in groundwater basins characterized by point and diffuse recharge, *Hydrology and Earth System Sciences Discussions*, 11, 2407–2014.

Stute, M., and P. Schlosser, 1993, Principles and applications of the noble gas paleothermometer, in *Climate change in continental isotopic records*, American Geophysical Union, 89–100.

Sun, X., et al., 2022, Morphological distribution and formation mechanisms of antimony in the Shallow groundwater of the Xikuangshan antimony mine in Hunan, China, *Frontiers Environments Science*, 10.

Tian, H., et al., 2020, Long-term trajectory of nitrogen loading and delivery from Mississippi River Basin to the Gulf of Mexico, *Global Biogeochemical Cycles*, 34.

USGS, 2025, *Arsenic and drinking water*, USGS.

Van't Dack, L., et al., 1983, Equilibrium constants for trace elements in natural waters, *Reviews in Analytical Chemistry*, 7(4), 297–401.

Walter, K.M., et al., 2007, Methane bubbling from northern lakes: Present and future contributions to the global methane budget, *Philosophical Transactions of the Royal Society*, 365, 1657–1676.

Whiticar, M.J., et al., 1986, Biogenic methane formation in marine and freshwater environments: CO_2 reduction vs. acetate fermentation: Isotopic evidence, *Geochimica et Cosmochimica Acta*, 50, 693–709.

Wolin, M.J., and T.L. Miller, 1987, Bioconversion of organic carbon to CH_4 and CO_2, *Geomicrobiological Journal*, 5(3–4), 239–259.

Wood, W.W., 1999, Use and misuse of the chloride-mass balance method in estimating groundwater recharge, *Groundwater, Technical Commentary*, 2–3.

Wood, W.W., and W.E. Sanford, 1995, Chemical and isotopic methods for quantifying ground-water recharge in a regional, semiarid environment, *Groundwater*, 33(3), 458–468.

Ziwa, G., et al., 2020, Geochemistry, mineralogy and microbiology of cobalt in mining-affected environments, *Minerals*, 11(1), 22.

Part II

Isotopes

7 Isotopic Definitions and Standards

7.1 INTRODUCTION

The term isotope, derived from Greek and meaning "equal place," refers to a species of a chemical element whose nuclei contain differing numbers of neutrons but occupy the same position on the periodic table. Isotopes are conventionally denoted in the form ${}^{A}_{Z}X$, where the superscript, A, represents the atomic mass number (number of protons plus neutrons), the subscript, Z, represents the atomic number (number of protons), and X represents the element. For example, carbon isotopes may be represented as ${}^{14}_{6}C$, ${}^{13}_{6}C$, and ${}^{12}_{6}C$. In isotope hydrogeology, the Z subscript is usually omitted. For example, we commonly us ^{13}C rather than ${}^{13}_{6}C$.

Isotopic abundances vary greatly between the chemical reservoirs of the atmosphere, biosphere, hydrosphere, and lithosphere and between the solid, dissolved, and gaseous phases in each reservoir. Isotopic fractionation and radioactive decay are pervasive in natural systems. The isotopic compositions of water, dissolved solutes, and gases are affected in a predictable manner during phase changes, by chemical and biological processes, and during radioactive decay. These changes make the use of isotopes powerful and often underutilized tools in hydrogeologic investigations.

More than 1,000 stable and unstable isotopes exist among the 92 naturally occurring elements. Stable isotopes are elements whose nucleus remains stable over geologic time. The nucleus of unstable isotopes undergoes spontaneous transformations or decay. Unstable isotopes are often referred to as radioactive because the decay process involves the transformation of a parent element into a daughter and element (e.g., $^{14}C \rightarrow {}^{14}N$) by the emission of particles and radiant energy.

Only a few isotopes are sufficiently abundant in groundwater for routine laboratory analysis. Isotopes most commonly used in environmental studies include the light elements hydrogen, carbon, oxygen, nitrogen, sulfur, and occasionally helium. Chlorine has been used for dating very old groundwater. The average terrestrial abundances of the most frequently used isotopes in groundwater investigations are listed in Table 7.1. In all instances the most abundant species of each element is the light isotope.

Absolute abundances are commonly reported as atom percent and fractional abundance. For example, the atom percent of ^{18}O is calculated as:

$$^{18}O = \left[{}^{18}O / \left({}^{16}O / {}^{18}O\right)\right]100 \tag{7.1}$$

and the fractional abundance may be calculated as:

$$^{8}F = {}^{18}O / \left({}^{16}O / {}^{18}O\right) \tag{7.2}$$

Measuring the absolute isotopic abundances of stable isotopes in water and aquifer minerals is extremely difficult. In most scientific investigations it is conventional, and just as useful, to determine the relative isotopic composition of a sample with respect to a standard of known isotopic composition. Concentrations are therefore not expressed as the absolute abundance ratio as with solutes but rather as the abundance relative to a standard. Relative abundances are described using delta (δ) units, which are defined as:

$$\delta = \left(\frac{R_{\text{sample}}}{R_{\text{standard}}} - 1\right) \times 10^{3} \tag{7.3}$$

DOI: 10.1201/9781003646129-9

TABLE 7.1
Average Terrestrial Abundance of Isotopes Used in Groundwater Investigations

Element	Isotope	Abundance %	Comments
Hydrogen	^{1}H	99.985	
	^{2}H or D	0.015	
	^{3}H or T	10^{-14} to 10^{-16}	half-life 12.3 years
Carbon	^{12}C	98.89	
		^{13}C	1.11
	^{14}C	$\approx 10^{-10}$	half-life 5730 years
Chlorine	^{35}Cl	75.77	
	^{36}Cl		half-life 3×10^{5} years
	^{37}Cl	24.23	
Oxygen	^{16}O	99.76	
	^{17}O	0.037	
	^{18}O	0.10	
Nitrogen	^{14}N	99.34	
	^{15}N	0.336	
Sulfur	^{32}S	95.02	
	^{33}S	0.75	
	^{34}S	4.22	

where R is the number of heavy isotopic nuclides relative to the number of light isotopic nuclides (i.e., $^{13}C/^{12}C$). The delta unit expresses the difference in isotopic composition in parts per thousand between the sample and the standard. In some older literature, del and the symbol Δ have been used rather than δ. Parts per thousand, known as permil and usually expressed in the symbolic form of ‰, are used because the abundance ratio difference between the sample and standard is so small. For example, a carbon-13 analysis might be expressed as $\delta^{13}C = -5.2$ ‰.

Analysis of stable isotopes are commonly undertaken in conjunction with other solutes and physical hydrogeology parameters and factors. It is sometimes useful to use a dual-isotope approach. For example, Otero et al. (2007) and Zhang et al. (2014) analyzed both d^{34}S and d^{18}O when sorting out the origin of dissolved sulfate.

Important Takeaway – Unlike solutes, where the absolute abundance is measured, stable isotopic compositions are analyzed relative to a standard of known isotopic composition. Because the difference between the sample and standard are very small, results are reported in permil (‰).

7.2 ISOTOPIC FRACTIONATION AND ISOTOPIC EXCHANGE

The nuclear structure of an element determines its physical properties, and the electron structure is largely responsible for its chemical behavior. Because all isotopes of an element contain the same number and arrangement of electrons, they tend to have similar chemical behaviors. It is this similar chemical behavior that makes stable isotopes so useful. For example, all sulfur isotopes participate in the precipitation of gypsum from a calcium-sulfate rich solution. Similarities in chemical behaviors are not, however, limitless. Mass differences between isotopic species of the same element result in different vibrational energies, which in turn cause differences in their physiochemical properties. These isotopic effects primarily manifest themselves by partitioning of isotopes during physical and biochemical reactions and phase changes. Isotopic partitioning is commonly referred to as isotopic fractionation. Isotopic fractionation occurs because the atomic vibrational frequency of an element is inversely dependent upon mass. The heavier isotopes of a particular element have greater mass and a corresponding lower vibrational frequency relative to the lighter isotopes of the same element. Chemical bonds formed with heavy isotopes also have lower vibrational frequencies than do those formed with light isotopes. These lower vibrational frequencies make chemical bonds between heavy isotopes stronger than those between lighter isotopes. During phase changes and chemical reactions, the molecules containing light isotopes are more energetic and tend to react more often than do the molecules containing the heavy isotopes, and the products of phase changes and chemical reactions tend to become enriched in the light isotopes.

Isotopic effects fall into two main groups: kinetic and equilibrium. Kinetic effects are not reversible and cause the lighter isotopes to accumulate in the reaction products. Equilibrium effects are reversible and cause the heavy isotopes to accumulate where they can form the strongest bonds. For example, in the $H_2O_{(l)} - H_2O_{(v)}$ system, the heavy isotopes accumulate in the $H_2O_{(l)}$ phase, where they can form stronger bonds than in the $H_2O_{(v)}$. During water evaporation the volatile gaseous molecules are preferentially enriched in the light and more energetic ^{16}O and ^{1}H isotopes, and the remaining liquid is preferentially enriched in the heavy isotopes ^{18}O and ^{2}H. Therefore, the vapor phase has $\delta^{18}O$ and $\delta^{2}H$ contents that are more negative than the parent liquid and are considered to be enriched in the light isotopes, and the liquid is depleted in the light isotopes or enriched in the heavy isotopes. Usually, but not always, enriched is used to mean more of the light isotope and depleted is used to mean less of the light isotope. The terms enriched or depleted are commonly used in groundwater investigation with only an inference regarding what is enriched or depleted. For clarification, it is important to state if the produce or reactant is either enriched or depleted in either the light or heavy isotope.

The amount of enrichment or depletion accompanying fractionation is often of great interest in hydrogeology in that it may provide insight into the origin and history of groundwater and its contained solutes and gases. For example, when the vapor source is an ocean or other large body of water the extent of isotopic depletion of light isotopes in the water body accompanying evaporation is insignificant, but the enrichment of the vapor in light isotopes may be consequential. In all large water bodies, the parent liquid is isotopically buffered. When the vapor source is a smaller body of water, the soil zone, or atmospheric conditions such as virga rain, both isotopic depletion of the source term and enrichment of the vapor may be hydrologically important. Virga is rain that completely or mostly evaporates prior to reaching the ground.

Isotopic fractionation can occur with or without chemical partitioning and may occur under equilibrium and non-equilibrium conditions. Isotopic fractionation is the relative partitioning of the heavier and lighter isotopes between two coexisting phases or components of a system. Chemical partitioning includes all reactions that involve a change in the concentration of the dissolved or gaseous constituents. Chemical partitioning includes dissolution, precipitation, and redox reactions.

The two principal phenomena leading to isotopic fractionation are:

1. isotope exchange and
2. kinetic processes, which are primarily dependent on differences in chemical reaction rates of isotopic molecules.

Isotopic exchange is fractionation without a change in mass (i.e., no chemical partitioning) or without a phase change. This type of fractionation encompasses processes in which isotopic equilibrium is approached or attained between different chemical substances, between different phases, or between individual molecules. Isotopic exchange may accompany ordinary chemical reactions, such as those described by the law of mass action. Although the resulting isotopic compositions of reactants and products are not primarily controlled by isotopic exchange, chemical reactions that proceed to equilibrium are often accompanied by isotopic fractionation. These processes include mineral dissolution and precipitation, evaporation and condensation, crystallization and melting, absorption and desorption, and diffusion and thermodiffusion. During redox equilibrium reactions the heavy isotopes tend to accumulate in the species with the higher oxidation state. For example, during the oxidation of a sulfide mineral to sulfate, the sulfate becomes enriched in ^{34}S relative to residual sulfide mineral. During equilibrium reactions the denser species become isotopically heavy relative to the less dense species. For example, the abundance of $\delta^{18}O$ and $\delta^{13}C$ follow the pattern $CO_2 < HCO_3^- < CaCO_3$.

During phase changes or during reactions involving a common element, the denser species tend to become enriched in the heavy isotopes. For example, during H_2O phase changes isotopic enrichment follows the pattern $\delta^{18}O_{(s)} > \delta^{18}O_{(l)} > \delta^{18}O_{(v)}$, where the subscripts s, l, and v represent solid, liquid, and vapor, respectively. During cloud condensation resulting in rain, the liquid phase becomes enriched in the heavy isotopes ^{18}O and ^{2}H, and the remaining cloud vapor becomes enriched in the light isotopes ^{16}O and ^{1}H. As the storm travels along the storm path, both the cloud vapor and the precipitation become progressively more isotopic enriched in the lighter isotopes due to the rainout effect.

Kinetic effects are the second main phenomena effecting isotopic fractionation. Many complex chemical and biochemical processes do not proceed to chemical or isotopic equilibrium. Kinetic effects are especially common during biochemical reactions when fractionation does not proceed to complete equilibrium. Kinetic fractionation accompanying biologic processes often complicates our understanding and analysis of isotopic data.

7.3 FRACTIONATION FACTORS

The quantum theory is the theoretical basis for explaining the physiochemical properties of isotopic fractionation. For many simple molecules, the vibrational frequencies determined by statistical mechanical methods may be used to calculate the isotopic fractionation that can take place under different thermal conditions (Urey, 1947; Bigeleisen and Mayer, 1947; Melander, 1960). For reactions of hydrogeochemical interest, isotopic fraction factors are usually determined experimentally. Isotopic fractionation obeys the law of mass action and may be expressed in terms of temperature-dependent isotopic equilibrium constants. The equilibrium constants are known as isotope fractionation factors (K) or more commonly, α. K should not to be confused with the mass action equilibrium constant K_{eq}. To help avoid confusion, let us consider an isotopic exchange reaction that proceeds to equilibrium and in which the products and reactants are chemically but not isotopically identical. As discussed in Chapter 2, chemical reactions and their corresponding chemical equilibrium expressions may be symbolically expressed as:

$$aA + bB = cC\ dD \tag{7.4}$$

and

$$Keq = \frac{[C]^c [D]^d}{[A]^a [B]^b} \tag{7.5}$$

where a, b, c and d are the number of moles of the chemical constituents A, B, C, and D, respectively. The law of mass action (Equation 7.4) expresses the chemical equilibrium between the reactants and products in terms of a thermodynamic equilibrium constant (Equation 7.5). In Equation 7.5 the brackets specify the effective molar concentrations or chemical activity.

Consider the following dynamic reactions that only involve the exchange of isotopes and where there is no change in the mass or chemical composition of either reactants or products:

$$CO_{2(g)} + H_2O = CO_{2(g)} + H_2O \tag{7.6}$$

$$HS^- + H_2S^- = HS^- + H_2S^- \tag{7.7}$$

Here the thermodynamic equilibrium constant (K_{eq}) for each reaction is 1 and is of little geochemical interest. Symbolically, isotopic exchange for reactions 7.6 and 7.7 may be written as:

$$aA_1 + bB_2 = aA_2 + bB_2 \tag{7.8}$$

where A and B are molecules having any element as a common constituent and the subscripts 1 and 2 represent the light and heavy isotope of the molecule, respectively. The isotopic exchange reactions for Equations 7.6 and 7.7 may be written as:

$$C^{16}O_2 + 2H_2{}^{18}O = C^{18}O_2 + 2H_2{}^{16}O \tag{7.9}$$

$$H^{32}S^- + H_2{}^{34}S = H^{34}S^- + H_2{}^{32}S^- \tag{7.10}$$

When considering Equations 7.9 and 7.10, it is important to keep in mind that both reactants and products exist in the solution prior to isotopic exchange and that the only difference between the reactants and products is their isotopic compositions. The exchange reactions only describe the exchange of light and heavy isotopes between reactants and products. There is no change in the molar concentration of either reactants or products. Equations 7.0 and 7.10 do not imply a complete exchange between the CO_2 and H_2O. H_2O may be a system buffer, and only a small fraction of the H_2O may undergo exchange. In a similar fashion, only a fraction of the system's CO_2 may undergo exchange.

When complete equilibrium exchange occurs, the isotopic equilibrium constant for the reaction in Equation 7.10 may be expressed as:

$$\alpha = \frac{\left[H^{34}S^-\right]\left[H_2{}^{32}S\right]}{\left[H^{32}S^-\right]\left[H_2{}^{34}S\right]} \tag{7.11}$$

The equilibrium isotope fractionation factor (α) for the exchange of sulfur isotopes between HS- and H_2S may be written as:

$$\alpha_{HS^- - H_2S} = \frac{\left({}^{34}S/{}^{32}S\right)H^-S}{\left({}^{34}S/{}^{32}S\right)H_2S} \tag{7.12}$$

The general form of the isotopic fractionation factor between two phases A and B may be defined as:

$$\alpha_{A-B} = \frac{R_A}{R_B} \tag{7.13}$$

where, as previously described, R is the ratio of the number of heavy isotope nuclides relative to the number of light isotope nuclides, such as $^{34}S/^{32}S$. In terms of laboratory-measured δ values, Equation 7.13 may be arranged to:

$$\alpha_{A-B} = \frac{1000 + R_A}{1000 + R_B} \tag{7.14}$$

The isotope fractionation factor is related to the chemical equilibrium constant in the following way:

$$\alpha = Keq^{1/n} \tag{7.15}$$

TABLE 7.2
Comparison of δ, Δ, α, and $10^3\alpha_{A\text{-}B}$ Values

δ_A	δ_B	$\Delta_{A\text{-}B}$	$10^3\alpha_{A\text{-}B}$	α
1.00	0	1.00	1.00	1.00000
5.00	0	5.00	4.99	1.00500
10.00	0	10.00	9.95	1.01000
12.00	0	12.00	11.93	1.01200
15.00	0	15.00	14.89	1.01500
20.00	0	20.00	19.80	1.02000
10.00	5.00	5.00	4.96	1.00498
20.00	15.00	5.00	4.91	1.00493
30.00	25.00	5.00	4.87	1.00488
30.00	20.00	10.00	9.76	1.00980
30.00	15.00	15.00	14.67	1.01478
30.00	10.00	15.00	19.61	1.01980

Source: Data from Friedman and O'Neil, 1977).

where n is the number of atoms exchanged and $K_{eq} \equiv \alpha$ in monatomic reactions. Isotope exchange reactions are usually written such that only one atom is exchanged and $K = \alpha$.

Isotopic fractionation factors are usually close to 1 and may be expressed as 1.00X, where X, the permil fractionation, may be either positive or negative. For example, in the reaction:

$$1/3\mathrm{C}^{16}\mathrm{O}_3^{2-} + \mathrm{H}_2^{18}\mathrm{O} = 1/3\mathrm{C}^{18}\mathrm{O}_3^{2-} + \mathrm{H}_2^{16}\mathrm{O} \tag{7.16}$$

at 20 °C the fractionation factor is about 1.030 (1.00X or 1.00 + 0.030), meaning that carbonate in the product will be 30 ‰ richer in ^{18}O than the water. It is useful to note that $10^3\ln(1.00X) = X$. Therefore, in the example above where $\alpha = 1.030$, the value of $10^3\ln\alpha = 30$. That is, the fractionation $10^3\ln\alpha$ is 30 ‰. Permil fractionation may sometimes be approximated by the A value where:

$$\mathrm{A}_{A-B} = \delta_A - \delta_B = 10^3\ln\alpha_{A-B} \tag{7.17}$$

Subtracting δ values can give a good approximation of the permil fractionation when the A and δ are less than 10 (Table 7.2). The magnitude of fractionation has also been represented as the isotopic enrichment factor ε:

$$\varepsilon_{A-B} = \delta_A - \delta_B = (\alpha - 1)10^3 \tag{7.18}$$

For small values of ε, $\varepsilon = 10^3\ln\alpha$.

At isotopic equilibrium, the magnitude of fractionation, as a function of temperature, can be expressed over large temperature ranges by the polynomial function:

$$\ln\alpha = A + \frac{B}{T^2} \tag{7.19}$$

or the more complicated polynomial function:

$$\ln\alpha = A + \frac{B}{T} + \frac{C}{T^2} \tag{7.20}$$

where T is in Kelvin and A, B, and C are coefficients that are usually determined experimentally. It is apparent from Equations 7.19 and 7.20 that $\ln\alpha$ is inversely proportional to temperature and that $\ln\alpha = f(T)$. As the temperature increases, $\ln\alpha$ or the amount of absolute fractionation decreases. Recall that $10^3\ln\alpha$ is the permil fractionation. It has been found that smooth, often nearly linear curves are obtained when values of $10^3\ln\alpha$ are plotted against values of T^{-1} or T^{-2}. For graphical purposes, the relationship $\ln\alpha = f(T)$ is often expressed as:

$$10^3 \ln\alpha = A + 10^6 BT^{-2} \tag{7.21}$$

or in the more complicated form:

$$10^3 \ln\alpha = A + 10^3 BT^{-1} + 10^6 CT^{-2} \tag{7.22}$$

Equations 7.21 and 7.22 are commonly used for calculating temperatures at which minerals crystalize and for estimating the temperatures of hydrothermal systems. The equations may also be used to aid in the interpretation of the isotopic history of a groundwater.

Friedman and O'Neil (1977) compiled and graphically illustrated some fractionation factors of geochemical interest. Figure 7.1 illustrates oxygen and hydrogen fractionation in the vapor-water system. In this system fractionation increases as the temperature decreases, that is the fractionation factor becomes larger at lower temperatures. For example, at 30 °C water vapor is enriched with respect to the liquid phase by approximately 8.75 and 72 ‰ in ^{18}O and ^{2}H, respectively. At 10 °C enrichment of the water vapor increases to approximately 10.5 and 95 ‰ in ^{18}O and ^{2}H, respectively. Fractionation of ^{13}C between calcite and $CO_{2(g)}$, illustrated in Figure 7.2.

It is important to keep in mind that fractionation alters the isotopic composition of both phases A and B (Equation 7.13), but measurable effects on the two phases may be quite different. When the mass of either participant is substantially greater than the mass of the other participant, the larger participant may act as an isotopic buffer. For example, during the evaporation of seawater, the vapor phase is greatly depleted in ^{18}O and ^{2}H with respect to the parent seawater (Figure. 7.1). However, the overall isotopic composition of seawater is relatively unaffected because so few molecules of seawater are involved. Similar buffering effects can be observed where groundwater and its contained solutes interact with the relatively inexhaustible isotopic reservoirs of minerals in the aquifer matric and in biologic and atmospheric environments.

In most groundwater systems the rock–water ratio greatly favors the aquifer matrix and the matrix may be thought of as an isotopic buffer. Over eons of geologic time, however, groundwater flowing through an aquifer may leach out, via isotopic exchange, isotopes in the aquifer matrix that are in contact with groundwater. Such effects tend to be more pronounced in relatively insoluble aquifers, such as batholithic terrains, than they are in carbonate terrains, whose mineral solubilities are considerably greater. For example, Taylor (1974, 1977) demonstrated that pervasive exchange between thermal waters and silicate reservoirs under epizonal batholithic conditions have depleted the $\delta^{18}O$ content of aquifers in some regions, thus limiting their potential for isotopic exchange with modern circulating groundwater. A similar depletion is not possible in soluble terrains such as carbonates, because the isotopically depleted mineral surface tends to dissolve and is carried away as aqueous ions.

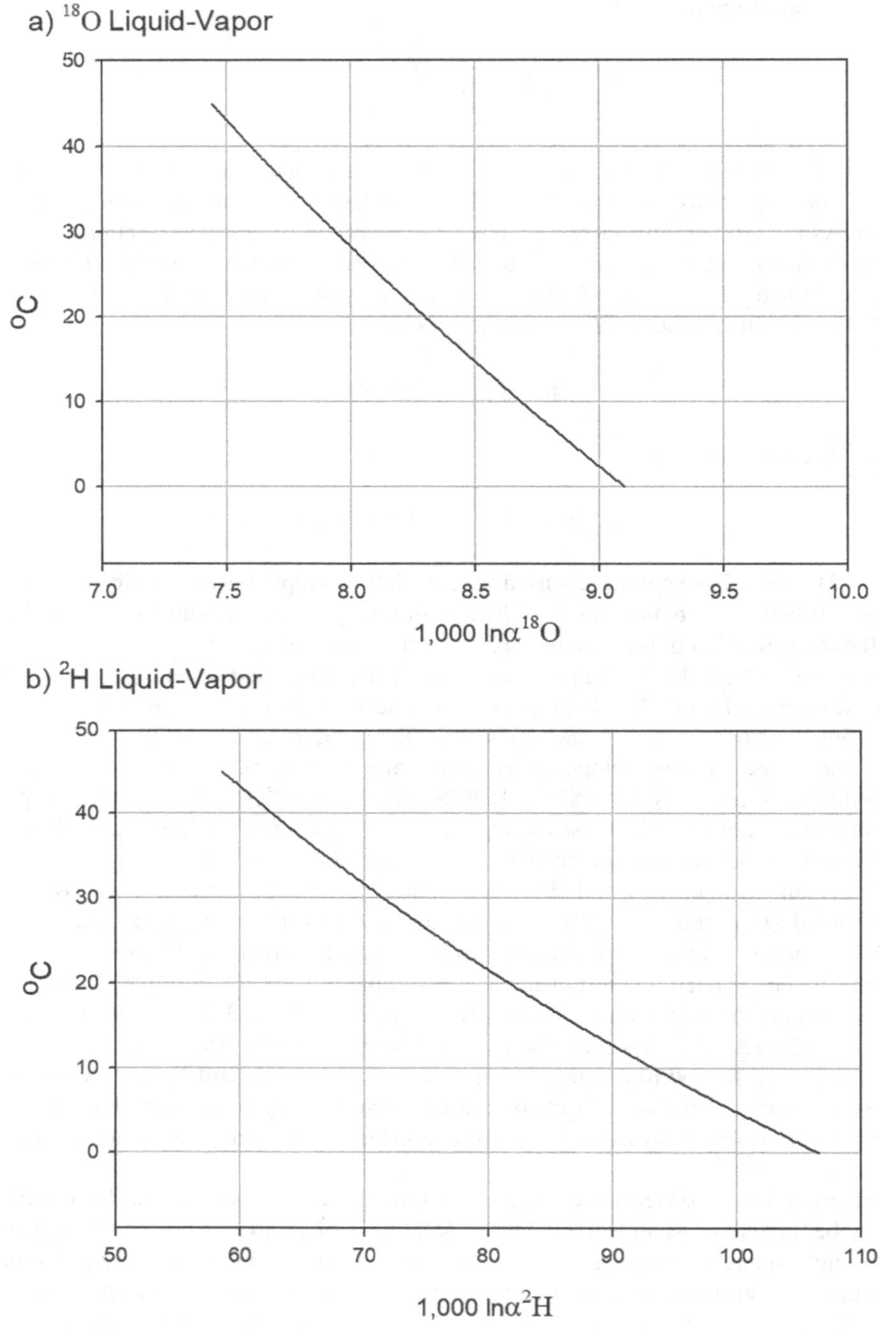

FIGURE 7.1 The temperature dependence of deuterium and ^{18}O isotopic fractionation between water and vapor.

7.4 ISOTOPIC EQUILIBRIUM

Rayleigh fractionation (pronounced "raylee") describes the thermodynamic equilibrium exponential portioning of isotopes between two reservoirs as one reservoir decreases in size. The original Rayleigh equation was derived by Lord Rayleigh in 1900 to describe fractional distillation of mixed

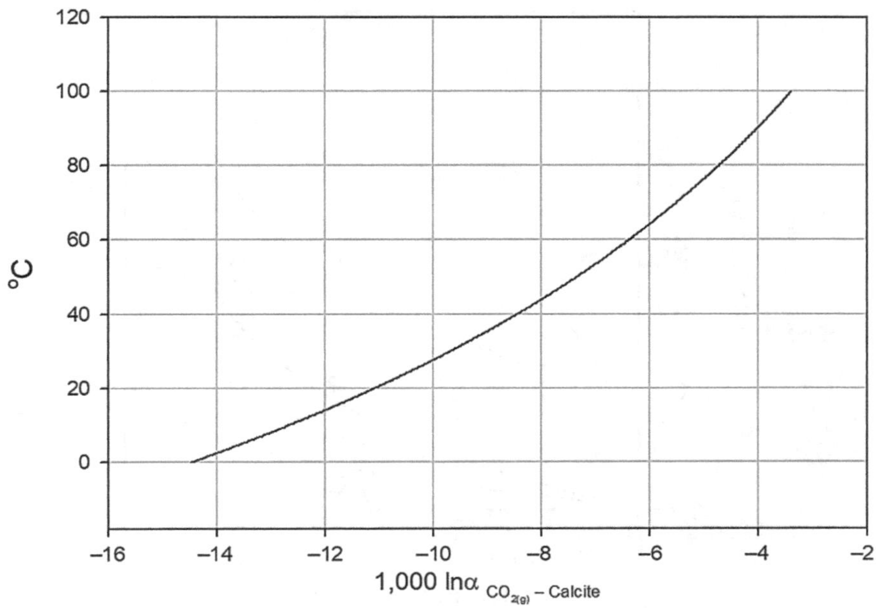

FIGURE 7.2 The $\delta^{13}C$ fractionation between $CO_{2(g)}$ and calcite.

liquids. The evolving isotopic composition of the residual material (i.e., reactant) can be described by:

$$\frac{R}{R^o} = \left(\frac{X}{X^o}\right)^{\alpha-1} \tag{7.23}$$

where R = ratio of isotopes (e.g., $\frac{^{18}O}{^{16}O}$) in the reactant, R^o is the initial ratio in the reactant, X is the concentration of the more abundant isotope (e.g., ^{16}O), and X^o is the initial concentration of the abundant isotope. Because the concentration of $X >> X_h$ (the heavier isotope), X is approximately equal to the amount of original material in the phase. If the $f = \frac{X}{X^o}$ = fraction of material remaining, then:

$$R = R^o \ln f^{(\alpha-1)} \tag{7.24}$$

Equation 7.24 can also be expressed as:

$$\delta = \delta^o \ln f^{(\alpha-1)} \tag{7.25}$$

where δ^o is the initial isotopic value. Equation 7.24 is valid for α values near 1, and δ^o values near 0 and less than 10.

Rayleigh fractionation is particularly useful for describing the isotopic evolution in open systems such as evaporation where the isotopic species removed are in thermodynamic and isotopic equilibrium with the remaining liquid (Figure 7.3). Rayleigh fractionation can also be used to describe closed system fractionation. In the figure, the fractionation curves bend upward because during evaporation the residual water becomes progressively enriched in the heavy isotopes. In the case of large water bodies, such as the ocean and many lakes, where the residual reservoir is essentially

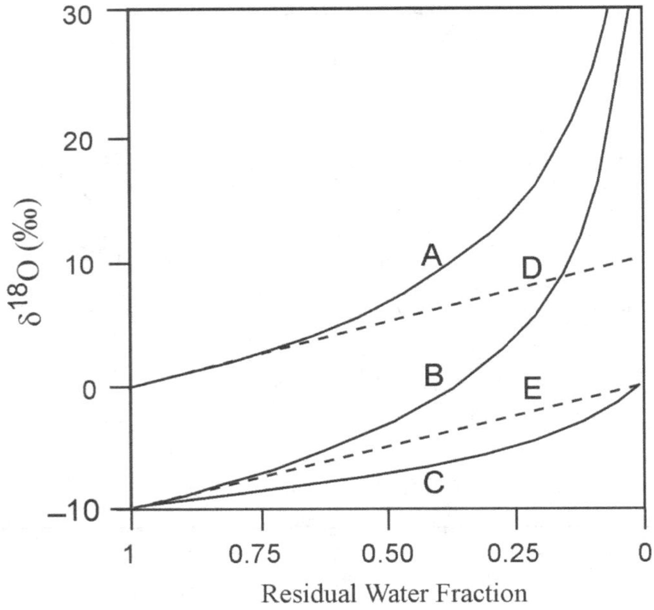

FIGURE 7.3 Isotopic exchange under open and closed Rayleigh fractionation of water evaporation with a fractionation factor $\alpha = 1.01$ and an initial liquid $\delta^{18}O = 0$ ‰ (modified after Gat and Gonfiantini 1981). The $\delta^{18}O$ of the residual water (solid line A), the instantaneous vapor removed (solid line B), and the accumulated vapor removed (solid line C) increase during evaporation. The $\delta^{18}O$ of a two-phase water closed system (dashed line D) and vapor (dashed line E) also increase during evaporation, but to a lesser extent.

infinite, the residual water fraction will approximate 1, and there will be no overall change in the isotopic composition of the liquid phase. The vapor phase will, however, become isotopically enriched in the lighter isotopes. The isotopic composition of an evaporating small water body will exhibit mass fractionation as the residual water fractionation becomes smaller. The usefulness of evaporation fractionation is further discussed in Chapter 8. Regardless of the size of the initial water body, most of the vapor phase will ultimately condense and fall out as precipitation as the cloud vapor moves across the ocean and inland. As the precipitation condenses, the heavier isotopes preferentially move from the air mass as a liquid and fall out of the sky. The residual air mass becomes progressively lighter, thus, subsequently, precipitation also becomes progressively lighter due to the rainout effect. A plot of isotopic precipitation has a downward slope because the residual is vapor and not liquid water.

At a specific fraction of the original vapor, the $d^{18}O$ of the residual vapor in the air mass can be calculated as:

$$\delta^{18}O_v \sim \delta_o{}^{18}O_v + \varepsilon_{l\text{-}v} \ln f \tag{7.26}$$

where $\delta^{18}O_v$ is the initial vapor d value and $\varepsilon_{l\text{-}v}$ is the liquid-vapor enrichment factor, and f is the fraction of the original vapor mass. The $\delta^{18}O$ of the precipitation can be calculated as:

$$\delta^{18}O_l \sim \delta_o{}^{18}O_v + \varepsilon_{l\text{-}v} \tag{7.27}$$

Relative humidity is the most important factor affecting the isotopic evaporation from an open surface water body. High humidity during evaporation results in small change in both $\delta^{18}O$ and δ^2H. At 95 percent humidity, the d value (deuterium excess) is similar to evaporation of the last 85 percent fraction of the water and results in an evaporation line with a slope of <8 on a $\delta^{18}O$–$\delta^{2}H$ plot. Evaporation under 100 percent humidity is essentially equivalent to closed-system evaporation, and on a plot of the data. The humidity factor is often important when evaluating the isotopic composition of paleo-groundwater when climate change is being considered.

Important Takeaway – Isotopic fractionation accompanies many common solution and dissolution reactions in water and during phase changes. Fractionation may occur without a change in the mass of reactants and products. Isotopic fractionation may be calculated in a manner similar to calculating chemical equilibrium. Chemical equilibrium is calculated using a temperature dependent equilibrium constant (K), and isotopic fractionation is calculated using a temperature-dependent fractionation factor (α).

7.5 RADIOACTIVE ISOTOPES

During radioactive decay the number of protons and neutrons in the nucleus change, transforming the parent radioactive element into a different stable or sometimes unstable (radioactive) daughter element. Radioactive decay or disintegration of the nucleus is spontaneous and random. Particles given off during decay are of three general types:

Alpha (α) – positively charged helium nuclei,
Beta (β) – electrons traveling at close to the velocity of light, and
Gamma (Γ) – electromagnetic radiation.

The rate of decay is described by the exponential law of radioactivity:

$$N = N_o \exp^{(-Bt)} \tag{7.28}$$

where N is the number of remaining radioactive parent atoms, N_o is the original number of atoms when time (t) was equal to 0, and B is the unique decay constant for the particular isotopic species. It is usually helpful to express the rate of decay of a particular species in terms of half-life ($t_{1/2}$):

$$t_{1/2} = \ln\frac{2}{B} \tag{7.27}$$

where the half-life is the time required for one-half of radionuclides to decay. The half-lives of the radioisotopes commonly used or having potential application in hydrogeology are ^{3}H, ^{14}C, and ^{36}Cl (Table 7.1.)

7.6 ISOTOPE REPORTING STANDARDS

Stable isotopic analyses are reported relative to arbitrary standards rather than as an absolute abundance, as is the case in solute analyses. The selection of standards is a very important issue in isotope geochemistry. Without internationally accepted standards, it would be impossible to compare the results of different laboratories and investigations. The International Atomic Energy Agency (IAEA) in Vienna, Austria, and the U.S. National Bureau of Standards (NBS) have coordinated the preparation, calibration, and distribution of internationally acceptable standards (IAEA, 1995, 2023a, 2023b). Many of the most common laboratory standards are listed in Table 7.3. In addition to the commonly used standards for reporting laboratory results, the IAEA distributes numerous lesser used standards for laboratory analysis. Many of the substitute reference standards start with the prefix V to indicate Vienna. For example, VSMOW2 is an acronym for the second iteration of

TABLE 7.3
The Evolution of Internationally Accepted Reference Standards for Reporting Isotopic Analysis

Acronym	Name	Substance
CDT	Canyon Diablo Troilite	meteoric sulfur
PDB	Pee Dee Formation *Belemnitella americana*	carbonate mineral
SLAP2	Standard Light Arctic Precipitation	water
SMOW	Standard Mean Ocean Water	water
VSMOW2	Vienna Standard Mean Ocean Water	water

Hydrogen – Oxygen
→ SLAP → SLAP2
SMOW → *VSMOW*
→ VSMOW → VSMOW2
Carbon – Oxygen
→ NPS19 → IAEA-063
PDB → VPDB
→ LSVEC
Sulfur
CDT → *VCDT* →UAEA-S-1

Standards in *italics* are the commonly used literature reference standards.

Vienna Standard Mean Ocean Water. Regardless of the reporting standard used, it is important to report the reference standard in the written communication of the results.

7.6.1 Oxygen

The commonly used reporting standard for $\delta^{18}O$ is VSMOW, whose isotopic composition is close to the average of ocean water. VSMOW2, artificially prepared and distributed by the IAEA, is a replacement for the earlier standards SMOW (Craig, 1961) and VSMOW, which are no longer available. The $\delta^{18}O$ value of VSMOW is 0 ‰.

Less frequently used standards include NBS-1, SLAP2, and PDB/NBS-20. The relationship between SMOW and NBS-1 is:

$$({}^{18}O/{}^{16}O)\ \text{SMOW} = 1.008\ ({}^{18}/{}^{16}O)\ \text{NBS-1} \tag{7.30}$$

SLAP2 (Standard Light Arctic Precipitation) is often used for analysis of ^{18}O depleted samples from high latitudes and arctic ice. The relationship between VSMOW2 and SLAP2 (IAEA, 2023b) is:

$$\delta^{18}O\ \text{SLAP2}/\text{VSMOW2} = -55.5\% \tag{7.31}$$

Knowledge of the $\delta^{18}O$ of a solid carbonate sample is often useful. Such measurements are usually made with reference to the carbonate standard PDB and then reported with respect to SMOW. The PDB standard is exhausted, but it has been calibrated against the VPDB laboratory reference standard.

The relationship between PDB and VSMOW (Friedman and O'Neil, 1977) is:

$$\delta^{18}O(VSMOW) = 1.03086\delta^{18}O(PDB) + 30.86\% \tag{7.32}$$

7.6.2 Hydrogen

Analysis of deuterium (^2H) from water samples is usually reported with respect to the standard VSMOW. Laboratory analyses may be performed using the water standards VSMOW2, SLAP2, or another IAEA standard. As in the case of the stable oxygen isotope ratios, the $^2H/^1H$ in SMOW (Craig, 1961) and VSMOW2 are very similar and are approximately equal to the theoretical composition of seawater. The abundance ratios are:

$$SMOW = 158 \pm 2 \times 10^{-6} \tag{7.33}$$

$$VSMOW2 = 155.76 \times 0.05 \times 10^{-6} \tag{7.34}$$

Numerical relationships between several of the standards are:

$$\delta^2H \text{ NBS-1 / SMOW} = -47.6\% (\text{Craig, 1961}) \tag{7.35}$$

$$\delta^2H \text{ SLAP/VSMOW} = -428\% (\text{Gonfiantini, 1978}) \tag{7.36}$$

VSMOW has a δ^2H value of 0 ‰.

Reference standards are usually not used for tritium (^3H) measurements because absolute concentrations may be determined by counting the decays of ^3H atoms. One tritium unit (TU) corresponds to 7.2 disintegrations per minute (dpm) per liter of water or 3.2 pCi/L. Dpm is a measure of the number of beta particles released during radioactive decay. This corresponds to an isotopic abundance of 10^{-18} ^3H atoms for each atom of ^1H.

7.6.3 Carbon

The reporting standard for the stable carbon isotopes $^{13}C/^{12}C$ is VPDB. The original PDB standard was carbon dioxide prepared from the shell of belemnite (*Belemnitella americana*) collected from the Peedee Formation (Upper Cretaceous) in South Carolina. The supply of PDB and the replacement VPDB are exhausted. The National Bureau of Standards distributed the substitute $\delta^{13}C$ reference standards NBS-19, NBS-20, and NBS-21 which are now exhausted. Craig (1957) determined the relationship of these standards to PDB as:

$$\delta^{13}C \text{ NBS-20 / PDB} = -1.06 \tag{7.37}$$

$$\delta^{13}C \text{ NBS-21 / PDB} = -27.79 \tag{7.38}$$

Friedman and O'Neil (1977) suggest the latter value is closer to −28.10 for NBD-21. The IAEA now distributes IAEA-603 and LSVEC reference standards for carbonate material. The reference standards are +2.46 and −46.6 ‰, respectively, relative to VPDB for $\delta^{13}C$ and −2.37 and −26.7 ‰, respectively, for $\delta^{18}O$. Regardless of the laboratory standard used, the results are reported relative to PDB and sometimes VPDB.

Unlike tritium measurements, ^{14}C analyses are reported with respect to a standard, and the absolute abundance of ^{14}C atoms in a water sample is not determined. The ^{14}C content of the sample is compared to the ^{14}C content of wood grown in 1860 in an environment relatively free of fossil carbon dioxide (i.e., prior to the industrial revolution). The ^{14}C content of a water sample is defined as the ratio of the measured disintegrations per minute per gram of carbon (dpm/g) of the sample versus the standard. The dpm are typically measured by counting the number of beta particles released during the isotopic decay. This ratio is referred to as the activity (*A*) of the sample. Because the concept of activity is used in relationship to solutes, gases, and isotopes, it is important to keep the context in mind when using the term. Analytical results are expressed in percent modern carbon (pmc) where:

$$A = \frac{\left({}^{14}C/{}^{12}C\right)_{sample}}{\left({}^{14}C/{}^{12}C\right)_{standard}} \times 100\ \text{(pmc)} \tag{7.39}$$

NBS oxalic acid is used for laboratory calibrations. The ^{14}C activity of modern carbon is now 13.56 dpm/gm (Choppin et al., 2002), or about 95 percent of the NBS oxalic acid.

7.6.4 Sulfur

The troilite phase (FeS) from the Canon Diablo iron meteorite (CDT) is the original reference standard for the sulfur stable isotope ratio $^{34}S/^{32}S$. In the standard the abundance ratio is 22.22, which is similar to the isotopic composition of mafic igneous rocks (Schneider, 1970). Most terrestrial and water samples have a $\delta^{34}S$ value that is within ± 40 ‰ of the standard. As with all original standards, CDT is exhausted, and substitute laboratory standards are used. Laboratory results are reported relative to VCDT using the IAEA-S-1 reference standard which has a $\delta^{34}S$ value of −0.30 ‰ relative to VCDT.

7.6.5 Nitrogen

Atmospheric nitrogen is used for both the working and reference standards for the stable isotope ratio $^{15}N/^{14}N$. A nitrogen gas sample, NBS-14, is used for laboratory calibration purposes.

7.6.6 Chlorine

Chlorine has stable and numerous radiogenic isotopes. Chlorine has atomic mass numbers ranging from 32 to 40. The two stable isotopes, ^{35}Cl (75.77 %) and ^{37}Cl (24.23 %), have the relative proportions of 3:1. The standard for the stable isotope ratio $^{35}Cl/^{37}Cl$ ($\delta^{37}Cl$) is Standard Mean Ocean Chloride (SMOC). Ocean chloride is thought to be homogeneous in oceans worldwide, where the chloride isotopic compositions are within approximately 0.15 ‰. NIST SRM 975 is internationally accepted as the standard for SMOC. The radioactive isotope chlorine-36 is measured using atomic mass spectrometry (AMS).

Important Takeaways – The original laboratory isotopic standards are exhausted, and the IAEA and NBS continuously develop and distribute substitute reference laboratory standards. The reference substitute standards are calibrated relative to the original standards, and laboratories report results referencing the original standards (i.e., VSMOW, VPDB, and VCDT). This standard referencing is essential for worldwide cross-correlation of results.

Because the supply of reference standards from the NBS and IAEA are limited and costly, for routine laboratory analysis laboratories use internal working standards that they create and calibrate relative to the reference standards purchased from the IAEA and NBS.

Problems

1. A laboratory reported the $\delta^{18}O$ of a water sample is –37.3 ‰ relative to PDB. What is the $\delta^{18}O$ relative to VSMOW?
2. Why are the isotopes ^{18}O and ^{16}O used in isotopic analysis, and what are the practical limitations of using ^{17}O?
3. How would a laboratory develop internal working standards using a standard purchased from the IAEA?
4. What is the difference between isotope equilibrium constant (K) and α isotope fractionation factors (α and ε), and how are they used?

Chapter Notation

Superscript

a,b,c,d	moles of species A, B, C, and D, respectively
A	atomic mass number (number or protons plus number of neutrons)
B	decay constant

Subscript

g	gas
Z	atomic number (number of protons)
[]	effective molar concentration
‰	per mill
α	isotopic fractionation factor
$\alpha_{A\text{-}B}$	isotopic fractionation factor between substances or phases A and B
ε	isotopic enrichment factor
$\delta_{A\text{-}B}$	isotopic enrichment factor between substances or phases A and B
δ	delta unit used for reporting isotopic abundance in the ‰ difference relative to the standard
δ_A	per mil isotopic fractionation of substance A
δ_B	per mil isotopic fractionation of substance B
a,b,c,d	moles of species A, B, C, and D, respectively
A	chemical activity of a ^{14}C sample
A,B,C,D	any charged or uncharged chemical molecule or species or any phase
A,B,C	experimentally determined coefficients
dpm/g	disintegration per minute per gram
exp	exponent
$f(T)$	function of temperature
K	thermodynamic equilibrium constant
K_{eq}	thermodynamic equilibrium constant
ln	natural logarithm (logarithm to the base e); ln a = 2.303 log a
$\ln\alpha$	natural log of the isotopic fractionation factor
n	number of atoms exchanged
N	number of radioactive parent atoms
N_o	number of atoms when time = 0
NBS-1	National Bureau of Standards isotopic standard #1
NBS-1A	National Bureau of Standards isotopic standard #1A
NBS-14	National Bureau of Standards isotopic standard #14
NBS-20	National Bureau of Standards isotopic standard #20
NBS-21	National Bureau of Standards isotopic standard #21

pCi/L	picocurie per liter
pmc	percent modern carbon; 1890 reference date
PDB	Pee Dee Belemnitella (carbon isotopic standard)
R	ratio of heavy isotope nuclides to number of light isotope nuclides
R_A	ratio of heavy isotope nuclides to number of light isotope nuclides in species A
R_B	ratio of heavy isotope nuclides to number of light isotope nuclides in species B
SMOW	Standard Mean Ocean Water (isotopic standard)
SLAP	Standard Light Arctic Precipitation (isotopic standard)
VSMOW	Isotopic standard
$t_{1/2}$	half life
T	temperature
TU	tritium unit; 1 TU = 3.2 pCi L^{-1}
X	permil fractionation
X	any element

REFERENCES CITED

Bigeleisen, J., and M.G. Mayer, 1947, Calculation of equilibrium constants for isotopic exchange reactions, *Journal of Chemical Physics*, 15, 261–267.

Choppin, G.R., et al., 2002, Radionuclides in nature in radiochemistry and nuclear chemistry, 3rd ed, *Butterworth-Heinemann*, 5, 94–122.

Craig, H., 1957, Isotopic standards for carbon and oxygen and correction factors for mass-spectrometric analysis of carbon dioxide, *Geochimica et Cosmochimica Acta*, 12, 113–149.

Craig, H., 1961, Standard for reporting concentrations of deuterium and oxygen-18 in natural water, *Science*, 33, 1833–1834.

Friedman, I., and J.R. O'Neil, 1977, *Compilation of stable isotope fractionation factors of geochemical interest – data of geochemistry part 440-KK*, U.S. Geological Survey Professional Paper 440-KK, 12.

Gat, G.R., and R. Gonfiantini, 1981, *Stable isotope hydrology deuterium and oxygen-18 in the water cycle*, International Atomic Energy Agency Technical Report, Series 210.

Gonfiantini, R., 1978, Standards for stable isotope measurements in natural compounds, *Nature*, 271, 534–536.

IAEA, 1995, Reference and intercomparison material for stable light elements, *International Atomic Emergency Agency, Techdoc*, 825.

IAEA, 2023a, *Reference products for environment and trade*, International Atomic Emergency Agency.

IAEA, 2023b, *Reference sheet for international measurement standards VSMOW2 and SLAP2*, International Atomic Emergency Agency.

Melander, L., 1960, *Isotope effects on reaction rates*, Ronald Press, New York.

Otero, N., et al., 2007, Using dual-isotope data to trace the origin and processes of dissolved sulphate: A case study in Calders stream (Llobregat basin, Spain), *Aquatic Geochemistry*, 13, 109–126.

Schneider, A., 1970, The sulfur isotope composition of basaltic rocks, *Contributions to Mineralogy and Petrology*, 254, 95–124.

Taylor, H.P., Jr., 1974, A low-^{18}O, late Precambrian granite batholith in the Seychelles Islands, India Ocean: Evidence for the formation of ^{18}O-depleted magmas and interaction with ancient meteoric groundwaters, *Geological Society of America, Abstracts with Programs*, 6, 981–982.

Taylor, H.P., Jr., 1977, Water/rock interactions and the origin of H_2O in granite batholiths, *Journal Geological Society (London)*, 133, 509–558.

Urey, H.C., 1947, The thermodynamic properties of isotopic substances, *Journal Chemical Society (London)*, 562–581.

Zhang, D., et al., 2014, Using dual isotopic data to track the sources and behaviors of dissolved sulfate in the western North China Plain, *Applied Geochemistry*, 52, 43–56.

8 Hydrogen and Oxygen Isotopes

8.1 INTRODUCTION

Hydrogen has two stable isotopes, ^1H and ^2H, whose terrestrial isotopic abundance's average 99.984 and 0.015 percent, respectively. Oxygen has three stable isotopes, ^{16}O, ^{17}O and ^{18}O, whose terrestrial isotopic abundance's average 99.76, 0.037, and 0.10 percent, respectively (Table 7.1). Water molecules have nine different stable isotopic configurations whose atomic masses range from 19 to 22. Only the molecules $^1H_2{}^{16}O$, $^1H^2H^{16}O$, and $^1H_2{}^{18}O$, which represent 99.78, 0.02, and 0.20 percent of all water molecules, respectively, occur in sufficient amounts to be useful in hydrology.

The study of the stable isotopic ratios $^2H/^1H$ and $^{18}O/^{16}O$ in natural waters began in the 1930s (Gilfillan, 1934; Harada and Titani, 1935; Lewis and Cornish, 1933; Tes, 1939) and resumed in earnest after World War II (Craig et al., 1956; Dansgaard, 1953, 1954, 1964; Ehhalt et al., 1963; Epstein, 1956; Epstein and Mayeda, 1953; Friedman, 1953; Kobayakawa and Horibe, 1960; Craig, 1961). It was during the post-war period that the fundamental principles on which all modern research is based were established.

The physiochemical behaviors of the various isotopic forms of water depend on their chemical activities. Chemical activity is related to vapor pressure, which in turn is a function of molecular vibrational frequency. The lightest water molecule, $^1H_2{}^{16}O$, has a higher vapor pressure than does the heaviest molecule. Differences in chemical activities tend to concentrate water molecules containing the heavy isotopes ^2H and ^{18}O in the liquid phase of the hydrologic reservoirs (e.g., ocean, atmosphere, precipitation, surface water, and groundwater). Such concentration is due to isotopic fractionation accompanying phase changes. Phase changes affecting water molecules include precipitation (vapor to liquid), evaporation (liquid to vapor), sublimation (solid to vapor), and freezing (liquid to solid).

Other oxygen isotopic fractionations in systems such as sulfate–water and calcite–water generally have minimal impact on the isotopic composition of water due to the large isotopic buffering effect of the oxygen in water. The significance of this is that the stable isotopic composition of most groundwater is not affected by mineral–water interactions and the isotopic composition of groundwater is usually the same as the isotopic composition of the recharge water. What this means is that in groundwater systems where precipitation does not undergo significant sublimation or evaporation on the land surface or in the soil, the isotopic composition of the groundwater provides information about the precipitation recharge conditions.

The stable isotopic compositions of hydrogen and oxygen are expressed in terms of the delta notation where:

$$\delta^2H = \left[\frac{\left({}^2H/{}^1H\right)_{sample}}{\left({}^2H/{}^1H\right)_{standard}} - 1 \right] \times 10^3\,(‰) \tag{8.1}$$

$$\delta^{18}O = \left[\frac{\left({}^{18}O/{}^{16}O\right)_{sample}}{\left({}^{18}O/{}^{16}O\right)_{standard}} - 1 \right] \times 10^3\,(‰) \tag{8.2}$$

DOI: 10.1201/9781003646129-10

Hydrogen and oxygen isotopic standards are discussed in Chapter 7.

Important Takeaway – Variations in the stable isotopic composition of water are extremely small and are reported as permil and not as percent differences relative to the standard Vienna Standard Mean Ocean Water (VSMOW). These differences are reported using the δ notation.

The isotopic composition of groundwater is usually the same as the recharge water and is generally not affected by rock–water interactions.

8.2 GLOBAL METEORIC WATER LINE (GMWL)

In his seminal paper, Harmon Craig (1961) analyzed 400 samples from worldwide locations of rivers, lakes, and precipitation for their stable isotopic ratios. About 40 percent of the samples were from North America. The purpose of Craig's research was to establish the isotopic relationships of meteoric water. Except for evaporated water in closed basins, the plotted trend of the data is represented by the equation:

$$\delta^2H = 8\delta^{18}O + 10\ ‰ \tag{8.3}$$

Equation 8.3 was originally called the Craig Line and is now known as the widely used Global Meteoric Water Line (GMWL). In the equation, the value 8 is the line slope and +10 is the deuterium intercept on the y-axis. The y-intercept is also known the deuterium excess. Craig found that the GMWL is consistent with Rayleigh liquid–vapor equilibrium distillation between –10 and 100 °C. Warmer precipitation corresponds to less negative and colder precipitation temperatures correspond to more negative isotopic values on the GMWL.

In his seminal work, Dansgaard (1964) analyzed 1,126 worldwide stable isotopic results from 84 International Atomic Energy Agency–World Meteorological Organization (IAEA–WMO) precipitation network stations. The precipitation network included all continents, except for inland Asia, and included numerous oceanic islands. Most of Dansgaard's stations were located near an ocean, and each dataset included at least one full year of measurements. He found that local precipitation often deviates from the GMWL, and that the data commonly deviate from Rayleigh distillation in both the line slope and the y-intercept. He suggested that a more correct form the of δ^2H and $\delta^{18}O$ relationship in coastal precipitation should be:

$$\delta^2H = s\delta^{18}O + d\ (‰) \tag{8.4}$$

where s is the line slope and d is the deuterium or ^2H excess. The slope of the line is primarily a function of isotope fractionation between water and vapor at the ocean water–vapor interface and between vapor–water during condensation accompanying cloud formation and precipitation (Merlivant and Jouzel, 1979). Under most evaporation and condensation conditions, the d-parameter is predominantly controlled by the humidity at the ocean surface. As a general rule, the deuterium excess will increase, that is the d-parameter value will become smaller, with decreasing humidity (Gat, 1983). Variations in the d-parameter in groundwater relative to local modern precipitation may be an indication of paleo-groundwater recharge, which is discussed later in this chapter.

Most precipitation is returned to the atmosphere by evapotranspiration processes. Dansgaard (1964) suggested that values of $s < 8$ and $d < 10$ may indicate evaporation of some water during or after precipitation. Convective rainstorms may also influence the d-parameter. Changes in d values due to convective storms typically occur in flat land where precipitation is induced by differences between land surface and air condensation temperatures.

It is important to note that the GMWL is only representative of average annual modern meteoric data with respect to worldwide coastal reporting stations. Adjustments, usually empirical, must be made for regional and local conditions and for paleo-groundwater. Because the GMWL is only technically valid for coastal zone precipitation, it is reasonable to ask how the GMWL can be applied to hydrology and hydrogeology investigations of terrestrial surface and groundwater. The short answer is that it is the best readily available approximation for local precipitation, and the isotopic compositions of surface and groundwaters usually do not stray too far from the composition of local precipitation.

The way we use the GMWL is to plot the precipitation and surface and groundwater δ^2H and $\delta^{18}O$ data relative to a straight-line segment of the GMWL that is in the range of the local data. For example, if the local data had $\delta^{18}O$ values ranging from −16 to −12 ‰, it might be reasonable to plot a segment of the GMWL with values ranging from −20 to −8 ‰. The GMWL is used as a reference when sufficient local long-term precipitation data are not available, which is usually the case. The δ^2H vs. $\delta^{18}O$ relationship for precipitation in a particular region often has a slope similar to the GMWL, but the data may fall above or below the line due to vapor–precipitation fractionation in the cloud. North American precipitation data often plot slightly sub-parallel to the GMWL. In other words, the d-excess is less than 10 ‰, and the slope is approximately the same as the GMWL. Precipitation data from Australia, the Middle East and Central Africa often plot above the GMWL (Gat, 1980). When sufficient local precipitation data are available it is sometimes useful to calculate a local meteoric water line (LMWL) or even a local groundwater line (LGWL) from the data and then compare this to a similar segment of the GMWL. A few years of local precipitation data are seldom sufficient to develop a LMWL. Stable isotopic data have been used by Benjamin et al. (2004), Klaus et al. (2015), and Yeh et al. (2014) to develop LMWLs. If sufficient data are available, a LMWL or LGWL equation can be calculated using regression techniques.

8.3 METEORIC AND GROUNDWATER

Almost all surface and groundwater have a meteoric origin that originated by the evaporation of an ocean reservoir. After evaporation, the isotopic composition of precipitation and groundwater may have been affected by rainout, evaporation of surface water, and other factors. The isotopic composition of ocean water is considered be 0 ‰ for both δ^2H and $\delta^{18}O$ with respect to Vienna Standard Mean Ocean Water (VSMOW). In earlier literature the reference standard is Standard Mean Ocean Water (SMOW). See Chapter 7 for a discussion of reporting standards. The power of using δ^2H and $\delta^{18}O$ in aqueous investigation is based on the fact that the isotopic evolution of water in the hydrologic cycle occurs in a predictable fashion, and this evolution is based on observable meteorological and hydrological events and processes (Figure 8.1). This evolution is based on temperature-dependent isotopic fractionation accompanying phase changes and on the isotopic behavior of water with respect to soil and bedrock minerals. When water evaporates, the more volatile isotopic forms, such as ${}^1H_2{}^{16}O$, evaporate preferentially relative to the heavier isotopic forms, and when water condenses the heavier isotopes preferentially condense.

Isotopic fractionation during a phase change is described by the Rayleigh equation, also known as the Rayleigh distillation:

$$R / R^o = \text{f ln}^{\alpha-1} \tag{8.5}$$

or

$$\delta = \delta^o \text{lnf}^{(\alpha-1)} \tag{8.6}$$

where R is the ${}^{18}O/{}^{16}O$ or ${}^2H/{}^1H$ ratio of the remaining vapor, R^o is the ${}^{18}O/{}^{16}O$ or ${}^2H/{}^1H$ ratio of the vapor before condensation, f is the fraction of vapor remaining, and α is the isotope fractionation

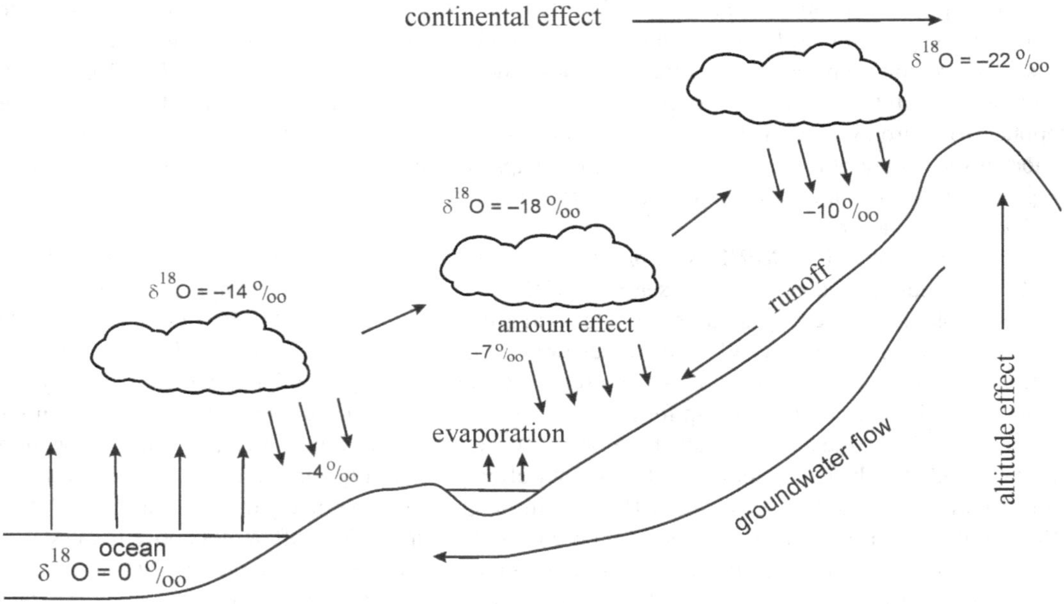

FIGURE 8.1 The stable isotopic evolution from seawater to mountain precipitation. Inland changes in the isotopic composition are due to isotopic fractionation by Rayleigh distillation that amplifies the amount, altitude, and continental effects, which result in progressively more negative isotopic compositions.

factor. Due to Rayleigh distillation, the vapor phase becomes enriched in ^1H and ^{16}O and the parent (residual) phase becomes enriched in ^2H and ^{18}O (Figure 8.2). What this means is that the δ^2H and $\delta^{18}O$ compositions of the vapor phase are negative with respect to the parent phase. The actual isotopic composition of the vapor usually does not precisely conform to the predicted values based on evaporation equilibrium. Craig and Gordon (1965) found that this nonideal behavior is due to kinetic factors.

Unlike the vapor phase, the isotopic composition of ocean water and most rivers and lakes is generally not changed by evaporation due to the isotopic buffering capacity of most large water bodies. Buffering occurs because the volume of the water body is great with respect to the volume of the vapor phase in the overlying air mass. On the other hand, the isotopic composition of a water body subjected to excessive evaporation, such as a playa lake or an evaporation pond, is often affected by isotopic fractionation.

The isotopic composition of water is usually not affected by changes in solute chemistry making it a powerful isotopic tool for the evaluation of groundwater flow systems. The δ^2H and $\delta^{18}O$ data may often be used to identify groundwater from specific aquifers (i.e., fingerprint the water), and the δ^2H and $\delta^{18}O$ composition may provide independent information regarding the origin of the groundwater (i.e., magmatic, connate, meteoric, evaporated surface water), recharge conditions (i.e., climate and paleoclimate, season, and elevation), subsurface thermal conditions, and groundwater mixing patterns.

Application of hydrogen and oxygen stable isotope geochemistry in groundwater investigations is based on several principles:

1. Almost all groundwater is recharged from meteoric sources.
2. Groundwater in different aquifers is often recharged from precipitation that condensed from air masses with different temperatures or from different areas.
3. The stable isotopic composition of groundwater is not generally affected by changes in solute chemistry.

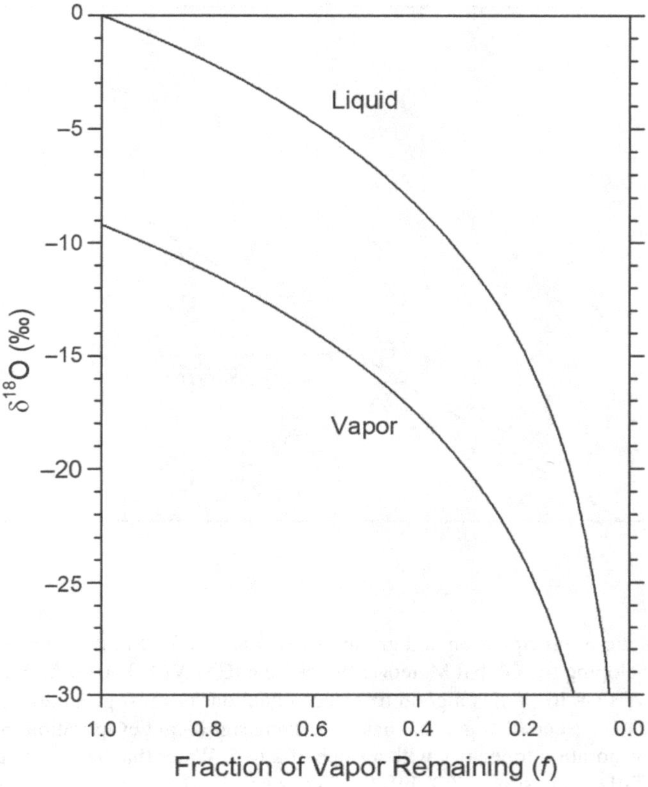

FIGURE 8.2 Fractionation of oxygen isotopes during water condensation from vapor at 25 °C using the Rayleigh distillation model (Equation 8.6), where $\alpha = 1.0092$ and the initial vapor has a $\delta^{18}O$ of −9.2 ‰. As precipitation proceeds and the fraction of the vapor remaining becomes smaller, both the residual vapor and the liquid (i.e., precipitation) become isotopically more negative. This model explains why freshwater is isotopically negative relative to seawater.

4. The isotopic composition of groundwater reflects the isotopic composition of the recharge water unless it has been altered by evaporation, subsurface thermal heating, or other unusual conditions.
5. Groundwater whose origin is not meteoric, such as marine connate or magmatic, has an isotopic composition that is understood and is not related to the GMWL.

The location of isotopic compositions of groundwater with respect to the GMWL (Figure 8.3) often provides insight into recharge altitude, recharge season, and recharge during cooler or warmer climates (Line A–B), evaporation (Line C–D), and geothermal heating (Line E–F). Because meaningful differences in isotopic compositions can be measured in tenths of a mil for $\delta^{18}O$ and a few mils for $\delta^{2}H$, great care should be exercised when evaluating the data. It is often useful to work with mean isotopic values, and considerable caution is needed when evaluating small datasets. Individual datum usually should not be micro-analyzed, particularly when the differences between the isotopic values are small.

Important Takeaway – Plotting $\delta^{2}H$ and $\delta^{18}O$ data relative to the GMWL is the starting point for most stable isotopic investigations. Determining the slope, ^{2}H intercept, and deviation of the data relative to the GMWL often provide insight into the thermal history of the recharge water and hence insight into the groundwater flow history.

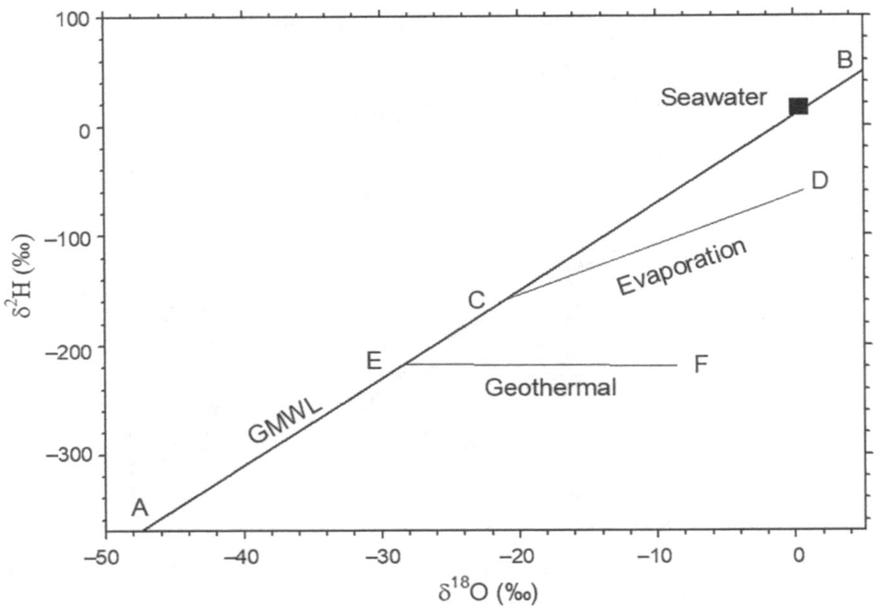

FIGURE 8.3 Schematic of precipitation and groundwater δ^2H and δ^{18}O in the range of data values used by Craig (1961) when developing the Global Meteoric Water Line (GMWL). The GLMW (line A–B), defined by the equation $\delta^2H = 8\ \delta^{18}O + 10$ (‰), is a good reference equation for most groundwater investigations. Line C–D illustrates the plotting trend of water that has undergone significant evaporation. Such evaporated water typically falls on an evaporation trajectory with a slope of 2 to 5. Water that has been geothermally heated to ~100 °C or more may fall on a positive δ^{18}O shift line such as E–F. The horizontal shift in δ^{18}O is the result of isotopic fractionation between the aquifer minerals and the water. Simultaneous hydrogen fractionation does not occur because aquifer minerals contain no or only limited amounts of hydrogen.

8.3.1 Isotopic Trends in Meteoric Water

Since 1964 it has been well documented that the stable isotopic composition of precipitation and groundwater may be effected by geographic and meteorological factors including air temperature, latitude, land surface elevation (altitude effect), distance from the coast (rainout effect), precipitation intensity (amount effect), season, climate and paleoclimate, and storm tracks (Clark and Fritz, 1997; Dansgaard, 1964; Gat, 2001; Merlivant and Jouzel, 1979; Rozanski et al., 1993; Siegenthaler and Oeschger, 1980; Yurtsever, 1975). Stable isotopes have also been used to evaluate interannual temperature changes (Bowen, 2008; Cai and Tian, 2016). It is important to bear in mind that all of the factors affecting the stable isotopic composition of precipitation and hence the isotopic composition of groundwater are surrogates for temperature and humidity conditions accompanying isotopic fractionation. Our task is to sort out the precipitation and hydrogeologic conditions that resulted in the isotopic compositions. Many of the effects discussed here are interrelated, and sometimes it can be difficult to sort out one effect from another. It is often necessary to consider the interactions of hydrogeologic, hydrogeochemical, and radiogenic reactions, climate and weather patterns, and geologic factors to fully interpret the stable isotopic data.

8.3.1.1 Air Temperature

Dansgaard (1964) found the following empirical relationships between air temperature and the stable isotopic composition of precipitation in the North Atlantic:

$$\delta^2H = 5.6T - 100\ (‰) \tag{8.7}$$

$$\delta^{18}O = 0.695T - 13.6\ (‰) \tag{8.8}$$

where T is the mean annual air temperature in °C. Similar correlations have been found for Central Asia (Wang, 2016):

$$\delta^{18}O = 0.78T - 16.01\,(‰) \tag{8.9}$$

and for southwest Oregon (Ersek et al., 2010):

$$\delta^{18}O = 0.9T - 15.96\,(‰) \tag{8.10}$$

However, Li et al. (2023) found the $\delta^{18}O$–T relationship for 28 North America stations to only be: 0.09–0.49 ‰ per °C

For Antarctica, Dansgaard (1964) found that:

$$\delta^{18}O = 6.04 - 51\,(‰) \tag{8.11}$$

The relationships expressed by Equations (8.7 to 8.11) are not completely linear, and on a worldwide scale, all values approach $\delta = 0$ ‰ at high air temperatures.

Equations 8.7 and 8.8 were developed from data representing very large areas and should not be used to calculate absolute temperatures for the condensation of precipitation. The equations do, however, provide first approximations, and in the absence of local data may be qualitatively useful for estimating relative seasonal differences in precipitation temperatures.

8.3.1.2 Amount, Rainout, and Continental Effects

As a storm progresses, precipitation becomes isotopically lighter due to Rayleigh distillation. The condensation of water vapor during a storm event enriches the water phase (precipitation) in ^{2}H and ^{18}O. Progressive precipitation results in the preferential removal of ^{2}H and ^{18}O from the air mass and the enrichment of the remaining vapor phase in ^{1}H and ^{16}O. Thus, the $\delta^{2}H$ and $\delta^{18}O$ of the air mass and any subsequent precipitation become progressively more negative as the storm progresses (Figure 8.1). The isotopic response to this moisture loss has been described as the amount or the rainout effect and the continental effect. Dansgaard (1964) was the first to describe the relation between $\delta^{18}O$ and the amount of precipitation. The amount or rainout effect is based on the idea that as the storm progresses during a precipitation event, the atmospheric moisture is enriched in ^{2}H and ^{18}O and the cloud vapor becomes enriched in ^{1}H and ^{16}O. As the precipitation condenses the heavier isotopes are preferentially removed from the air mass as a liquid and fall out of the sky. The residual air mass becomes progressively isotopically lighter; thus, subsequent precipitation also becomes progressively lighter due to the so-called rainout effect. It is sometime useful to think of the amount effect as the location-dependent fractionation accompanying a single, relatively short-lived, and often intense precipitation event, and the rainout effect as the isotopic depletion of cloud moisture as the air mass moves inland and loses water during precipitation over a large area. The physical process responsible for the amount effect have been investigated by Risi et al. (2008), Lee and Fung (2007), Mix et al. (2019), and Tindall et al. (2009).

The continental effect is the rainout effect applied to a very large geographic region where the isotopic composition is affected by both the rainout effect and re-evaporation of precipitation and surface water. Polyakov and Kolesnikov (1978) found the following empirical continental effect for regions having an annual average air temperature of less than about 20 °C:

$$\delta^{2}H = 8\delta^{18}O + 10 - 0.7A^{2}\,(‰) \tag{8.12}$$

where A is the distance from the ocean water vapor source, in thousands of kilometers. In coastal regions and islands, $A = 0$.

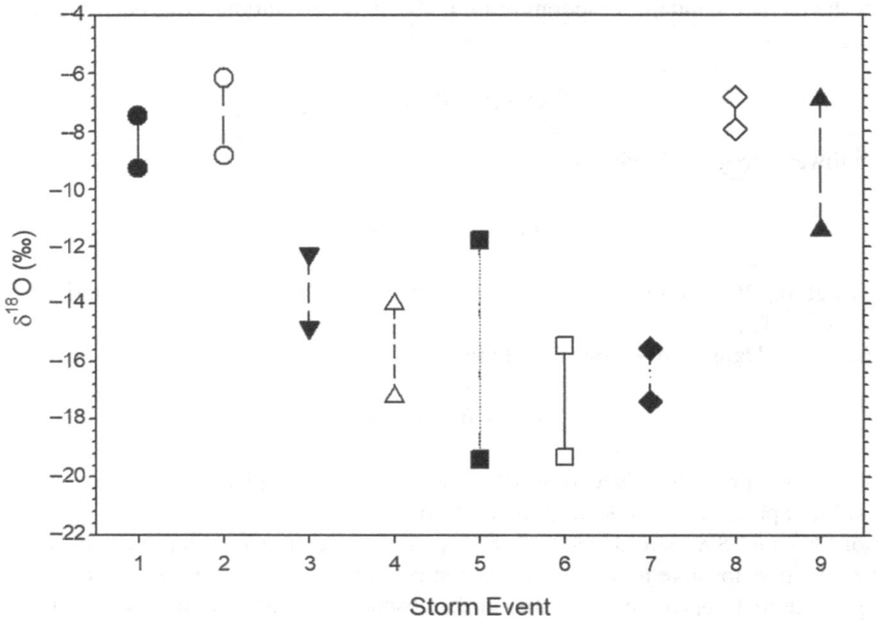

FIGURE 8.4 Intra-storm stable isotopic changes in precipitation during single day storm events in central Utah (elevation 2,280 m amsl). The daily variations ranged from 1.11 to 7.62 ‰ and averaged 3.25 ‰. Sampling events ranged from 5 to 11 hours apart. Unpublished data from Alan Mayo, David Tingey, and Kevin Rey.

Although $\delta^{18}O$ variations as great as −15 ‰ have been observed in the intra- and inter-storm precipitation (Pionke and DeWalle, 1992; Gat et al., 2002; Han et al., 2020), there is little evidence for seasonal or longer amount effects in groundwater (Easto and Dettman, 2016; Hager and Foelsche, 2015). Intra-storm daily and intra-storm duration variability in $\delta^{18}O$ are illustrated in Figures 8.4 and 8.5. The average intra-storm single day variation ranged from −1.1 to −7.62 ‰ and averaged −3.25 ‰ for samples collected 5 to 11 hours apart (Figure 8.4). Daily intra-storm $\delta^{18}O$ differences for storms ranging in duration from 2 to 4 days varied from 0.3 to −19.6 ‰ and averaged 3.85 ‰ over the duration of the storms (Figure 8.5).

Land surface evaporation usually does not appreciably change the isotopic composition of recharge water. Fractionation accompanying the partial evaporation of falling rain (virga effect), sublimation of snow, and evaporation of soil zone moisture may alter the isotopic composition of groundwater recharge. Sublimation of snow and evaporation of rain during its descent, which is common in arid environments during summer months, will enrich the recharge water in the heavier ^{18}O and ^{2}H isotopes (Gat, 1983). Groundwater recharged by snowmelt may have more positive values than the snow (Zentner, 2001). Changes in the isotopic composition of recharge water from rain are relatively small and rarely exceed 1 ‰ in $\delta^{18}O$ (Bath, 1983; Gat, 1983). The largest changes occur in arid and semi-arid regions. Where vegetation covers the recharge area, the fractionation accompanying infiltration is usually insignificant.

8.3.1.3 Latitude Effect

Worldwide mean annual $\delta^{18}O$ precipitation data are illustrated in Figure 8.6a and for North American in Figure 8.6b. Low latitude precipitation is generally isotopically heavier than high latitude precipitation because of a combination of factors including the:

1. rainout effect induced by the poleward movement of air masses,
2. a general decrease in air temperature with increasing latitude, and
3. recycling of water into the atmosphere by re-evaporation of water from rain droplets and from the evaporation of surface terrestrial water.

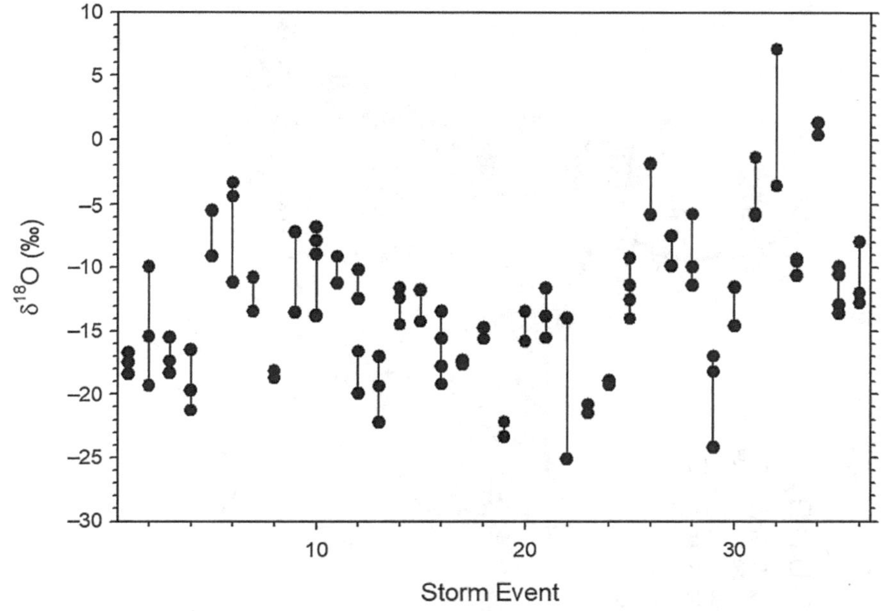

FIGURE 8.5 Intra-storm stable isotope variations in central Utah precipitation during storms with duration of 2 to 4 days (elevation 2,280 m amsl). Each filled circle represents a single day. Nearly every month of the year is represented in the data, and all of the data are the average daily compositions of consecutive days during the storm event. Winter months include snow and rain. The maximum storm differences vary from 0.3 to 10.6 ‰ and average 3.85 ‰. Unpublished data from Alan Mayo, David Tingey, and Kevin Rey.

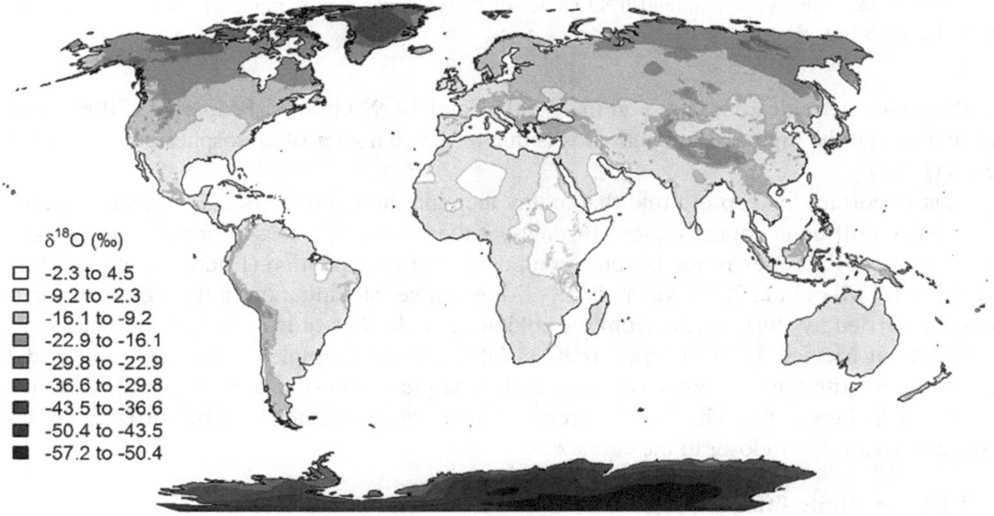

FIGURE 8.6a The average annual $\delta^{18}O$ values of global precipitation. Used with permission of waterisotopes.org.

The $\delta^{18}O$ latitude effect over North America is about 0.5 ‰ per degree of latitude (Yurtsever, 1975); however, it is more pronounced in the western Great Basin and the central and northern Rocky Mountains. High latitude locations have less negative $\delta^{18}O$ values in the summer and more negative values in the winter. Tropical areas have the reverse condition. The boundaries between

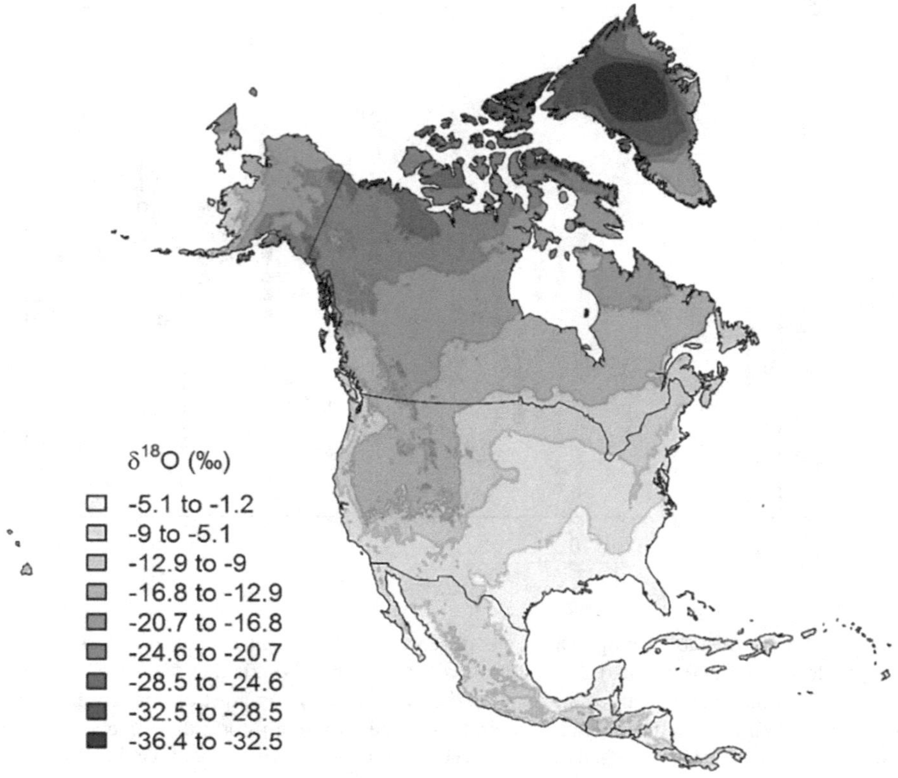

FIGURE 8.6b The average annual $\delta^{18}O$ values of North American precipitation. Used with permission of waterisotopes.org.

high and low latitude zones are nominally 30 °S and 30 °N latitude (Feng et al., 2009). The difference between the high and low latitudes is due to the character of atmospheric general circulation cells (AGC).

It is important to keep in mind that at any location the latitude effect is not static. Storms may arrive from different source regions throughout the course of a season or year, and there are corresponding variabilities in the isotopic composition of precipitation (Friedman et al., 2002; Wang et al., 2012; Mix et al., 2019; Mayo, 2023). For example, precipitation in the western United States may be carried by storm tracks from the cold arctic (>30 °N latitude), the warm South Pacific, and the gulfs of Mexico and California (<30 °S latitude). An example of the storm tracks dynamics resulting in latitudinal differences in the stable isotopic compositions of precipitation source terms is shown in Figure 8.7. The latitude trends reflect colder ocean water in northern latitudes and warmer ocean water closer to the equator.

8.3.1.4 Altitude Effect

The altitude or topographic effect is based on the idea that high-elevation precipitation condenses at lower temperatures than does nearby low-elevation precipitation. This phenomenon is observed by snow falling in the mountains and rain in the adjacent lowland during a storm event. What the altitude effect means is that during a given storm event in mountainous terrain, the stable isotopic composition of high elevation precipitation will plot more negatively relative to the GMWL than will the lower elevation precipitation. Typical $\delta^{18}O$ and $\delta^{2}H$ precipitation altitude gradients range from −0.15 to −0.5 ‰ and −1.2 to −4 ‰ per 100 m of elevation increase, respectively (Gat, 1980;

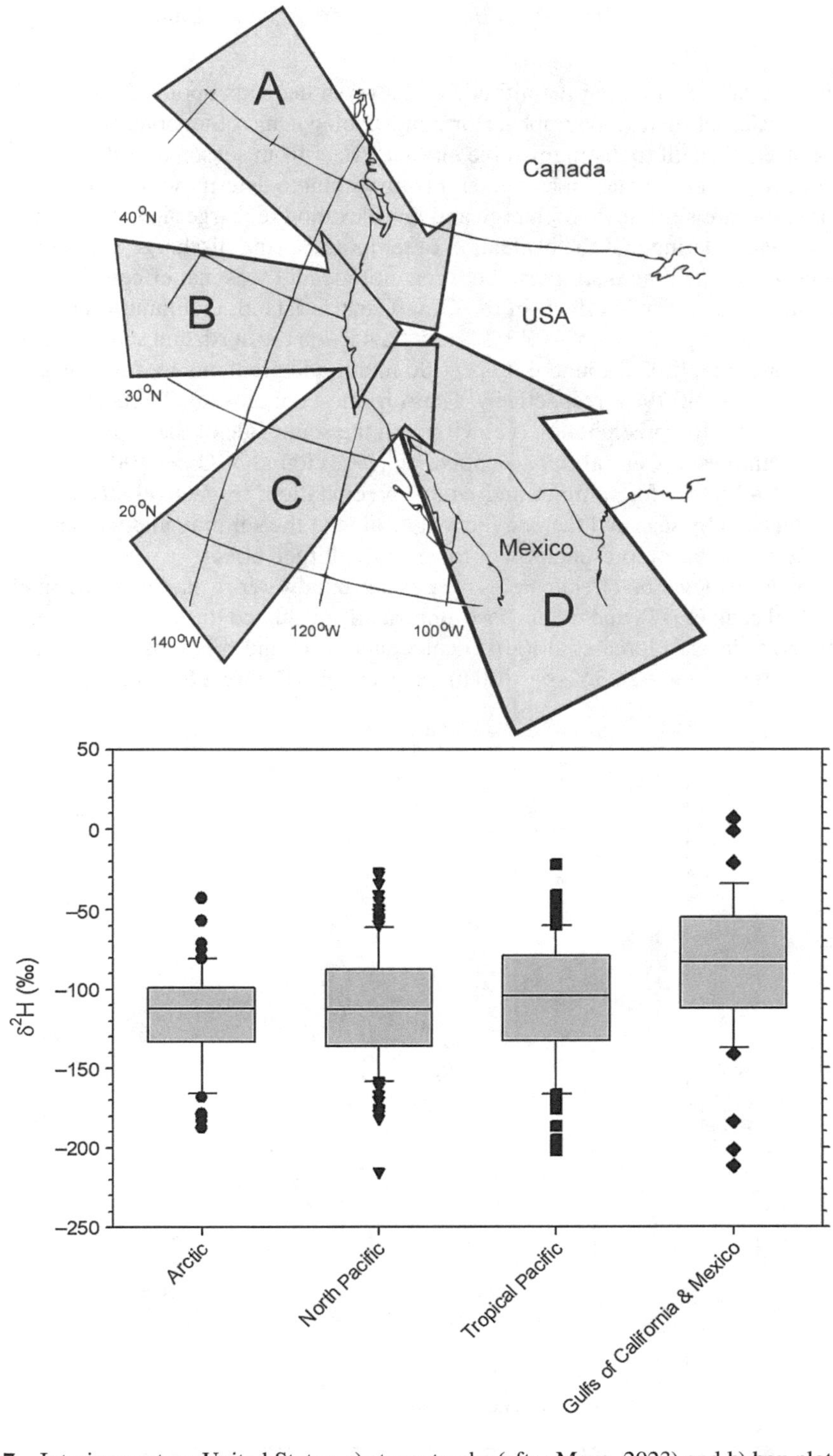

FIGURE 8.7 Interior western United States: a) storm tracks (after Mayo, 2023) and b) box plots of the isotopic compositions of storm track precipitations (unpublished data from Alan Mayo and David Tingey). The storm tracks are (A) Arctic, (B) North Pacific Ocean, (C) South Pacific Ocean, and (D) Gulfs of Mexico and California. Arrow sizes are not qualitatively meaningful. The median δ^2H values for Arctic, North Pacific, South Pacific, and Gulfs are −117.8, −112.8, −104.8, and −86.3 ‰, respectively. The top and bottom of each box is the upper and lower quartile, and the line in the middle is the average values. Maximum values are shown by whiskers and outlier days lie beyond the whiskers.

Eriksson, 1965, 1983). Evaluating the altitude effect often depends upon sorting out the influence of regional and microclimate, topography, storm morphology, and other conditions. In groundwater samples it is often difficult to distinguish the altitude effect from seasonal or paleoclimatic effects. If recharge occurs in both winter and summer it is important to determine if the percentages of seasonal contributions are similar for both high and low elevation recharge sites. Independent methods such as groundwater dating and the evaluation of temporal spring discharge recession characteristics may be useful for helping distinguish between altitude and seasonal effects.

As examples of the altitude effect, in the Gran Canaria Island, Gonfiantini et al. (1976) found $\delta^{18}O$ per 100 m altitude gradients of −0.13 ‰ and −0.24 ‰ on the north and south sides of the island, respectively. Kern et al. (2020) found $\delta^{18}O$ per 100 m altitude gradients for the Adriatic-Pannonian region of −0.12 ‰ and −0.79‰, respectively. The Adriatic-Pannonian differences were attributed to the evaporation of falling precipitation (i.e., virga) on the south side of the island. In a precipitation study in the Apennines of Central Italy, Zuppi et al. (1974) found $\delta^{18}O$ per 100 m altitude gradients of −0.17 ‰ and −0.54 ‰ for summer and winter precipitation, respectively. Because the altitude effect was influenced by seasonal factors, the mean value of the summer and winter data, −0.34 ‰, was considered as the best representation of the overall altitude effect.

The precipitation elevation effect carries over as a groundwater recharge elevation effect. Eriksson (1965), Stahl et al. (1974), and others have empirically evaluated the effect of altitude on groundwater. In the Himalayas, Florea et al. (2017) calculated $\delta^{18}O$ and $\delta^{2}H$ lapse rates for groundwater as −0.25 ‰ per 100 m and −2.23 ‰ per 100 m, respectively (Figure 8.8). Although the quantitative

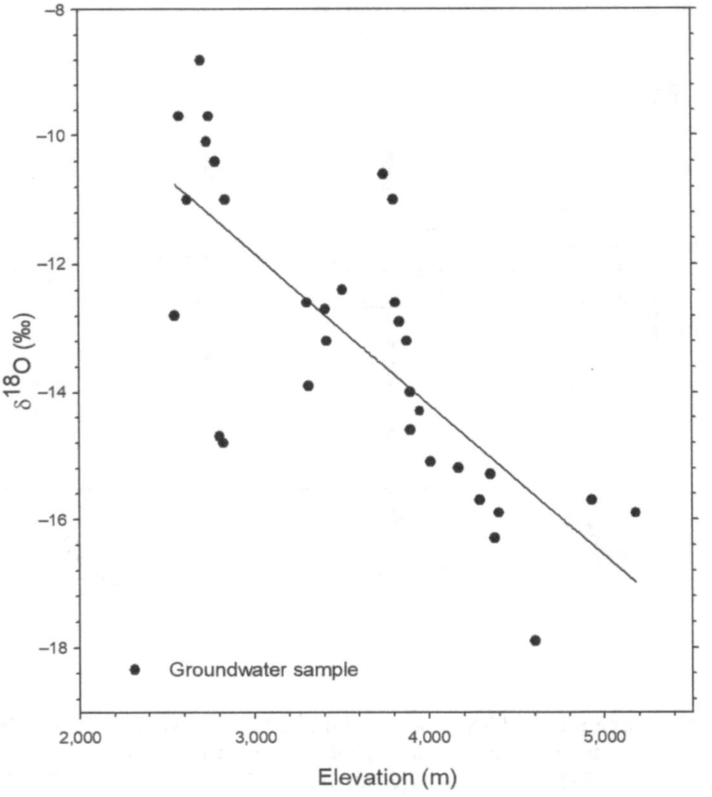

FIGURE 8.8 Elevation effect for groundwater samples collected in the Koshi watershed in the Southern Himalayas of Nepal. The $\delta^{18}O$ calculated altitude groundwater lapse rate for the entire 40 km long watershed is −0.25 ‰/100 m. Data from Florea et al. (2017).

effect of altitude must be determined independently for each investigation, the following relationships between stable isotopic compositions of groundwater and recharge altitude are good starting points:

$$\delta^{18}O = \frac{-0.2 \text{ to} - 0.4‰}{100 \text{ m elevation increase}} \quad (8.13)$$

$$\delta^{2}H = \frac{-2 \text{ to} - 4‰}{100 \text{ m elevation increase}} \quad (8.14)$$

where m = meters.

Elevation effects are usually evaluated by plotting the stable isotopic composition vs. the spring or well elevation. It is also possible to sample a mountain stream at various elevations and plot the isotopic compositions for each elevation. The elevation lapse rate can then be calculated from a best fit line fitted through the data. It may be necessary to adjust the data for the downstream increase in the stream flow.

8.3.1.5 Seasonal Trends

Seasonal variations observed in a local precipitation data are related to two factors: 1) seasonal differences in air temperature (Wu et al., 2022; Wang et al., 2021; Feng et al, 2009; Jackisch et al., 2022; Sun et al., 2019; Araguás et al., 1998) and 2) storm tracks (Friedman et al., 1992; Kurita, 2014; Vuille and Werner, 2005). Because seasonal temperature and storm tracks are often interrelated, it is not always easy to sort out the differences. Precipitation stations located in mid-continental settings commonly have seasonal isotopic variations (Gat, 2001; IAEA, 2021). For example, in Vienna, Austria, and in Ottawa, Canada, the average seasonal variations in $\delta^{18}O$ are as great as 18 ‰ between winter and summer precipitation. Based on a limited dataset, Benjamin et al. (2004) found several permil variability in $\delta^{18}O$ between winter and summer precipitation events at a single station in the interior of the western United States.

Seasonal isotopic trends in precipitation may be manifested in stream and river flow data, and temporal variations in the isotopic composition of a nearby stream may be useful when weather station data are not available. Equations 8.7 and 8.8 can be a useful starting point for distinguishing summer vs. winter recharge. For example, in Central Utah there is a correlation between long-term seasonal isotopic compositions in precipitation and the isotopic trend in river flows (Figure 8.9). The figure shows a) the $\delta^{2}H$ compositions of 265 precipitation events in central Utah between 2000 and 2012, and b) the monthly average isotopic compositions of undammed streams in the same watershed. There is a strong correspondence between the increasing isotopic depletion of stream waters as the contributions from winter and springtime precipitation and snowmelt increases. At the end of the spring freshet, the stream isotopic composition is more positive as the base flow and summer monsoonal precipitation dominate the stream discharge.

Interpretation of the seasonal recharge effect is helped by knowledge or an estimation of the seasonal variation in the isotopic composition of precipitation. Using precipitation data is based on the idea that the isotopic composition of recharge water is similar to the average composition of precipitation and that the isotopic composition varies greatly by season. Long-term stable isotopic data correlated with precipitation data from nearby weather stations provide the most reliable background information. Isotopic data from a single precipitation event or even a season of events usually should not be considered representative of local long-term precipitation.

Seasonal recharge effects are most easily recognized in shallow circulating, local groundwater systems where groundwater discharge rates respond to annual or semi-annual recharge cycles (Jasechko et al., 2014). The terms local, intermediate, and regional groundwater systems are used as defined (Toth, 1963). Local system have discharge rates that fluctuate seasonally, have a TDS that is consistent with groundwater flow in shallow alluvium and bedrock, have discharge temperatures

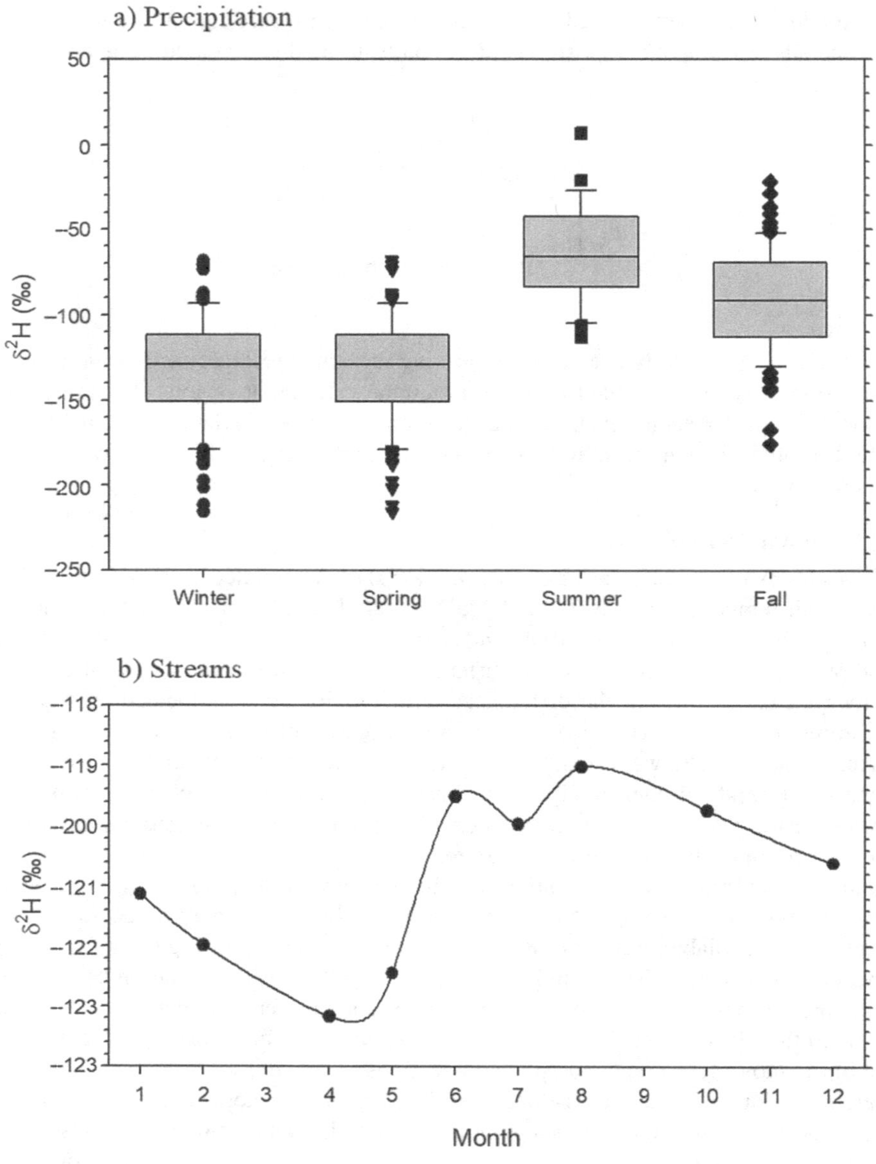

FIGURE 8.9 Seasonal isotopic effects on a) precipitation and b) stream flows in central Utah, during the 2004 water year. The stream flows represent six undammed streams in the same watershed. The most negative values of the streams correspond with the spring snowmelt freshet, and the most positive values correspond with the summertime monsoon season. The average precipitation δ^2H values for winter, spring, summer, and fall are −133.3, −133.3, −64.0, and −91.5 ‰, respectively. Unpublished data from Alan L. Mayo and David Tingey.

that approximate mean annual air temperature, and contain modern concentrations of ^3H and ^{14}C. Regional systems have a steady groundwater flux, may have elevated TDS, have a slightly warm to considerably thermal groundwater temperature, do not contain ^3H, and have ^{14}C that is representative of older groundwater. Local system recharge waters may move as a slug along the flow path and may retain the isotopic composition of major recharge events. On the other hand, the long residence times and the slow movement of deep circulating, intermediate, or regional groundwater systems tend to mix recharge waters from different seasons and years. Such mixing results in blended isotopic compositions of discharge waters, which masks seasonal trends. In tectonically complex terrain it is common for local, intermediate, and regional groundwater systems to discharge at the same location. A large flux of springtime recharge and groundwater discharge often means that it is possible to sort out the seasonal trend.

In the absence of precipitation data, the seasonal effect in groundwater may be evaluated on the basis of differences between mean summer and winter temperatures (Equations 8.7 to 8.11). Summer recharge has less negative δ^2H and $\delta^{18}O$ values than does winter recharge. For example, in the western region of Ireland, Bowen and Williams (1973) found that mean $\delta^{18}O$ of karstic groundwater was −7.2 ‰ and the $\delta^{18}O$ of average annual precipitation was −5.4 ‰. They concluded that the recharge occurred during the winter when precipitation was depleted in ^{18}O and that most summer precipitation enriched in the heavy isotopes was lost to evaporation.

The potential impact of the seasonal effect may need to be considered when designing an isotopic sampling program. This is particularly true when the groundwater to be sampled may include mixed discharge water from local, intermediate, and regional groundwater flow systems. Mixing is common where regional or intermediate groundwater systems share a common discharge location with local groundwater systems. Such mixing often occurs in the topographic lows of basins and in regions where the geology is structurally complex. The isotopic composition of a mixed discharge will have solute and isotopic signatures that are the mean weighted compositions of the discharge contributions from each system. In regions dominated by winter and springtime recharge the isotopic composition of late summer and fall discharge water will tend be similar to the base flow component of the regional groundwater system, and the early spring and early summer discharge will reflect mixed base flow and the previous season's recharge.

8.3.1.6 Evaporation

In the case of large water bodies, such as the ocean and many lakes, where the residual reservoir is essentially infinite, the residual water fraction will approximate 1, and there will be no overall change in the isotopic composition of the liquid phase. The vapor phase will be isotopically enriched in the lighter isotopes. The isotopic composition of evaporating small water bodies will exhibit fractionation and become enriched in ^2H and ^{18}O as the residual water fraction becomes smaller.

Relative humidity is the most important factor affecting the isotopic evaporation from an open surface water body. High humidity during evaporation results in small changes in both $\delta^{18}O$ and δ^2H. At 95 percent humidity, the *d* value (deuterium excess) is similar to evaporation of the last 85 percent fraction of the water and results in an evaporation line with a slope <8 on a $\delta^{18}O$-δ^2H plot. Evaporation under 100 percent humidity is essentially equivalent to closed-system evaporation and the data plot along a slope of 8. The humidity factor is often important when evaluating the isotopic composition of paleo-groundwater where climate change is being considered.

The isotopic composition of surface water, which has undergone excessive evaporation, is subject to isotopic enrichment and will plot on an evaporation trajectory such as line C–D with respect to the GMWL (Figure 8.3). Point C in the figure represents the isotopic composition prior to evaporation. The slope of the evaporation trajectory, typically in the range of 2 to 5, is due to the more rapid vaporization of ^1H relative to the vaporization of ^{16}O. Because evaporation trajectories slope, an evaporation line should not have a d-parameter excess that is greater than +10 ‰. However, in the eastern Mediterranean region Nir (1967) reported a precipitation d-parameter excess as great as +22 ‰ due to evaporation effects. Groundwater that has received recharge from the infiltration of evaporated water will also have an isotopic composition that plots along line C–D. In the unusual circumstances where recharge water has been subjected to soil zone evaporation that substantially increases the groundwater TDS, the groundwater may have stable isotopic compositions that plot along an evaporation line.

Important uses of evaporated isotopic signatures are to identify groundwater that has been recharged 1) from the leakage of an evaporative body of water, such as a wastewater disposal pond,

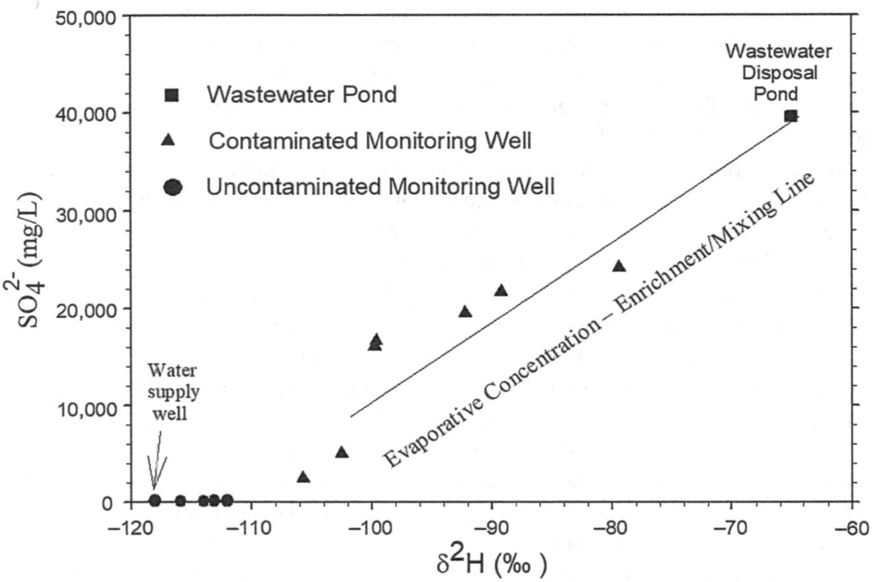

FIGURE 8.10 Stable isotope mixing of wastewater disposal pond water and background groundwater. Many of the contaminated monitoring wells have δ^2H and SO_4^{2-} concentrations that are mixtures of water leaking from the evaporative wastewater disposal pond and natural background groundwater. The purpose of the pond, located in western Utah, is to dispose of wastewater by evaporation, which makes the water about 60 ‰ more positive in δ^2H than the background groundwater.

and 2) from precipitation whose moisture has been derived from an evaporative water body. It is a common industrial and mining practice to dispose of wastewater in evaporation ponds. Most ponds are designed to prevent leakage, but over time pond linings may fail and captive water may escape to the surrounding environment. Figure 8.10 illustrates how stable isotopes have been used to detect leakage from an evaporative water body. The down-gradient monitoring wells contain a mixture of evaporative pond and background water, whereas up-gradient monitoring wells do not contain any evaporative water.

Figure 8.11 illustrates how the δ^2H and $\delta^{18}O$ composition of Antelope Island groundwater, Great Salt Lake (GSL), Utah, was used to identify an evaporative source of recharge water. The 43 mi^2 island rises 730 m above the surrounding hypersaline lake. GSL water is highly evaporated, with δ^2H and $\delta^{18}O$ values of about –70 and –4 ‰, respectively. The stable isotopic composition of Antelope Island groundwater plots along a mixing line of GLS water and non-island groundwater found in the nearby Weber delta. Weber delta groundwater is recharged from Wasatch Mountain streams flowing from the east. δ^2H and $\delta^{18}O$ data plot along an evaporation line, indicating the recharge water contains a component of precipitation recharge derived from evaporated GSL water.

In shallow, closed basin environments it is tempting to attribute elevated groundwater TDS and stable isotopes that plot along an evaporation trajectory to near-surface evaporation. For example, at Pilot Valley, Utah–Nevada, and at California-Nevada, there are strong correlations between evaporated shallow groundwater and elevated TDS. In Pilot Valley the groundwater TDS and isotopic signatures change abruptly from fresh to saline to evaporative and to isotopically less negative at the alluvial fan–playa interface. The coincidental abrupt isotopic and TDS evolution at about 900 m from the reference point suggests a commom cause (Figure 8.12). However, the elevated TDS and evaporative water occur below a reasonable evaporation depth, and a more complex explanation is required. Groundwater evolves from fresh to hypersaline near the alluvial fan/playa interface, where there is an inverted salinity gradient and a groundwater pressure ridge. The pressure ridge

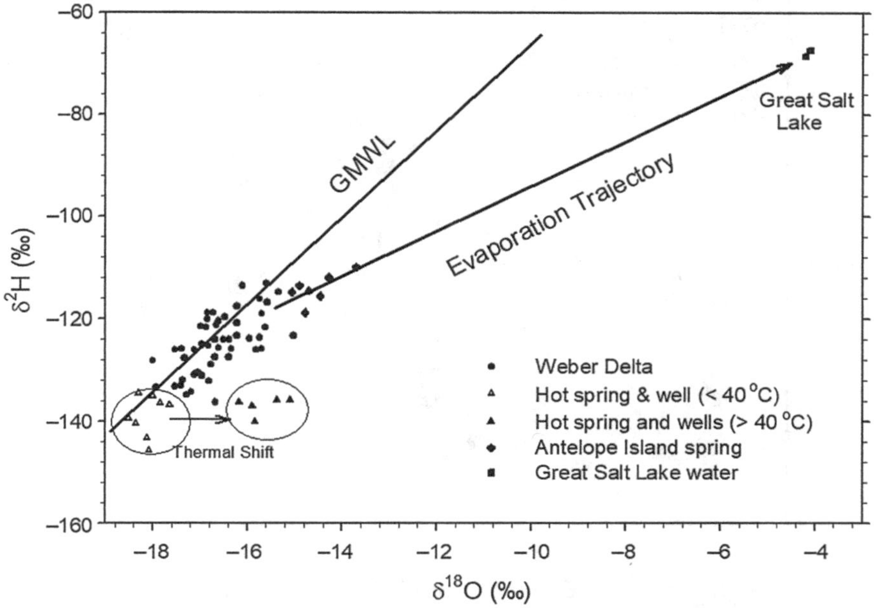

FIGURE 8.11 Plot of δ^2H and $\delta^{18}O$ of various groundwaters in the vicinity of Antelope Island, Great Salt Lake (GSL), Utah. Modified after Mayo and Klauk, 1991. Island groundwaters are a mixture of recharge water derived from evaporated GSL water and groundwater recharge with Pacific Ocean moisture. Thermal groundwater in the nearby Wasatch Range with discharge temperatures >40 °C exhibit a stable $\delta^{18}O$ thermal shift.

and inverted salinity interface are due to: 1) osmotic pressure established between high hydraulic conductivity oolitic sand and the overlying low hydraulic conductive lake mud at the fan/playa interface and 2) the collision between fresh groundwater flow driven by a steep hydraulic head with hypersaline groundwater flow driven by a nearly flat hydraulic head (Mayo et al., 2020). Isotopic fractionation up to −8 ‰ is attributed to hyperfiltration in the clay sequences in the playa sediments, and the elevated TDS is attributed to the dissoluton of NaCl and $CaSO_4$ in the playa sedimments.

At California-Nevada, shallow groundwater flows to the terminal sump where the TDS increases to as much as 30,000 mg/L due to dissolution of evaporite minerals in the ancestral terminal lake. The δ^2H evolves independently due to shallow groundwater evaporation (Figure 8.13). Both Pilot Valley and California-Nevada demonstrate the importance of not accepting the first and easiest solution to chemically and isotopically complex problems. It is important to keep in mind that although the concentrating effect of evapotranspiration processes in the soil zone – and to a lesser extent from the infiltration of highly evaporated surface water – can locally result in elevated TDS, salinization is not necessarily accompanied by isotopic enrichment.

8.3.1.7 Paleoclimate

Stable isotopic data may be used to aid in the identification of paleo-groundwater either independently or in conjunction with radiocarbon methods. Such interpretations of the paleoclimate effect are possibly due to the generally cooler and more humid global climatic conditions during Pleistocene glaciation. Cooler ancient climate caused many paleo-recharge waters to be depleted in ^2H and ^{18}O with respect to modern recharge waters. The depletion is due to the differences in paleo- and modern evaporation and condensation temperatures at the oceanic source areas (Dansgaard, 1964) and to differences in the air temperatures at the recharge areas. More negative isotopic values, which correspond to precipitation during periods of glaciation, have been verified in ice cores from Antarctica (EPICA, 2004), the Byrd Station, Antarctica, the Camp Century Station, Greenland

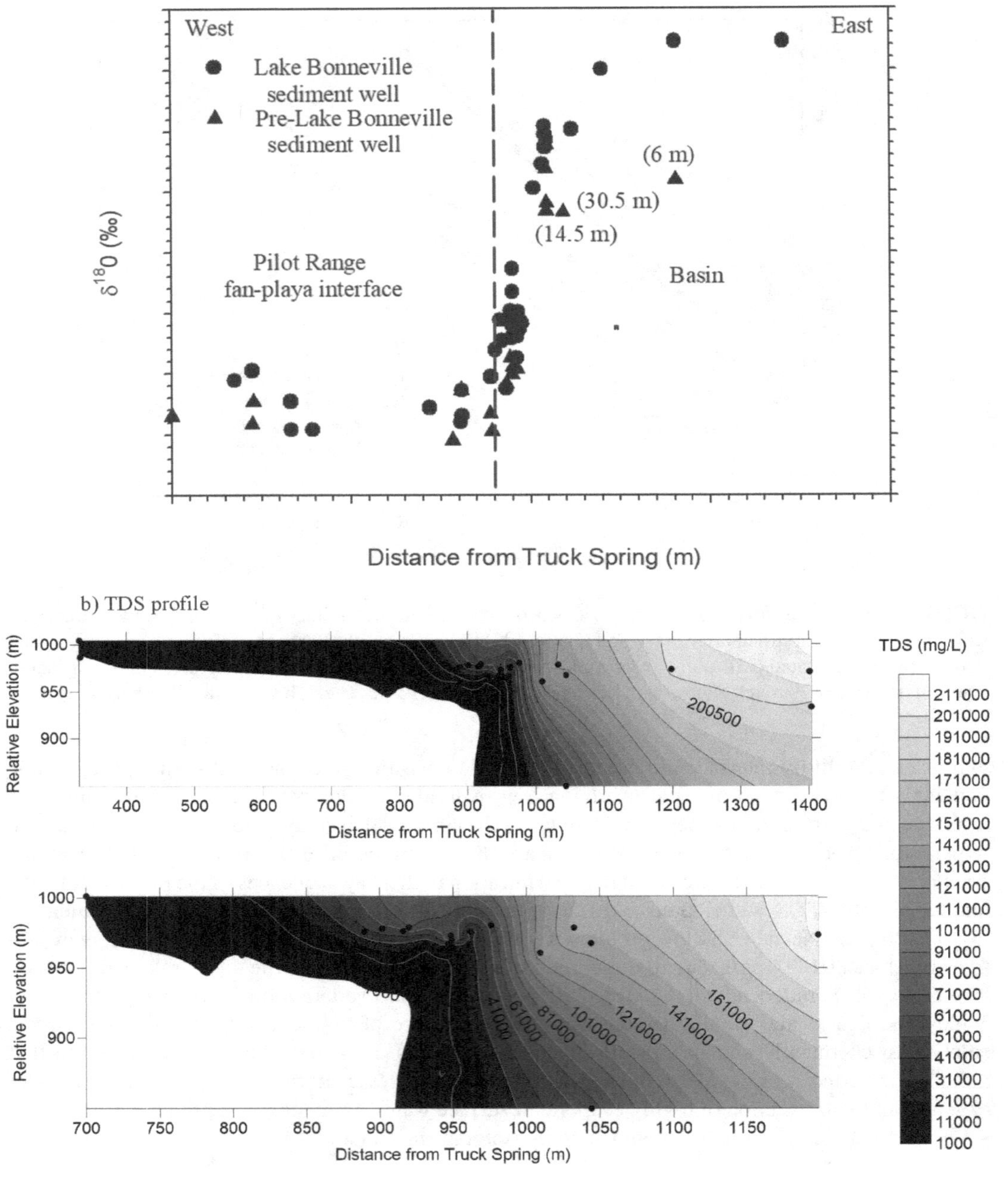

FIGURE 8.12 The complex $\delta^{18}O$ and TDS evolution of Pilot Valley, Utah-Nevada closed basin shallow groundwater at the the boundary between fresh water recharge from the west and hypersaline playa groundwater flow from east to west (modified after Mayo et al., 2020; Mayo and Tingey, 2021). The two waters collide at about 900 m from the reference point. The coincidiental abrupt isotopic a) and TDS b) evolution at about 900 m from the reference point suggests a commom cause. However, both the isotopic and TDS evolution occur below a reasonable evaporation depth, and a more complex explanation is needed. Isotopic fractionation up to −8 ‰ is attributed to hyperfiltration in the clay sequences in the playa sediments, and the elevated TDS is attributed to the dissoluton of NaCl and $CaSO_4$ in the playa sediments.

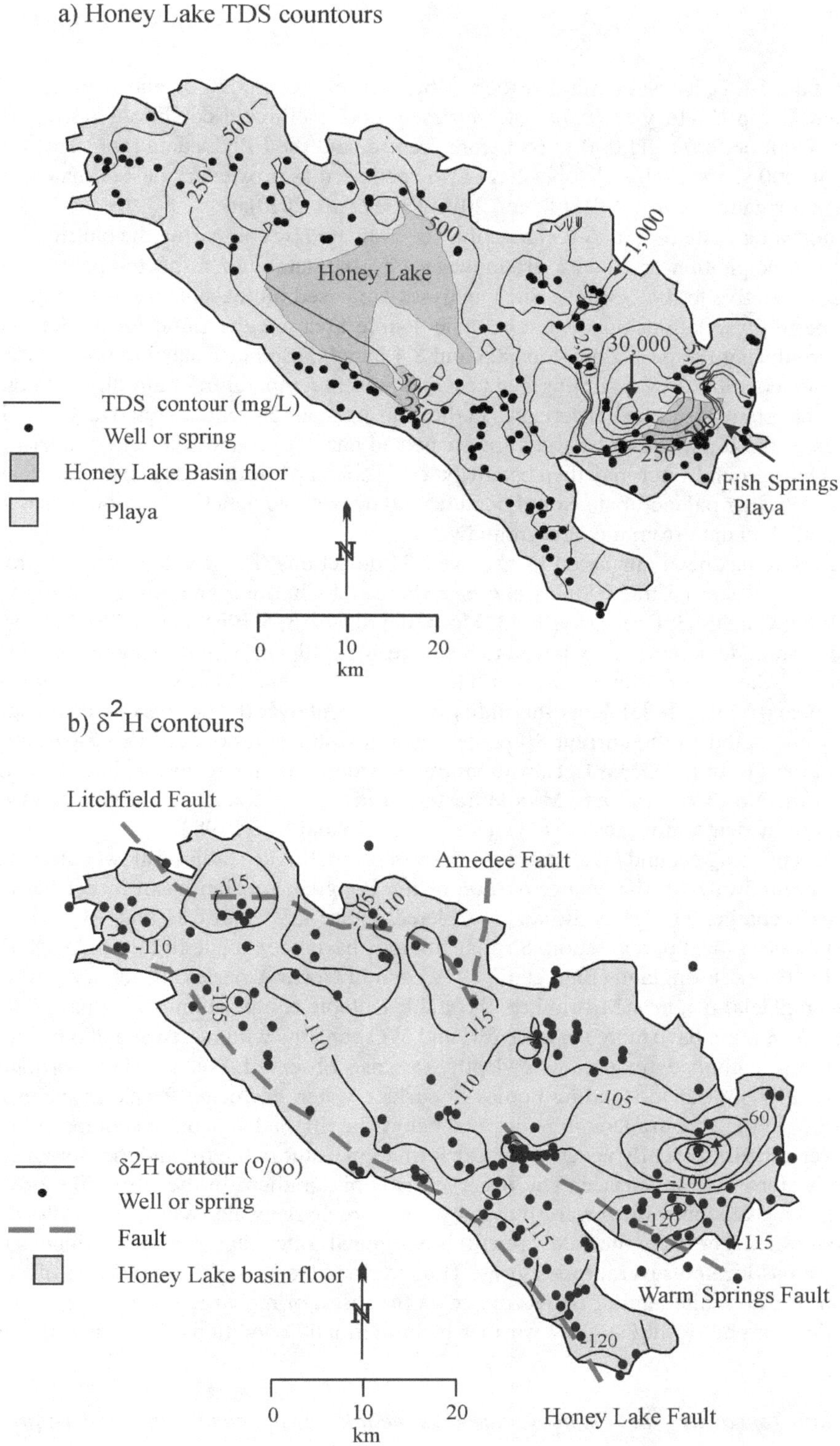

FIGURE 8.13 TDS and δ²H distribution in the California-Nevada, closed basin. Throughout most of the valley the a) TDS and b) δ²H are in the range 250–500 mg/L and −105 to −110 ‰, respectively (after Mayo et al., 2010). Shallow groundwater flows to the terminal lake in the east; where the TDS of shallow groundwater increases to as much as 30,000 mg/L due to dissolution of evaporite minerals in the ancestral terminal sump, and the δ²H evolves due to shallow groundwater evaporation.

(Johnson et al., 1972), ice wedges in Northern Siberia (Meyer et al., 2000), and elsewhere. The Byrd Station and Camp Century ice core data represent cooler climatic conditions during Wisconsin glaciation about 66,000 to 11,000 years before present, and the EPICA data represent climate over the past 740,000 years. Stable isotopes have also been used as proxies for past climate as related to speleothem formation (Cross, 2015; Duan, 2016; Pape et al., 2010).

In groundwater systems, paleo-recharge may be evaluated by comparing the plotting locations of present-day precipitation or modern groundwater isotopic data with suspected paleo-groundwater isotopic data relative to the GMWL. Such analyses are based on the starting assumption that temperature–humidity relationships at the oceanic source area are the same for modern conditions as they were during the past. Equations 8.3 and 8.4 may be used as a starting point for evaluating paleoclimatic conditions by assuming that past and present temperature–humidity relationships are reasonably constant. Observed differences in the plotting locations of data relative to the GMWL or differences in the d-parameter between the suspected ancient groundwater and modern precipitation or modern groundwater may then be attributed to colder or sometimes warmer ancient climatic conditions. Warmer paleoclimate would be indicated by isotopic data having more positive isotopic compositions than data from modern groundwater.

Paleoclimate has been attributed to observed ^{2}H depletions (i.e., d values <10) in groundwater systems in the western United States and elsewhere (Al-Charideh and Kattaa, 2016; White and Chuma, 1987; Gat, 1983; Ferronsky, 1983; Mosher et al., 1983; Schlegal et al., 2009; Werner et al., 2018). Deuterium depletions are believed to be the result of higher humidity and colder temperature accompanying the evaporation of seawater during the Pleistocene (Merlivant and Jouzel, 1979; van der Straaten and Mook, 1983). Paleo-humidity of 90 percent over the oceans during cooler climatic periods as compared to the current 80 percent would result in a cool climate d-parameter of +5 rather than the +10 of the GMWL. Low d-parameter values are not always an indication of ancient climate. In the North African and Middle Eastern arid zones, modern recharge waters have been characterized by deuterium excess as low as +5 (Ouijano and Dray, 1983).

Examples of using groundwater stable isotopes in paleoclimate studies follow. Salati et al. (1974) attributed groundwater in the deeper portion of the Potiguar artesian basin in northern Brazil to Pleistocene recharge. The groundwater is depleted about −25 ‰ in δ^2H and −3 ‰ in δ^{18}O with respect to modern local precipitation. Similar findings have been made in North Africa (Edmunds and Wright, 1979), in England (Bath et al., 1979,) and in North America (Bradbury et al., 1985). In a Wisconsin glacial till, near Milwaukee, the stable isotopic compositions of formation waters plot along the GMWL but have more negative δ^2H and δ^{18}O contents with increasing depth (Figure 8.14). Corresponding tritium extinction with depth was also observed. Near surface formation water was isotopically similar to standing bodies of surface water, corrected for excessive evaporation. Groundwater in the Niagara Dolomite, located below the till, had an isotopic composition intermediate between the isotopically negative deeper formation water in the till and the isotopically heavy formation water near the surface. The observed isotopic gradient in the till is interpreted as the result of mixing of slowly downward moving younger recharge water with late Pleistocene glacial till connate water. The dolomite below the till is a regional water supply aquifer, which is recharged at outcrop areas located several miles away. The intermediate isotopic composition of the groundwater from the dolomite aquifer is interpreted as the result of recharge from Pleistocene meteoric water, which formed during slightly warmer global climatic conditions than when the glacial ice existed.

Important Takeaway – In summary, care must be taken when comparing modern precipitation or modern groundwater isotopic data with isotopic data from suspected old groundwater or from other regional groundwater flow systems. It is important to keep in mind that the stable isotopic composition of older groundwater often represents the integrated mean of recharge events that may have occurred over long periods of time. Isotopic data from a single or a few recent precipitation events do not represent the annual or long-term

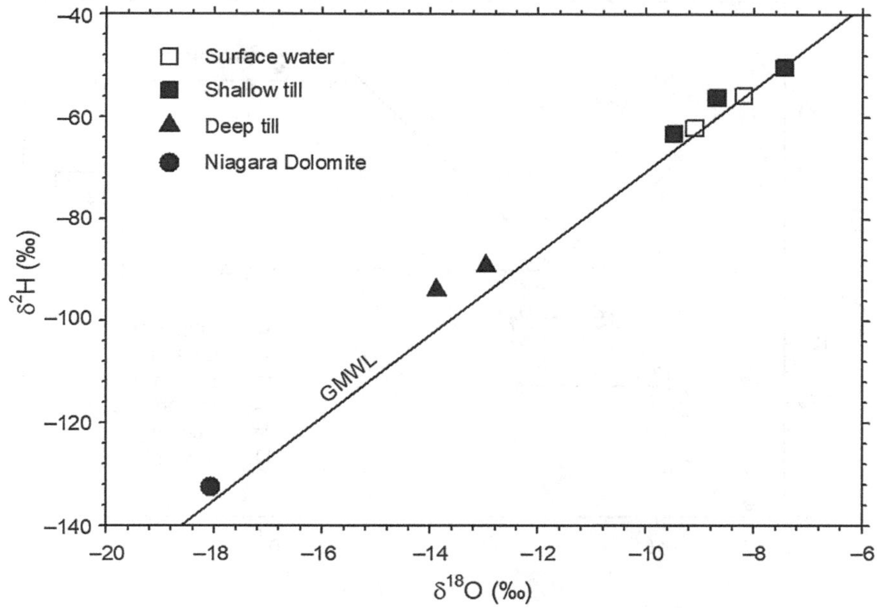

FIGURE 8.14 Plot of stable isotopes illustrating the effect of paleoclimate on groundwater recharge. Modern surface and shallow till groundwater plot in a close cluster along the GMWL. Deep Niagara Dolomite groundwater in Wisconsin has a very negative plotting location, suggesting groundwater recharge during colder recharge conditions. The deep till groundwater plots intermediate between the end member waters and is interpreted as the result of the mixing of slowly downward-moving younger recharge water with late Pleistocene glacial till connate water. Tritium data, not shown, support this idea. Unpublished data.

isotopic composition of recent precipitation (Siegenthaler and Oeschger, 1980). It is also important to keep in mind that groundwater recharge events may be seasonal and that changes in climate may also change the ratio of summer vs. winter recharge. Whenever possible the occurrence of older groundwater should be verified by radioisotope methods such as carbon-14 or chlorine-36. In addition to the effects on δ^2H and $\delta^{18}O$ data, paleoclimate may also affect the $\delta^{13}C$ composition of the total dissolved inorganic carbon that are used to interpret ^{14}C groundwater ages.

8.3.2 Thermal Groundwater

Subsurface groundwater heating, due to the geothermal gradient or close proximity to a cooling igneous body, may change the oxygen isotopic composition of groundwater according to:

$$\ln\alpha = A + \frac{B}{T^2} \tag{8.15}$$

or the more complicated polynomial function:

$$\ln\alpha = A + \frac{B}{T} + \frac{C}{T^2} \tag{8.16}$$

These equations are represented as Equations 7.19 and 7.20 in Chapter 7. Similar changes in the hydrogen isotopic composition seldom occur. The reason for the different isotopic behaviors is the fact that in most groundwater–rock interactions, the aquifer matrix is the oxygen isotopic buffer and

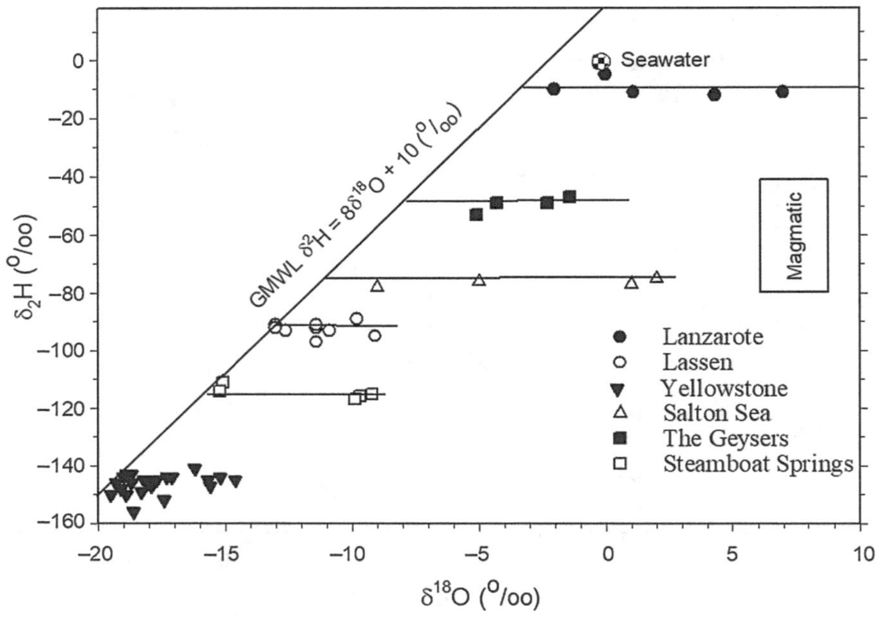

FIGURE 8.15 $\delta^{18}O$ shifts in worldwide geothermal waters. Data sources: Lanzarote (Arana and Pnichi, 1974), Lassen (Craig, 1963), Yellowstone (Bergfeld et al., 2014), Salton Sea (Craig, 1966), the Geysers (Lowenstern et al., 1999), and Steamboat Springs (Craig, 1963).

the groundwater is the hydrogen isotopic buffer. In other words, most of the oxygen molecules are contained in aquifer minerals and most of the hydrogen molecules are contained in the groundwater. What this means is that the oxygen in the water can exchange with oxygen in the aquifer matrix, but the matrix contains little exchangeable hydrogen. Accordingly, thermal groundwater may have isotopic compositions that are related to but do not lie on the GMWL. Thermal groundwater whose temperatures have, at some time in their flow history, been elevated above about 100 °C to 150 °C often have more positive $\delta^{18}O$ compositions relative to cooler groundwater of the same origin (Craig, 1963; Giggenbach, 1978; Dotsika et al., 2021). Kharaka and Carothers (1986) suggest that positive oxygen isotope shifts are generally not attained in deep sedimentary basins at temperatures below 150 °C. The positive oxygen isotope shift is primarily the result of oxygen isotope exchange between warm silicate minerals in the aquifer and groundwater (Craig et al., 1956). Minimal ^{18}O wall rock exchange has been reported in rocks of carbonate composition (Shieh and Taylor, 1969; Nabelek et al., 1984).

A typical trend of thermal groundwater is shown as line E–F in Figure 8.3, where E represents the original isotopic composition of the groundwater. Notable geothermal shifts are illustrated on Figure 8.15. The magnitude of the oxygen shift is reported as $\delta^{18}O$ whether or not fractionation equilibrium has occurred. A theoretical equilibrium delta can be calculated using experimentally determined fractionation factors and Equations 8.16 and 8.17. The actual $\delta^{18}O$ of a groundwater system can be determined by subtracting the $\delta^{18}O$ of the thermal groundwater from the anticipated $\delta^{18}O$ of the groundwater prior to warming. A delta of a few permil should be observed before assuming subsurface thermal conditions.

Oxygen isotope shifts are uncommon at low temperatures because equilibration rates are infinitely slow at low temperatures. A threshold or critical temperature exists below which no measurable shift occurs even though the water has had a very long contact time with the aquifer matrix and isotopic fractionation is theoretically possible. The critical temperatures for various aquifer

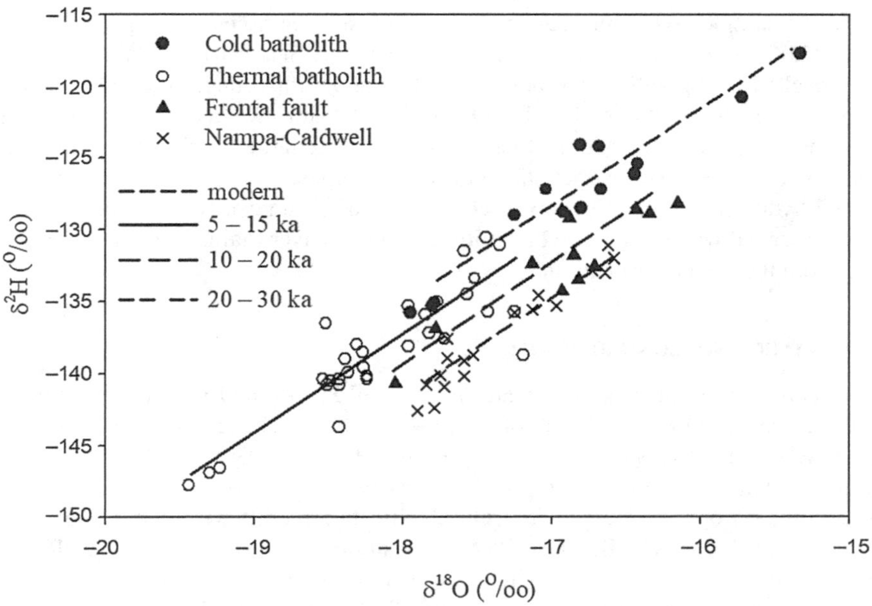

FIGURE 8.16 Local groundwater lines developed for modern and paleo-groundwater systems in the Boise, Idaho, area (modified after Schlegel et al., 2009). ^{14}C ages range from 5 to 20 ka. The more positive isotopic compositions of the younger waters are attributed to a 9 percent sea surface humidity increase, a 6–7 °C increase in sea surface temperature, and a 4–5 °C local temperature increase from the last glacial maximum to the present time.

minerals are poorly understood. The potential for an observable shift is greatest when the difference between the $\delta^{18}O$ of the source water (meteoric) and the $\delta^{18}O$ of the aquifer minerals is large. Continental meteoric waters commonly have $\delta^{18}O$ compositions at least 5 ‰ smaller than most silicate and aluminosilicate aquifer minerals, which is large enough for oxygen isotope shifts to theoretically occur. If the aquifer matrix has been previously depleted in ^{18}O due to long-term contact with thermal groundwater, such as found in hydrothermal alteration zones, an oxygen isotope shift may not be observed in very hot groundwater.

Although hydrogen-bearing silicate minerals may undergo substantial ^{2}H depletion in deep thermal systems (Taylor, 1974, 1977; Kharaka and Carothers, 1986), thermal groundwater often does not exhibit $\delta^{2}H$ shifts for two reasons: 1) most groundwater has low $\delta^{2}H$ values relative to the $\delta^{2}H$ of most aquifer minerals, and 2) hydrogen is scarce in the solid phase of most crystalline rock environments. In clay-rich sedimentary environments, negative hydrogen isotopic shifts are possible in that isotopic fractionation factors for most water-clay minerals reactions are less than 1 (Friedman and O'Neil, 1977; Savin, 1980).

The stable isotopic compositions of cold and thermal groundwater systems in the southern Idaho Batholith in the Snake River Plain have been compared by Mayo and Mitchell (1988) and Schlegel et al. (2009). The systems are adjacent to a boundary fault zone, which separates the two hydrogeologic provinces near Boise, Idaho. Elevations of the groundwater systems range from 900 to 2,750, 825 to 850 and 760 to 825 m for the batholith, fault zone, and Snake River Plain (Nampa-Caldwell), respectively (Figure 8.16). An unexpectedly wide range of d-parameter values were calculated for the various groundwater regimes. Calculated *d*-parameter values were +9.0 for cold groundwater systems in the batholith, +7.1 for thermal systems in the batholith, −1.5 for thermal systems in the boundary fault zone, and −6.0 for thermal systems in the Snake River Plain. Positive $\delta^{18}O$ shifts, due to thermal equilibration with host rocks, were observed in 6 of 83 datasets. Radiocarbon ages were calculated for 21 samples. Maximum estimated mean residence times were approximately 12,000, 17,000, and 23,000 years for the batholith, fault zone, and Snake River Plain groundwater systems, respectively. Mixing with anthropogenic carbon and with younger groundwater was evident in many analyses. Based on the δ^2H and $\delta^{18}O$ and the ^{14}C data, two conclusions were made. First, thermal systems in the batholith and the Snake River Plain recharged during different paleoclimatic conditions. Their waters have distinct stable isotopic compositions, and the groundwater systems have minimal hydraulic communication. Second, thermal groundwater in the fault zone may be a mix of parent thermal waters from the batholith and Snake River Plain, with the largest component contributed from the Snake River Plain.

8.3.3 Magmatic and Connate Water

The isotopic composition of magmatic water is not well known and may vary. Magmatic waters are believed to have $\delta^{18}O$ compositions of +6 to +9 ‰ and δ^2H compositions of −40 to −80 ‰ (Figure 8.15; White, 1974; Taylor, 1977; Truesdell and Hulston, 1980). Mitchell et al. (2022) found a wider range of values. The $\delta^{18}O$ composition must be generally greater than +5 ‰ because at high temperatures isotopic exchange with ^{18}O rich silicate minerals will be minimal. Most silicate minerals in intrusive igneous bodies have $\delta^{18}O$ compositions of +7 to +10 ‰ (Taylor, 1977). The δ^2H composition has been estimated from an analysis of water in fluid inclusions, volcanic emanations, and hydrous minerals in igneous rocks (Befus et al., 2000; Taylor, 2001).

Connate water of marine origin should have a stable isotopic composition similar to seawater, which is about 0 ‰ in both $\delta^{18}O$ and δ^2H. Except for those aquifers that have active hydraulic communication with the ocean, the occurrence of marine connate waters is not common in shallow sedimentary basins. The occurrence of connate water in deep sedimentary basins is yet to be resolved. Clayton et al. (1966) found 95 oil-field brines from the Illinois, Michigan, and Alberta basins and the Gulf Coast to be dominated by meteoric water. The deuterium content of most of this water had not been affected by isotopic fractionation, although thermal oxygen exchange had taken place. Kharaka et al. (1979) and Kharaka and Carothers (1986) found an unusual stable isotopic trend in deep groundwater believed to be connate in the northern Gulf of Mexico basin. Connate water was suggested because the formation fluids are geopressured (i.e., they have fluid pressure gradients greater than hydrostatic) and the calculated fluid potential was greater than the elevation of any recharge area in the basin.

Important Takeaway – The δ^2H and $\delta^{18}O$ data may often be used to identify groundwater from specific aquifers. The δ^2H and $\delta^{18}O$ composition may provide independent information regarding the origin of the groundwater (i.e., magmatic, connate, meteoric, evaporated surface water), recharge conditions (i.e., climate and paleoclimate, season, and elevation), subsurface thermal conditions, and groundwater mixing patterns.

Problems

1. Well and evaporation pond water samples from a mineral extraction facility had permil $\delta^{18}O$ and δ^2H compositions of:

Wells		Evaporation pond	
δ^2H (‰)	$\delta^{18}O$ (‰)	δ^2H (‰)	$\delta^{18}O$ (‰)
−15.4	−131	−7	−89
−15	−129	−6.6	−89.5
−15.5	−129.5	−5	−78.5
−13.5	−12.5		

Wells	
−12.5	−104
−10.6	−104
−11.8	−110.5
−10.5	−107.8
−6.3	−87.5
−6.6	−89.5
−9.2	−104.3

Plot the data relative to the GMWL. How would you interpret the results?

2. Snow samples collected in January and February at a weather station in Utah (elevation 4,300 feet) had the average $\delta^{18}O$ and $\delta^{2}H$ compositions of −18.7 ‰ and −136.2 ‰, respectively.
   a. What is a reasonable estimate of the average isotopic values at 335 m on a nearby mountain peak?
   b. The average January–February temperature is −3.3 °C, and the average July temperature is 24.5 °C. What would you expect for the average stable July isotopic values?
3. The stable isotopic data from groundwater samples collected in the San Luis Valley, Colorado, are plotted relative to the GMWL. The valley aquifers include a shallow unconfined, a upper confined, and a lower confined. How would you interpret the data represented in the following figure (Figure 8.17)?
4. What criteria would you use to determine that the isotopic compositions of water samples represent thermal and meteoric groundwater?
5. The average isotopic composition of rain from a storm in Indianapolis, Indiana, was $\delta^{2}H$ = −39.06 ‰ and $\delta^{18}O$ = −6.25 ‰. The storm originated in the Gulf of Mexico and passed

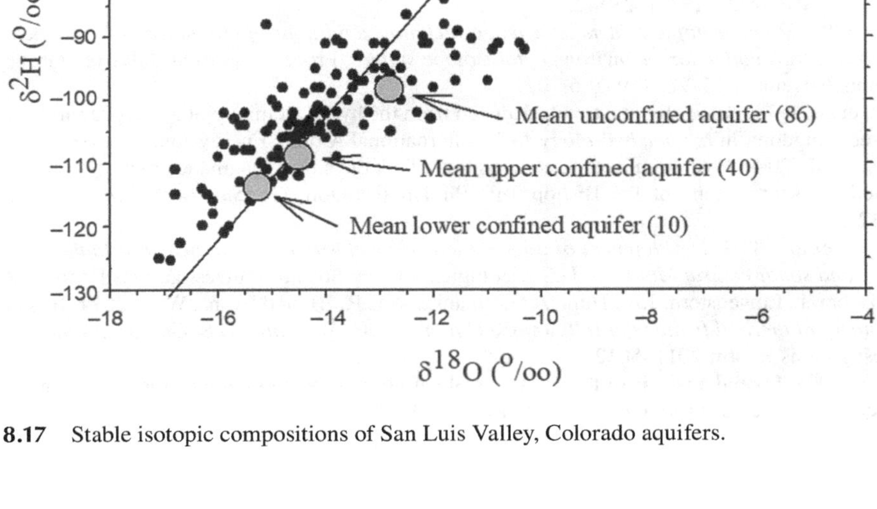

FIGURE 8.17 Stable isotopic compositions of San Luis Valley, Colorado aquifers.

over New Orleans, Louisiana, which is located about 1,325 km to the south. Assuming the storm continuously rained along the storm path, what was the possible isotopic composition of the precipitation in New Orleans?

6. The average δ^2H values of summer and winter precipitation are –70 ‰ and –120 ‰, respectively. The base flow δ^2H of a spring –120 ‰ and the average δ^2H of the spring water is –90 ‰.
   a. What types of groundwater systems (local, intermediate, regional), discharge at the spring, and how do you know?
   b. What is the contribution of summer precipitation to the spring?
7. A city in Southern California stores water in a small open reservoir (elevation 100 m amsl) cut into the top of a 30 m high hill. The water originates in the Sierra Nevada Mountains about 650 km to the north (elevation 1,200 to 1,800 m amsl). Numerous springs discharge in the yards of residents at the bottom of the hill. The reservoir owner claims the spring discharge from a local groundwater system, and the home owners claim the reservoir leaks. Construct a graph of the GMWL and show how you would determine who is correct.

8.4 CHAPTER NOTATION

‰	per mill
°C	temperature in degrees Centigrade
α	isotopic fractionation factor
δ	delta unit used for reporting isotopic abundance in ‰ difference relative to a standard
a	distance from the ocean
d	‰ deuterium excess
f	fraction of vapor remaining
m	meter
GMWL	global meteoric water line, defined as $\delta^2H = 8\ \delta^{18}O + d$, where d is usually about 10 ‰
R	ratio of remaining heavy isotope nuclides to number of light isotope nuclides
T	temperature

REFERENCES CITED

Al-Charideh, A., and V. Kattaa, 2016, Isotope hydrology of deep groundwater in Syria: Renewable and non-renewable groundwater and paleoclimate impact, *Hydrology Journal*, 14, 79–98.

Araguás, L., et al., 1998, Stable isotope composition of precipitation over southeast Asia, *Journal Geophysical Research*, 103(D22), 28721–28742.

Bath, A.H., 1983, *Stable isotopic evidence for paleo-recharge conditions of groundwater in Paleoclimates and paleowaters: A collection of environmental isotope studies*, Proceedings of an Advisory Group Meeting, Vienna, November 25–28, 169–186.

Bath, A.H., et al., 1979, Paleo climatic trends deduced from the hydrochemistry of a Triassic sandstone aquifer, United Kingdom, in *Isotope hydrology 1978*, International Atomic Energy Agency (IAEA), 2.

Befus, K.S., et al., 2000, Hydrogen isotope composition of a large silicic magma reservoir preserved in quartz-hosted glass inclusions of the Bishop Tuff Plinian Eruption, *Geochemistry Geophysics Geosystem*, 21(12).

Benjamin, L., et al., 2004, *Development of a local meteoric water line for southeastern Idaho, western Wyoming, and south-central Montana*, U.S. Geological Survey Scientific Investigations Report 2004-5126.

Bergfeld, Deborah, Lowenstern, J.B., Hunt, A.G., Shanks, W.C.P., III, and Evans, W.C., 2014, *Gas and isotope chemistry of thermal features in Yellowstone National Park, Wyoming*, U.S. Geological Survey Scientific Investigations Report 2011–5012.

Bowen, G.J., 2008, Spatial analysis of the intra-annual variation of precipitation isotope ratios and its climatological corollaries, *Journal Geophysical Research*, 13.

Bowen, R., and P.W. Williams., 1973, Geohydrologic study of the Gort Lowland and adjacent areas of western Ireland using environmental isotopes, *Water Resources Research*, 9, 753–758.

Bradbury, K.R., et al., 1985, Groundwater movement through till, northwestern Wisconsin, USA, in *Hydrology of rocks of low permeability*, 17th International Congress International Association. Hydrologist, Tucson, AZ.

Cai, Z., and L. Tian, 2016, Atmospheric controls on seasonal and interannual variations in precipitation isotope in East Asian Monsson region: Atmospheric controls on seasonal and interannual variations in the precipitation, *American Meteorological Society*, 1339–1352.

Clark, I.D., and P. Fritz, 1997, *Environmental isotopes in hydrogeology*, CRC Press.

Clayton, R.N., et al., 1966, The origin of saline formation waters 1, Isotopic composition, *Journal Geophysical Research*, 71, 3869–3882.

Craig, H., 1961, Isotopic variations in meteoric waters, *Science*, 133, 1702–1703.

Craig, H., 1963, The isotopic geochemistry of water and carbon in geothermal areas in nuclear geology of geothermal areas, *Consiglio Nazionale Delle Ricerche*, 17–53.

Craig, H., 1966, *Isotopic Composition and Origin of the Red Sea and Salton Sea Geothermal brines: Science*, 154, 1544–1548.

Craig, H., et al., 1956, *Isotopic geochemistry of thermal waters*, National Academy Science National Resource Council, Publication 400, 29–38.

Craig, J., and L.I. Gordon, 1965, Deuterium and oxygen-18 variations in the ocean and the marine atmosphere, in *Stable isotopes in oceanographic studies and paleotemperature, Spoleto, July 26–27, 1965*, Consiglio Nazionale delle Ricereche, Laboratorio di Geologia Nucleare, Pisa, 1–222.

Cross, M., 2015, Great Basin hydrology, paleoclimate, and connections with the North Atlantic: A speleothem stable isotope and trace element record from Lehman Caves, NV, *Quaternary Science Reviews*, 127, 186–198.

Dansgaard, W., 1953, The abundance of O^{18} in atmospheric water and water vapour, *Tellus*, 5, 461–469.

Dansgaard, W., 1954, The abundance of O^{18} in fresh water, *Geochimica et Cosmochimica Acta*, 6, 241–260.

Dansgaard, W., 1964, Stable isotopes in precipitation, *Tellus*, 6, 435–468.

Dotsika, E., et al., 2021, Chemical and isotopic characterization of the thermal fluids emerging along the North – Northeastern Greece, *Scientific Reports*, 16291

Duan, W., 2016, The transfer of seasonal isotopic variability between precipitation and drip water at eight caves in the monsoon regions of China, *Geochimica et Cosmochimica Acta*, 183, 250–266.

Easto, C.J., and D.L. Dettman, 2016, Isotope effects in hydrogeologic and climate reconstructions of monsoon climates: Implications of some long-term data sets for precipitation, *Chemical Geology*, 430, 78–89.

Edmunds, W.M., and E.P. Wright, 1979, Groundwater recharge and paleoclimate in the Sirte and Kufra Basins, Libya, *Journal Hydrology*, 40(3–4), 215–241.

Ehhalt, D., et al., 1963, Deuterium and oxygen 18 in rain water, *Journal Geophysical Research*, 68(13), 3775–3780.

EPICA Community Members, 2004, Eight glacial cycles from an Antarctic ice core, *Nature*, 429, 623–628.

Epstein, S., 1956, Variations of the O18/O16 ratios of fresh water and ice, *National Academy Science Nuclear Science Series Report*, 19, 20–25.

Epstein, S., and T. Mayeda, 1953, Variation on ^{18}O content of waters from natural sources, *Geochimica et Cosmochimica Acta*, 4, 213–214.

Eriksson, E., 1965, Deuterium and oxygen-18 in precipitation and other natural waters, some theoretical considerations, *Tellus*, 17, 498–512.

Eriksson, E., 1983, Stable isotopes and tritium in precipitation, in *Guidebook on nuclear techniques in hydrology*, 1983 ed, International Atomic Energy Agency (IAEA), Vienna, Technical Report Series, 91, 19–33.

Ersek, V., et al., 2010, Variations of $\delta^{18}O$ in rainwater from southwestern Oregon, *Journal Geophysical Research*, 115.

Feng, X., et al., 2009, Seasonality of isotopes in precipitation: A global perspective, *Journal of Geophysical Research Atmospheres*, 4(D8).

Ferronsky, V.L., 1983, Relationships between climatic changes and variations in isotopic composition of groundwater, precipitation and organic soil matter in the Quaternary period, in *Paleoclimates and paleowaters: A collection of environmental isotope studies*, Proceedings Advisory Board Meeting, International Atomic Energy Agency (IAEA), November 25–28, 1980, 13–36.

Florea, L., et al., 2017, Stable isotopes of river water and groundwater along altitudinal gradients in the High Himalayas and the Eastern Nyainqentanghla Mountains, *Journal of Hydrology Regional Studies*, 4, 37–48.

Friedman, I., 1953, Deuterium content of natural waters and other substances, *Geochimica et Cosmochimica Acta*, 4, 89–103.

Friedman, I., and J.R. O'Neil, 1977, *Data of geochemistry, compilation of stable isotope fractionation factors of geochemical interest data of geochemistry*, 6th ed, U.S. Geological Survey Professional Paper 440-KK.

Friedman, I., et al., 1992, Stable isotope composition of waters in Southeastern California – 1. Modern precipitation, *Journal of Geophysical Research Atmospheres*, 97(D5), 795–5812.

Friedman, I., et al., 2002, Stable isotope composition of waters in the Great Basin, United States 1. Air-mass trajectories, *Journal of Geophysical Research Atmospheres*, 107(D19).

Gat, J.R., 1980, The isotopes of hydrogen and oxygen in precipitation, in *Handbook of environmental isotope geochemistry*, Elsevier, 1(A), 21–74.

Gat, J.R., 1983, *Precipitation, groundwater and surface waters: Control of climate parameters on their isotopic composition and their utilization as paleoclimatologic tools in paleoclimates and paleowater: A collection of environmental isotope studies*, Proceeding of Advisory Group Meeting, International Atomic Energy Agency (IAEA), Vienna, November 25–28, 1980, 3–12.

Gat, J.R., 2001, *Environmental isotopes in the hydrogeological cycle – Principles and applications, technical documents I hydrology*, International Atomic Energy Agency (IAEA), 2.

Gat, J.R., et al., 2002, Inter- and intra-storm variability of the isotope composition of precipitation in Southern Israel: Are local or large-scale factors responsible? in *Conference international conference on study of environmental change using isotope techniques, Vienna (Austria), 23–27 Apr 2001*, International Atomic Energy Agency (IAEA), 41–53.

Giggenbach, W.F., 1978, The isotopic composition of waters from the El Tatio geothermal field, Northern Chile, *Geochimica et Cosmochimica Acta*, 42(7), 979–988.

Gilfillan, E.S. Jr., 1934, The isotopic composition of sea water, *Journal American Chemical Society*, 56(2), 406–408.

Gonfiantini, R., et al., 1976, *Environmental isotopes and hydrogeochemistry of groundwater of Gran Canaria in interpretation of environmental isotope and hydrochemical data in groundwater hydrology*, International Atomic Energy Agency (IAEA), Vienna, 159–170.

Hager, G., and U. Foelsche, 2015, Stable isotope compositions of precipitation in Austria, *Australian Journal Earth Science*, 108(2), 2–13.

Han, T., et al., 2020, Sub-hourly variability of stable in precipitation in the marginal zone of East Asian Monsoon, *Water*, 12(8).

Harada, M., and T. Titani, 1935, Isotopic composition of rain and snow water, *Chemical Society Japan*, 10, 206–263.

IAEA, 2021, *Global network of isotopes in precipitation (GNIP)*, International Atomic Energy Agency (IAEA), accessed March 21, 2021.

Jackisch, D., et al., 2022, Precipitation stable isotopic signatures of tropical cyclones in Metropolitan Manila, Philippines, show significant negative isotopic excursions, *Natural Hazards and Earth Systems Science*, 22(1), 213–226.

Jasechko, S., et al., 2014, The pronounced seasonality of global groundwater recharge, *Water Resources Research*, 50, 8845–8867.

Johnson, S.J., et al., 1972, Oxygen isotope profiles through the Antarctic and Greenland ice sheets, *Nature*, 235, 429–434.

Kern, Z., et al., 2020, Isotopic altitude and continental effects in modern precipitation across the Adriatic–Pannonian region, *Water*, 12(6).

Kharaka, Y.K., and W.W. Carothers, 1986, Oxygen and hydrogen isotope geochemistry of deep basin brines, in *Handbook of environmental isotope geochemistry*, Elsevier, 2(B), 305–360.

Kharaka, Y.K., et al., 1979, *Geochemistry of formation waters from Pleasant Bayou No. 2 well and adjacent areas in Coastal Texas*, Proceedings of 4th Conference, Geopressured-Geothermal Resources Austin, Texas.

Klaus, J., et al., 2015, Where does stream water come from in low-relief forested watersheds? A dual-isotope approach, *Hydrology and Earth Systems Science*, 19(1), 125–135.

Kobayakawa, H.Y., and Y. Horibe, 1960, Deuterium abundance of natural waters, *Geochimica et Cosmochimica Acta*, 20(3–4), 273–283.

Kurita, N., 2014, Atmospheric circulation controls on the inter-annual variability in precipitation isotope ratio in Japan, *Climate Past Discussion*, 10(5), 3989–4032.

Lee, J-E., and I. Fung, 2007, Amount effect of water isotopes and quantitative analysis of post-condensation processes, *Hydrological Processes*, 22(10), 1–8.

Lewis, G.N., and R. Cornish, 1933, Separation of the isotopic forms of water by fractional distillation, *Journal American Chemical Society*, 55(6), 5061–5062.

Li, Y., et al., 2023, Re-evaluation of the effects of precipitation amount and temperature on precipitation $\delta^{18}O$ at the monthly and interannual timescales, *Journal of Water and Climate Change*, 14(6), 1816–1830.

Mayo, A.L., 2023, Precipitation cycles relative to storm tracks, ENSO and PDO, and drought – continental interior central western U.S., *American Journal of Climate Change*, 12, 335–358.

Mayo, A.L., R. Henderson, and D. Tingey, 2010, Chemical evolution of shallow playa groundwater in response to post-pluvial isostatic rebound, Honey Lake Basin, California-Nevada, USA, *Hydrogeology Journal*, 18, 725–747.

Mayo, A.L., and R.H. Klauk, 1991, Contributions to the solute and isotopic groundwater geochemistry, Antelope Island, Great Salt Lake, Utah, *Journal of Hydrology*, 127, 307–335.

Mayo, A.L., and J.C., Mitchell, 1988, Comparison of stable and unstable isotopic compositions from thermal groundwaters in the southern Idaho batholith and adjacent Snake River Plain, *Geological Society of America, Abs. w/Programs*, 20(6), 430.

Mayo, A.L., and D. Tingey, 2021, Shallow groundwater chemical evolution, isotopic hyperfiltration, and salt pan formation in a hypersaline endorheic basin: Pilot Valley, Great Basin, USA, *Hydrogeology Journal*, 29, 2219–2243.

Mayo, A.L., et al., 2020, Shallow ground water flow and inverted fresh/saline water interface in a hypersaline endorheic basin – Great Basin, USA, *Hydrogeology Journal*, 28, 2877–2902.

Merlivant, L., and J. Jouzel, 1979, Global climatic interpretation of the deuterium-oxygen 18 relationship for precipitation, *Journal American Chemical Society*, 84(C8), 5029–5033.

Meyer, H., et al., 2000, Paleoclimate studies on Bykovsky Peninsula, Northern Siberia – hydrogen and oxygen isotopes in ground ice, *Polarforschung*, 70, 37–41.

Mitchell, S.J., et al., 2022, Isotopic signatures of magmatic fluids and seawater within silicic submarine volcanic deposits, *Geochimica et Cosmochimica Acta*, 326, 214–233.

Mix, H., et al., 2019, Evaluating the roles of rainout and post-condensation processes in a land falling atmospheric river with stable isotopes in precipitation and water vapor, *Atmosphere*, 10(2).

Mosher, H., et al., 1983, *Stable isotope studies on paleowater and other, in paleoclimates and paleowaters: a collection of environmental isotope studies*, Proceedings Advisory Board Meeting, International Atomic Energy Agency (IAEA), Vienna, November 25–28, 1980, 201–204.

Nabelek, P.I., et al., 1984, Contrasting fluid/rock interaction between the Notch Peak granitic intrusion and argillites and limestones in western Utah: Evidence from stable isotopes and phase assemblages, *Contribution to Mineralogy and Petrology*, 86, 25–34.

Nir, A., 1967, Development of isotope methods applied to groundwater hydrology, in isotope techniques in the hydrologic cycle, *American Geophysical Union, Monograph*, 1, 109–116.

Ouijano, L., and M. Dray, 1983, Origin of groundwater, in *Guidebook on nuclear techniques in hydrology*, 1983 ed, *International Atomic Energy Agency, Technical Report Series*, 91, 273–284.

Pape, J.R., et al., 2010, Controls on oxygen isotope variability in precipitation and cave drip waters, Central Texas, USA, *Journal Hydrology*, 385(3–4), 203–215.

Pionke, H.B., and D.R. DeWalle, 1992, Intra- and inter-storm ~80 trends for selected rainstorms in Pennsylvania, *Journal Hydrology*, 138(1–2), 131–143.

Polyakov, V.A., and L.N. Kolesnikov, 1978, *Regional aspects of isotopic composition variation in precipitation in VII Vses*, Symposium Isotope Geokhim. Tez. Dokl. Moskva, 104–105.

Risi, C., et al., 2008, Influence of convective processes on the isotopic com-position ($\delta^{18}O$ and δD) of precipitation and water vapor in the tropics: 2 Physical interpretation of the amount effect, *Journal of Geophysical Research*, 113(D19).

Rozanski, K., et al., 1993, Isotopic patterns in modern global precipitation, climate change in continental isotopic record, *Geophysical Monograph Series*, 78, 1–37.

Salati, E., et al., 1974, *Environmental isotopes in hydrogeological study of Northeastern Brazil in isotopes techniques in groundwater hydrology*, 1st Proceedings of Symposium, International Atomic Energy Agency (IAEA), Vienna, 1974, 259–282.

Savin, S.M., 1980, Oxygen and hydrogen isotope effects in low temperature mineral-water interactions, in *Handbook of environmental isotope geochemistry*, Elsevier, 1(A), 283–328.

Schlegel, M.E., et al., 2009, Paleo-climate of the Boise area, Idaho from the last glacial maximum to the present based on groundwater δ^2H and $\delta^{18}O$ compositions, *Quaternary Research*, 71, 172–180.

Shieh, Y. N., and H.P. Taylor, 1969, Oxygen and hydrogen isotope studies of contact metamorphism in the Santa Rosa Range, Nevada, and other areas, *Contributions to Mineralogy and Petrology*, 20, 306–356.

Siegenthaler, U., and H. Oeschger, 1980, Correlation of ^{18}O in precipitation with temperature and altitude, *Nature*, 285, 314.

Stahl, W., et al., 1974, *Origin of artesian and thermal waters determined by oxygen, hydrogen and carbon isotope analyses of water samples from the Sperkhios Valley, Greece in Isotopes technique in groundwater hydrogeology*, 1st Proceedings of Symposium, International Atomic Energy Agency (IAEA), Vienna, 1974, 317–339.

Sun, C., et al., 2019, Controls on the isotopic composition of precipitation in the south-central United States, *Journal of Geophysical Research: Atmospheres*, 124, 8320–8335.

Taylor, B., 2001, *Hydrogen isotope composition of magmatic water: Review of variations due to source, igneous environment, and degassing processes*, AGU Spring Meeting Abstracts.

Taylor, H.P. Jr., 1974, The application of oxygen and hydrogen isotope studies on problems of hydrothermal alteration and ore deposition, *Economic Geology*, 69, 843–883.

Taylor, H.P. Jr., 1977, Water/rock interactions and the origin of H_2O in granitic batholiths, *Journal Geological Society (London)*, 133, 509–558.

Tes, R.V., 1939, Isotopic composition of rain water, *Akad. Nauk, USSR*, 23, 674.

Tindall, J.C., et al., 2009, Stable water isotopes in HadCM3: Isotopic signature of El Nino-Southern Oscillation and the tropical amount effect, *Journal of Geophysical Research*, 114(D4), 111.

Toth, J., 1963, A theoretical analysis of groundwater flow in small drainage basins, *Journal Geophysical Research*, 68, 4795–4812.

Truesdell, A.H., and J.R. Hulston, 1980, Isotopic evidence on environments of geothermal systems, in *Handbook of environmental isotope geochemistry*, Elsevier, 1(A), 179–226.

van der Straaten, C.M., and W.G. Mook, 1983, *Stable isotopic composition of precipitation and climatic variability in paleoclimates and paleowaters: a collection of environmental isotope studies*, Proceedings Advisory Board Meeting, International Atomic Energy Agency (IAEA), November 25–28, 1980, 53–64.

Vuille, M., and M. Werner, 2005, Stable isotopes in precipitation recording South American summer monsoon and ENSO variability: Observations and model results, *Climate Dynamics*, 25(4), 401–413.

Wang, J., et al., 2021, Characteristics of stable isotopes in precipitation and their moisture sources in the Guanling Region, Guizhou Province, *Journal of Chemistry*, (4), 1–12.

Wang, S., 2016, Factors controlling stable isotope composition of precipitation in arid conditions: An observation network in the Tianshan Mountains, central Asia, *Tellus B: Chemical and Physical Meteorology*, 68(1).

Wang, S.-Y., et al., 2012, Multidecadal drought cycles in the great basin recorded by the great salt lake: Modulation from a transition-phase teleconnection, *Journal of Climate*, 25(5), 1711–1721.

Waterisotopes.org, 2023, accessed November 23, 2023.

Werner, M., et al., 2018, Reconciling glacial Antarctic water stable isotopes with ice sheet topography and the isotopic paleothermometer, *Nature Communications*, 9.

White, A.F., and N.J. Chuma, 1987, Carbon and isotopic mass balance models of Oasis Valley-Fourtymile Canyon groundwater basin, southern Nevada, *Water Resources Research*, 23(4), 571–582.

White, D., 1974, Diverse origins of hydrothermal ore fluids, *Economic Geology*, 69, 954–973.

Wu, H., et al., 2022, Temporal variations of stable isotopes in precipitation from Yungui Plateau: Insights from moisture source and rainout effect, *Journal of Hydrometeorology*, 23, 39–51.

Yeh, H., et al., 2014. Identifying seasonal groundwater recharge using environmental stable isotopes, *Hydrology*, 6, 203–215.

Yurtsever, Y., 1975, *Worldwide survey of stable isotope in precipitation*, International Atomic Energy Agency (IAEA), Report Section Isotope Hydrology.

Zentner, M.A., 2001, *Time and mechanism of groundwater recharge in a mountainous terrain in a temperate climate*, unpublished MS thesis, Brigham Young University.

Zuppi, G.M., et al., 1974, *Isotopes du milieu et circulations d'eaux sulfurees dans le Latium in Isotopes technique in groundwater hydrogeology*, 1st Proceedings of Symposium, International Atomic Energy Agency (IAEA), Vienna, 1974, 341–361.

9 Carbon-13

9.1 INTRODUCTION

The lithosphere is the earth's great carbon reservoir (Table 9.1), although carbon is abundant in the atmosphere, hydrosphere (mostly ocean water), terrestrial biosphere, soil, and fossil fuels (coal, oil, and gas). Approximately 99.9 percent of the potentially available carbon is sequestered in ancient sedimentary carbonate and argillaceous rocks. An additional 180 million gigatons (Gt) of carbon is in the upper and lower mantle (Suarez et al., 2019). Most lithospheric carbon is in rocks of marine sedimentary origin. This carbon is potentially available but is relatively immobile (Ruby, 1964). There is a wide range in published worldwide annual carbon budgets, and the data in Table 9.2 should be considered as a general approximation of the relative amounts. The annual carbon imbalance of +4 Gt is the result of the introduction of anthropogenic carbon to the atmospheric that is only partially offset by oceanic and terrestrial uptake. The worldwide carbon cycling budget is relatively unimportant to the groundwater carbon cycle. The interactions of carbon between carbon reservoirs regulate the concentration and isotopic composition of groundwater carbon. The carbon species CO_2, HCO_3^-, and CO_3^{2-} are the most important carbon compounds in groundwater. In the normal pH range of groundwater, HCO_3^- is the dominant species. The partial pressure of soil gas CO_2 and the dissolution of carbonate minerals in the soil zone and aquifer skeleton are the two principal factors controlling the concentration of carbon in most groundwater. The concentration of soil zone P_{CO_2} is largely regulated by biologic activity.

Carbon isotopes of interest in environmental studies are the stable isotopes ^{12}C and ^{13}C and the radioactive isotope ^{14}C. The stable isotopes have isotopic abundances of 98.89 and 1.11 percent, respectively (Table 7.1). Carbon-14 is naturally present in only trace amounts in water that recharged during the past 75,000 years or so and is not measurable in older water. A key element in using $\delta^{13}C$ and ^{14}C in groundwater investigations is understanding the numerous processes than can contribute carbon and their isotopic compositions to groundwater. The final $\delta^{13}C$ of a groundwater is the result of the fractionation of the carbon source term(s). As is discussed later, most groundwater carbon is from soil zone CO_2-carbonate mineral interactions. Numerous other carbon sources and interactions that can be sources of groundwater carbon need to be understood and considered when evaluating groundwater carbon isotopes.

Knowledge of the groundwater carbon isotopic compositions can provide valuable insights into the origin and chemical evolution of carbon species in groundwater and are essential for modeling radiocarbon groundwater ages. Even though it is possible to measure ^{14}C contents in groundwater that recharged up to 75,000 years ago, the practical upper limit for ^{14}C dating is considerably less (Chapter 11).

The stable isotopic composition of carbon is expressed in terms of the delta notation, where:

$$\delta^{13}C = \left[\frac{\left(^{13}C/^{12}C\right)_{\text{sample}}}{\left(^{13}C/^{12}C\right)_{\text{standard}}} - 1 \right] \times 10^3 \left(‰\right) \tag{9.1}$$

Stable isotopic compositions are referenced with respect to the standard PDB; however, in the literature, isotopic compositions have sometimes been reported relative to other standards.

DOI: 10.1201/9781003646129-11

TABLE 9.1
The Range of $\delta^{18}O$ Values of Stored Carbon in the Primary Carbon Reservoirs

Reservoir	Estimated total carbon (Gt)[1]	$\delta^{13}C$ (‰)
Atmosphere	700–800	−7
Hydrosphere		
Surface ocean	670	0.5 to 2.5
Ocean inorganic	37,400	
Deep ocean	38,400	−1 to 0
Organic	1,000	
Terrestrial biosphere		
Living biomass	2,000	
C3 (trees)[2]		−20 to −37 (−27 mode)
C4 (grasses)		−12 to −20 (−11 mode)
Cam		−10 to −20 (mode ~ −14)
Dead biomass	1,200	−13 to −25
Soil	1,500–2,300	−13 to −25
Fossil fuels	4,000–10,000	−20 to −80
Lithosphere	60–100 million	0 to −10

[1] Gigaton (Gt) = one billion metric tons, metric ton = 2,205 pounds

Source: Data from Falkowski et al. (2000), Gruber et al. (1999), Kohn (2010), O'Leary (1988), and Winter and Holtum (2002).

TABLE 9.2
Annual Global Carbon Flux in Gigatons/Year (TG/y)

From	Destination	Atmosphere: Annual input (GT)	Atmosphere: Annual output (GT)
Atmospheric	Plants and soil carbon		123
	Ocean		92
Terrestrial			
Plants	Atmosphere	60	
Soil	Atmosphere	60	
Fossil fuels, cement, land use	Atmosphere	9	
Oceanic	Atmosphere	90	
Total		219	215
Net terrestrial uptake		3	
Net ocean uptake		2	
Net atmosphere increase		4	

Source: Data from Riebeek, 2011.

9.2 CARBON ISOTOPIC FRACTIONATION

The hydrologic usefulness of stable carbon isotopes is based on the fact that the $\delta^{13}C$ of carbon in each of the major carbon reservoirs (carbonate rocks, oceans, atmosphere, terrestrial biosphere, soil, and fossil fuels) are measurably different from each other (Table 9.1, Figure 9.1) and that the isotopic composition of groundwater carbon is affected by several well-understood processes. The typical $\delta^{13}C$ contents of the principal carbon reservoirs of interest in hydrogeology are about 0 ‰ for ocean water and marine carbonate rock, –7 ‰ for atmospheric CO_2, and –10 to –32 ‰ for living and dead terrestrial organic matter.

Isotopic differences between the carbon reservoirs, including groundwater, are the result of isotopic fractionation, isotopic exchange, and mixing between:

1. inorganic and organic carbon in the atmosphere, the oceans, and the photosynthetic cycle,
2. inorganic isotopic exchange between aqueous carbon species, and
3. aqueous and solid carbon species in the soil zone and aquifer skeleton.

9.2.1 Atmospheric, Oceanic, and Biologic Carbon

Atmospheric carbon, mostly CO_2, is a critical link in the chain of organic and inorganic carbon that cycles through the atmosphere, biosphere, and hydrosphere (Figure 9.2). Atmospheric carbon is moderated by the dynamic equilibrium between atmospheric CO_2 and near-surface mobile carbon in the oceans and terrestrial environments. Ocean water bicarbonate is the great buffering and exchange reservoir for atmospheric carbon. Ocean water buffers the isotopic composition of atmospheric carbon for two reasons. First, the quantity of mobile carbon in the atmosphere is small relative to the quantity in the oceans (Table 9.1). Second, the residence time of atmospheric carbon is short, about 5–10 years, compared with the long residence time of dissolved ocean carbon, about 1,000 to 1,500 years. Although plants are the primary sink of atmospheric CO_2, CO_2 exchange between the atmosphere and the oceans is a major factor in the atmospheric CO_2 cycle. Plants, root respiration, and bacterial decay

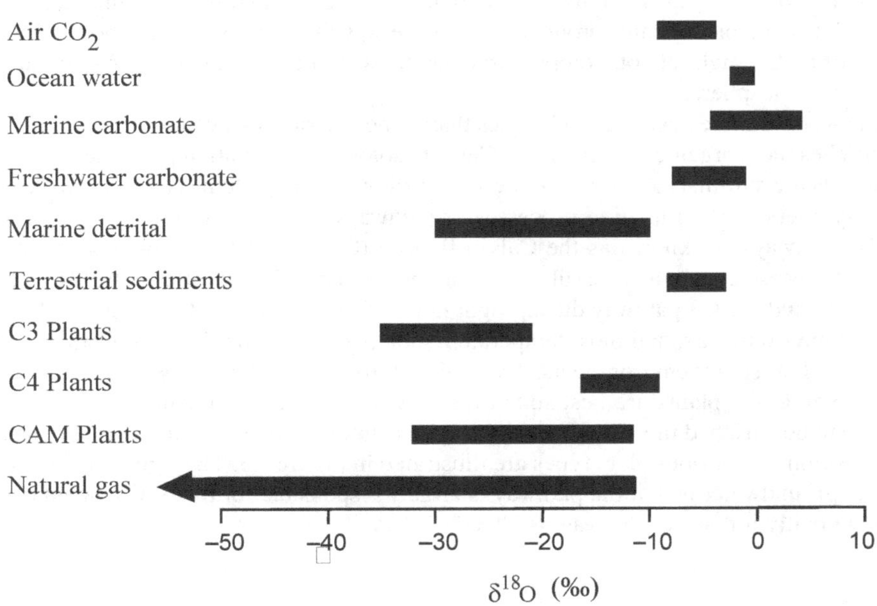

FIGURE 9.1 Typical $\delta^{13}C$ ranges of carbon sources of interest to hydrogeochemistry.

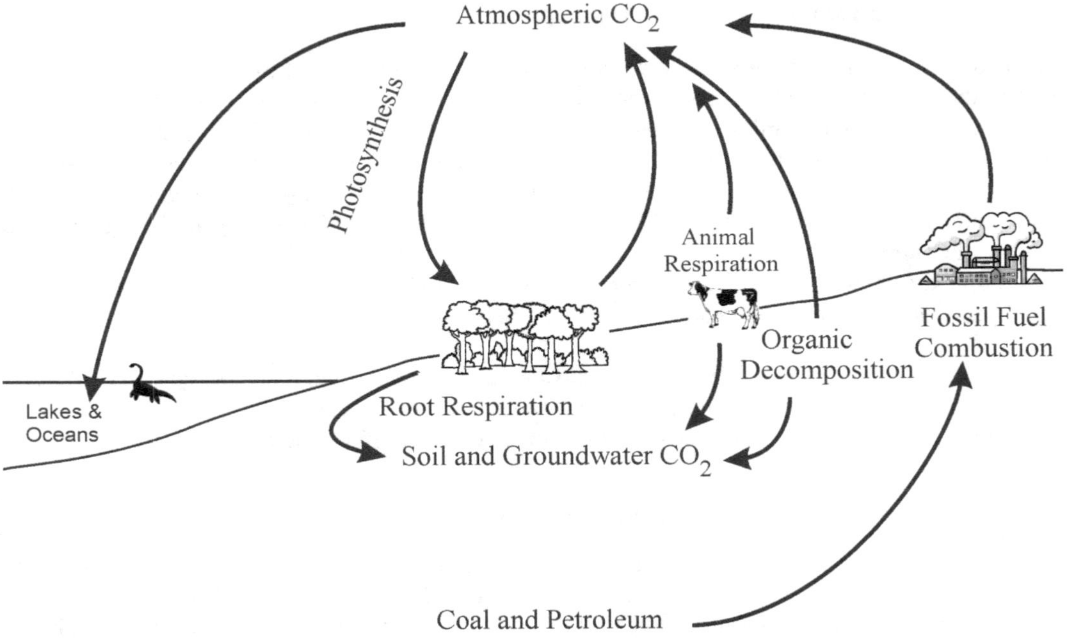

FIGURE 9.2 The carbon cycle.

of dead organic matter in the soil account for more than 50 percent of annual natural return of CO_2 to the atmospheric (Table 9.2). Burning fossil fuels, cement production, and other anthropogenic sources contribute about 9 Gt annually to the atmosphere (Riebeek, 2011).

Carbon isotopic fractionation between atmospheric CO_2 and dissolved oceanic carbon enriches seawater bicarbonate in ^{12}C and depletes atmospheric CO_2 in ^{12}C. The $\delta^{13}C$ of ocean surface water ranges between 0.5 and 2.5 ‰. This narrow range of $\delta^{13}C$ in dissolved oceanic carbon and the isotopic evolution of biologically produced atmospheric CO_2, whose $\delta^{13}C$ is about –7 ‰, can be accounted for by the high rate of isotopic exchange between dissolved oceanic carbon and the carbon gas in the atmosphere.

Most plants have more negative $\delta^{13}C$ values than atmospheric CO_2 because biologic fractionation greatly enriches most organic matter in ^{12}C. Carbon isotopic fractionation processes accompanying photosynthesis are summarized here and are reviewed in detail by Deines (1980) and Vogel (1993). Two photosynthetic carbon fixation processes or pathways regulate the $\delta^{13}C$ content of terrestrial plants. The pathways are known as the Calvin-Benson (C3) and Hatch-Slack (C4) cycles. A third type of plant process known as Crassulacean acid metabolism (CAM) follows the C4 pathway during dark hours and the C3 pathway during light hours (Osmond, 1975). C3 plants are best adapted for cool, wet environments, and most temperate region plants have a C3 metabolism. C4 plants are best suited for hot, sunny environments. Tropical, subtropical, and arid region plants tend to have C4 metabolism. CAM plants are best suited for very hot, dry environments. Plants living in arid regions tend to be enriched in ^{13}C with respect to plants living in more humid regions. The ranges of $\delta^{13}C$ compositions for various plant types are illustrated in Figure 9.3. The significance of metabolic pathways to groundwater is that the pathway is largely responsible for the $\delta^{13}C$ content of soil zone CO_2, which greatly influences the water's HCO_3^- $\delta^{13}C$.

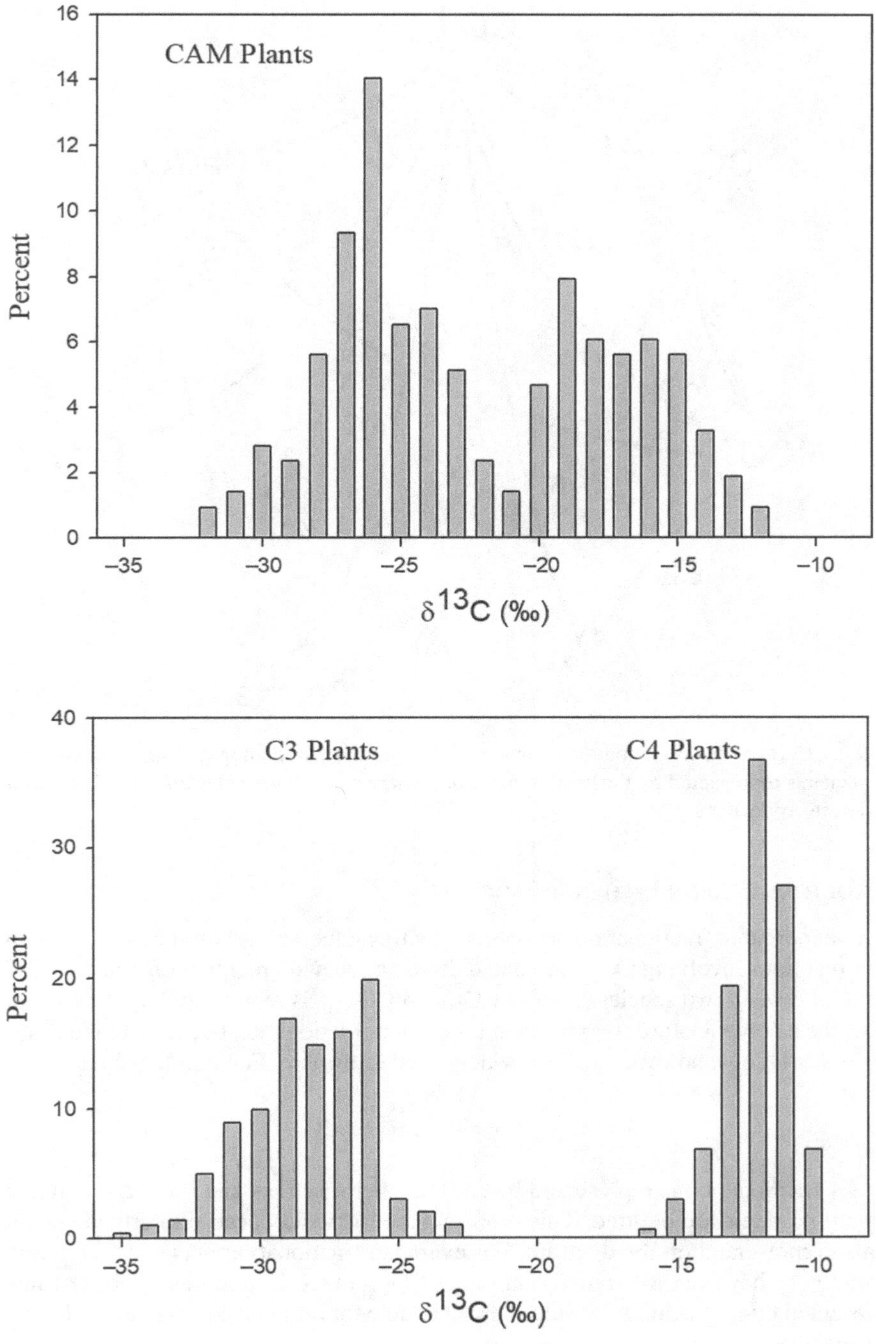

FIGURE 9.3 Carbon isotopic composition of photosynthetic plants. Data from Hoefs (2010) and Deines (1980).

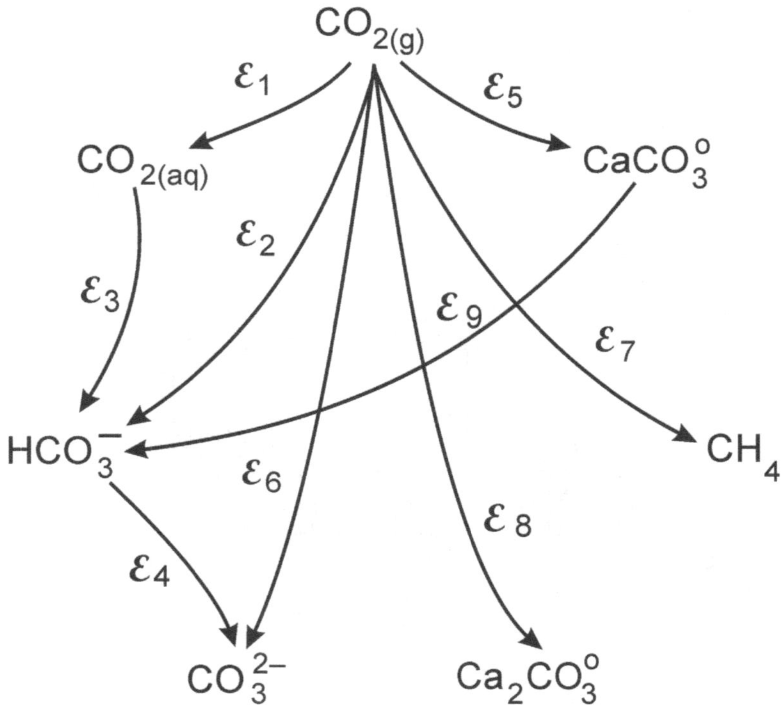

FIGURE 9.4 Carbon fractionation pathways of interest in hydrogeologic investigations. Fractionation between species is represented by the epsilon notation, where $\varepsilon = 10^3\ln\alpha$. Selected isotopic fractionation factors are listed in Appendix D.

9.2.2 Aqueous Carbon Exchange Processes

The principle inorganic fractionation processes effecting aqueous carbon species may be envisioned in terms of reactions involving CO_2 and related dissociation and precipitation reactions (Figure 9.4). Fractionation between two species, such as $CO_{2(g)} \rightarrow CO_{2(aq)}$, is represented by the ε (epsilon) notation. Use of the ε nomenclature is convenient for carbon fractionation because it facilitates radiocarbon dating calculations; however, ε is not widely used elsewhere. For small epsilon values,

$$\varepsilon = 10^3 \ln\alpha \tag{9.2}$$

In Figure 9.4 fractionation is represented by end-member reactants and products, and intermediate reactants and products are omitted. This representation is valid because intermediate stable products are also represented on the diagram. For example, fractionation between $CO_{2(g)}$ and CO_3^{2-} is represented by ε_9; however, the transformation of $CO_{2(g)}$ to CO_3^{2-} requires a series of intermediate steps represented by ε_1, ε_3, and ε_4. ε values are available for all intermediate and end-member reactions in Appendix D.

9.3 SOURCES OF GROUNDWATER CARBON

Most groundwater acquires about 50 percent of its carbon from $CO_{2(g)}$ in the soil zone and 50 percent from the dissolution of carbonate minerals in the soil zone and aquifer skeleton (Chapter 6). In the typical range of groundwater pH, the evolution of CO_2 into HCO_3^- can be represented as:

$$CO_{2(g)} + H_2O = CO_{2(aq)} + H_2O \tag{9.3}$$

$$CO_{2(aq)} + H_2O = H_2CO_3 \tag{9.4}$$

$$H_2CO_3 = HCO_3^- + H^+ \tag{9.5}$$

The H^+ in Equation 9.5 dissolves soil zone and aquifer carbonate minerals:

$$H^+ + CaCO_3 = Ca^{2+} + HCO_3^- \tag{9.6}$$

At low or high pH the carbon will be in the form of H_2CO_3 and CO_3^{2-}, respectively. The $\delta^{13}C$ of groundwater that has acquired carbon from the dissolution of carbonate minerals in the presence of CO_2 will be the mass weighted average of the $\delta^{13}C$ contributions represented by Equations 9.5 and 9.6. Carbon contributions from the dissolution of marine carbonate minerals typically have a $\delta^{13}C$ of about 0 ± 3 ‰, and the contribution from organic carbon in the soil will usually have a $\delta^{13}C$ in the range −14 to −28 ‰. Assuming a 50/50 contribution of carbon from each source, the $\delta^{13}C$ of most groundwater will be about −10 ± 2 to 3 ‰.

Isotopic fractionation between carbon in soil gas CO_2, aqueous carbon species, and solid carbonate minerals is common, and the final groundwater $\delta^{13}C$ content may differ from that anticipated by the simple mixing of soil zone CO_2 and carbon from carbonate minerals. In many groundwater investigations the 1 to 2 ‰ fractionation difference is of little consequence and may be neglected. However, subtle differences in the stable isotopic compositions can be important for radiocarbon dating and in isotopic investigations where two or more groundwaters are used to identify or characterize groundwater systems, groundwater mixing, recharge locations, climatic variations, and paleoclimate. In these situations, the effects of isotopic fractionation between the dissolved carbon species and the other carbon species encountered in the flow path must be accounted for.

The carbon history of groundwater systems may be complex, and numerous factors in addition to soil zone CO_2 and carbonate minerals can be involved. Groundwater whose isotopic composition is beyond the range of typical values or which has unusually low or high HCO_3^- and P_{CO_2} concentrations may have acquired appreciable carbon from other sources. The P_{CO_2} may be readily calculated by means of computer codes such as WATEQF (Plummer et al., 1976) and PHREEQC (Parkhurst and Appelo, 2013) and commercially available groundwater chemistry programs. Locally, groundwater may acquire carbon from atmospheric CO_2; however, in almost all systems, carbon is from other sources. CO_2 and CH_4 are released by the reduction of shallow organic matter (biogenic), during catagenesis of organic carbon accompanying sediment burial (thermocatalytic), and by the chemical weathering of non-carbonate minerals. CO_2 is released from magmatic fluids and by the release of CO_2 from carbonate minerals during metamorphism.

Important Takeaway – Typical groundwater acquires about 50 percent of its carbon from soil zone CO_2 gas and about 50 percent from the dissolution of carbonate minerals. The isotopic composition of soil zone gas is determined by the plant type, C3, C4, or CAM. Assuming a 50/50 contribution of carbon from each source, the $\delta^{13}C$ of most groundwater is about −10 ± 2 to 3 ‰.

9.3.1 Groundwater CO_2

In most groundwater investigations it is safe to ignore atmospheric contributions of carbon. Usually only a small fraction of groundwater carbon is directly from atmospheric CO_2 because the P_{CO_2} in the atmosphere is about $10^{-3.5}$ atm and the P_{CO_2} in most groundwater is about 10^{-2} to $10^{-1.5}$ atm. However, in areas of little or no vegetation, particularly in arid regions with sandy soil, groundwater

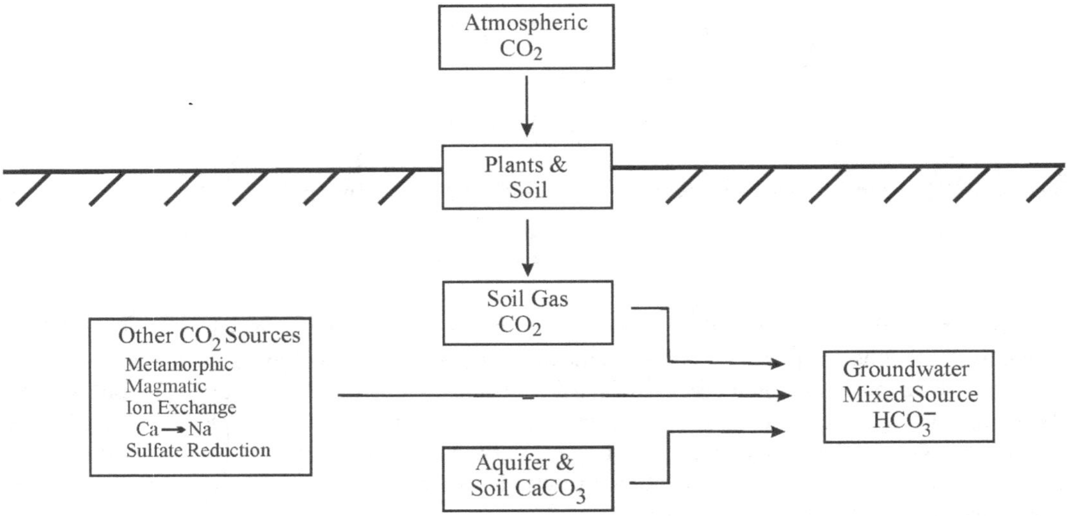

FIGURE 9.5 Pathways for the contribution of carbon to groundwater. In most groundwater systems about 50 percent of the carbon is from soil gas CO_2 and 50 percent from the dissolution of $CaCO_3$ in the soil and aquifer. Because many groundwaters acquire about 50 percent of their HCO_3^- from soil zone $CO_{2(g)}$ ($\delta^{13}C \approx -20$ ‰) and about 50 percent from the carbonate matrix ($\delta^{13}C \approx -0$ ‰), the resulting groundwater will commonly have a HCO_3^- $\delta^{13}C \approx -10$ ‰.

P_{CO_2} may be low and atmospheric CO_2 may be an important carbon source. When atmospheric CO_2 is an important or major carbon source, the $\delta^{13}C$ of groundwater will be in the range of –7 to –3 ‰. Although –7 ‰ is at the end of the range of values expected for mixed biogenic and dissolved carbonate mineral sources, groundwater having a major component of atmospheric CO_2 can often be identified by the low HCO_3^- and TDS concentrations and the calculated P_{CO_2}.

As previously discussed, in most groundwater systems approximately one-half of the total dissolved carbon is contributed by biogenic CO_2 in the soil zone (Figure 9.5), where the partial pressure of soil gas CO_2 may be as great as 10^{-2} to $10^{-1.5}$ atm. This gas is the product of both root respiration and decomposition of dead organic matter, and the CO_2 is considered to be biogenic. Plant roots and some soil biota respire CO_2, and CO_2 is released when organic soil material is degraded by microbial oxidation (Gillon et al., 2012). It is of great importance to hydrogeology that photosynthetic pathways are generally linked to climatic and that $\delta^{13}C$ of soil zone CO_2 is directly related to the photosynthetic pathways. Knowledge of the variations in the $\delta^{13}C$ of soil CO_2 is important for modeling radiocarbon groundwater ages (Chapter 11) and are often useful for evaluating paleoclimatic conditions of ancient groundwater recharge.

Soil zone root respiration CO_2 has approximately the same isotopic composition as the local plant cover and is isotopically similar to plant carbon regardless if it is from C3, C4, or CAM plants (Park and Epstein, 1960, 1961; Bottrell, 2019). Decomposed wood, peat, and lignite also have about the same $\delta^{13}C$ composition as fresh wood (Craig, 1954; Deines, 1980). However, CO_2 from microbial oxidation of soil organic matter is depleted in ^{13}C (Bottrell, 2019). Cerling et al. (1991) suggest this difference may be due to the diffusion coefficients of $^{12}CO_2$ and $^{13}CO_2$.

The three terrestrial plant groups have large variations in soil gas $\delta^{13}C$ values, and the bimodal distribution of the isotopic composition of soil gas carbon mirrors the distribution of the carbon isotopic composition of carbon in plants. C3 plants are the most common plant type and include most agricultural crops, many vegetables, and most trees. They are named after a 3-carbon compound (3-phosphoglycerate) that is a product of photosynthesis. They are sensitive to high temperatures and water stress. C4 plants are named after a 4-carbon compound. They do well in hot and dry environments. Lerman (1972) reports average soil gas CO_2 values of −13 ± 4 ‰ for C4 plants, −27 ± 5 ‰ for C3 plants and a spread of values of about −17 ‰ for the CAM plants. The bimodal distribution of $\delta^{13}C$ in CAM plants suggests a predominance of either C4 or C3 cycle metabolism in individual plant species. Galimov (1966) and Rightmire and Hanshaw (1973) found that root respiration and decay of dead organic matter in moderate climates with mixed plant types have soil CO_2 with a $\delta^{13}C$ of −25 ± 5 ‰. Smith and Epstein (1971) and Park and Epstein (1960) found that tropical, subtropical, and arid climates should have a soil CO_2 with a $\delta^{13}C$ of −12 ± 5 ‰. Reported mean $\delta^{13}C$ values of soil CO_2 in desert regions range from −16.7 to −18.5 ‰ (Rightmire and Hanshaw, 1973; Kunkler, 1969); however, Hart (2009) found average soil values of −11 ‰ ± 2‰ for desert soils in Utah. Because it is not practical to measure the $\delta^{13}C$ of soil gas CO_2 in most groundwater investigations, the soil gas $\delta^{13}C$ must be estimated based on the local plant communities. Such estimation becomes particularly difficult when paleo-groundwater is involved, and the climate and, hence, plant community at the time of recharge may have been different than today's.

9.4 SOIL ZONE AND AQUIFER CARBON PROCESSES

The combined carbon contributions from soil zone CO_2 and dissolved carbonate minerals often yield groundwater HCO_3^- concentrations exceeding 200 ppm (3.3 meq L^{-1}). Soil zone chemical reactions of significance in the dissolution of carbonate minerals involve organic acid from the decomposition of organic matter in the soil and carbonic acid formed from root respiration CO_2 (Chapter 6). The general forms of these reactions are:

$$\underset{\text{(organic acid)}}{R-COOH} + CaCO_3 = HCO_3^- + Ca^{2+} + R-COO^- \tag{9.7}$$

$$CO_2 + H_2O + CaCO_3 = 2HCO_3^- + Ca^{2+} \tag{9.8}$$

where COOH is the carboxyl group and R is the remainder of the molecule. It is important to remember that Equations 9.7 and 9.8 are the simplified form of several independent reactions. A more comprehensive treatment of these processes is found in Chapter 5.

Soluble carbon is found in the soil zone, carbonate rocks, argillaceous sediments, and fossil fuels such as coal and petroleum. The carbon in each of these carbon reservoirs has characteristic $\delta^{13}C$ contents, and that portion of the dissolved carbon in groundwater acquired from these sources will have a $\delta^{13}C$ representative of the reservoir source term.

The $\delta^{13}C$ contents of marine carbonate rocks of Cambrian to Tertiary age reflect the isotopic composition of ancient seawater. Reported isotopic values cluster about 0 ‰, and a default value of 0 ‰ is often used when local data are not available. The range of $\delta^{13}C$ values of marine carbonate rocks have been reported as +3 to −3 ‰ (Craig, 1963), +4 to −4 ‰ (Truesdell and Hulston, 1980), +4 to −1 ‰ (Giggenbach et al., 1983), and +2 to −2 ‰ (Oehlert and Swart, 2014). Muller and Mayo (1986) investigated the spatial distribution of $\delta^{13}C$ over a 500 km^2 area of a Grand Canyon regional carbonate aquifer. They found a mean $\delta^{13}C$ value of −2.11 ‰ and a range of values from +0.28 to −3.61 ‰ for weathered and unweathered samples.

Freshwater carbonates are enriched in ^{12}C relative to marine carbonates. Keith and Weber (1964) report an average $\delta^{13}C$ value of −4.93 ± 2.75 ‰ for 183 freshwater carbonates. The more negative $\delta^{13}C$ values of freshwater carbonates are largely due to the contribution of CO_2 liberated by the oxidation of plant debris in the soil and by plant respiration.

In a few groundwater systems, organic sediments and other factors are an import source of groundwater carbon. The carbon isotopic compositions of marine and nonmarine organic-rich detrital sediments reflect the environmental conditions under which organic matter was living prior to deposition. Plankton is the principal source of organic carbon in most marine detrital rocks, although terrestrial sources and inorganic precipitation can contribute some carbon. The $\delta^{13}C$ of marine detrital rocks ranges from –10 to –30 ‰; however, 90 percent of the values fall in the narrow range of –20 to –27 ‰ (Deines, 1980). There is a relationship between surface water temperature (and hence latitude) and the $\delta^{13}C$ of detrital marine sediments. In an extensive study of particulate organic $\delta^{13}C$ in the Atlantic Ocean sediments, Verwega et al. (2021) found relatively high values (≥ –22 ‰) in the low latitudes, with more negative values in the Arctic Ocean (~-24 ‰) and even lower values (≤ –28 ‰) in the southern oceans.

Organic materials in terrestrial sedimentary rocks exhibit a wider range of $\delta^{13}C$ values than do marine sediments. Lake bed sediments do not display the peaked $\delta^{13}C$ frequency distribution found in marine detrital sediments, and the sediments may be seasonally dependent. Kämpf (2020) found winter $\delta^{13}C$ values of –3.2 to –7.1 ‰, and summer values –5.9 to –7.7 ‰ in lake bed sediments.

Diagenesis and lithification do not appreciably alter the $\delta^{13}C$ composition of organic rich sediments. Factors controlling the isotopic composition of kerogen in sediments include:

1. the $\delta^{13}C$ content of the organic matter buried in place,
2. the $\delta^{13}C$ content of transported organic matter, most notably in aquatic sediments, soils, and peat, and
3. isotopic changes in the organic matter before it is incorporated into the sediment.

During diagenesis, complex carbon compounds that were synthesized by photosynthetic processes become unstable and disintegrate mechanically and chemically. A portion of the decay products are oxidized to CO_2 and H_2O, some are used by other living organism, and some degrade into brown-colored, base-soluble organic matter known as humic acid (insoluble in most acids) and fluvic acids (soluble in most acids). Humic acids are responsible for the brown color of groundwater associated with organic-rich terrestrial sedimentary rocks.

With increasing compaction and dewatering, the degradation of carbon molecules are accompanied by an increase in atomic weight. If compaction and dewatering proceed far enough, the carbon molecules form non-extractable amorphous solids termed "kerogen." Further burial and increased compaction may lead to catagenesis, during which additional CO_2 and H_2O may be lost and the carbon compounds may be converted to hydrocarbons. During the final stages of catagenesis, methane (CH_4) may evolve. Organic carbon in sedimentary rock may be further reduced to elemental carbon (graphite) during metamorphism.

The anaerobic biogenic decomposition of organic material (methanogenesis) in an aquifer can yield substantial quantities of CO_2 and CH_4. Isotopic fractionation between CO_2 and CH_4 results in the accumulation of isotopically light carbon in CH_4. The isotopically heavier CO_2 may evolve to carbonic acid, which in turn may dissolve additional isotopically heavy carbonate minerals. In a Maryland aquifer, Chapelle and Knobel (1985) found that this process was responsible for the downgradient enrichment of ^{12}C where the $\delta^{13}C$ evolved from –13.0 to –5.3 ‰. The general fermentation reactions for the liberation of CO_2 may be represented as:

$$\underset{\text{(cellulose)}}{C_6H_{10}O_5} + H_2O = \underset{\text{(glucose)}}{C_6H_{12}O_6} \tag{9.9}$$

$$\underset{\text{(glucose)}}{C_6H_{12}O_6} + 6H_2O = 6CO_2 + 12H_2 \tag{9.10}$$

$$\underset{\text{(glucose)}}{C_6H_{12}O_6} + 2H_2O = \underset{\text{(acetic acid)}}{2CH_3CO_2^-} + 2H^+ + 2CO_2 + 4H_2 \tag{9.11}$$

The products of fermentation reactions may be utilized by methanogenic bacteria to produce additional CO_2 and methane:

$$2CH_3CO_2^- + H^+ = CH_4 + CO_2 \quad (9.12)$$
(acetic acid)

$$CO_2 + 4H_2 = CH_4 + 2H_2O \quad (9.13)$$

The carbon contained in coal, petroleum, natural gas, oil shale, and tar sands are enriched in ^{12}C with respect to atmospheric CO_2. The isotopic compositions of fossil fuels reflect their biogenic origin. The $\delta^{13}C$ composition of coal is similar to modern land plants, and there is only about a 1 to 2 ‰ shift in $\delta^{13}C$ during the coalification process (i.e., transformation of lignite to anthracite). During coalification, reduced volatile organic matter is oxidized and CO_2 is released. The general form of the chemical reaction may be represented as:

$$\tfrac{1}{2}\, C_nH_2O = \tfrac{1}{2}\, CO_2 + C_nH_{2n}^{2+} \quad (9.14)$$

where C_nH_2O represents reduced organic material and n is the number of molecules. Coalification has been found to be an important source of CO_2 in the groundwaters in the Atlantic and Coastal Plain regions and in the Powder River Basin in Wyoming and Montana (Flores et al., 2008; Winograd and Farlekas, 1974). As in the case of sulfate reduction, the released CO_2 will be isotopically similar to the original organic matter. Because only CO_2 is released and bicarbonate ions are not first-order products of this process, the final isotopic composition of the groundwater will be controlled by Equation 9.8. It may be isotopically difficult to distinguish such groundwater from groundwater whose carbon is from soil zone CO_2 and the dissolution of carbonate minerals in the soil zone and in the aquifer skeleton.

Considerable volumes of CH_4 also evolve during coal maturation and during the thermocatalytic breakdown of organic matter in organic-rich sediments. Most of this gas migrates away from the source area and either escapes to the atmosphere or is trapped as natural gas deposits. Some CH_4 may be carried away by groundwater. Locally, methane may substantially alter the isotopic composition of groundwater because isotopically heavy carbon accumulates in CO_2 during fractionation between CO_2 and CH_4. The oxidation of methane to CO_2 may be represented as:

$$CH_4 + 2O_2 = CO_2 + H_2O \quad (9.15)$$

The $\delta^{13}C$ of CH_4 in coal gas exhibits a wide range of values (−12 to −71 ‰). The $\delta^{13}C$ composition of crude oil ranges from −21 to −32 ‰, and most crude oils have a $\delta^{13}C$ content of about −28 ‰. The carbon isotopic composition of methane gas associated with petroleum varies from −25 to −65 ‰ (Deines, 1980).

Natural gas is a term used to describe the mixture of hydrocarbon and non-hydrocarbon compounds found in the subsurface. Natural gas occurrences are not necessarily associated with coal or petroleum. For example, it is estimated that 11 to 75 Gt of methane is released to the atmosphere each year by thawing permafrost in the pan-Arctic region (Elder et al., 2021). Methane (CH_4) is the most abundant hydrocarbon compound in natural gas. As previously described, CH_4 is important in many groundwater investigations because it may be dissolved in the water or it may undergo isotopic fractionation with dissolved carbon-bearing species, particularly CO_2. Other hydrocarbon constituents include ethane, propane, n-butane, and isobutene (Chapter 5). The following notations often refer to the carbon compounds in natural gas:

C1 = methane (CH_4)
C2 = ethane (C_2H_6)
C3 = propane (C_3H_8)
C4 = n-butane (C_4H_{10})
C5 = isobutane (C_4H_{10})
C2+ = concentration of C2 to C5 relative to C1

$$C2+=\left(1-\frac{C1}{Cn}\right)\times 100 \tag{9.16}$$

where *n* is the sum of all hydrocarbons.

9.5 OTHER INORGANIC CARBON SOURCES

Several less common sources of inorganic carbon may locally be the driving mechanism of groundwater carbon. Such sources include magmatic, metamorphism of carbonate rocks, sulfate reduction, and connate water.

9.5.1 Magmatic Carbon

Gases emanating from geothermal and volcanic fumaroles and cooling lava flows may contain appreciable amounts of CO_2 and smaller amounts of CH_4 (about 1 percent). Much of this gas was thought to be from juvenile sources (discharge from the upper mantle) by many early investigators. The isotopic compositions of subcrustal minerals such as carbonatites and diamonds have been investigated to determine isotopic composition of subcrustal carbon and, by extension, the juvenile carbon in groundwater. Carbonatites are igneous rocks containing more than 50 percent calcite and dolomite and are often associated with deposits of rare earth elements (REE). Although there are 37 known carbonatite deposits in North America (Millonig and Groat, 2013), the only working REE mine in the United States is in California's Mojave Desert. The mine produces about 16 percent of the world's REE supply.

The $\delta^{13}C$ content of carbonatites range from –2.0 to –8.0 ‰ (Conway and Taylor, 1969; Galimov et al., 1975; Santos and Clayton, 1995). Deines and Gold (1969) found an average $\delta^{13}C$ value of –5.1 ± 1.4 ‰ for 22 carbonatites and –4.7 ± 1.2 ‰ for carbonates from 13 kimberlites. Jones et al. (2013) found primary igneous carbonatites have $\delta^{13}C$ ranging from –3.1 to –7.7 ‰. Regional differences in the isotopic values may indicate that carbonatites are not reliable indicators of deep-seated carbon (Faure, 1977). Moreover, the $\delta^{13}C$ variations and values of igneous carbonate minerals are difficult to distinguish from marine and nonmarine sedimentary carbonates. Diamonds occurring in kimberlite pipes and other rocks are believed to originate in the mantle. The isotopic composition of colorless diamonds falls within the $\delta^{13}C$ range of –2.0 to –10.0 ‰ (Faure, 1977); however, colored diamonds have values ranging from –5.0 to –32.3 ‰ (Koval'skiy and Cherskiy, 1973), which overlap with the values found in organic carbon. The wide variability of isotopic compositions may be due the heterogeneity of carbon in the mantle or due to high temperature isotopic fractionation occurring in the crust, or both.

The great isotopic variation found in subcrustal carbon species means that the isotopic composition of carbonatites and diamonds are not a reliable independent guide to distinguish mantle carbon from crustal carbon. Fuex and Baker (1973) estimated the average $\delta^{13}C$ value of crustal rocks at about –7.0 ‰ on the basis of geochemical balance calculations. If most of the carbon in the crust was originally the product of mantle outgassing, the average composition of crustal rocks should reflect the average composition of the mantle. The Furex and Baker estimate is consistent with the values obtained from carbonatites and diamonds if allowances are made for isotopic fractionation. Because the estimated isotopic composition of mantle carbon appears to be in the range of groundwater isotopic values from atmospheric CO_2 and values obtained by the stoichiometric mixing of carbon from organic sources and marine carbonate rocks, it is difficult to identify juvenile carbon in groundwater based on the $\delta^{13}C$ content.

Most researchers now believe that little if any of the gases escaping from geothermal and fumaroles are juvenile and that the isotopic composition of dissolved groundwater carbon or gas is of little use in identifying juvenile gas or groundwater. Craig et al. (1956) and Craig (1963) demonstrated that geothermal waters have $\delta^{13}C$ contents that are similar to the isotopic composition of local precipitation, and most thermal groundwater is believed to have a meteoric origin. Ellis and Mahon (1964) have shown that some chemical constituents of geothermal fluids (Cl, B, and Li) could be leached at reasonable geothermal temperatures from volcanic rocks, and juvenile sources need not be called upon to explain the high concentrations of these trace species. CO_2 collected from cooling lava typically has $\delta^{13}C$ values between –14 and –28 ‰ (Faure, 1977), and the isotopic values of CO_2 emanations from regions not containing limestone, such as Steamboat Springs (Nevada), Lassen Park (California), and the Geysers (California) range between –8 and –11 ‰. Craig (1963) noted that the $\delta^{13}C$ composition of CO_2 in fumaroles in areas where limestone is present is similar to that of the carbonate rocks. The $\delta^{13}C$ of CO_2 collected in geothermal areas generally ranges between –2 to –5 ‰ in $\delta^{13}C$.

9.5.2 Metamorphic Carbon

During metamorphism of carbonate rocks, CO_2 is released, and the isotopic composition of both the carbon and oxygen in the rock are isotopically fractionated. Both the δ13C and the δ18O of the carbonate rock become more negative, and the isotopic composition of the released CO2 is positive with respect to the unaltered rock. In a study of a contact metamorphic zone of an argillaceous limestone, Deines and Gold (1969) found the isotopic composition of calcite as follows:

TABLE 9.3
Carbon-13 compositions of various carbonate rocks

Source	$\delta^{13}C$ ‰	$\delta^{18}O$ ‰
Unaltered limestone	+1.13	+24.12
Marble, contact aureole	–0.58	+18.4
Calcite within intrusion	–3.91	+11.06

Shieh and Taylor (1969) attribute similar depletions in dolomite and calcite in a contact metamorphic zone to decarbonization reactions such as:

$$\text{Calcite} + \text{Quartz} \rightarrow \text{Wollastonite} + CO_2 \tag{9.17}$$

$$2\ \text{Dolomite} + \text{Quartz} \rightarrow 2\ \text{Calcite} + \text{Forsterite} + 2CO_2 \tag{9.18}$$

$$3\ \text{Dolomite} + \text{K-feldspar} + H_2O \rightarrow \text{Phlogopite} + 3\ \text{Calcite} + 3CO_2 \tag{9.19}$$

$$5\ \text{Dolomite} + 8\ \text{Quartz} + H_2O \rightarrow 3\ \text{Calcite} + \text{Tremolite} + 7CO_2 \tag{9.20}$$

The significance of these reactions to hydrogeology is that substantial quantities of CO_2 are released during metamorphism of carbonate minerals, and this CO_2 is enriched in ^{13}C. Because most of the world's carbon is sequestered in thick sequences of carbonate and carbonate-cemented rocks, heating of the deepest portions of these sequences by the geothermal gradient may release carbon to groundwater systems in tectonically active areas. Shieh and Taylor (1969) estimated that the $\delta^{13}C$ of CO_2 released by decarbonization reactions is about 6 ‰ more positive than the parent carbonate minerals. However, CO_2 in equilibrium with calcite at temperatures above 200 °C should only be about 2.8 ‰ more positive (Bottinga, 1969).

In a non-thermal, CO_2 evolving groundwater system, Mayo and Muller (1997) found the HCO_3^- in CO_2 evolving springs ($P_{CO2} \approx 1$ atm) was about 6 ‰ more positive in $\delta^{13}C$ than HCO_3^- found in nearby non-CO_2 evolving springs. The excess CO_2 was attributed to the thermal decomposition of carbonate minerals at depth. After liberation, the CO_2 then migrated upward along faults until it encountered a shallow groundwater system.

9.5.3 Sulfate Reduction Carbon

Sulfate reduction, which occurs in the absence of free oxygen, requires the catalyzing action of anaerobic bacteria and may be identified by the rotten egg smell of evolving hydrogen sulfide gas. Groundwater that has undergone sulfate reduction is usually depleted in SO_4^{2-} . The general form of the chemical reaction may be expressed as:

$$\underset{\text{(organic matter)}}{2(CH_2O)} + SO_4^{2-} = H_2S + 2HCO_3^- \tag{9.21}$$

The isotopic content of the released bicarbonate will be quite negative, having a $\delta^{13}C$ value similar to the original organic matter. The $\delta^{13}C$ content of carbon introduced into groundwater by sulfate reduction may become progressively diluted and less negative due to the dissolution of carbonate minerals by carbonic acid, which forms in response to the evolved CO_2. The final carbon isotopic composition of groundwater influenced by sulfate reduction will depend on the amounts and isotopic compositions of the dissolved carbon prior to sulfate reduction, the carbon introduced during sulfate reduction, and isotopic exchange between the new and previously existing carbon species.

9.5.4 Connate Water

Unaltered interstitial (connate) water should have the $\delta^{13}C$ composition of the original fluid (i.e., 0.5 to 2.5 ‰ for oceanic sediments and −10 ± 3.0 ‰ for typical groundwater). The $\delta^{13}C$ of interstitial water in deep-sea sediments ranges from +18 and −22 ‰ (Deines, 1980). $\delta^{13}C$ values of 0 to −22 ‰ have been attributed to varying degrees of mixing between marine connate water (0 ‰) and CO_2 produced by anaerobic fermentation of organic matter. The positive values, which occur with less frequency, are more difficult to explain. Nissenbaum et al. (1972) suggested that these values are consistent with presence of bacteria produced CH_4. Three possibilities have been suggested: 1) isotope exchange between CH_4 and CO_2 (Bottinga, 1969), 2) formation of CH_4 and CO_2 fermentation (Rosenfeld and Silverman, 1959), and 3) reduction of CO_2 by methane-forming bacteria (Nissenbaum et al., 1972; Games and Hayes, 1974, 1976). Most researchers prefer the latter possibility.

Important Takeaway – Sources of groundwater carbon include atmospheric CO_2, CO_2, and CH_4 released during the reduction of shallow organic matter (biogenic), catagenesis of organic carbon accompanying sediment burial (thermocatalytic), chemical weathering of non-carbonate minerals ($\delta^{13}C$ variable), CO_2 released from magmatic fluids, and the release of CO_2 from carbonate minerals during metamorphism. The $\delta^{13}C$ from these sources is variable, ranging from 0.0 ‰ for metamorphic CO_2 to as negative as ~-80 ‰ for some fossil fuel CH_4.

Problems (all $\delta^{13}C$ values are in VPDB)

1. A non-thermal groundwater in a carbonate aquifer has a $\delta^{13}C$ composition of –5 ‰.
   a. What are the possible explanations for the $\delta^{13}C$ value?
   b. Describe the isotopic evolution of the water.
2. A non-thermal groundwater has a $\delta^{13}C$ composition of –20 ‰.
   a. What geologic/biologic conditions could account for the isotopic composition?
   b. What water chemistry would you expect to find?
3. A non-thermal groundwater has a $\delta^{13}C$ composition of –40 ‰.
   a. What are the possible explanations for the $\delta^{13}C$ value?
   b. Describe the isotopic evolution of the water.
4. A groundwater exsolves methane gas with a $\delta^{13}C$ of –60 ‰. What methane source would you expect?
5. A soil zone recharge water is in equilibrium with soil zone gas having a $\delta^{13}C$ of –24‰. What would you expect the $\delta^{13}C$ of the DIC at pH 7.5?
6. If a carbonate rock with a $\delta^{13}C$ of –10 ‰ was flooded with the leakage from an acid storage pond, what would the $\delta^{13}C$ of the dissolved HCO_3^- be, assuming all of the H+ ions are from the acid?

Chapter notation

‰	permil
α	isotopic fractionation factor
δ	delta unit used for reporting isotopic abundances in ‰ difference relative to a standard
ε	carbon isotopic enrichment fractionation in ‰
atm	atmosphere
CAM	Crassulacean acid plant metabolism
km	kilometer
ln	natural logarithm (logarithm to the base e); ln a = 2.303 log a
lnα	natural log of the isotopic fractionation factor
meq L^{-1}	milliequivalent per liter
ppm	part per million
P_{CO_2}	partial pressure of $CO_{2(g)}$

REFERENCES CITED

Bottinga, Y., 1969, Calculated fractionation factors for carbon and hydrogen isotope exchange in the system calcite-carbon dioxide-graphite-methane-hydrogen-water vapor, *Geochimica et Cosmochimica Acta*, 33, 49–64.

Bottrell, S., 2019, Carbon-13 in groundwater from English and Norwegian crystalline rock aquifers: A tool for deducing the origin of alkalinity? *Sustain Water Resources Management*, 5, 267–287.

Cerling, T.E., et al., 1991, On the isotopic composition of carbon in soil carbon dioxide, *Geochimica et Cosmochimica Acta*, 55(11), 3403–3405.

Chapelle, F.H., and L.L. Knobel, 1985, Stable carbon isotopes of HCO_3^- in the Aquia Aquifer, Maryland: Evidence for isotopically heavy source of CO_2, *Ground Water*, 23(5), 592–599.

Conway, C.M., and H.P. Taylor Jr., 1969, O^{18}/O^{16} and C^{13}/C^{12} ratios of coexisting minerals in the Oka and Magnet Cove carbonatite bodies, *Journal Geology*, 77, 618–626.

Craig, H., 1954, Carbon-13 in plants and the relationships between carbon-13 and carbon-14 variations in nature, *Journal Geology*, 62, 115–149.

Craig, H., 1963, The isotopic geochemistry of water and carbon in geothermal areas, in *Nuclear geology of geothermal areas. Spoleto, 1963*, Consiglio Nazionale delle Richerche, Laboratorio di Geologia Nucleare, Pisa, 7–54.

Craig, H., et al., 1956, *Isotopic geochemistry of thermal waters*, National Academy of Science-National Research Council Publication 400, 29–38.

Deines, P., 1980, The isotopic composition of reduced organic carbon, in *Handbook of environmental isotope geochemistry*, Elsevier, 1(A), 329–406.

Deines, P., and D.P, Gold, 1969, The change in carbon and oxygen isotopic composition during contact metamorphism of Trenton limestone by the Mount Royal pluton, *Geochimica et Cosmochimica Acta*, 33, 421–424.

Elder, C.D., et al., 2021, Characterizing methane emission hotspots from thawing permafrost, *Global Biogeochemical Cycles*, 35(12).

Ellis, A J., and W.A.J. Mahon, 1964, Natural hydrothermal systems and experimental hot-water/rock interactions, *Geochimica et Cosmochimica Acta*, 28, 1323–1357.

Falkowski, P., et al., 2000, The global carbon cycle: A test of our knowledge of earth as a system, *Science*, 290(5490), 291–296.

Faure, G., 1977, *Principles of isotope geology*, Wiley.

Flores, R.M., et al., 2008, Methanogenic pathways of coal-bed gas in the Powder River Basin, United States: The geologic factor, *International Journal of Coal Geology*, 76(1–2), 52–75.

Fuex, A.N., and D.R. Baker, 1973, Stable carbon isotopes in selected granitic, mafic and ultramafic rocks, *Geochimica et Cosmochimica Acta*, 37, 2509–2521.

Galimov, E.M., 1966, Carbon isotopes in soil CO_2, *Geochemistry International*, 3, 889–898.

Galimov, E.M., et al., 1975, Carbon isotopic composition in carbonatites and carbonate like rocks in connections with the source of carbonatite material, *Geochemistry International*, 11, 503–510.

Games, L.M., and J.M. Hayes, 1974, Carbon in groundwater in the Columbus, Indiana landfill, in *Solid Waste disposal by land burial in southern Indiana*, Indiana University Water Resources Research Center, Bloomington, IN, 81–110.

Games, L.M., and J.M. Hayes, 1976, On the mechanisms of CO_2 and CH_4 production in natural anaerobic environments. Carbon, nitrogen, phosphorus, sulfur and selenium cycles, *Environmental Biogeochemistry*, 1, 51–73.

Giggenbach, W., et al., 1983, Geothermal systems, in *Guidebook on nuclear techniques in hydrology*, 1983 ed, International Atomic Energy Agency (IAEA), 359–377.

Gillon, M., et al., 2012, Controls on 13C and 14C variability in soil CO_2, *Geoderma*, 189–190, 431–441.

Gruber, N., et al., 1999, Spatiotemporal patterns of carbon-13 in the global surface oceans and the oceanic Suess effect, *Global Biogeochemical Cycles*, 13(2), 307–335.

Hart, R., 2009, *Isotopic evaluation of carbon dioxide in soil gas in Utah for a more accurate input variable in groundwater age determining models*, unpublished MS thesis, Brigham Young University, 2248.

Hoefs, J., 2010, Geochemical fingerprints: A critical appraisal, *European Journal of Mineralogy*, 22.

Jones, A.P., et al., 2013, Carbonate melts and carbonatites, *Review in Mineral and Geochemistry*, 75(1).

Kämpf, L., 2020, Stable oxygen and carbon isotopes of carbonates in lake sediments as a paleo flood proxy, *Geology*, 48, 3–7.

Keith, M.L., and J.N. Weber, 1964, Isotopic composition and environmental classification of selected limestone and fossils, *Geochimica et Cosmochimica Acta*, 28, 1787–1816.

Kohn, J.M., 2010, Carbon isotope composition of terrestrial C3 plants as indicators of (Paleo) ecology and (Paleo) climate, *Proceedings of the National Academy of Sciences of the United States of America,* 107.

Koval'skiy, V.V., and N.V., Cherskiy, 1973, Possible sources and isotopic composition of carbon in diamonds, *International Geology Reviews*, 15, 1224–1228.

Kunkler, J.L., 1969, *The sources of carbon dioxide in the zone of aeration of the Bandelier Tuff, near Los Alamos, New Mexico*, U.S. Geological Survey Professional Paper 650-B, B185–B188.

Lerman, J.C., 1972, *Carbon-14 dating: origin and correction of isotope fractionation errors in terrestrial living matter*, Proceedings 8th International Conference Radiocarbon Dating, New Zealand, 612–624.

Mayo, A.L., and A.B. Muller, 1997, Contributions of external CO_2 gas to a shallow groundwater system, *Journal of Hydrology*, 194, 286–304.

Millonig, L.J., and L.A. Groat, 2013, Carbonatites in western North America – occurrences and metallogeny in economic geology, *Society of Economic Geologist (Special Publication)*, 17, 245–264.

Muller, A.B., and A.L. Mayo, 1986, Carbon-13 variation in limestone on an aquifer-wide scale and its effects on groundwater radiocarbon dating models, *Radiocarbon*, 28(3), 1041–1054.

Nissenbaum, A., et al., 1972, Early diagenesis in a reducing fjord, Saanich Inlet, British Columbia, I. Chemical and isotopic changes in major components of interstitial water, *Geochimica et Cosmochimica Acta*, 36, 1007–1027.

Oehlert, A., and P. Swart, 2014, Interpreting carbonate and organic carbon isotope covariance in the sedimentary record, *Nature Communication*, 5, 4672.

O'Leary, M.H., 1988, Carbon isotope in photosynthesis, *BioScience*, 38(5), 328–336.

Osmond, C.B., 1975, Environmental control of photosynthetic options, in *Crassulacean plants in environmental and biological control of photosynthesis*, Springer, 299–309.

Park, R., and S. Epstein, 1960, Carbon isotope fractionation during photosynthesis, *Geochimica et Cosmochimica Acta*, 21, 110–126.

Park, R., and S. Epstein, 1961, Metabolic fractionation of C3 and C4 in plants, *Plant Physiology*, 36(2), 133–138.

Parkhurst, D.L., and C.A.J. Appelo, 2013, *Description of input and examples for PHREEQC version 3-A computer program for speciation, batch-reaction, one-dimensional transport, and inverse geochemical calculations*, U.S. Geological Survey Techniques and Methods, 6, A43.

Plummer, L.N., et al., 1976, *WATEQF - A FORTRAN IV*, USGS Water Resources Investigations 76-13.

Riebeek, H., 2011, *The carbon cycle: NASA Earth Observatory*, earthobservatory.nasa.gov/features/CarbonCycle

Rightmire, G.T., and B.B. Hanshaw, 1973, Relationship between the carbon isotope composition of soil CO_2 and dissolved carbonite species in groundwater, *Water Resources Research*, 9(4), 958–967.

Rosenfeld, W.D., and S.R. Silverman, 1959, Carbon isotope fractionation in bacterial production of methane, *Science*, 130, 1658–1659.

Ruby, W.W., 1964, Geologic history of sea water, in *The origin and evolution of atmospheres and oceans*, Wiley, 1–73.

Santos, R.V., and R.N. Clayton, 1995, Variations of oxygen and carbon isotopes in Carbonatites: A study of Brazilian alkaline complexes, *Geochimica et Cosmochimica Acta*, 59(7), 339–1352.

Shieh, Y.N., and H.P. Taylor, 1969, Oxygen and carbon isotope studies of contact metamorphism of carbonate rocks, *Journal Petrology*, 10, 307–331.

Smith B.N., and S. Epstein, 1971, Two categories of $^{13}C/^{12}C$ rations for higher plants, *Plant Physiology*, 47, 380–384.

Suarez, A., et al., 2019, Earth catastrophes and their impact on the carbon cycle, *Elements*, 15.

Truesdell, A.H., and J.R. Hulston, 1980, Isotopic evidence on environments of geothermal systems in Handbook on environmental isotope geochemistry, in *The terrestrial environment*, Elsevier, 1(A), 179–226.

Verwega, M.-T., et al., 2021, Description of global marine particulate organic carbon-13 isotope data set, *Earth System Science Data*, 13(10), 4861–4880.

Vogel, J.C., 1993, Variability of carbon isotope fractionation during photosynthesis, in *Stable isotopes and plant carbon-water relations*, Academic Press, 29–46.

Winograd, I.J., and G.M. Farlekas, 1974, Problems in ^{14}C dating of water from aquifers of deltaic origin: An example from the New Jersey coastal plain, *Isotope Techniques in Groundwater Geohydrology*, II, 83–107.

Winter, K., and J.A.M. Holtum, 2002, How closely do the delta 13d values of crassulacean acid metabolism plants reflect the proportion of CO2 fixed during day and night? *Plant Physiology*, 129(4), 1843–1851.

10 Sulfur-34 and Nitrogen-15

10.1 SULFUR ISOTOPE FRACTIONATION AND THE SULFUR CYCLE

Sulfur has four stable and numerous short-lived radiogenic isotopes. The stable isotopes ^{32}S, ^{33}S, ^{34}S, and ^{36}S have abundances of 95.02, 0.75, 4.21, and 0.02 percent, respectively (Table 7.1). ^{32}S and ^{34}S are the most widely used sulfur isotopes in environmental investigations because of their isotopic abundances. The stable isotopic composition of sulfur is expressed in terms of the delta notation:

$$\delta^{34}S = \left[\frac{\left(^{34}S/^{32}S\right)\text{sample}}{\left(^{34}S/^{32}S\right)\text{standard}} - 1 \right] \times 10^3\ (‰) \tag{10.1}$$

where the reporting standard is Vienna Cañon Diablo troilite (VCDT).

Sulfur occurs in valance states ranging from −2 to +6. The highly soluble evaporite minerals gypsum and anhydrite contain oxidized +6 sulfur, whereas common sulfide minerals such as pyrite (FeS_2) contain reduced −2 sulfur. Under oxidizing groundwater conditions, sulfate (SO_4^{2-}) is the dominant species, and sulfide, either as $H_2S_{(aq)}$ (pH ~ <7) or HS^- (pH ~ >7), prevails under reducing conditions (Garrels and Christ, 1965).

Sulfur isotopes are a powerful but often complicated tool for evaluating the origin and history of groundwater sulfur. The $\delta^{34}S$ of primary sulfur falls in a narrow range of values that center about 0 ‰ (Figure 10.1), as evidenced by the isotopic compositions of meteorites and deep-seated primary sulfur in basic sills, magmatic deposits, and carbonatites (Hutchison et al., 2019; Mitchell and Krouse, 1975; Thode et al., 1962). The multiple sulfur valance states allow it to participate in numerous geochemical and biologic reactions and process. Once primary sulfur is cycled into the atmosphere, lithosphere, and hydrosphere it is subjected to a variety of abiotic and biotic sulfur fractionating processes, many of which are redox regulated. The parameters used to calculate several sulfur fractionation factors of interest in hydrogeochemistry are listed in Appendix D.

Volcanic sulfur and igneous rocks have a wider range of $\delta^{34}S$ values than primary sulfur due to inorganic reactions that tend to enrich oxidized and to deplete reduced sulfur in ^{34}S. Sulfur in evaporate minerals, sedimentary rocks, and hydrocarbons that have been subjected to biologic activity have a much wider range of $\delta^{34}S$ values than abiotic evaporitic sulfur (Figure 10.1).

Groundwater acquires sulfur from a mixture of fractionated sources including atmospheric dry fall and precipitation, the dissolution of calcium sulfate and calcium carbonate minerals in evaporite and non-evaporite sedimentary rocks, the oxidization of sulfide minerals, the biologic oxidation and reduction of organic material, and anthropogenic contamination including sewage, fertilizers, detergents, and industrial and mine wastewater (Knöllerand and Schubert, 2010; Ślósarczyk et al., 2022; Tostevin et al., 2016). Abiotic and biotic processes fractionate sulfur compounds accompanying simple mineral dissolution, and during the oxidation, reduction, and disproportionation (simultaneous oxidation and reduction) of sulfur compounds. As examples. the abiotic precipitation of gypsum from seawater fractionates sulfur 0 to 5 ‰ (Raab and Spiro, 1991), the oxidation of H_2S to SO_4^{2-} at low temperatures fractionates sulfur about −5.2 ‰ (Fry et al., 1988), and microbes fractionate sulfur −65 to 0 ‰ during metabolization (Table 10.1).

Isotopically, most groundwater sulfate belongs to one of two groups:

1. ^{34}S enriched sulfate from the dissolution of evaporite and CAS containing sediments and
2. ^{34}S depleted sulfate from reduced sulfur sources such as sulfide minerals, soil organic material, and anthropogenic environments.

DOI: 10.1201/9781003646129-12

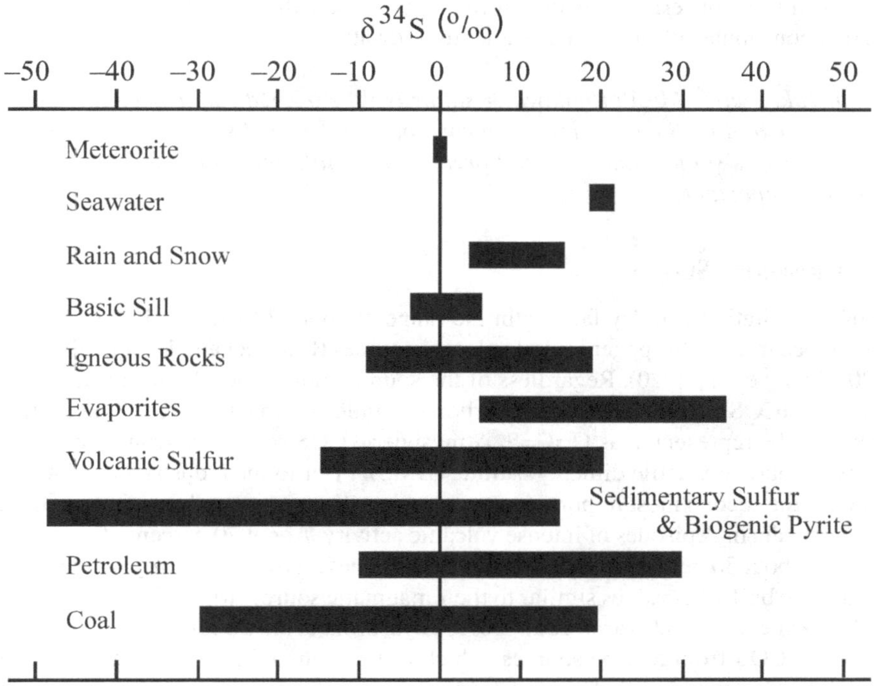

FIGURE 10.1 Range of $\delta^{34}S$ values of natural sulfur sources (data from Jensen and Nakai, 1961; Seal et al., 2000; Spoelstra et al., 2021; Thode, 1991; Tostevin et al., 2014; Zhu et al., 2016).

TABLE 10.1
Range of Observed $\delta^{34}S$ Values for Some Natural Processes

Process	Range of observed $\delta^{34}S$ (‰) fractionation
Assimilatory sulfate reduction	–2.8 to –0.9
Dissimilatory sulfate reduction	–65.6 to 0
Sulfite reduction	–41 to +0.3
Sulfide oxidation	–18.0 to +3
Sulfur disproportionation	Sulfate: –0.6 to +20.2
	Sulfide: –8.6 to –5.5
Thermochemical sulfate reduction	+10 to +25
Gypsum formation	0 to +4.2

The primary sulfur compound in groundwater is sulfate (SO_4^{2-}), whose isotopic composition reflects the isotopic composition of the source material. CAS (carbonate-associated sulfur) is sulfate in fluid inclusions, adsorbed phase sulfate, and sulfate incorporated in carbonate mineral crystal lattices. CAS is commonly found as aragonite and calcite in both ocean and terrestrial sedimentary environments. CAS is responsible for much of the small sulfur concentrations associated with groundwater in carbonate terrains.

Small sulfide concentrations occur in some reduced groundwater including thermal systems, groundwater flowing in organic rich sediments, and mine and other anthropogenic environments.

Because our primary concern is aqueous sulfur species, we will focus on those sulfur sources and processes that contribute sulfur to surface and groundwater.

> *Important Takeaway – Most groundwater sulfur is ^{34}S enriched sulfate from the dissolution of evaporite and CAS containing sediments, but ^{34}S depleted sulfate from reduced sulfur sources such as sulfide minerals, soil organic material, and anthropogenic sources may be locally important.*

10.1.1 Atmospheric Sulfur

The $\delta^{34}S$ of precipitation and dry fall are in the range of about 2 to 15 ‰ (Figure 10.1), reflecting the mixing of oceanic, anthropogenic, and volcanic sources (Cortecci and Longinelli, 1970; Hattori et al., 2020; Zheng et al., 2020). Regardless of the source, most atmospheric sulfur is converted to carbonyl sulfide (COS). In some literature carbonyl sulfide is known as OCS. The chemical bonding of COS may be represented as O=C=S. Atmospheric COS originates from fossil fuel combustion and ocean-sourced volatile dimethyl sulfide (DMS). Prior to the industrial revolution, volcanic emissions were the second most important atmospheric sulfur source and remain as major contributors, particularly during episodes of intense volcanic activity. About 70 percent of sulfur in volcanic gas is SO_2, and about 30 percent is H_2S (Martin, 2018). These gases are readily oxidized into sulfate aerosols that have bulk $\delta^{34}S$ values similar to their magmatic source, where $\delta^{34}S$ is –2 to 6 ‰ (Liotta et al., 2012; Sakai et al., 1982; Taran et al., 2018; Toyoshima et al., 2022).

Tropospheric COS from marine sources, which is a mixture of direct oceanic COS, DMS, and carbon disulfide (CS_2), has a $\delta^{34}S$ of about 14 to 15 ‰. The $\delta^{34}S$ of oceanic DMS is about 19 to 21 ‰; however, when transformed into the long-lived atmospheric COS, the $\delta^{34}S$ is fractionated to about 13 ‰ (Amrani et al., 2013; Angert et al., 2019; Davidson et al., 2021; Serio et al., 2023). The $\delta^{34}S$ of ocean waters CS_2 is about 15 ‰.

Sulfate in continental precipitation is generally depleted in ^{34}S with respect to seawater SO_4^{2-} due to dilution with sulfur from anthropogenic sources, the oxidation of biogenic H_2S, volcanic gases, and terrestrial dust. Near-costal precipitation SO_4^{2-} from modern sea spray has a $\delta^{34}S$ of about 20 ‰. The chloride content of precipitation may sometimes be used to estimate the percentage contribution of marine aerosol sulfate in coastal precipitation, and by extension the contribution of marine aerosol sulfate to groundwater recharge. The coastal Cl-SO_4 groundwater relationship shown in Equation 6.46 may be used to calculate the sea spray sulfate contribution:

$$A = \frac{SO_4^{2-}\text{ seawater}}{SO_4^{2-}\text{ observed}} \times 100 = \frac{\left(Cl^-\right)\text{observed}}{7.1\left(SO_4^{2-}\right)\text{observed}} \times 100 \tag{10.2}$$

where A is the percent of sea spray sulfate. This relationship assumes that all chloride comes from marine aerosols and that the value of the Cl^- / SO_4^{2-} ratio in the marine aerosol component in precipitation is 7:1. The parameter A may then be plotted against the $\delta^{34}S$ of the sulfate in the sample. At A = 100 percent, $\delta^{34}S$ should be about 20 ‰. Because sulfate in non-coastal-zone continental precipitation and in dry fall tends to have a terrestrial origin rather than a saline water origin (Berner and Berner, 1987; Munger, 1982), Equation 10.2 should be used with caution.

Human-induced perturbations to atmospheric sulfur make isotopic methods a powerful tool in the investigation of anthropogenic atmospheric contributions (Grey and Jensen, 1972; Jongebloed et al., 2023; Robinson and Robins, 1970). These perturbations, which affect dry fall and precipitation $\delta^{34}S$, have a global reach and may be particularly important near population and industrial centers. The major human contributions of gaseous sulfur oxide are from the burning of fossil fuels, with lesser but locally important contributions from agricultural activities and bio-mass burning. Chen et al. (2016) found that burning coal enriches the flue gas in ^{34}S and depletes the PM2.5 in ^{34}S.

PM2.5 is the fugitive particulate matter with a diameter of 2.5 micrometers or less. The $\delta^{34}S$ in flue gas increased 2.55 ‰, and in the PM2.5 it decreased by 5.92 ‰. Elsewhere, polluted air has been found to have $\delta^{34}S$ values ranging from 2.5 to 4.5 ‰ (Chung et al., 2019; Panettiere et al., 2000).

Desert dust is the primary global atmospheric aerosol (Kok et al., 2021). Locally, aerosols from gypsum processing and other industrial sources contribute sulfur to the atmosphere and precipitation (Mayo and Klauk, 1991; Rightmire et al., 1974). Commercial gypsum processing dust has a typical $\delta^{34}S$ value of about 20 ‰. Reported desert dust $\delta^{34}S$ values include 1 to 5 ‰ in India (Dasari et al., 2022), 13 to 26 ‰ in the central Andes (Olson et al., 2021), and 10.1 ± 9.9 in the Sahara (Harris et al., 2012).

Important Takeaway – The $\delta^{34}S$ of precipitation and dry fall is in the range of about 2 to 15 ‰, reflecting the mixing of oceanic, anthropogenic, and volcanic sources. Although atmospheric sulfur contributions to groundwater are generally small, locally the sources may be significant, particularly in non-carbonate terrain.

10.1.2 Calcium Sulfate and Calcium Carbonate Minerals

Sulfur isotopes are not fractionated during the abiotic dissolution of gypsum, anhydride, or CAS. Most groundwater acquires abiotic sulfate from CAS minerals and not from evaporite layers. What this means is that the $\delta^{34}S$ of abiotic groundwater sulfate will have approximately the same $\delta^{34}S$ composition as the bulk $\delta^{34}S$ of the soil zone and aquifer skeleton.

Most sulfur enters the ocean as SO_4^{2-} from surface and groundwater discharges. The origin of the sulfur includes the weathering of:

1. sulfide-bearing sedimentary rocks (black shale and carbonates),
2. evaporite rocks of marine origin, and
3. sulfide minerals in volcanic and other igneous rocks.

The $\delta^{34}S$ content of ocean sulfate and sulfur in associated sediments are the result of several competing processes (Holser, 1977; Jørgensen et al., 2019). Sulfur is removed from ocean water by the formation of evaporite rocks such as gypsum and anhydrite in marginal marine basins, inclusion in carbonate sediments (CAS), precipitation of biochemical barite in pelagic water, and deep water bacterial reduction of sulfate to sulfide, which precipitates as pyrite (Present et al., 2020).

Evaporite mineral precipitation does not appreciably change the bulk $\delta^{34}S$ composition of ocean water sulfate because the ocean water is the sulfur buffer. Thode and Monster (1965) and Raab and Spiro (1991) have shown the absolute fractionation during the precipitation of gypsum from sulfate-bearing brines is less than −2 ‰. The $\delta^{34}S$ compositions of modern marine sulfate lies in a narrow range, about 20 ‰. The sulfur in marine CAS is slightly fractionated during mineral formation. In high CAS-concentrated sediments, the abiotic CAS-SO_4^{2-} fractionation is reported as 1.0 ± 0.3 1 ‰ for aragonite and 3.1 ± 0.6 and 1.7 ± 0.4 for low and high activity ratios of SO_4^{2-}/CO_3^{2-} (Barkan et al., 2020).

The isotopic composition of sulfate minerals from marine sediments has varied systematically over geologic time (Figure 10.2). This variation not only implies a systematic temporal change in the sulfur cycle, but it may also be reflected in the isotopic composition of groundwater, which acquires sulfur from marine sediments. For example, groundwater acquiring sulfate from Permian age marine sediments would have a $\delta^{34}S$ of about 12 ‰, whereas a similar groundwater flowing through Tertiary age marine sediments would have a $\delta^{34}S$ of about 20 ‰. It is also significant to note the isotopic composition of the oceans has fluctuated about the value 20 ‰ over geologic time. This fluctuation and knowledge of the ^{18}O content of marine sulfate has been used as evidence for

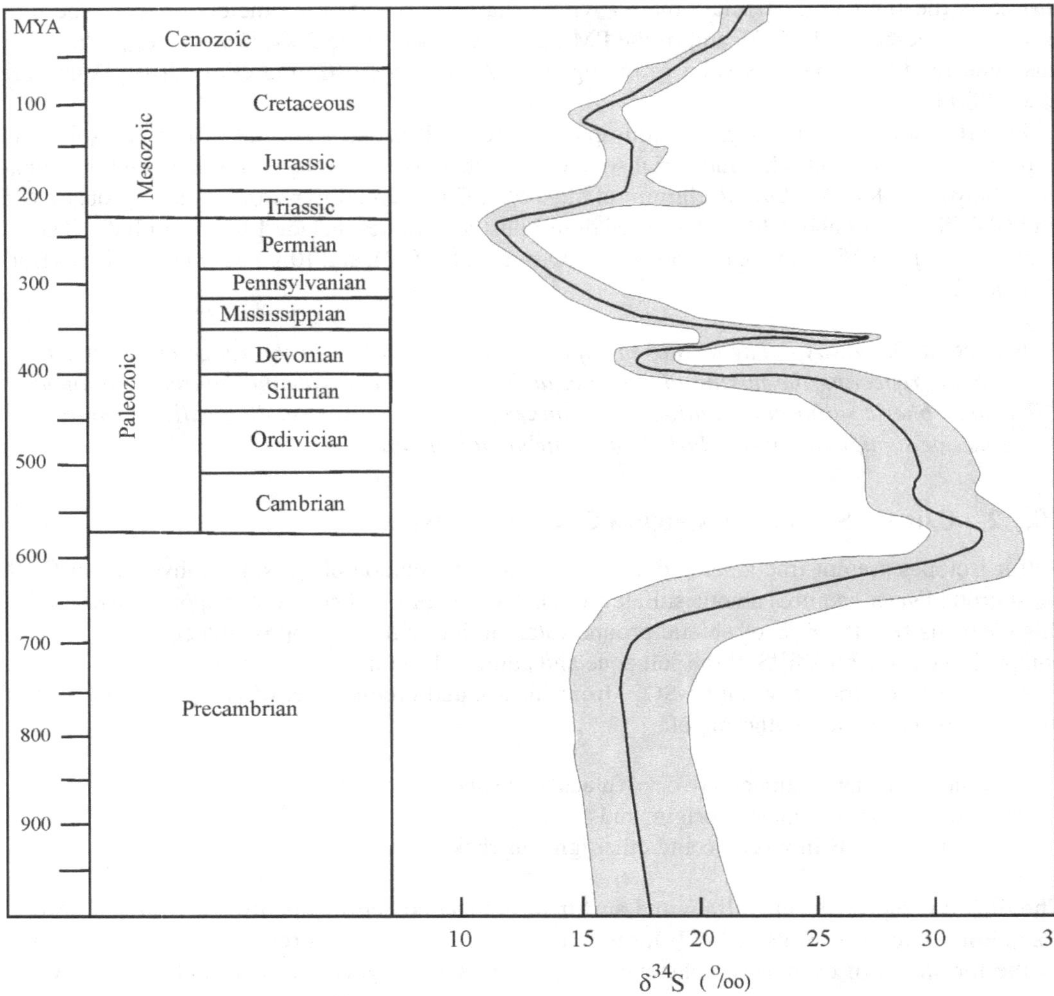

FIGURE 10.2 Sulfur isotopes in evaporites during the past 1 billion years. The shaded area represent the range of values, and the solid line is the typical value. Data from Claypool et al., 1980; Holser and Kaplin, 1966; Paytan et al., 2020.

a stabilized oceanic salinity at 3.5 × 10^9 years ago (Ferronsky and Polyakov, 1982; Vinogradov, 1975). The geologic time fluctuations in the $\delta^{34}S$ composition of marine evaporite rocks have been attributed to climatic and tectonic perturbations (Yao and Paytan, 2020).

> *Important Takeaway – most sulfur in groundwater will have a $\delta^{34}S$ similar to the evaporite minerals and CAS source minerals (~10 to 20 ‰). When possible, it is important to validate the source $\delta^{34}S$ to confirm that the sulfur is from evaporite and CAS minerals and to determine the δS34 of the source terms.*

10.1.3 Sulfate-Sulfide Redox and Biologic Fractionation

The large range of $\delta^{34}S$ values observed in non-siliceous rock reservoirs (Figure 10.1) is mostly due to biologic fractionation accompanying sulfate and sulfide reduction and oxidation. Equilibrium and non-equilibrium (kinetic) fractionation tend to enrich oxidized sulfur species in ^{34}S and enrich reduced sulfur species in ^{32}S. Microorganisms play an important role in the distribution of sulfur isotopes. The organisms thrive in both aerobic and anaerobic environments, where their metabolic functions fractionate sulfur compounds. These processes are described in Chapter 6 and are illustrated in Figure 10.3.

Only minor isotopic fractionation accompanies sulfate assimilation by microorganisms (Kaplan and Rittenberg, 1994), and little of this sulfide is available to the environment until the death and decay of the organism. The most important biologic factor in sulfur isotopic fractionation is dissimilatory sulfate reduction accompanying anaerobic bacteria such as the microorganism *Desulfovibrio desulfuricans*, which lives in the substrate of ocean and lake sediments and elsewhere. Dissimilatory sulfate reduction involving oxidation of organic carbon can be written as:

$$2CH_2O + SO_4^{2-} \rightarrow 2HCO_3^- + H_2S \quad (10.3)$$

The first step in the dissimilatory process, the entrance of sulfate into the cell, is accompanied by a small isotopic shift. The bacteria then split oxygen from the sulfate ion and excrete H_2S, which is enriched in ^{32}S relative to the original sulfate. During dissimilatory reduction the sulfate ion performs the same function as oxygen does for aerobic organisms; that is, it is an electron acceptor. This process leads to a very large range of isotopic fractionations accompanying dissimilatory reduction (Canfield, 2001; Kemp and Thode, 1968).

Sulfide-bearing species in groundwater may be chemically and isotopically altered by oxidation and reduction processes. Only a minor amount of groundwater sulfide is released by the dissolution of sulfide minerals because of the very low solubility of such minerals. Sulfide derived from the

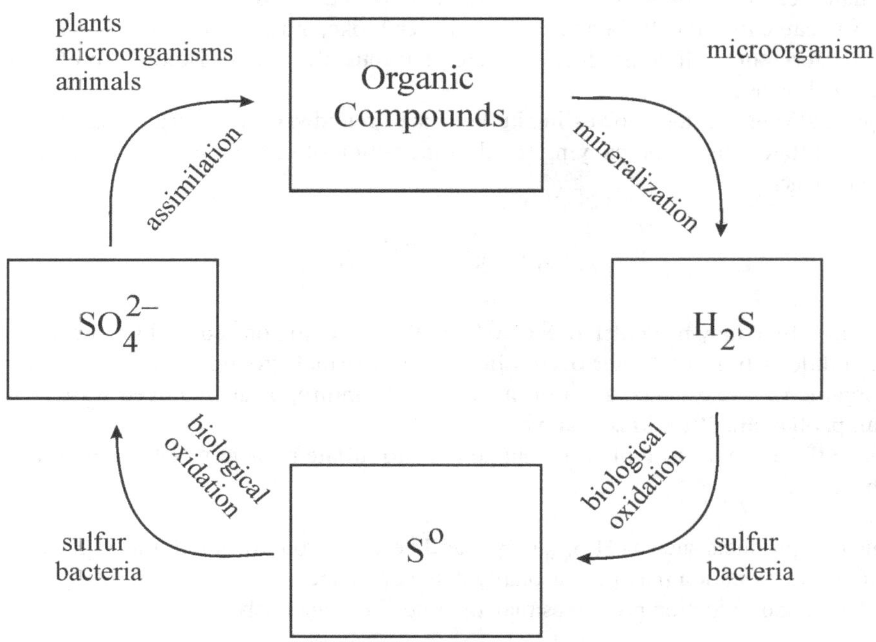

FIGURE 10.3 Microbiological conversion in the organic sulfur cycle (modified after Kushkevych et al., 2021).

dissolution of sulfide minerals may be oxidized to elemental sulfur and to sulfate without biological participation. Abiotic conversion of sulfide to sulfate by oxidation may be represented by the reaction:

$$\text{sulfide} \xrightarrow{\text{(oxidation)}} SO_4^{2-} \tag{10.4}$$

In this situation, as Schoen and Rye (1970) found at Yellowstone, Wyoming, isotope fractionation is negligible. At concentrations of only a few milligrams per liter, groundwater sulfide, H_2S, and HS^-- are stable under reducing conditions.

Sulfide-oxidizing microorganisms enrich ^{34}S in the reaction products as much as 12.5 ‰ relative to the reactants (Pellerin et al., 2019). The extent of isotopic sulfate fractionation during reduction accompanying bacterial metabolism is variable and is inversely proportional to the rate of sulfate reduction, which in turn is regulated by temperature and the availability of sulfate. Harrison and Thode (1957) demonstrated that ^{32}S–O bonds are broken more easily than are ^{34}S–O bonds and that the H_2S produced by reduction of SO_4^{2-} may be enriched in ^{32}S by as much as 22 ‰ relative to the sulfate. Kaplan and Rittenberg (1994) found ^{32}S enrichments in H_2S of up to 46 ‰ relative to sulfate. Isotopic fractionation in natural environments is more extensive than suggested by laboratory experiments. Sulfide minerals in recent sediments may be enriched in ^{32}S by as much as 50 ‰ relative to the associated marine sulfate. Isotopic fractionation due to biological oxidation is generally much smaller than that achieved during reduction processes.

Seasonal effects on isotopic fractionation may be complex (Knossow et al., 2015; Mayuko et al., 2021). Krouse (1980) found an increase in the conversion of SO_4^{2-} to H_2S in springtime discharges with the onset of warmer weather. This was attributed to an increase in the activity of sulfate-reducing bacterial. An opposite trend was noted in the City of Camrose, Alberta, water-sewer system. During the winter the water was covered by ice, and the oxygen utilized by sewage oxidation processes was derived from inorganic ions such as SO_4^{2-}. This led to the production of copious quantities of sulfide ions that were greatly depleted in ^{34}S ($\delta^{34}S = -55$ ‰). The unreacted sulfate ions became isotopically heavier. When the ice broke, most of the sulfide was re-oxidized by anaerobic bacteria and sulfide production decreased; thus, the overall isotopic composition of the water became less negative.

Most groundwater sulfide evolves in situ near the groundwater discharge area by the bacterial reduction of sulfate ions accompanying the decomposition of organic material. This process may be represented as:

$$\text{evaporitic } SO_4^{2-} \xrightarrow{\text{(reduction)}} \text{sulfide} \tag{10.5}$$

The resulting sulfide may be depleted 0 to 50 ‰ with respect to non-reduced sulfate. Although less common, sulfide-rich groundwater occurs in reduced thermal groundwater, groundwater flowing through organic rich sediments, and in mine and other anthropogenic-sourced wastes (Ball et al., 1998; Giampaoli et al., 2013; Massmann et al., 2003).

Assessing the extent of isotopic fractionation due to sulfate reduction occurring in the aquifer is difficult because:

1. reduction products such as $H_2S_{(aq)}$ may readily escape from the aquifer as $H_2S_{(g)}$,
2. isotopic fractionation may be seasonally dependent, and
3. oxidation and reduction processes may operate simultaneously.

If all fractionation products remain in the groundwater system, the measured $\delta^{34}S$ would be the accumulated isotopic product; however, if fractionation products are rapidly lost, the measured isotopic value would be the instantaneous isotopic product (Krouse, 1980). For example, some or all of the H_2S evolved during reduction may escape into the atmosphere, or it may precipitate with iron as pyrite or a similar mineral. If the H_2S is rapidly lost, the measured isotopic composition would be the instantaneous product. On the other hand, if only a portion of the reduction products are lost, the measured isotopic composition would be somewhere between the instantaneous and accumulation products. Loss of reduction products will be accompanied by an overall decrease in the concentration of dissolved sulfur species.

Important Takeaway – Abiotic sulfate reduction has little effect on the isotopic composition; however, biotic reduction may result in fractionation of 50 ‰ or more. Assessing the extent of fractionation from the sulfur in a water sample may be difficult because reaction products may have escaped, the fractionation may be seasonal, and both oxidation and reduction may occur simultaneously.

10.1.4 Anthropogenic Sulfur

Anthropogenic sulfur has a wide range of $\delta^{34}S$ values, reflecting the diversity of the original sulfur-bearing material (Table 10.2). For example, if a landfill contains a significant amount of construction dry wall (mostly $CaSO_4$), the leachate $\delta^{34}S$ can be as great as 20 ‰, but if the leachate source is biodegradable organic matter, the leachate $\delta^{34}S$ will be less positive. Although there is considerable overlap in the $\delta^{34}S$ values in anthropogenic sulfur, it is possible to identify the groundwater sulfur sources by analyzing the site-specific isotopic composition of potential end members. For example, Jakóbczyk-Karpierz and Ślósarczyk (2022) analyzed the sulfate $\delta^{34}S$ and $\delta^{18}O$ of 65 possible anthropogenic sources. They found the isotopic compositions of synthetic fertilizers ($\delta^{34}S$ = 3.0 ‰ ± 6.4 ‰, $\delta^{18}O$ = 12.4 ‰ ± 2.3 ‰), landfill leachates ($\delta^{34}S$ = 16.4 ‰ ± 4.9 ‰, $\delta^{18}O$ = 10.8 ‰ ± 2.2 ‰) and ore mine drainage ($\delta^{34}S$ = 8.4 ‰ ± 3.9 ‰, $\delta^{18}O$ = −1.8 ‰ ± 0.8 ‰) to be most distinctive, and the values in wastewater ($\delta^{34}S$ = 2.3 ‰ ± 2.6 ‰, $\delta^{18}O$ = 3.2 ‰ ±1.5 ‰), organic fertilizers ($\delta^{34}S$ = 3.5 ‰ ± 2.5 ‰, $\delta^{18}O$ = 2.0 ‰ ± 0.8 ‰), and leachates from heaps ($\delta^{34}S$ = 4.0 ‰ ± 3.3 ‰, $\delta^{18}O$ = 5.4 ‰ ± 2.3 ‰) to be less distinctive. Based on the isotopic values and sulfate concentrations, they concluded that wastewater was the main source of iron in the rivers and sulfur in the Gliwice aquifer, Poland.

Small sulfate concentrations in groundwater can sometimes be attributed to sulfate in dry fall and precipitation (Bottrell et al., 2008; Tostevin et al., 2016). Li et al. (2007) found that in the up-gradient region of an aquifer, where the SO_4^{2-} concentration was less than 70 mg/L, air pollution from coal combustion ($\delta^{34}S$ = 2.9 to 5.1 ‰) was a major source of groundwater sulfur and nitrogen. The $\delta^{34}S$ in rain was 1.7 to 5.6 ‰, which overlapped the values found in coal. Down gradient, the SO_4^{2-} concentration was greater than 130 mg/l and sulfur contributions from fertilizer ($\delta^{34}S$ average −0.8 ‰) and household detergent ($\delta^{34}S$ average 15.9 ‰) became increasingly important.

Example E10.1 Sulfur isotope mixing

Problem: A groundwater has a $\delta^{34}S$ value of +10 ‰. Two potential sources of sulfate are identified: river water with $\delta^{34}S$ = −2 ‰ and mine drainage with $\delta^{34}S$ = +25 ‰. What is the contribution of each source if the groundwater is a mixture of the two?

Solution: Let x be the fraction of river water and $1 - x$ the fraction of mine drainage.

Set up the equation: $(+10‰) = x(-2‰) + (1 - x)(+25‰)$

Solve for x: $+10 = -2x + 25 - 25x$

Simplify: $27x = 15$

Therefore, $x = 15/27 = 0.556$ or 55.6% river water

The mine drainage contributes $1 - 0.556 = 0.444$ or 44.4%.

TABLE 10.2
Range of $\delta^{34}S$ and Sulfate $\delta^{18}O$ Values in Anthropogenic Sulfur Sources (‰)

	$\delta^{34}S$	$\delta^{18}O$ in SO_4^{2-}	Source
Acid mine drainage/mine tailing	−17.3	−12.5	8
	−10.5		6
	−1.1 to 3.9	−4.5 to −8.6	1
Automobile exhaust	12.1 to 17		9
Copper ores	−4.3 to 2.4		9
Detergent	15.2 to 17.2		10
Heap leachate	4.0 ± 3.3	5.4 ± 2.3	8
Landfill leakage	16.4 ± 4.9	10.8 ± 2.2	8
	−3.3 to 9.1		2
	8.87		11
	10 to 17		5
Livestock manure	3.5 ± 2.5	2.0 ± 0.8	8
	−0.9 to 6.8		4
	−0.5 to 6		12
	4.7 ± 1.5		3
Chicken	1.1 to 10.2		4
Cattle	3.3 to 11.8		4
Swine	6.3 to 37.2		4
Organic fertilizer	3.5 ± 2.5	2.0 ± 0.8	8
Sewage	10.8 ± 2.3		8
	4.5 to 8.4		7
	9.5 to 12.35	8.1 to 17.9	13
Smelter plume	−3.8 to 3.4		9
Synthetic fertilizer	3 ± 6.4	12.4 ± 2.3	8
	−0.5 to 19.9		4
	10	7.6	14
	10.8 ± 2.4		3
	4.9 ± 6.7	12.6 ± 3.3	14
	5.6 to 7.7		10
	−1.8 to 21.4	8.9 to 15.6	12
Urban wastewater	2.3 ± 2.6	3.2 ± 1.5	8
	3.5 to 6.4		4

1 Bernhard and Spangenberg, 2005
2 Boris et al., 2007
3 Bottrell et al., 2008
4 Cravotta, 1997
5 Fritz et al., 1994
6 Gałuszka et al., 2020
7 Hunt, et al., 1995
8 Jakóbczyk-Karpierz and Ślósarczyk, 2022
9 Letolle and Olive, 1983
10 Li et al., 2006
11 Pujiindiyati and Syafalni, 2012
12 Vitòria et al., 2004
13 Vengosh et al., 2013
14 Zhang, 2013

10.2 NITROGEN CYCLE AND ISOTOPIC FRACTIONATION

Nitrogen has two stable isotopes, ^{14}N and ^{15}N, whose atmospheric isotopic abundances are 99.33 and 0.37 percent, respectively (Table 7.1). Several radioactive nitrogen isotopes exist, but their half-lives are too short for use in environmental studies. The stable isotopic composition of nitrogen is expressed in terms of the delta notation:

$$\delta^{15}N = \left[\frac{\left(^{15}N / ^{14}N \right) \text{sample}}{\left(^{15}N / ^{14}N \right) \text{standard}} - 1 \right] \times 10^3 \ (‰) \tag{10.6}$$

where the atmospheric $^{15}N/^{14}N$ ratio is the standard. Similar to other isotopic standards, the nitrogen standard has been assigned a value of 0 ‰.

10.2.1 Nitrogen Sources

Isotopic nitrogen is extensively used in biological studies but is not routinely investigated in groundwater investigations because aqueous nitrogen concentrations are usually quite small. Despite the usually low concentrations, isotopic fractionation accompanying nitrogen transformations make nitrogen stable isotopes useful in some aqueous investigations. Beginning in the late 1960s and early 1970s nitrogen isotopes began to be used to distinguish fertilizer nitrate from natural aqueous nitrogen (Kohl et al., 1971). Nitrogen isotopes are now used in investigations of nitrogen-contaminated surface and groundwater, $N_{2(g)}$ exsolving groundwater, and gas field reservoirs.

Most aqueous nitrogen is nitrate (NO_3^-), but nitrogen gas (N_2), ammonium (NH_4^+), ammonia (NH_3), and nitrite (NO_2^-) are important aqueous species. Groundwater nitrate originates from both natural and contamination sources in the surface, soil zone, and shallow subsurface environments. Natural nitrogen sources include precipitation and the ammonification–nitrification of dead organic matter in the soil. In some groundwater systems the natural nitrogen content is overwhelmed by anthropogenic nitrogen from artificial fertilizers, barnyard waste, domestic wastewater, industrial wastes, and atmospheric and groundwater pollution from industrial and power generating facilities, automobile combustion exhaust, and nitrogen spills. $NO_{2(g)}$ is a major component of the total gas composition in gas fields and in some groundwaters that exsolve gas. In aqueous systems, ammonia, often in the form of ammonium, occurs in environments such as peat bogs, sewer treatment facilities, landfills, and some groundwater systems.

The typical range of $\delta^{15}N$ and $\delta^{18}O$ in natural and anthropogenic nitrogen compounds are listed in Table 10.3, and the range of N_2 and NO_3^- $\delta^{15}N$ values are shown on Figure 10.4. Even though the atmosphere only contains about 2 percent of the total global nitrogen, the mobility of atmospheric nitrogen makes the atmosphere the nitrogen isotopic buffer. Most atmospheric nitrogen is $N_{2(g)}$, but the atmosphere also contains small percentages of ammonium and nitrogen oxides (NOx), whose isotopic compositions vary widely. Nitrogen oxides dissolve in precipitation-forming nitrate, which is an important source of soil nitrogen. Despite the fact that the isotopic composition of natural nitrogen in the atmosphere is relatively stable, seasonal variations have been observed in the $\delta^{15}N$ content of precipitation (Li et al., 2019). Freyer and Aly (1976) observed that rain over western Germany had NO_3^- $\delta^{15}N$ values of −5 to 5 ‰ and NH_4^+ $\delta^{15}N$ values between −5 and −10 ‰. They suggested that ^{15}N-depleted ammonia originates from evaporation of NH_3 in fields. Wada et al. (1975) interpreted atmospheric ^{15}N depletion as a consequence of burning fossil fuels. In the North China Plain, an intensive agricultural region undergoing rapid economic development, seasonal precipitation variations in NO_3^- $\delta^{15}N$ and NH_4^+ $\delta^{15}N$ were −10.0 to 20 ‰ and −12.7 to 12.9 ‰, respectively (Zhang et al., 2008). They concluded that in the rural areas the nitrogen was mostly from agricultural sources and in the Beijing area from multiple pollution sources. For example, at two rural sites

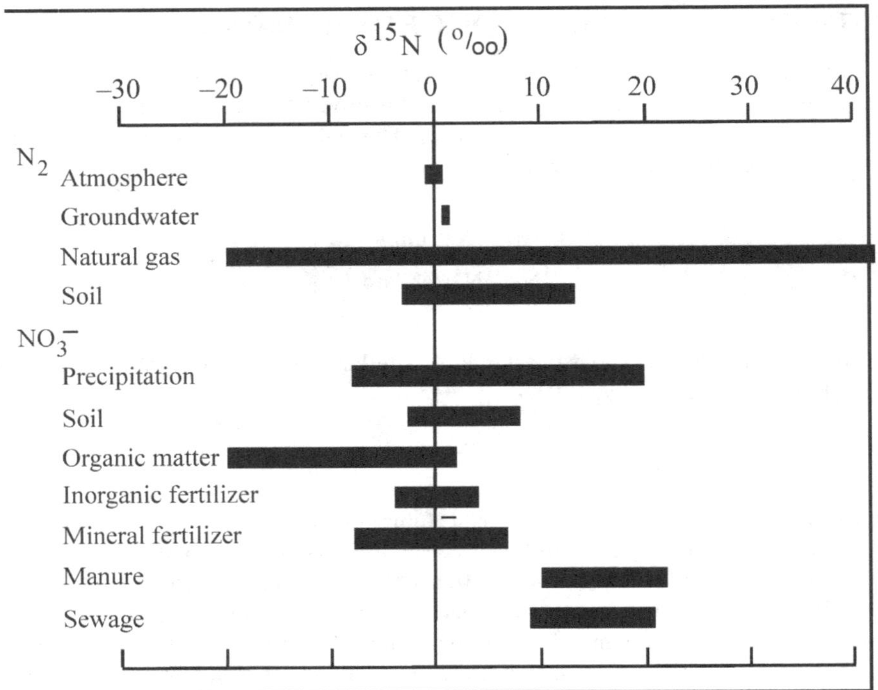

FIGURE 10.4 Range of $\delta^{15}N$ values of natural and anthropogenic sulfur sources.

the NO_3^- $\delta^{15}N$ and NH_4^+ $\delta^{15}N$ were −2.8 ± 4.06 ‰ and −1.19 ± 3.74 ‰, respectively, and at two urban sites in Beijing it was 4.23 ± 4.34 ‰ and 0.45 ± 4.39 ‰, respectively.

Although the percentage of the total global nitrogen in the soil zone is small (≈ 10^{-4}), soil and regolith nitrogen are the primary nitrogen contributors to unpolluted groundwater (Qian et al., 2021; Wu et al., 2021). There is a strong global relationship between soil nitrogen isotopes and mean annual temperature (MAT) and mean annual precipitation (MAP). In both hot and dry environments, soil organic matter is enriched in $\delta^{15}N$ relative to wet environments (Craine et al., 2015a). It has also been observed that cultivated soil has slightly more positive $\delta^{15}N$ content than uncultivated soil (Broadbent et al., 1980). This difference is due to the use of atmospheric nitrogen-based artificial fertilizers.

The $N_{2(g)}$ $\delta^{15}N$ in groundwater and in the soil is readily distinguishable from fugitive natural gas nitrogen. The $\delta^{15}N$ of $N_{2(g)}$ associated with soil gas ranges from −3 to 13 ‰, and natural gas ranges from −20 to 45 ‰ (Table 10.3). Despite the large variations in natural gas values, the range of values in individual gas fields are small, and $\delta^{15}N$ has been used to evaluate natural gas migration and mixing patterns (Stahl et al., 1975). Decaying plant material is the main nitrogen source in natural soils. Most soils have NO_3^- and NH_4^+ isotopic compositions ranging from 3 to 8 ‰ and −5 to 15 ‰, respectively. In most soil zones the nitrogen concentration decreases with increasing depth. In some soils $\delta^{15}N$ becomes more negative with increasing depth, but in most soil the $\delta^{15}N$ becomes more positive with increasing depth (Brenner et al., 2001; Mariotti et al., 1980; Zhang, 2013). In a comparison of plants and the underlying soil at 10 and 50 cm Amundson et al. (2003) found the average $\delta^{15}N$ of plants was −1.74 ‰ and the $\delta^{15}N$ of underlying soil at 10 cm, and soil at 50 cm was 3.56, and 4.35 ‰ (Figure 10.5).

At elevated concentrations nitrate can have serious health consequences. Nitrate concentrations in surface and shallow groundwater from natural biologic decomposition and precipitation seldom

TABLE 10.3
Range of δ¹⁵N and δ¹⁸O Values in Natural and Anthropogenic Nitrogen Sources

		$\delta^{15}N$ %	$\delta^{18}N$ %	Source
N_2				
	Atmosphere	0		
	Groundwater	0.85 to 0.99		2, 5
	Natural gas	−20 to 45		12
	Soil	−3 to 13		9
NO_3^{-1}				
	Precipitation	−10 to 20		4, 14
	Soil	3 to 8	−10 to 10	3, 6, 7, 10
	Organic matter	−20 to 2		9
	Inorganic fertilizer	−4 to 4	8 to 22	1, 4, 11
	Mineral fertilizer	−8 to 7		10
	Manure/animal waste	10 to 22	−10 to 17	1, 7, 8
	Sewage	9 to 21		1, 8
NH_4^+				
	Precipitation	−12.7 to 12.9		4, 14
	Soil	−5 to 15		9
	Organic matter	2.4 to 4.1		10
	Inorganic fertilizer	0 to 4	−8 to 12	4, 7, 11
	Manure/animal waste	8 to 11		10
	Household waste	5 to 9		10
NO_x (NO & NO_2)				
	Lightning	0 to 4		13
	Automotive exhaust	−19.1 to 9.8		13

Sources:
[1] Aravena and Mayer, 2009
[2] Benson and Parker, 1961
[3] Bremner and Tabatabai, 1973
[4] Freyer and Aly, 1976
[5] Klots and Benson, 1963
[6] Kendall, 1998
[7] Kendall and Aravena, 2000
[8] Kreitler, 1975
[9] Letolle, 1980
[10] Nikolenko et al., 2018
[11] Shearer et al., 1974
[12] Stahl et al., 1975
[13] Walters et al., 2015
[14] Zhang et al., 2008

exceed a few milligrams per liter (Letolle, 1980; Matiatos et al., 2021). Elevated concentrations are often due to contamination from agricultural activities, industry, sewage, septic tanks, and landfills. Agricultural activities include the application of artificial fertilizer, biomass burning, and the disposal of animal wastes. Groundwater contaminated with artificial fertilizer, animal wastes, and sewage disposal may have nitrate concentrations in excess of 10 mg L^{-1} (Abascal et al., 2022; Filpse and Bonner, 1985; Letolle and Olive, 1983). Sewage treatment facilities transform nitrogen

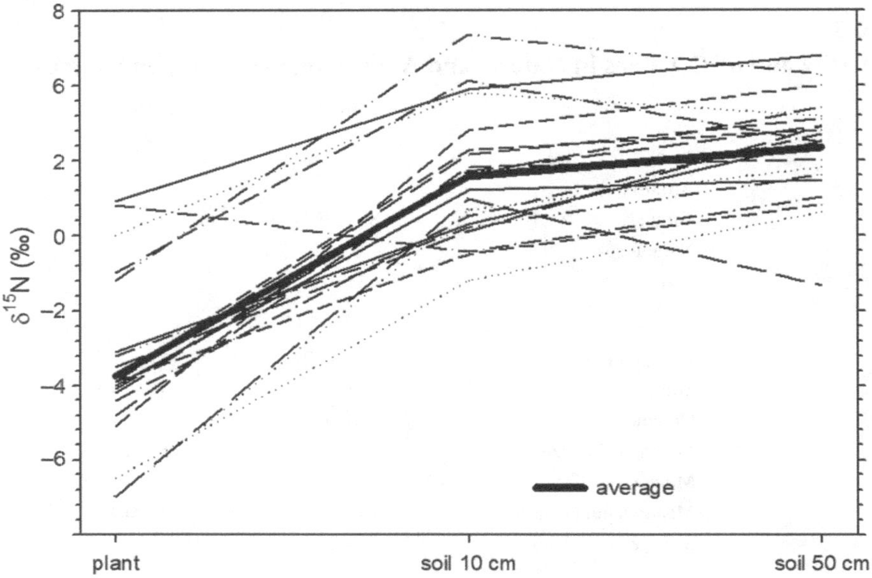

FIGURE 10.5 Comparison of plant and corresponding soil depth bulk $\delta^{15}N$ from 21 samples collected from the western United States, Hawaii, Tanzania, and France. The vegetation included desert, Joshua tree, pinyon-juniper pine, oak, mixed forest, mixed conifer, various grasslands, and heather. The average values are −1.74, 3.56, and 4.35 ‰ for plants, soil at 10 cm, and soil at 50 cm, respectively (data from Amundson et al., 2003).

compounds into the nitrate, but they do not reduce the total nitrogen content. The nitrogen isotopic composition of sewage discharge has $\delta^{15}N$ in NH_4^+ and NO_3^- of 5 to 9 ‰ and 9 to 21 ‰, respectively, which are isotopically similar to animal wastes. Animal tissue and manure are usually enriched in ^{15}N relative to the animal feed.

Nitrogen from human and animal excretory wastes is mainly in urea. Isotopic enrichment is due to catabolic pathways that systematically eliminate, via urea, small nitrogen molecules that are enriched in ^{14}N. Urea is hydrolyzed to ammonia and converted to ammonium and nitrate. During the process ammonia is readily volatilized and may be lost to the atmosphere:

$$\underset{\text{(Urea)}}{CO(NH_2)_2} \rightarrow NH_3 \rightleftharpoons NH_4^+ \rightarrow NO_3^- \tag{10.7}$$

Nitrate derived from urea and urea-based fertilizers should be depleted in ^{15}N with respect to animal feed. However, Kreitler (1975) observed ^{15}N enrichment in nitrate from soils and groundwater associated with cattle feedlots. Similar observations were made by Watkins et al. (1972) for forest soils fertilized with urea pellets and by Kreitler (1979) for calcareous soils treated with ammonium-based fertilizers. This apparent contradiction is the result of the volatile loss of ammonia from urea and urea-based fertilizers. Ammonia volatilization favors ^{14}N in the vapor phase. Artificial fertilizers are impoverished in ^{15}N relative to soil organic matter because they are synthesized from atmospheric $N_{2(g)}$. The isotopic compositions of inorganic fertilizer range from 0 to 4 ‰ for NO_3^- and −4 to 4 ‰ for NH_4^+. NO_3^- and NH_4^+ from the oxidation of manure and animal waste have compositions ranging from 10 to 22 ‰ and 8 to 11 ‰, respectively.

Important Takeaway – Most aqueous nitrogen is nitrate (NO_3^-), but nitrogen gas (N_2), ammonium (NH_4^+), ammonia (NH_3), and nitrite (NO_2^-) are important aqueous species. Groundwater nitrate originates from both natural and contamination sources in the surface, soil zone, and shallow subsurface. Natural nitrogen sources include precipitation and the ammonification–nitrification of dead organic matter in the soil. In some groundwater systems the natural nitrogen content is overwhelmed by anthropogenic nitrogen from artificial fertilizers, barnyard waste, domestic wastewater, industrial wastes, and atmospheric and groundwater pollution from industrial and power-generating facilities, automobile combustion exhaust, and nitrogen spills. Each source has a range of isotope values that can be used to identify the aqueous nitrogen source.

10.2.2 Nitrogen Isotopic Fractionation

Nitrogen isotopic fractionation accompanies both inorganic and biological redox processes. Some theoretical and experimentally determined fractionation factors are listed in Table 10.4 and Appendix D. Nitrogen species may be transformed by nitrogen fixation, ammonification (a.k.a. mineralization), nitrification, and denitrification (Figure 10.6). Inorganic fractionation is equilibrium based,

TABLE 10.4
$\delta^{15}N$ Fractionation Factors of Interest in Groundwater Investigations

	$N_{2(g)}$	$NO^-_{2(aq)}$	$NO^-_{3(aq)}$	$NH^-_{3(aq)}$
Equilibrium				
$N_{2(g)}$			1.0772[1]	1.00
			1.0678	
$NO^-_{3(aq)}$	1.02[6]			
$NO_{(g)}$		1.026[1]	1.086[4]	
			1.097[4]	
$NO_{2(aq)}$			1.061[4]	
	1.0059 to 1.0295[5]		1.0774[4]	
			1.0664[4]	
			1.059[4]	
$NH^+_{4(aq)}$	1.0309 to 1.0327[2]	1.03 to 1.035[6]	1.0228 to 1.0241[3]	0.980 to 0.983[1]
Biological kinetic				
$NH^+_{4(aq)}$		1.02 to 1.035[6]	1.0189 to 1.0251[3]	
		0.9663[4]		
$NO^-_{3(aq)}$		1.01 to 1.03[6]		
		1.0294 ± 0.0246[5]		
		1.024 ± 0.009[5]		

Sources:
[1] Cejudo and Schiff, 2018
[2] Kobayashi et al., 2019
[3] Haas et al., 2022
[4] Letolle, 1980
[5] Mariotti et al., 1981
[6] Richet et al., 1977

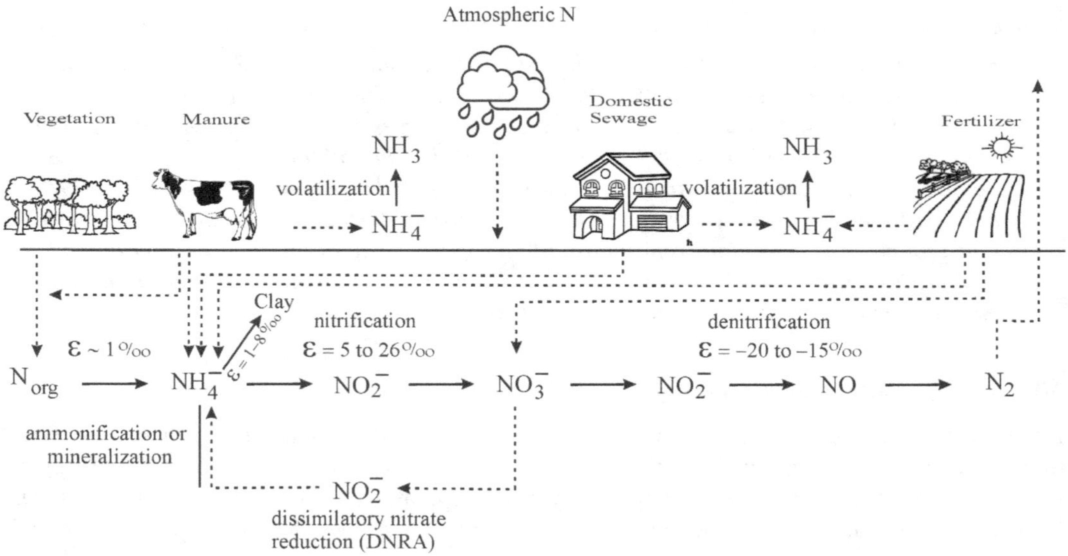

FIGURE 10.6 Terrestrial nitrogen transformations and associated isotopic values. Isotopic fractionation is represented by the ε notation.

and the reactions are reversible. The principal inorganic nitrogen isotopic fractionation processes include:

1. diffusion through porous and aqueous materials,
2. dissolution of molecular nitrogen in water,
3. evaporation of volatile nitrogen compounds,
4. oxidation of reduced nitrogen, and
5. ion exchange.

Biologic fractionation is usually unidirectional and is often kinetically impeded, where the magnitude of fractionation may be considerably less than that predicted for equilibrium conditions.

Nitrogen fixation is the reduction of $N_{2(g)}$ to ammonia and ammonium (NH_3 and NH_4^+), nitrite (NO_2^-), and nitrate (NO_3^-). Common nitrogen fixation processes can be represented as:

$$N_2 + O_2 \rightarrow 2NO \quad \text{(lightning)} \tag{10.8}$$

$$N_2 + 3H_2 \rightarrow 2NH_3 \quad \text{(anthropogenic)} \tag{10.9}$$

$$N_2 + 8H^+ + 8e \rightarrow 2NH_3 + H_2 \quad \text{(BNF)} \tag{10.10}$$

Anthropogenic atmospheric nitrogen, collectively known as NO_x ($NO + NO_2$), is transformed into nitric acid (HNO_3) and ultimately into ammonium and nitrate by complex oxidation and hydrolysis pathways (Chen et al., 2024). Biological nitrogen fixation (BNF) is the primary natural mechanism for terrestrial fixed nitrogen. Not all N_2 in aqueous systems becomes reduced. Craig (1968) observed fractionation of nitrogen gas that had been injected into a flowing stream of CO_2. This carrier diffusion effect has been invoked for the decrease in the nitrogen content with migration distances for natural gas accumulations (Stahl et al. 1975). Small fractionations accompanying the molecular dissolution of nitrogen in water have also been observed. Dissolved N_2 in water is enriched 0.85 to 0.99 ‰ (ε = 1.085 to 1.099) with respect $N_{2(g)}$ (Klots and Benson, 1963; Miyake and Wada, 1971).

Lee et al. (2015) measured the temperature dependent fractionation as 0.91, 0.73, −0.04 and −0.42 ‰ at 5, 20, 40, and 60 °C, respectively. They suggest that the temperature dependence is kinetic-based and may explain why small negative $\delta^{15}N$ values occur in some spring water.

Ammonification is the aerobic and anaerobic biologic breakdown of organic matter into ammonia and ammonium. Ammonification can be simplified to:

$$R-NH_2 + H_2O \rightarrow NH_3 + R-OH \tag{10.11}$$

$$NH_3 + H_2O \rightarrow NH_4^+ + OH^- \tag{10.12}$$

and is often summarized as:

$$RNH_2 + 2H_2O \rightarrow NH_4^+ + ROH \tag{10.13}$$

where R is an alkyl group.

Ammonification is commonly followed by nitrification, which is the oxidization of ammonia and ammonium into nitrite and nitrate:

$$NH_3 + O_2 \rightarrow NO_2^- + 3H^+ + 2e \tag{10.14}$$

$$NO_2^- + H_2O \rightarrow NO_3^- + 2H^+ + 2e \tag{10.15}$$

$$2NH_4^+ + 3O_2 \rightarrow 2NO_2^- + 2H_2O + 4H^+ \tag{10.16}$$

$$2NO_2^- + O_2 \rightarrow 2NO_3^- \tag{10.17}$$

During this process the nitrate is depleted 2 to 35 ‰ with respect to ammonium (Delwiche and Steyn, 1970; Freyer and Aly, 1974; Haas et al., 2022).

Denitrification, the reduction of NO_3^- to N_2O or N_2,_ can be represented as:

$$NO_3^- \rightarrow NO_2^- \rightarrow NO \rightarrow N_2 \tag{10.18}$$

Freyer and Aly (1974) investigated isotopic fractionation during the evaporation of volatile nitrogen compounds in ammonium fertilizer. They found that escaping NH_3 became depleted in ^{15}N with respect to aqueous NH_4^+. The escape of NH_3 is a nonreversible reaction and cannot be compared with evaporation of water, because during nitrogen evaporation isotopic exchange between a gas and a liquid phase does not occur. The resulting NH_4^+ can lost to the aqueous environment by exchange with clay particles in soils. London (1961) reported enrichment of 5 to 20 ‰, and Cejudo and Schiff (2018) reported enrichment of 19 and 30 ‰ at pH 8.5 and 9.2, respectively. Nitrite and nitrate ions are not involved in soil ion exchange reactions.

The biologic nitrogen cycle involves nitrogen fractionation in the soil zone and lower atmosphere. The principal chemical processes in the biochemical nitrogen cycle are ammonification, nitrification, and denitrification. Kinetic affects accompanying organic processes are dependent on the relative speed of chemical reactions involving the isotopes ^{15}N and ^{14}N. The speed of the reaction is influenced by temperature, humidity, mineral content of the soil, and the oxygen, water, and organic content of the soil. The first step in the biologic nitrogen cycle is the fixation of atmospheric molecular nitrogen. Meints et al. (1975) found that organic matter is depleted several permil with respect to nitrate when nitrate is the nitrogen source for plants. In a global study, Craine et al. (2015b) found that organic matter is depleted −7.8 to +8.3, with a mean value of 0.9 ‰. After plant death, decomposition of organic matter in the soil converts the biologically fixed nitrogen to NH_4^+

by ammonification. Complex organic matter converted to NH_4^+ by ammonification is depleted 5 to 14 ‰ in ^{15}N with respect to the original organic material (Feigin et al., 1974; Yu et al., 2021).

Denitrification, the transformation of NH_3^- to molecular nitrogen (N_2), is associated with bacteria such as *Thiobacillus denitrificans*. Denitrification involves several intermediate steps that are accompanied by the formation of nitrogen oxides. Complicated kinetic factors accompanying denitrification often results in incomplete isotopic fractionation; thus, a wide range of isotopic fractionations for $NO_3^- \rightarrow NO_2^-$ have been observed ranging from −20 to −1 ‰. The mean overall fractionation for $NO_3^- \rightarrow NO_2^- \rightarrow NO \rightarrow N_2$ ranges from −20 to −15 ‰ (Cook et al., 1970; Granger et al., 2008; Wellman et al., 1968).

Dissimilatory nitrate reduction (DNRA) is the reduction of nitrate to ammonium via an intermediate nitrite step. DNRA is a reduction process catalyzed by anaerobic microorganisms most commonly associated with low-oxygen-bearing marine sedimentary environments. There is some evidence that DNRA occurs in some terrestrial ecosystems (Pandey et al., 2020).

Important Takeaway – Nitrogen species may be transformed by nitrogen fixation, ammonification (a.k.a. mineralization), nitrification, and denitrification. Isotopic fractionation accompanies each step. Locally evaluating flow path nitrogen fractionation may be important when sorting out the nitrogen isotopic compositions.

10.2.3 Nitrogen in Groundwater

Aqueous nitrate is very mobile and may be transported considerable distances. Only two processes, dilution and denitrification, can reduce groundwater nitrogen concentrations. Ammonium exchange with clay can slightly reduce the nitrogen content of some waters, but the ammonium concentration is generally small compared to the nitrate concentration. Both dilution and denitrification may affect the isotopic composition of the nitrogen. The isotopic composition of diluted groundwater will depend on the extent of mixing and the isotopic compositions of the mixed waters. The change of a groundwater system from oxidizing to reducing may, in some situations, lead to denitrification. As previously described, fractionation accompanying denitrification enriches the remaining NO_3^- in ^{15}N. The absence of sufficient quantities and types of organic matter in groundwater systems often inhibit the growth of denitrifying bacteria. In unconfined groundwater systems, N_2O or $N_{2(g)}$ liberated by denitrification may escape to the unsaturated zone.

Despite the fact that there is considerable overlap in the $\delta^{15}N$ of nitrogen sources, nitrogen isotopes are frequently used to identify the nitrogen source in contaminated surface and groundwater. Water contamination is usually indicated by elevated nitrogen concentrations. The focus is usually on analyzing $\delta^{15}N$ of nitrate and sometimes ammonium in the contaminated water relative to the local $\delta^{15}N$ compositions of possible sources. Evaluating the fractionation processes leading to the isotopic compositions is usually a secondary concern but may be critical in systems that chemically and isotopically evolve along the flow path. In aqueous nitrogen isotopic investigations, surface water, particularly streams and rivers, may be the primary interest; however, agricultural and septic tank wastewater contamination is usually manifested in groundwater systems. The migration of contaminated water is largely the result of surface flow even though groundwater-supported base flow may be the nitrogen source. The $\delta^{15}N$ of some sources such as sewage treatment outflow and industrial discharge can be measured directly at the source. The $\delta^{15}N$ of agricultural and rural domestic waste can be more difficult to characterize. When available local wells are a good option, sometimes it is necessary to analyze the upstream reaches of tributary creeks and streams that can be tied directly to the agricultural and rural domestic sources. Examples of $\delta^{15}N$ investigations involving river loading, atmospheric deposition, and groundwater follow.

Example E10.2 Nitrogen Fractionation

Problem: A groundwater sample has a NO_3^- $\delta^{15}N$ value of +12 ‰. If fertilizer nitrate with a NO_3^- $\delta^{15}N$ value of 0‰ is the primary source of nitrate contamination, what percentage of the nitrate in the groundwater is likely to have originated from denitrification processes, given that denitrification can enrich the remaining nitrate in the sample, potentially producing a NO_3^- $\delta^{15}N$ of +20‰?

Solution: A two-end member mixing model can be used to estimate the contribution from each source. If we let x = the fraction of nitrate from a fertilizer with a $\delta^{15}N$ = 0 ‰ and $1 - x$ = the fraction of nitrate from the denitrification source $\delta^{15}N$ = +20 ‰:

$$x + (1 - x) = 12\ ‰$$

$$x\,0\ ‰ + (1 - x)\bullet 20\ ‰ = 12\ ‰$$

$$0 + 20 - 20x = 12$$

$$20 - 20x = 12$$

$$20x = 20 - 12$$

$$20x = 8$$

$$x = 0.4$$

and the denitrification fraction is 60 percent.

Example 10.3 Upper Elbe River, Czech Republic

The upper one-third of the Elbe River in Western Europe is located in the Czech Republic. The Elbe is Europe's fourth-largest river, draining 148,270 km^2. Except during peak runoff, river flows are supported by groundwater discharge. Nitrate concentrations in some reaches exceed the European Union standard. Historically, inorganic fertilizer has been considered to be the primary nitrate source. Nitrate concentrations, flow, and stable isotopic data were used to evaluate the contributions associated with commercial fertilizer, mineralized soil nitrate, and animal and human waste (Mayo et al., 2019). The relationship between city size and nitrate loading was also examined. More than 50 percent of the basin is used for agriculture, and more than 30 percent is forest (Figure 10.7). Except for the capital city, Prague, all cities have a population of less than 200,000, and many residents live in small farming villages that do not have a central sewage treatment facility. Nitrogen sources were analyzed by direct sampling of sewage treatment and industrial outfalls and by sampling upstream reaches of tributary creeks where a single nitrogen groundwater interflow source could be identified. Samples from 138 river, stream, and creek locations were analyzed for NO_3^- $\delta^{15}N$. Approximately 76 percent of the samples had $\delta^{15}N$ that was consistent with animal and human waste, 20 percent with mineralized soil, and only 4 percent with commercial fertilizer (Figure 10.8). An analysis of nitrate loading and city size in each sub-basin suggests human wastewater including individual septic systems rather than artificial fertilizer is the major source of nitrate in the upper Elbe River basin.

Example 10.4 Sichuan Basin, China

Li et al. (2006) investigated nitrate-contaminated groundwater in four rural areas located near cities in the 330,000 km^2 Sichuan Basin, China. In the basin, agriculture use of sulfur and nitrogen-based

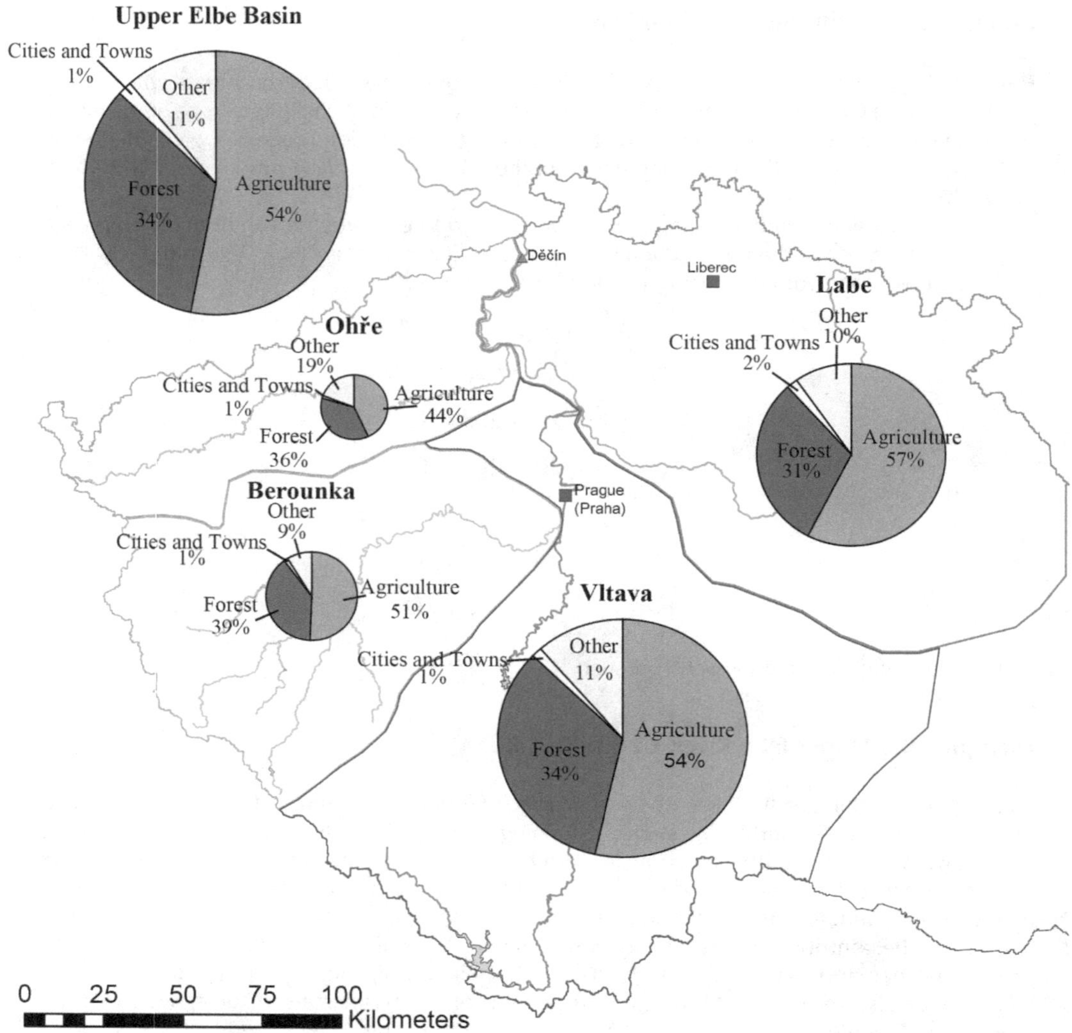

FIGURE 10.7 Land use in the upper Elbe River basin Czech Republic. Sub-basins pie chart sizes are sized proportional to sub-basin catchment area size (after Mayo et al., 2019).

chemical fertilizer is intensive. More than 50 percent of the samples collected had nitrate concentrations > 10 mg/L, and the $\delta^{15}N$ varied greatly by source and location:

	NO_3^-	$\delta^{15}N$ (‰) N_{tot}	NH_4+
Farmyard wells	9.7 ± 4.7		9.7 ± 6.1
Farmland wells	3.7 ± 2.1		−1.2 ± 3.0
Springs	−0.2 ± 3.7		−8.5 ± 4.0
Rain	−7.6 ± 1.9		−6.6 ± 4.0
Inorganic fertilizer		1.0 ± 2.6	

The isotopic data suggested multiple nitrogen sources. The rain had elevated ammonium concentrations (3.0 ± 2.0 mg/L) that varied by season. The highest concentrations occurred during

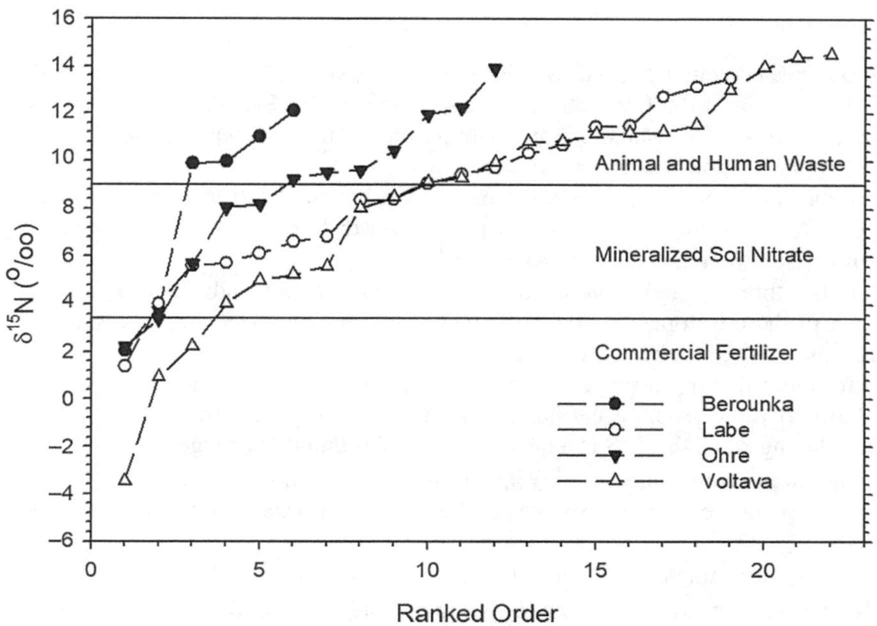

FIGURE 10.8 Ranked order of $\delta^{15}N$ values of upper Elbe River water organized by sub-basin. Approximate source term isotopic boundaries are shown for commercial fertilizer, mineralized soil nitrate, and animal and human waste (after Mayo et al., 2019).

May and June, the seeding time, when fertilizer was applied. NO_x from automotive exhaust was considered to be the primary nitrate rain source, and the primary ammonium source is the vaporization of fertilizers and animal waste. Springs had the lowest nitrogen concentrations. The near-neutral nitrate $\delta^{15}N$ suggested soil organic matter as a primary source.

The isotopic differences in farm-water nitrogen were particularly interesting. Nitrate concentrations were typically elevated and ammonium concentrations were low, sometimes less than the detection limit of 0.05 mg/L. Animal wastes and commercial fertilizer were identified as the primary agricultural nitrogen sources. The data suggests that the farmyard and farmland well waters have different sources. Farmyard wells had nitrate $\delta^{15}N$, consistent with animal enclosures, and farmland wells had $\delta^{15}N$, suggesting commercial fertilizer.

Example 10.5 Riparian Groundwater, Delaware

Inamdar et al. (2024) evaluated the effects of milldams and road salt salinization on riparian nitrogen processes in Delaware. One of the two milldams is located downstream of a major highway that is heavily salted during the winter. Groundwater Na^+ at the non-salted milldam site was 33mg/L compared 199 mg/L at the salted site. The mean NO_3^- $\delta^{15}N$ in riparian groundwater at the salted site was −0.79 ‰ (−9.62 to 4.82 ‰) and was 9.07 ‰ (−9.9 to 29.96 ‰) at the non-salted site. The NO_3^- $\delta^{18}O$ for the two sites were also different (average 13.07 and 5.88 ‰, respectively). Stream waters at both sites had NO_3^- $\delta^{15}N$, consistent with local precipitation. Groundwater NO_3^- $\delta^{15}N$ values at the non-salted site were higher in the upland area and declined down slope. ^{15}N enrichment at the upland riparian edge was attributed to denitrification along the flow path. The more depleted values at the salted site were attributed to suppressed denitrification, potentially greater DNRA, and anaerobic mineralization in riparian soils where persistent anoxic conditions and salinization were prevalent.

Problems

1. A groundwater sample has a $\delta^{34}S$ value of +20 ‰. Potential sulfur sources include evaporite deposits ($\delta^{34}S$ = +30 ‰), mine drainage ($\delta^{34}S$ = −10 ‰), and a wastewater treatment plant ($\delta^{34}S$ = +15 ‰). Make a graph with a mixing line to determine the likely contributions from each potential source?
2. A groundwater sample has a $\delta^{34}S$ value of +5 ‰. Potential sulfur sources include bedrock deposits ($\delta^{34}S$ = +20 ‰) and agricultural fertilizer ($\delta^{34}S$ = −10 ‰). What is the likely contribution from each potential source?
3. Sulfur has three stable isotopes: ^{32}S (95.02 %), ^{33}S (0.76 %), and ^{34}S (4.21 %). The atomic masses of these isotopes are 31.972 amu, 32.970 amu, and 33.967 amu, respectively. Calculate the average atomic mass of sulfur.
4. Groundwater flowing through an area with sulfate-reducing bacteria shows a change in $\delta^{34}S$. Initially, the groundwater had a $\delta^{34}S$ of +10 ‰. After the water flows through a sulfate-reducing zone, the $\delta^{34}S$ changes to +20 ‰. Explain this change.
5. A groundwater sample collected from a well in an agricultural area has a NO_3^- $\delta^{15}N$ value of +8 ‰. What are the likely sources of the nitrate, and what is the contribution from each source?
6. A groundwater sample in an agricultural area has a NO_3^- $\delta^{15}N$ value of +8 ‰. The NO_3^- $\delta^{15}N$ value of a known manure source is +15 ‰, and the NO_3^- $\delta^{15}N$ value of a known fertilizer source is +4 ‰. Assuming a simple mixing model, what is the percentage contribution of each source to the groundwater nitrate? Solve the problem both graphically and numerically.
7. Groundwater samples show a range of $\delta^{15}N$ values (e.g., −5 ‰ to +20 ‰) depending on location and depth. What processes (e.g., nitrification, denitrification) are likely occurring, based on the isotopic shifts observed?

REFERENCES CITED

Abascal, E., et al., 2022, Global diagnosis of nitrate pollution in groundwater and review of removal technologies, *Science of the Total Environment*, 810.

Amrani, A., et al., 2013, Sulfur isotope homogeneity of oceanic DMSP and DMS, *Proceedings National Academy of Science USA*, 46, 18413–18418.

Amundson, R., et al., 2003, Global patterns of the isotopic composition of soil and plant nitrogen, *Global Biogeochemical Cycles*, 17(1), 1031.

Angert, A., et al., 2019, Sulfur isotopes ratio of atmospheric carbonyl sulfide constrains its sources, *Scientific Reports*, 9(74).

Aravena, R., and B. Mayer, 2009, Chapter 7 – Isotopes and processes in the nitrogen and sulfur cycles, in *Environmental isotopes in biodegradation and bioremediation*, CRC Press, 203–246.

Ball, J.W., et al., 1998, *Chemical analyses of hot springs, pools, geysers, and surface waters from Yellowstone National Park, Wyoming and vicinity, 1974–1975*, U.S. Geological Survey Open-File Report 98-182.

Barkan, Y., et al., 2020, Sulfur isotope fractionation between aqueous and carbonate-associated sulfate in abiotic calcite and aragonite, *Geochimica et Cosmochimica Acta*, 280, 317–339.

Benson, B.B., and P.D.M. Parker, 1961, Nitrogen/argon and nitrogen isotope ratios in aerobic sea water, *Deep-Sea Research*, 7, 237–253.

Berner, E.K., and R.A. Berner, 1987, *The global water cycle*, Prentice-Hall.

Bernhard, D.B., and J.E. Spangenberg, 2005, Sulfur speciation and stable isotope trends of water-soluble sulfates in mine tailings profiles, *Environmental Science*, 39(15), 5650–5656.

Boris, U., et al., 2007, Biogeochemistry and isotope geochemistry of a landfill leachate plume, *Journal of Contaminant Hydrology*, 65(3–4), 245–268.

Bottrell, S., et al., 2008, isotopic composition of sulfate as a tracer of natural and anthropogenic influences on groundwater geochemistry in an urban sandstone aquifer, Birmingham, UK, *Applied Geochemistry*, 23(8), 2382–2394.

Bremner, J.M., and M.A. Tabatabai, 1973, Nitrogen-15 enrichment of soils and soil derived nitrate, *Journal of Environmental Quality*, 2, 363–365.

Brenner, D.L. et al., 2001, Soil N and ^{15}N variation with time in a California annual grassland ecosystem, *Geochimica et Cosmochimica Acta*, 65, 4171–4186.

Broadbent, F.E., et al., 1980, Spatial variability of N-15 and total N in some virgin and cultivated soils, *American Journal Soil Science Society*, 44, 524–527.

Canfield, D.E., 2001, Biogeochemistry of sulfur isotopes, *Reviews in Mineralogy and Geochemistry*, 43(1), 607–636.

Cejudo, E., and S.L. Schiff, 2018, Nitrogen isotope fractionation factors (α) measured and estimated from the volatilization of ammonia from water at pH 9.2 and pH 8.5, *Isotopes in Environmental and Health Studies*, 6, 642–655.

Chen, M., et al., 2024, A study of chemical processes of nitrate in atmospheric aerosol and snow based on stable isotopes, *Atmosphere*, 15(1), 59.

Chen, T., et al., 2016, Spatial and temporal variations of PM2.5 and its relation to meteorological factors in the urban area of Nanjing, China, *International Journal Environmental Research and Public Health*, 13(9), 921.

Chung, C.H., et al., 2019, Sulfur isotope analysis for representative regional background atmospheric aerosols collected at Mt. Lulin, Taiwan, *Scientific Reports*, 9.

Claypool, G.E., et al., 1980, The age curves of sulfur and oxygen isotopes in marine sulfate and their mutual interpretation, *Chemical Geology*, 28, 199–260.

Cook, F.D., et al., 1970, *Nitrogen isotope fractionation in the nitrogen cycle*, International Symposium on Hydrogeochemistry and Biochemistry, Tokyo, September 6–12, 1970, B1/6.

Cortecci, G., and A. Longinelli, 1970, Isotopic composition of sulphate in rainwater, Pisa, Italy, *Earth Planetary Science Letters*, 8, 36–40.

Craig, H., 1968, Isotope separation by carrier diffusion, *Science*, 159, 93–96.

Craine, J.M., et al., 2015a, Ecological interpretations of nitrogen isotope ratios of terrestrial plants and soils, *Plant and Soil*, 396, 1–26.

Craine, J.M., et al., 2015b, Convergence of soil nitrogen isotopes across global climate gradients, *Scientific Reports*, 5, 8280.

Cravotta, C.A., 1997, *Use of stable isotopes of carbon, nitrogen, and sulfur to identify sources of nitrogen in surface waters in the Lower Susquehanna River Basin, Pennsylvania*, U.S. Geological Survey Water-Supply Paper, 2497.

Dasari, S, et al., 2022, Sulfur isotope anomalies ($\delta^{33}S$) in urban air pollution linked to mineral-dust-associated Sulfate, *Earth Science & Technology Letters*, 9(7).

Davidson, C., et al., 2021, Tropospheric carbonyl sulfide mass balance based on direct measurements of sulfur isotopes, *Earth Atmospheric Planetary Sciences*, 118(6).

Delwiche, C.C., and P.L. Steyn, 1970, Nitrogen fractionation in soils and microbial reactions, *Earth Science & Technology Letters*, 4, 929–935.

Feigin, A.D., et al., 1974, Variations of the natural nitrogen-15 abundance in nitrate mineralization during incubation of several Illinois soils, *Soil Science Society of America Proceedings*, 38, 90–95.

Ferronsky, V.I., and V.A. Polyakov, 1982, *Environmental isotopes in the hydrosphere*, Wiley.

Filpse, W.J., and F.T. Bonner, 1985, Nitrogen-isotope ratios in nitrate in groundwater under fertilized fields, Long Island, New York, *Groundwater*, 23(1), 59–67.

Freyer, H.D., and A.I.M Aly, 1974, Nitrogen-15 variations in fertilizer nitrogen, *Journal of Environmental Quality*, 3, 405–406.

Freyer, H.D., and A.I.M. Aly, 1976, *Nitrogen-15 studies on identifying fertilizer excess in environmental systems in isotope ratios as pollutant source and behavior indicators*, International Atomic Energy Agency (IAEA), 21–33.

Fritz, S.J., et al., 1994, A geochemical and isotopic approach to delineate landfill leachates in a RCRA study, *Ground Water*, 32, 743–750.

Fry, B., et al., 1988, Sulfur isotope effects associated with oxidation of sulfide by O_2 in aqueous solution, *Chemical Geology, Isotope Geoscience Section*, 73(3), 205–210.

Gałuszka, A., et al., 2020, Trace elements and stable sulfur isotopes in plants of acid mine drainage area: Implications for revegetation of degraded land, *Journal of Environmental Sciences*, 94, 128–136.

Garrels, R.M., and C.L. Christ, 1965, *Solutions, minerals and equilibria*, Freeman-Cooper.

Giampaoli, S., et al., 2013, Hydrogen sulfide in thermal spring waters and its action on bacteria of human origin, *Microchemical Journal*, 108, 210–214.

Granger, J., et al., 2008, Nitrogen and oxygen isotope fractionation during dissimilatory nitrate reduction by denitrifying bacteria, *Limnology and Oceanography*, 53.

Grey, D.C., and M.L. Jensen, 1972, Bacteriogenic sulfur in air pollution, *Science*, 177, 1099–1100.
Haas, S., et al., 2022, Characterization of nitrogen isotope fractionation during nitrification based on a coastal time series, *Limnology and Oceanography*, 67, 1714–1731.
Harris, E., et al., 2012, Sulfur isotope fractionation during heterogeneous oxidation of SO_2 on mineral dust, *Atmospheric Chemistry and Physics*, 12(11).
Harrison, A.G., and H.G. Thode, 1957, The kinetic isotope effect in chemical reduction of sulphate, *Transactions Faraday Society*, 53, 1–4.
Hattori, S., et al., 2020, Constraining the atmospheric OCS budget from sulfur isotopes, *Earth, Atmospheric and Planetary Sciences*, 117(34), 20477–20452.
Holser, W.T. 1977, Catastrophic chemical events in the history of the ocean, *Nature*, 267, 403–408.
Holser, W.T., and I.R. Kaplin, 1966, Isotope geochemistry of sedimentary sulfates, *Chemical Geology*, 1, 93–135.
Hunt, C.D., et al., 1995, *Deer Island effluent characterization and pilot treatment plant studies*, Massachusetts Water Resources Authority, Technical Report 95-97.
Hutchison, W., et al., 2019, Sulphur isotopes of alkaline magmas unlock long-term records of crustal recycling on Earth, *Nature Communications*, 10, 4208.
Inamdar, S., et al., 2024, Riparian groundwater nitrogen (N) isotopes reveal human imprints of dams and road salt salinization, *Geophysical Research Letters*, 51.
Jakóbczyk-Karpierz, S., and K. Ślósarczyk, 2022, Isotopic signature of anthropogenic sources of groundwater contamination with sulfate and its application to groundwater in a heavily urbanized and industrialized area (Upper Silesia, Poland), *Journal of Hydrology*, 612C.
Jensen, M.L., and N. Nakai, 1961, Sources and isotopic composition of atmospheric sulphur, *Science*, 134, 2102–2104.
Jongebloed, U.A., et al., 2023, Sulfur isotope quantifies the impact of anthropogenic activities on industrial-era Arctic sulfate in a Greenland ice core, *Environmental Research Letters*, 18.
Jørgensen, B., et al., 2019, The biogeochemical sulfur of marine sediments, *Frontiers in Microbiology*, 10, 849.
Kaplan, I.R., and C.C. Rittenberg, 1994, Microbiological fractionation of sulfur isotopes, *Journal General Microbiology*, 34, 195–212.
Kemp, A.L.W., and H.G. Thode, 1968, The mechanism of the bacterial reduction of sulphate and sulphite from isotope fractionation studies, *Geochimica et Cosmochimica Acta*, 32, 71–91.
Kendall, C., 1998, Tracing Nitrogen sources and cycling, in *Catchments in isotope tracers in catchment hydrology*, Elsevier Science, 519–576.
Kendall, C., and R. Aravena, 2000, Nitrate isotopes in groundwater systems, in *Environmental tracers in subsurface hydrology*, Springer, 261–295.
Klots, C.E., and B.B. Benson, 1963, Isotope effect in the solution of oxygen and nitrogen in distilled water, *Journal of Chemical Physics*, 38, 890–893.
Knöllerand, M., and M. Schubert, 2010, Interaction of dissolved and sedimentary sulfur compounds in contaminated aquifers, *Chemical Geology*, 276(3–4), 284–293.
Knossow, N., et al., 2015, Annual sulfurs cycle in a warm monomictic lake with sub-millimolar sulfate concentrations, *Geochemical Transactions*, 16(7).
Kobayashi, K., et al., 2019, Dual nitrogen and oxygen isotope fractionation during anaerobic ammonium oxidation by anammox bacteria, *ISME Journal*, 13, 2426–2436.
Kohl, D.A., et al., 1971, Fertilizer nitrogen: contribution to nitrate in surface water in a Corn Belt watershed, *Science*, 174, 1331–1334.
Kok, J.F., et al., 2021, Contribution of the world's main dust source regions to the global cycle of desert dust, *Atmospheric Chemistry and Physics*, 21, 8169–8193.
Kreitler, C.W., 1975, *Determining the source of nitrate in groundwater by nitrogen isotope studies*, Bureau of Economic Geology, Report 83.
Kreitler, C.W., 1979, Nitrogen-isotope ratio studies of soils and groundwater nitrate from alluvial fan aquifers in Texas, *Journal of Hydrology*, 42(1–2), 147–170.
Krouse, H.R., 1980, Chapter 11 – Sulphur isotopes in our environment, in *Handbook of environmental isotope geochemistry*, Elsevier Science, 1, 435–471.
Kushkevych, I., et al., 2021, Microscopic methods for identification of sulfate-reducing bacteria from various habitats, *Intentional Journal Molecular Science*, 22(8), 4007.
Lee, H., et al., 2015, Kinetic nitrogen isotope fractionation between air and dissolved N_2 in water: Implications for hydrothermal systems, *Geochemical Journal*, 49(5), 571–573.
Letolle, R., 1980, Nitrogen-15 in the natural environment, in *Handbook of environmental isotope geochemistry*, Elsevier, 1, 405–443.

Letolle, R., and P. Olive, 1983, Isotopes as pollution tracers, in *Guidebook on nuclear techniques in hydrology*, International Atomic Energy Agency (IAEA), 411–422.

Li, X., et al., 2007, Nitrogen isotope study on nitrate-contaminated groundwater in the Sichuan Basin, China, *Water Air Soil Pollution*, 178, 145–156.

Li, X.-D., et al., 2006, Degradation of groundwater quality due to anthropogenic sulfur and nitrogen contamination in the Sichuan Basin, China, *Geochemical Journal*, 40, 309–322.

Li, Z., et al., 2019, Nitrate isotopic composition in precipitation at a Chinese megacity: Seasonal variations, atmospheric processes, and implications for sources, *Earth and Space Science*, 6, 2200–2213.

Liotta, M., et al., 2012, Sulfur isotopic compositions of fumarolic and plume gases at Mount Etna (Italy) and inferences on their magmatic source, *Geochemistry Geophysics, Geosystems*, 13(5).

London, M., 1961, *Separation of isotopes*, Newnes, London.

Mariotti, A., et al., 1981, Experimental determination of nitrogen kinetic isotope fractionation: Some principles; illustration for the denitrification and nitrification processes, *Plant Soil*, 62, 413–430.

Mariotti, A., et al., 1980, The abundance of natural nitrogen 15 in the organic matter of soils along an altitudinal gradient, *Catena*, 7(1), 293–300.

Martin, E., 2018, Volcanic plume impact on the atmosphere and climate: O- and S-isotope insight into sulfate aerosol formation, *Geosciences*, 8(6), 198.

Massmann, G., et al., 2003, Sulfide oxidation and sulfate reduction in a shallow groundwater system (Oderbruch Aquifer, Germany), *Journal of Hydrology*, 278, 231–243.

Matiatos, I., et al., 2021, Global patterns of nitrate isotope composition in rivers and adjacent aquifers reveal reactive nitrogen cascading, *Communications Earth & Environment*, 2(52).

Mayo, A.L., and R.H. Klauk, 1991, Contributions to the solute and isotopic groundwater geochemistry, Antelope Island, Great Salt Lake, Utah, *Journal Hydrology*, 127(1–4), 307–335.

Mayo, A.L., D.J. Ritter, J. Bruthans, and D. Tingey D., 2019, Contributions of commercial fertilizer, mineralized soil nitrate, and animal and human waste to the nitrate load in the Upper Elbe River Basin, Czech Republic, *HydroResearch*, 1, 25–35.

Mayuko, N., et al., 2021, Seasonal change in microbial sulfur cycling in monomictic Lake Fukami-ike, Japan, *Limnology and Oceanography*, 57.

Meints, V.W., et al., 1975, Natural ^{15}N abundance in soils, leaves and grain as influenced by long term addition of fertilizer N as several rates, *Journal of Environmental Quality*, 4, 486–490.

Mitchell, R.H., and H.R. Krouse, 1975, Sulfur isotope geochemistry of carbonatites, *Geochimica et Cosmochimica Acta*, 39, 1505–1513.

Miyake, Y., and E. Wada, 1971, The abundance of 15N/14N in marine environment's, *Records Oceanographic Works (Japan)*, 11, 1–6.

Munger, J.W., 1982, Chemistry of atmospheric precipitation in the north-central united states: Influence of sulfate, nitrate, ammonia and calcareous soil particulates, *Atmospheric Environment*, 16, 1633–1645.

Nikolenko, O., et al., 2018, Isotopic composition of nitrogen species in groundwater under agricultural areas: A review, *Science of The Total Environment*, 621, 1415–1432.

Olson, E., et al., 2021, Mineral dust and fossil fuel combustion dominate sources of aerosol sulfate in urban Peru identified by sulfur stable isotopes and water-soluble ions, *Atmospheric Environment*, 260, 18482.

Pandey, C.B., et al., 2020, DNRA: A short-circuit in biological N-cycling to conserve nitrogen in terrestrial ecosystems, *Science of the Total Environment*, 738, 139710.

Panettiere, P., et al., 2000, Chemistry and sulfur isotopic composition of precipitation at Bologna, Italy, *Applied Geochemistry*, 5(10), 1455–1467.

Paytan, A., et al., 2020, Chapter 9 – Sulfur isotope stratigraphy, in *Geologic time scale 2020*, Elsevier, 259–278.

Pellerin, A., et al., 2019, Large sulfur isotope fractionation by bacterial sulfide oxidation, *Science Advances*, 5(7).

Present, T.M., et al., 2020, Variability in sulfur isotope records of Phanerozoic seawater sulfate, *Geophysical Research Letters*, 47.

Pujiindiyati, E.R., and S. Syafalni, 2012, Study of sulphate origin in shallow groundwater in the vicinity of Bantar Gebang-Bekasi landfill using sulphur-34 and oxygen-18, *Atom Indonesia*, 33(2).

Qian, J., et al., 2021, Stable isotope analyses of nitrogen source and preference for ammonium versus nitrate of riparian plants during the plant growing season in Taihu Lake Basin, *Science of The Total Environment*, 763, 143029.

Raab, M., and B. Spiro, 1991, Sulfur isotopic variations during seawater evaporation with fractional crystallization, *Chemical Geology*, 86(4), 323–333.

Richet. P., et al., 1977, A review of hydrogen, carbon, nitrogen, oxygen, sulphur and chlorine stable isotope fractionation among gaseous molecules, *Annual Review of Earth and Planetary Sciences*, 5(1), 65–110.

Rightmire, C.T., et al., 1974, Distribution of sulfur isotopes of sulfates in groundwaters from the principal artesian aquifer of Florida and Edwards aquifer of Texas, United States of America, in *Isotope techniques in groundwater hydrology*, International Atomic Energy Agency (IAEA), II, 191–207.

Robinson, E., and R.C. Robins, 1970, Gaseous sulfur pollutants from urban and natural sources, *Journal Air Pollution Control Association*, 20, 233–235.

Sakai, H., et al., 1982, Chemistry and isotope ratios of sulfur in basalts and volcanic gases at Kilauea volcano, Hawaii, *Geochimica et Cosmochimica Acta*, 46(5), 729–738.

Schoen, R., and R.O. Rye, 1970, Sulphur isotope distribution in Solfataras, Yellowstone National Park, *Science*, 170(3962), 1082–1084.

Seal, R.S., et al., 2000, Stable isotope systematics of sulfate minerals, *Reviews in Mineralogy and Geochemistry*, 40(1), 541–602.

Serio, C., et al., 2023, Trend and multi-frequency analysis through empirical mode decomposition: An application to a 20-year record of atmospheric carbonyl sulfide measurements, *Journal of Geophysical Research Atmospheres*, 128(3).

Shearer, G.B., et al., 1974, The precision and determination of the natural abundance of nitrogen-15 in soils, fertilizers and shelf chemicals, *Soil Science*, 118, 308–316.

Ślósarczyk, K., et al., 2022, Identification of water contamination sources using hydrochemical and isotopic studies – the Kozłowa Góra Reservoir Catchment Area (Southern Poland), *Water*, 14(6).

Spoelstra, J., et al., 2021, Isotopic characterization of sulfate in a shallow aquifer impacted by agricultural fertilizer, *Groundwater*, 59, 658–670.

Stahl, W., et al., 1975, *Carbon and nitrogen isotope data of upper Carboniferous and Rotiegende natural gases from North Germany and their relationship to the maturity of the organic source material*, 7th International Meeting Organic Chemistry, Madrid.

Taran, Y., et al., 2018, *Sulfur and carbon isotopes in volcanic gases of the Kuril Island arc, NW Pacific*, 20th EGU General Assembly, EGU2018, Proceedings from the Conference, April 4–13, 2018, Vienna, Austria, https://ui.adsabs.harvard.edu/search/q=author:%22Chaplygin%2C+Ilya%22&sort=datedesc,bibcodedesc2018

Thode, H.G., 1991, Chapter 1 – Sulphur isotopes in nature and the environment: An overview, in *Assessment of natural and anthropogenic sulphur in the environment*, Wiley & Sons, 1–26.

Thode, H.G., and J. Monster, 1965, Sulfur-isotope geochemistry of petroleum, evaporites and ancient seas in Fluids in subsurface environments, *American Association Petroleum Geology Memoir,* 4, 367–377.

Thode, H.G., et al., 1962, Sulfur isotope abundances in rocks of the Sudbury District and their geological significance, *Economic Geology*, 57, 565–578.

Tostevin, R., et al., 2014, Multiple sulfur isotope constraints on the modern sulfur cycle, *Earth and Planetary Science Letters*, 396, 14–21.

Tostevin, R., et al., 2016, Sources of environmental sulfur in the groundwater system, southern New Zealand, *Applied Geochemistry*, 70, 1–16.

Toyoshima, S., et al., 2022, *A study on sulfur isotope ratios in Hakone Volcanic Gas*, Japan Geoscience Union Meeting 2022, SVC31–14.

Vengosh, A., et al., 2013, Isotopic imprints of mountaintop mining contaminants, *Environmental Science and Technology*, 47(17), 10041–10048.

Vinogradov, V.I., 1975, How old are the oceans? *Priroda*, 12, 50–57.

Vitòria, L., et al., 2004, Fertilizer characterization: Isotopic data (N, S, O, C, and Sr), *Environmental Science & Technology*, 38, 3254–3262.

Wada, E., et al., 1975, ^{15}N abundance in nitrogen of naturally occurring substances and global assessment of denitrification from isotopic view point, *Geochemical Journal*, 9, 139–148.

Walters, W., et al., 2015, The nitrogen stable isotope composition (δ15N) of vehicle emitted NOx, *Environmental Science and Technology*, 49(4).

Watkins, S.H., et al., 1972, Factors influencing ammonia losses from urea applied to northwestern forest soils, *Soil Science Society of America Proceedings*, 80(12), 1119–1121.

Wellman, R.O., et al., 1968, Nitrogen-15 microbial alteration of abundance, *Science*, 161, 269–270.

Wikipedia Contributors, 2024, *Sulfur isotope biogeochemisty*, Wikipedia.

Wu, H., et al., 2021, Identifying nitrate sources in surface water, regolith and groundwater in a subtropical red soil critical zone by using dual nitrate isotopes, *CATENA*, 198, 104994.

Yao, W., and A. Paytan, 2020, Possible triggers of the seawater sulfate S-isotope increase between 55 and 40 million years ago, *Chemical Geology*, 552, 119788.

Yu, L., et al., 2021, Determination of the nitrogen isotope enrichment factor associated with ammonification and nitrification in unsaturated soil at different temperatures, *Environmental Research*, 202, 111670.
Zhang, C.Y., 2013, Nitrogen isotope studies of nitrate contamination of the thick vadose zones in the wastewater-irrigated area, *Environmental Earth Science*, 68, 1475–1483.
Zhang, Y., et al., 2008, Nitrogen inputs and isotopes in precipitation in the North China Plain, *Atmospheric Environment*, 42(7), 1436–1448.
Zheng, F., et al., 2020, Chemical and sulfur isotopic characteristics of precipitation in a representative urban site, South China: implication for anthropogenic influences, *Air Quality Atmosphere & Health*, 13, 349–359.
Zhu, G., et al., 2016, Chemical and sulfur isotopic composition of precipitation in Beijing, China, *Environmental Science and Pollution Research*, 6(23), 5507–5515.

11 Groundwater Dating

11.1 INTRODUCTION

The concept of groundwater age is a confusing one. First, we must consider the idea of relative and absolute age. For example, a person will have an absolute age based on their birthday but will also have relative ages relative to their siblings. In groundwater we may be concerned with the absolute age of the water when using ^{14}C and with the relative age when using ^{3}H. A volume of water in an aquifer is the product of the mixture of many waters of possible diverse origins and histories. Except for the unusual occurrence of pure piston flow, a water sample has no unique absolute age. The calculated age of a groundwater sample or, more correctly, the integrated mean residence time of the recharge waters can often be estimated by means of radioactive isotopes and some non-radiogenic atmospheric gases.

Groundwater dating can be one of the most powerful tools in hydrogeochemistry; however, not all groundwater is amenable to dating, and considerable care must be exercised when attempting to date groundwater. Groundwater dating is not a routine process such as calculating the saturation index of calcite, where even the most inexperienced novice may obtain a correct answer by means of a computer program. For example, when attempting to radiocarbon date a groundwater the researcher must fully understand the geochemical evolution of the total dissolved inorganic carbon (TDIC) to know which models are applicable and to calculate a groundwater age only when geologic and hydrogeochemical conditions warrant. It is not uncommon to determine that a groundwater sample may not be dated after ^{14}C or other data have been collected.

Groundwater ages may be qualitatively described as ranging from modern to very old (Figure 11.1). At least a dozen isotopes, whose theoretical dating ranges are shown in the figure, are potentially useful for estimating the age of groundwater. From Figure 11.1 it is apparent that it is not possible to estimate the age of many groundwaters because of their young ages. Only the radioisotopes ^{3}H and ^{14}C are routinely used to date groundwater. Anthropogenic CFCs (chlorofluorocarbons) and SF_6 (sulfur hexafluoride) can provide modern groundwater ages for young piston flow systems, $^{3}H/^{3}He$ is useful for some young waters, and ^{36}Cl may be used for some very old groundwaters.

The radiogenic isotopic composition of precipitation is of particular interest to surface and groundwater investigations. With the advent of atmospheric thermonuclear testing in 1953, atmospheric bomb ^{3}H overwhelmed the natural stratospheric ^{3}H concentration. The long-lasting atmospheric bomb spike historically provided a time marker for many water-related investigations. Atmospheric bomb ^{3}H is now exhausted due to the rainout effect, and more recent precipitation contains natural background ^{3}H concentrations. The short 12.3-year half-life makes ^{3}H particularly useful for identifying recent groundwater recharge. Theoretically, ^{14}C can be used to date groundwater as young as 10 or 20 years. However, the practical lower limit of ^{14}C dating is about 200–300 years due to the dilution of atmospheric carbon with dead industrial CO_2, which has been released into the atmosphere since the industrial revolution, and to radioactive (live) carbon that was injected into the atmosphere by atmospheric testing of nuclear weapons. With conventional analytical laboratory methods, the upper limit of ^{14}C dating is about 35,000 years. Laboratory measurements of ^{14}C using enrichment techniques can extend the maximum range to about 75,000 years, and ^{14}C measurements performed using an accelerator can push the maximum age to about 100,000 years. Calculating such older ages is problematic due to laboratory procedural and measurement errors and the inherent errors in the numerical models, which are used to estimate ages. Chlorine-36 extends the possible range of groundwater dating from the ^{14}C upper limit to about 1 million years, which satisfies the needs of most groundwater investigations.

DOI: 10.1201/9781003646129-13

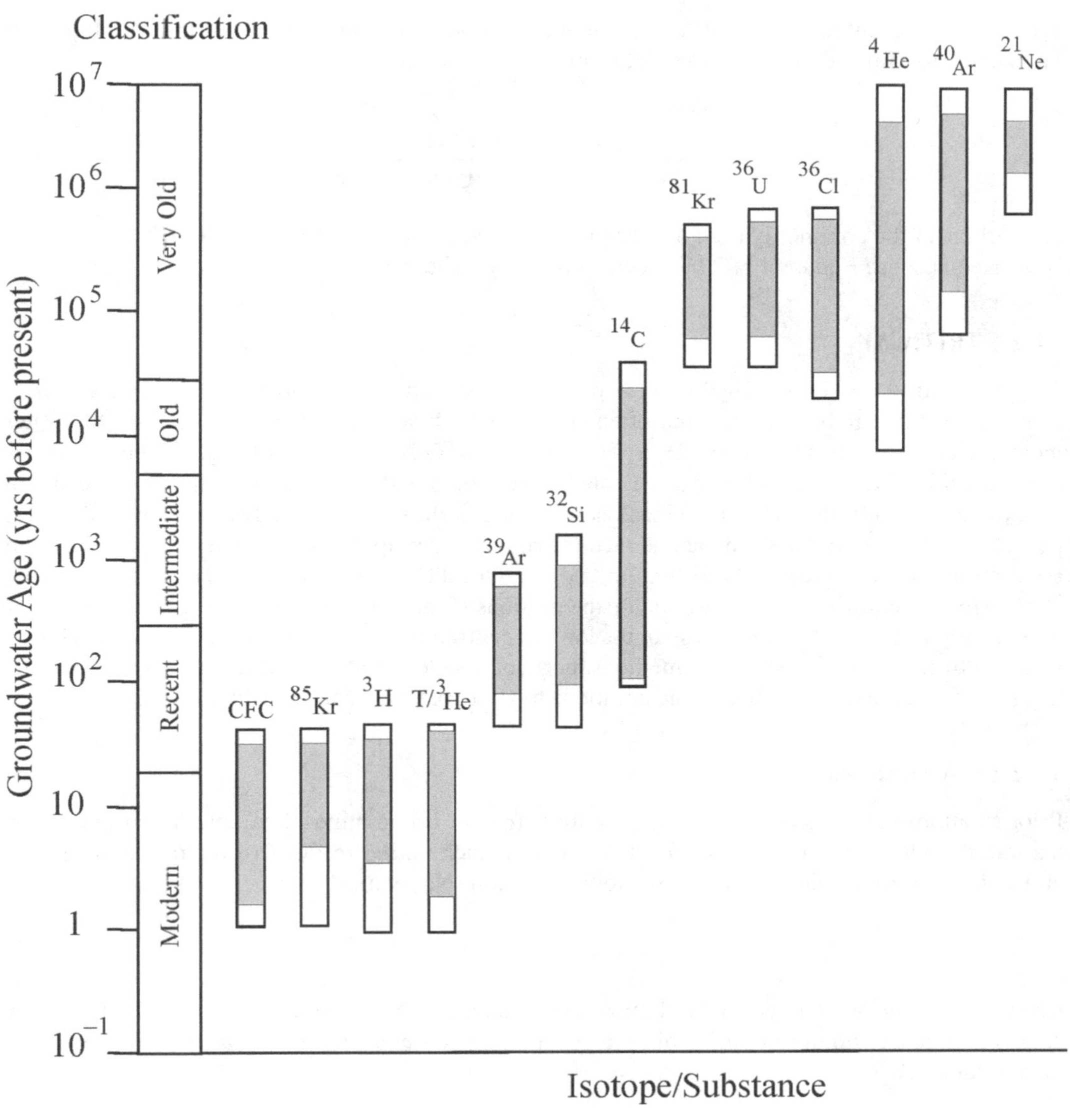

FIGURE 11.1 Range of isotopic and non-isotopic potential groundwater ages. The filled bars are the typical useful ranges and the open bars are ranges in selected applications. Water can also be classified as modern to very old. Only CFCs, ^{3}H, and ^{14}C are routinley used for groundwater dating. ^{3}H (T)/^{3}He and ^{36}Cl are useful in some circumstances.

Groundwater dating using radioisotopes is based on the idea that the mass of radioisotopes disintegrates at a constant rate and that the isotopic half-life is the time required for one-half of the mass to disintegrate into daughter products. At the end of the 1st half-life, one-half of the original activity (A_o) remains, and the remaining mass is designated by the remaining activity (A_t) of ^{14}C in the groundwater. At the end of the 2nd half-life (A_t) only represents 25% of A_o. This decay relationship can be expressed as:

$$A_t = A_o \left(\frac{1}{2}\right)^{\frac{t}{\text{half-life}}} \tag{11.1}$$

Where A_o is the initial activity, A_t is the remaining radioactivity (activity) after time (t). Equation 11.1 can be rearranged in the very useful form:

$$t = \frac{\text{half} - \text{life}}{\ln 2} \times \ln \frac{A_t}{A_o} = \frac{\text{half} - \text{life}}{-0.693} \times \ln \frac{A_t}{A_o} \quad (11.2)$$

For carbon-14 the commonly used half-life is 5,730 years, and it is 12.35 years for tritium. Caution is advised because Equation 11.2 has been incorrectly written elsewhere.

11.2 TRITIUM

Tritium is introduced into the hydrologic cycle from both natural and anthropogenic sources. Once formed, tritium may be oxidized and often combines with atmospheric oxygen to form the water molecule HTO, where H is either ^1H or ^2H, T is ^3H, and O is an isotope of oxygen. The short tritium half-life of 12.35 years (Fritz and Fontes, 1980) combined with the fact that tritium occurs in precipitation worldwide makes it an ideal isotope for determining if groundwater recharged during past 50 to 60 years. Because the abundance of tritium in precipitation is usually very low, tritium concentrations are expressed in tritium units (TU). One TU corresponds to a $^3H/^1H$ ratio of about 10^{-18}, which is equivalent to 7.2 atomic disintegrations (Beta particle releases) per minute per liter of water (dpm/ L), or 3.2 pCi/L. Despite the low concentrations of tritium in precipitation and hence in groundwater, modern laboratory methods, particularly electrolytic enrichment, make the routine measurement of extremely small concentrations to about 0.3 TU economically feasible.

11.2.1 Atmospheric Tritium

Prior to atmospheric thermonuclear testing most tritium in the atmosphere and hydrosphere was naturally produced in the upper atmosphere, in a manner similar to ^{14}C. Cosmic rays bombarding atmospheric nitrogen yield ^3H according to the reaction (Nir, et al., 1966):

$$^{14}N + n = {}^{12}C + {}^{3}H \quad (11.3)$$

where n is a neutron from cosmic radiation. Only about 3 to 5 percent of all neutrons in the upper atmosphere react with atmospheric nitrogen (). Tritium subsequently decays to helium releasing a beta particle (e):

$$^{3}H \rightarrow {}^{3}He + e \quad (11.4)$$

Tritium concentrations in water are measured by counting the released beta particles. In the laboratory the beta emissions are not directly counted. The water is mixed with a solution that emits a pulse of energy that can be measured by very sensitive equipment that shield the sample from atmospheric and lithospheric ^3H.

With the onset of atmospheric thermonuclear weapons testing in 1952 large quantities of anthropogenic ^3H entered the northern hemispheres atmosphere. Massive injections of tritium were placed into the atmosphere in 1954, 1955, 1958 and 1961–62. Prior to 1957, thermonuclear explosions were relatively small and were carried out at intermediate altitudes. Most of the ^3H released during the low altitude tests only reached the troposphere and fell out rapidly with precipitation. Prior to the nuclear atmospheric test ban treaty approximately 200 EBq (200×10^{18} Bq) of tritium was released into stratosphere from the high-altitude explosions in 1958 and in 1961–62. One Bq (Becquerel) is the amount of radioactive material needed for one nucleus decay per second.

Atomic industries such as nuclear research, power and research reactors, nuclear fuel reprocessing plants, weapons production, medical research, and others are also sources of anthropogenic environmental tritium. These sources release small amounts of tritium into the lower atmosphere and directly into the hydrologic cycle. Except for nuclear power plants such releases are relatively unimportant globally, but locally may be need to be considered. Tritium is the primary radionuclide released by nuclear power plants and annual releases are about 0.1 EBq per year (Eyrolle et al., 2018). The nuclear power plant accidents at Chernobyl in 1986 and Fukushima in 2011 were locally catastrophic, but did not change the atmospheric content of either carbon-14 or tritium (Florkowski et al., 1988; Kashiwaya et al., 2017).

Between 1952 and about 1985 tritium concentrations in the northern hemispheres atmosphere and hence precipitation was greatly influenced by atmospheric weapons testing. During this time the distribution of tritium in precipitation reflected the effects of weapons testing (Figure 11.2). Since 1963 the amount of tritium in northern hemisphere precipitation decreased almost exponentially while the concentration of tritium in precipitation in the southern hemispheric slightly increased as a result of the stratospheric transfer of the bomb produced tritium (Schell et al., 1974). Bomb produced stratospheric tritium, which had a long-term effect on the tritium content of precipitation, has now been exhausted and precipitation since about 1995 is approximately natural atmospheric levels.

In the northern hemisphere modern precipitation contains about 5 to 10 TU and the tritium concentration in precipitation in the southern hemisphere is slightly less (Figure 11.2). Tritium concentrations are storm and seasonally variable with elevated concentrations in the late winter and spring. For example, during a single year the tritium in precipitation varied between storm events by as much as 25 TU in Lindon-Provo, Utah, USA (Figure 11.3a), and there was a strong seasonal pattern (Figure 11.3b). During the summer to early spring the average concentration was 5–7 TU and was greater than 10 TU in April and May. Because the tritium concentration in precipitation varies greatly between storm events and during the course of a year, measured values from a few storm events do not represent local precipitation.

Tritium concentrations also have latitudinal, and continental vs. oceanic influences. Tritium is rapidly removed from the atmosphere because the water vapor only has a residence time of 10–30 days. Coastal precipitation typically has low tritium concentrations due to the exchange of atmospheric water vapor with the low tritium concentrations in ocean surface water (Schell et al., 1974; Chae and Kim, 2019). When storm systems move inland (continental effect) troposphere tritium concentrations increase by varying amounts as the tritium concentration is affected by the timing and flux of the stratosphere-troposphere exchange. The increase in tropospheric tritium in the continental interior is mostly due to the so-called spring leak, and thunderstorms. The spring leak occurs in the spring time between 30° N and 60° N when the tropospheric breakdown permits the mixing of troposphere air with tritium rich stratosphere air (Michel et al., 2018). Elevated tritium deposition is also associated with summer thunderheads that penetrate the troposphere and bring down precipitation with elevated ^{3}H concentrations. Evaporation and atmospheric exchange of inland surface water and saturated soil with elevated tritium can extend the length of the spring leak (Gat et al., 2001). The consequence of these processes combined with the variability in storm track trajectories results in the wide range of precipitation ^{3}H concentrations.

Topographic relief also exerts a great influence on the distribution of tritium in precipitation. Large mountain ranges have a profound effect on the circulation of air masses and induce precipitation and the accompanying tritium rainout. When an air mass moves away from the mountainous region it is depleted in tritium. Harms et al. (2016) found the tritium along the California coast was about 4.0 pCi/L but increased to almost 18 pCi/L in the nearby Sierra Nevada Mountains. Other examples of the mountain effect include the convergence of Pacific Ocean, Arctic, north Atlantic, and Gulf of Mexico air masses along the boundary of the eastern front of the Rocky Mountains and the Great Plains which results in elevated Great Plains region tritium concentrations at low latitudes compared with smaller concentrations at similar latitudes along the eastern and western coastal regions.

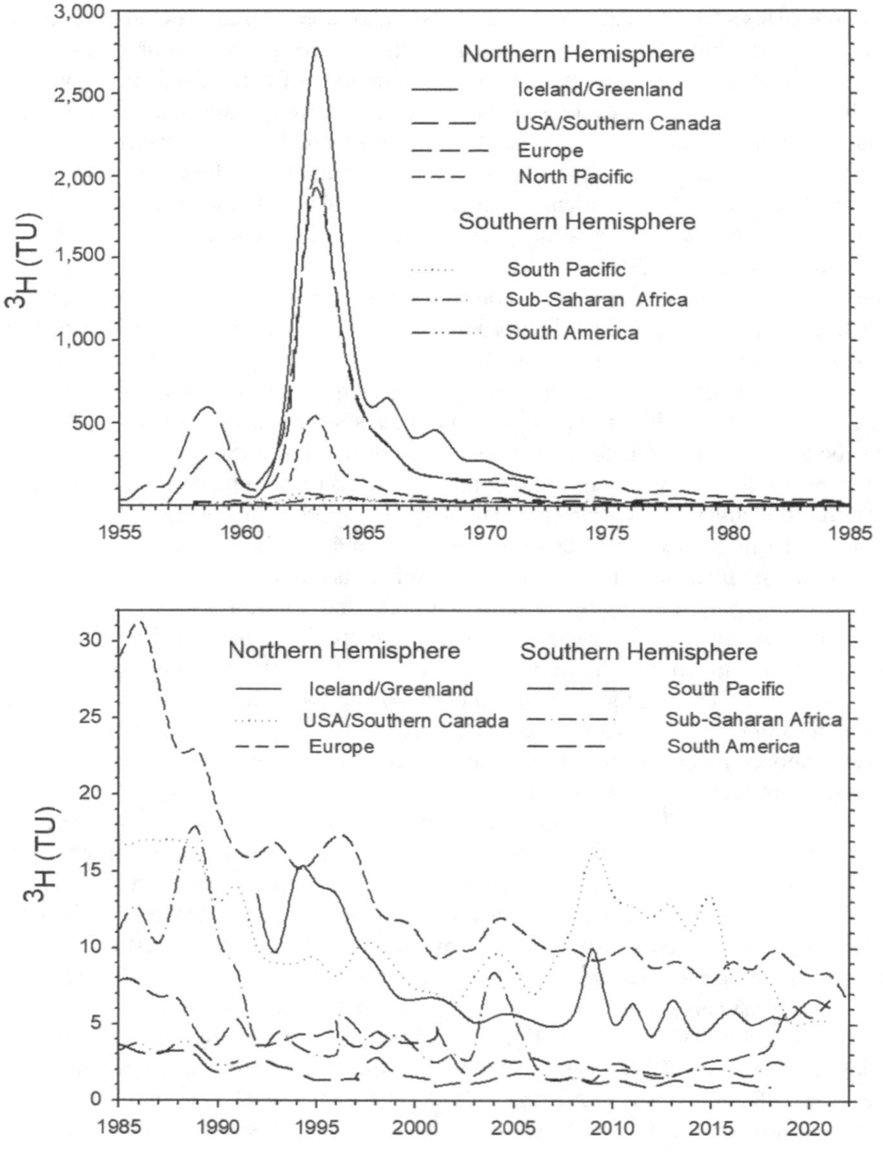

FIGURE 11.2 Average annual world-wide tritium concentrations in precipitation between 1955 and 2022 (Data from the International Atomic Energy Agency Global – Network Isotopes in Precipitation (IAEA GNIP, 2024).

Important Takeaway – Atmospheric tritium is produced by the interaction of cosmic rays and nitrogen gas. The tritium quickly oxidizes to water which falls to the earth as precipitation. Tritium in precipitation is reported in tritium units (TU). The natural average tritium concentration in precipitation is about 5–10 TU. The atmospheric weapons testing tritium spike has been exhausted since about 1995. Natural tritium concentration in precipitation is affected by the season, rainout, and mountain (elevation), and continental effects. Locally the tritium concentration in precipitation can vary as much as 25 TU. The results from short term and limited precipitation sampling campaigns should not be considered as representative of local precipitation averages and should only be used as anecdotal information.

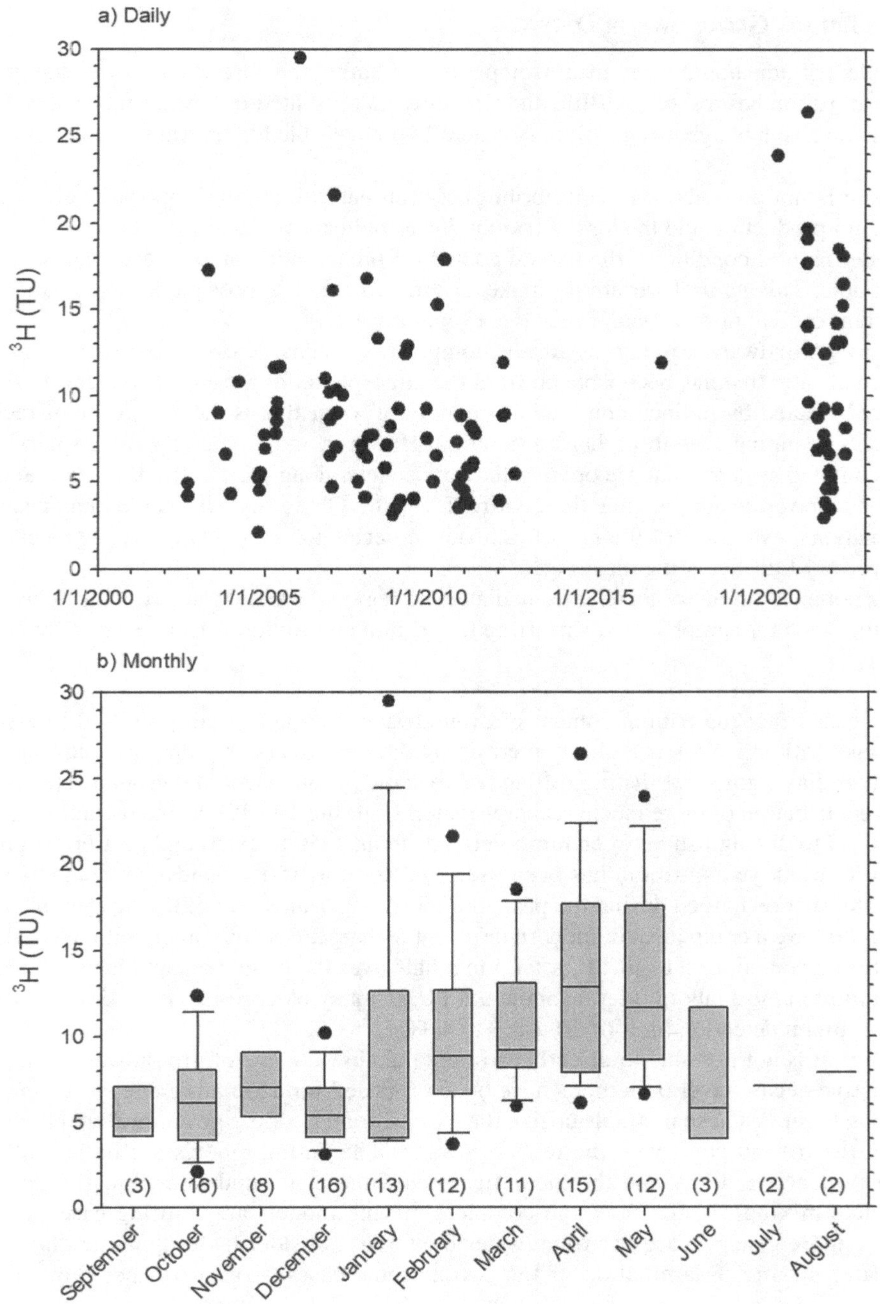

FIGURE 11.3 Daily (a) and monthly (b) tritium concentrations in Lindon-Provo, Utah, precipitation between 2004 and 2022. In the box plots the horizontal line is the median value, the filled boxes are the 50th percentile, the whiskers are the 25th and 75th percentile, and the filled circles are outliers. During a single year the concentrations varied by as much as 25 TU (a), and there is a strong season concentration trend (b). In the summer to mid spring the median concentration was 5–7 TU and was greater than 10 TU in April and May. The numbers in () are the number of samples. Unpublished data from Alan L. Mayo and David G. Tingey.

11.2.2 Tritium Groundwater Dating

In principle, tritium should be an ideal isotope for estimating the absolute age of young groundwater because tritium has a short half-life and is readily incorporated into water molecules. However, estimating an absolute age using tritium is generally not possible for several reasons:

1. There is not a steady-state equilibrium between natural and anthropogenic atmospheric tritium production and the loss of tritium due to radioactive decay.
2. Under natural conditions, the tritium content of precipitation at any particular location is variable. This natural variability makes it difficult if not impossible to predict the initial tritium content of a particular recharge event or events.
3. Many groundwater discharges are the composite of recharge over a period of years. Old groundwater that has been isolated from the atmosphere for most of its history and which may be dated by radiocarbon methods may have water that is the composite of recharge events spanning tens to perhaps a hundred. However, many young groundwaters are in unconfined systems that are open to the atmosphere along their entire flow path and may have received recharge along their entire flow path. Obviously, absolute tritium age determinations, by means of the law of radioactive decay, for such young groundwater would have very little scientific meaning.
4. It is not uncommon for local, intermediate, and regional groundwater systems to discharge at the same topographic low. Unsorting the tritium contributions from each source can be difficult.

Despite the fact that the tritium content of groundwater may not usually be used to estimate an absolute age, tritium is very useful for determining the relative age of young groundwater. Tritium has often used as a guide for distinguishing between old groundwater and groundwater containing at least a contribution of more recent recharge water. Until the mid-1990s, the tritium content could often be used to distinguish between pre-bomb (i.e. about 1950 to 1955) and post-bomb groundwater. In more recent years, tritium has been used to determine if a groundwater contains a proportion of water that recharged during the past 50 to 60 years. Water containing measurable tritium is considered to have a component of modern or recent recharge. Because precipitation has an average natural tritium content of 5 to 10 TU, after 4 to 5 half-lives the tritium content is usually below the detection limit in most laboratories, at about 0.3 TU. In many older research studies and papers, the tritium minimum detection limit (mdl) was 3 to 4 TU.

Although it is not possible to directly estimate the absolute age of groundwater using the law of radioactive decay, several methods have been proposed for estimating ages of young groundwater using tritium. These methods utilize the tritium content of the groundwater at the sampling point and the tritium content of the recharge water. The dating models are based on considerations of radioactive decay and the theoretical behaviors of groundwater flow (i.e., piston flow vs. complete mixing). Piston flow and complete mixing models are limiting cases and seldom occur. The more general case of groundwater flow must consider hydrodynamic dispersion and groundwater mixing. Examination of the two extreme cases of groundwater flow, piston and complete mixing, is instructive because they may be used to estimate the minimum and maximum groundwater age.

The tritium piston flow model assumes that recharge waters from specific recharge events follow each other sequentially along groundwater flow lines and that mixing of groundwater from different recharge events does not occur. In the piston flow model, the concentrations of radioactive isotopes at the discharge point may be expressed as:

$$A_t = A_o \exp(-x_o / VT_{1/2}) = A_o \exp(-tT_{1/2}) \tag{11.5}$$

where A_t is the concentration of the radioactive isotope at the sampling point located a distance, x_o, from the recharge zone, A_o is the concentration of the radioactive isotope at the recharge point, V is the rate of groundwater motion, t is the turnover or travel time of the groundwater, and $T_{1/2}$ is isotopic half-life. $T_{1/2} = \ln 2/T$, where T is the lifetime of an isotope (Nir, 1964). After algebraic transformation, the travel time of groundwater at the sampling point may be expressed as:

$$t = B^{-1}\left(\ln\frac{A_o}{A_t}\right) \quad (11.6)$$

where $B = T_{1/2}$. This model is generally considered to be inapplicable to groundwater systems because almost all groundwater has a mixed recharge origin. Additionally, the model assumes that A_o remains constant and is therefore not generally valid for tritium.

The complete mixing model assumes that recharge water instantaneously and completely mixes with existing groundwater and that the flux of groundwater flow (Q) within the reservoir remains constant. The concentration of the radioactive isotopes at the discharge point may be represented as:

$$A_o - A_t = \left(A_o - A_t\right)\exp(1 - \left(\frac{Q}{V}\right)t_{1/2} \quad (11.7)$$

where mean $t = \frac{Q}{V}$ and V is the volume of groundwater in the reservoir (Fontes, 1983). Any numerically consistent units of measurements may be used for Q, V and $t_{1/2}$. Dispersion models are often used to describe groundwater mixing of radioactive and non-radioactive tracers.

Important Takeaway – Although tritium dating models such as piston flow and complete mixing exist, it is seldom possible to calculate an absolute groundwater tritium age based on the law of radioactive decay. Nonetheless, the short tritium half-life (12.3 years) means that groundwater containing a component of tritium received recharge during the past 50 to 60 years.

11.3 ^{3}H/^{3}He DATING

The ^{3}H/^{3}He ratio is the basis for tritium/helium groundwater dating. The method was originally described by Tolstikhin and Kamensky (1969) and has subsequently been used by researchers to determine the age and recharge history of oceans, lakes, and groundwater (Bouchaou, 2008; Gil-Márquez et al., 2020; Manning et al., 2005; Poreda et al., 1988; Post et al., 2019; Shapiro et al., 1999; Solomon et al., 1993, 1998; Solomon and Sudicky, 1991). Groundwater can acquire helium from the exchange with the atmosphere, excess air, underground radiogenic decay, and the decay of groundwater ^{3}H into tritiogenic ^{3}He ($^3He_{tri}$). The relatively short half-life of tritium means that contributions of tritium from the mantle ($^3He_{rad}$) and the deep radiogenic decay of uranium and thorium are usually insignificant in ^{3}H/^{3}He dating investigations. The $^3He_{rad}$ concentration can be determined by an analysis of the radiogenic ^{4}He concentration. The atmosphere can also be a source of ^{3}He. The contribution of atmospheric helium ($^3He_{atm}$) in aqueous solutions can be determined by the neon (Ne) concentration. Determining the $^3He_{tri}$ in groundwater requires the analysis of the ^{3}H, ^{4}He, and Ne concentrations to sort out non-tritium sources (Gilmore et al., 2021; Solomon and Cook, 2000). After the contributions from non-tritiogenic sources have been identified, the ^{3}H/^{3}He groundwater age can be calculated:

$$r = \frac{1}{\gamma}\ln\left(1 + \frac{^3He_{tri}}{^3H}\right) \quad (11.8)$$

Where r is the age and is the tritium half-life (0.05626/yr). $^3H/^3He$ dating requires precise and accurate sampling and laboratory methods. Dating is most suitable for closed freshwater systems.

> *Important Takeaway – The $^3H/^3He$ ratio can be used to calculate the absolute age of some modern groundwaters that receive all of their ^3He from groundwater ^3H.*

11.4 RADIOCARBON

Radiocarbon is naturally produced in the atmosphere by the collision of secondary neutrons from cosmic ray spallation reactions with atmospheric nitrogen. The reaction may be represented as:

$$^{14}N + n = p + ^{14}C \tag{11.9}$$

where n is a neutron and p is a proton (Libby, 1955). Carbon-14 isotopes are not uniformly created in the polar, mid-latitude, and tropical regions of the atmosphere (Golubenko et al., 2022; Wu, 2017). Atmospheric carbon-14 is oxidized within a few hours to $^{14}CO_2$, which subsequently mixes with other CO_2 molecules in the troposphere and then enters the biosphere and hydrosphere. The latitudinal variation in the atmospheric neutron flux has a limited influence on the total troposphere ^{14}C activity and hence a limited effect on the concentration of ^{14}C in precipitation and organic matter. Unlike tritium, the north–south ^{14}C activity gradient and the natural atmospheric ^{14}C activity over a single hemisphere do not vary more than 3 to 5 percent from low to high latitudes or altitudes (Braziunas et al., 1995; Ferronsky and Polyakoc, 1982; Levin et al., 2010). The relative homogeneity of ^{14}C within the atmosphere of each hemisphere results from three factors:

1. ^{14}C in the troposphere mixes rapidly, in less than 10 years.
2. Most atmospheric ^{14}C is in the molecular form CO_2, which does not undergo substantial rainout, as is the case of ^3H, which is incorporated directly into water molecules.
3. The relatively long half-life of ^{14}C means that radioactive decay is not a factor. The natural troposphere ^{14}C content of the southern hemisphere is about 4 to 5 percent less than that of the northern hemisphere due to the large amount of CO_2 uptake from the southern hemisphere's vast ocean surface.

Loss of ^{14}C due to radioactive decay may be described by the reaction:

$$^{14}C = ^{14}N + e \tag{11.10}$$

Where e is a low-energy beta particle. Note that in redox chemistry e is often used to denote the transfer of electrons. Despite the fact that there are natural variations in atmospheric ^{14}C production, the transfer of atmospheric carbon to other carbon reservoirs, primarily the oceans, promotes the steady-state carbon composition of the atmosphere. Most of the atmospheric CO_2 is buffered by the oceans in a few tens of years, but a substantial portion remains in the atmosphere for centuries (Archer et al., 2009; Eby et al., 2009).

The carbon reservoirs of the hydrosphere, biosphere, and atmosphere contain less than 10 percent of the earth's total carbon but contain more than 97 percent of the earth's ^{14}C (live carbon). Almost all of the ^{14}C that has been deposited in the lithosphere has decayed away, and almost all lithospheric carbon is considered to be dead carbon (i.e., does not contain ^{14}C). The ^{14}C content in the hydrosphere is the results of the dynamic interaction between the ^{14}C in the atmosphere and biosphere with the dead carbon in the lithosphere. At some point in the distant past, all or almost all of the lithosphere's dead carbon began life as live carbon. The quantity of living organic matter has no effect on the ^{14}C content in the atmosphere because the concentration of ^{14}C in living mater is proportional to that in the atmosphere.

TABLE 11.1
Reporting ^{14}C Values

Value	Unit	Description
$\Delta^{14}C$	‰	Permil depletion or enrichment relative to the NBS 1 oxalic acid standard normalized for ^{13}C isotopic fractionation.
$\delta^{14}C$	‰	Permil depletion or enrichment relative to the NBS 1 oxalic acid standard not normalized for ^{13}C isotopic fractionation.
pM	%	Percent modern carbon relative to the NBS 1 oxalic acid standard normalized for ^{13}C isotopic fractionation.
pmc	%	Percent modern carbon relative to the NBS 1 oxalic acid standard not normalized for ^{13}C isotopic fractionation.
T	years	Radiocarbon age in years before 1950

$\Delta^{14}C$ and $\delta^{14}C$ are reported in permil because the differences between the sample and standard are so small.
Source: IAEA, 2013.

Carbon-14 concentrations are reported with a variety of laboratory reporting values (Table 11.1). The preferred unit in groundwater investigations is pmc, although $\delta^{14}C$ is sometimes used. Both pmc and $\delta^{14}C$ have not been corrected for ^{13}C fractionation. Laboratory carbon-14 analyses are routinely performed for archeological and paleo-botany dating, and these data are usually normalized in the laboratory for the samples' ^{13}C content. $\delta^{14}C$ is the ratio of $^{14}C/^{12}C$ in the sample relative to the $^{14}C/^{12}C$ in the standard. Normalized laboratory ^{14}C values ($\Delta^{14}C$) are corrected for the ^{13}C difference from $\delta^{13}C = -25$‰.

In groundwater studies the natural steady state content of atmospheric ^{14}C has been arbitrarily defined as 100 percent modern carbon (pmc), which is equal to 13.56 ± 0.07 disintegrations per minute per gram of carbon (dpm/g). One hundred percent modern carbon is believed to be the activity of ^{14}C in the atmosphere prior to the industrial revolution and has been arbitrarily set at about 1860 (Libby, 1955). Stuiver (1967) and Stuiver and Quay (1980) pointed out that the natural atmospheric production of ^{14}C may be affected by:

1. variations in the intensity of cosmic radiation due to fluctuations in solar activity,
2. variations in the magnetic flux of the earth's dipolar magnetic field, and
3. climatic changes, particularly during glacial periods.

Except for climate change, these factors are generally of minor importance in groundwater investigations because almost all unmixed groundwater is a mixture of recharge waters that fell as precipitation over a limited period of time (i.e., years to tens of years). During glacial periods the carbon content of the oceans decreased and the production rate of ^{14}C in the atmosphere varied. The production rate was sometimes the same as during modern times, but typically it was 25 percent greater than in modern times. About 40 ka it was as much as 1.7 times greater than the present (Beck et al., 2001; Cheng et al., 2018; Dinauer et al., 2020).

Since the industrial revolution, human activities have greatly altered the ^{14}C content in the atmosphere. Burning of fossil fuels, whose ^{14}C had completely decayed away prior to burning, decreased the $\Delta^{14}CO_2$ of the troposphere by nearly −25 ‰ (Figure 11.4a). In atmospheric science literature, the atmospheric ^{14}C content is often reported in $\Delta^{14}C$. Atmospheric dilution with dead carbon from the industrial revolution emissions is known as the Suess effect after the initial discoverer of the phenomena (Suess, 1955). Atmospheric weapons testing increased the $\Delta^{14}CO_2$ to nearly 800 ‰ by the mid-1960s (Figure 11.4b). This increase was rapidly reflected in the ^{14}C content

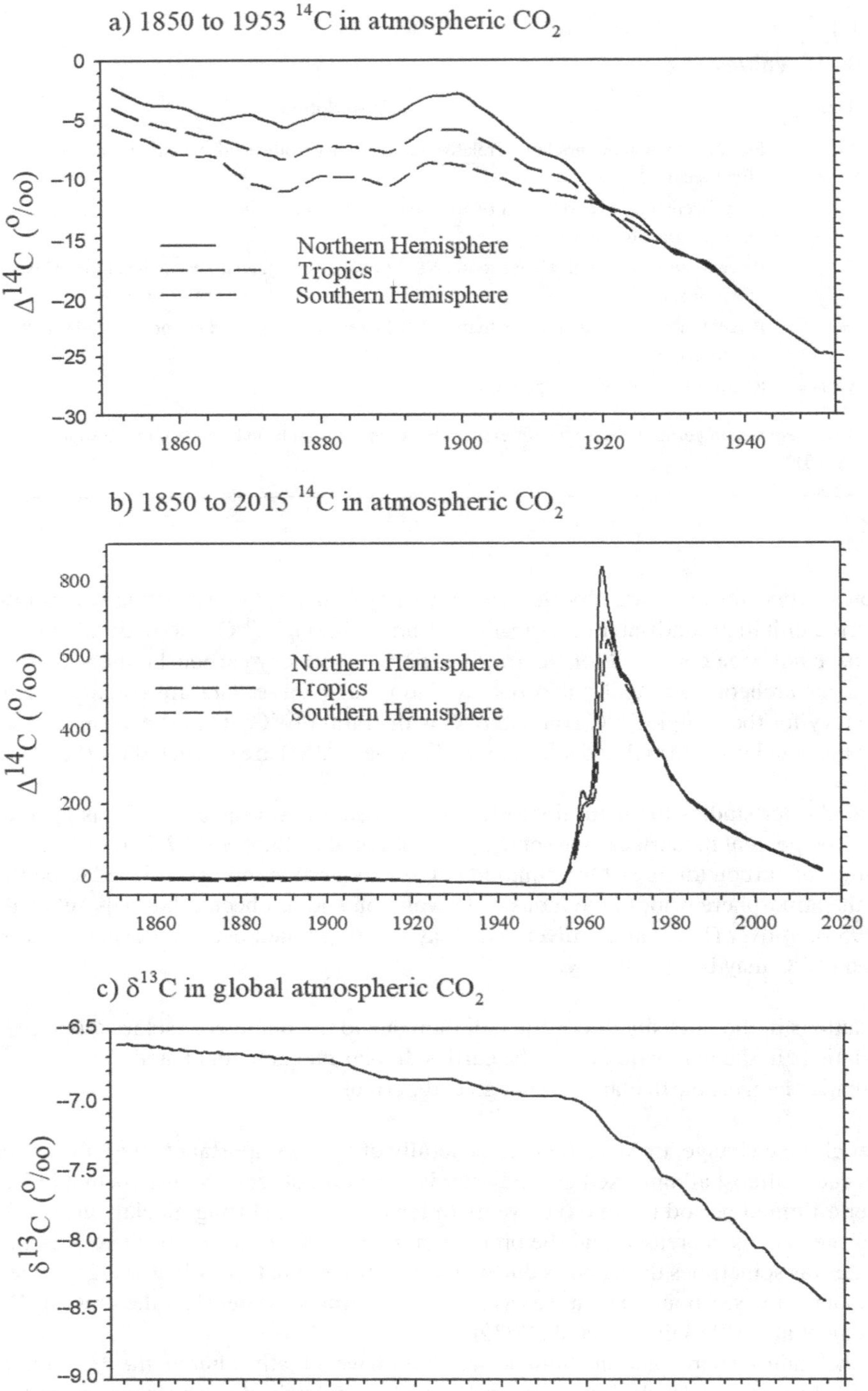

FIGURE 11.4 Annual mean ^{14}C values of industrial revolution and post-industrial revolution atmospheric forcing during a) 1850–1953, b) 1850–2015 ^{14}C, and c) 1850–2015 $\delta^{13}C$. $\Delta^{14}C$ is the ratio of $^{14}C/^{12}C$ in the sample relative to the $^{14}C/^{12}C$ atmospheric ratio prior to the industrial revolution. $\Delta^{14}C$ is reported in permil rather than percent. The northern hemisphere is defined as 30–90° N, the tropics are 30° N–30° S, and the southern hemisphere is 30–90° S. Data from Graven et al. (2017).

the hydrosphere and biosphere carbon reservoirs. Since the mid 1960 there has been a substantial decrease in atmospheric ^{14}C due to three factors:

1. the discontinuation of atmospheric weapons testing after the atmospheric test ban treaty,
2. absorption of ^{14}C by the terrestrial biosphere, and
3. most importantly, the CO_2 exchange between the atmosphere and the oceans. Approximately 80 percent of the deposition of atmospheric ^{14}C occurs in the oceans and only 20 percent is absorbed into the biosphere. Thus, the oceans are the major carbon sink, and only a small amount of atmospheric ^{14}C becomes available for groundwater recharge.

Important Takeaway – Carbon-14 is naturally created in the upper atmosphere by the collision of nitrogen molecules with cosmic rays. Carbon-14 is known as live carbon, and ^{12}C and ^{13}C are known as dead carbon. Prior to the industrial revolution, the concentration of natural atmospheric live carbon fluctuated by known amounts and is nominally considered to be 100 percent modern carbon (pmc). During the industrial revolution the atmosphere was flooded with dead CO_2, which suppress the atmospheric ^{14}C concentration to less than 100 pmc. This suppression is known as the Suess effect. Atmospheric weapons testing flooded the upper atmosphere with live carbon, and the concentration peaked at 600–800 pmc. Since about 1995, the bomb ^{14}C has been exhausted and the atmospheric ^{14}C now approaches 100 pmc.

11.4.1 Radiocarbon Dating

Th primary purpose of ^{14}C dating is to estimate the absolute age of the groundwater. Groundwater age can be used as stand-alone data and as a real-world check on the travel times predicted by groundwater flow models. When dating groundwater by radiocarbon methods, it is assumed that the natural production of atmospheric ^{14}C has been constant during the past 50,000 years. Although not rigorously true, this assumption is sufficiently valid for groundwater dating in light of the uncertainty in estimating other factors affecting the ^{14}C content of groundwater systems.

The principles behind radiocarbon dating are best illustrated by the dating of vegetal material. Plants obtain all of their carbon from atmospheric CO_2. Except for a minor enrichment due to fractionation, on the order of 2 pmc, the ^{14}C content of living plant tissue is the same as the atmosphere. When the plant dies, it is no longer resupplied with ^{14}C, and radioactive decay slowly reduces the amount of ^{14}C in the plant at the rate of about one-half every 5,730 ± 30 years. If the ^{14}C content of dead plant material is measured sometime later as 25 pmc (i.e. 3.337 dpm/g), the age of the plant material can be calculated using Equation 11.2:

$$t = \frac{5730 \ln\left(\frac{At}{Ao}\right)}{-0.693} \tag{11.11}$$

where (t) is the time it has taken to reduce the initial ^{14}C activity (A_o) to the measured ^{14}C level of the sample (A_t). In this example the age of our plant would have died about 11,460 years ago because the elapse of two complete half-lives is required to reduce the initial ^{14}C activity of 100 pmc to the final activity of 25 pmc. When the remaining activity (A_t) is 50, 25, 12.5 percent, Equation 11.2 is not necessary because the half-lives can simply be added. Except for a few long-lived plants, such as bristlecone pines (lifespan 5,000 years), sequoia trees (lifespan 3,500 years), and redwood trees (lifespan 2,000 years), the oldest and youngest part of the plant have essentially the same ^{14}C content. What this means is that for most plants the ^{14}C content at the time of the plant's death and a composite sample or any portion of the plant may be used for laboratory analysis.

It is apparent that the determination of the age of a sample requires knowledge or an estimate of its current ^{14}C activity (A_t) and its initial ^{14}C activity (A_o). For organic material, A_o is assumed to be ≈100 pmc, and A_t may be measured in the laboratory. The radiocarbon date is then corrected for systematic variations in atmospheric ^{14}C over the past 50,000 years. A calibration curve has been established for dating organic material by comparing the tree-ring ^{13}C of the wood of sequoia and bristlecone pines with their corresponding radiocarbon ages for the past 0–12,500 years. The ^{14}C calibration curve has been extended using marine records for 12,500 to 50,000 years ago (Lamarche and Harlan, 1973; Michael and Ralph, 1970; van der Plicht, 2007). For reasons discussed later it is important to tell the laboratory to not apply the ^{13}C calibration curve corrections to groundwater ^{14}C data unless the groundwater only acquired ^{14}C from organic sources and is organic carbon. Ask the laboratory to provide the analysis as pmc.

Radiocarbon dating of most groundwater is not as straightforward as dating a dead plant material. Recall that a plant has an A_o of approximately 100 pmc, corrected for systematic atmospheric ^{14}C variations, and that the measured A_t of dead material is less than A_o only because of radioactive decay. There are two principal reasons why the A_o of groundwater is not equal to the atmospheric ^{14}C activity of 100 pmc. First, as with dead plant tissue, some ^{14}C has undergone radioactive decay. Second and most importantly, almost all groundwater has acquired its carbon from mixed sources, where the live carbon has been diluted with dead carbon.

The crux of ^{14}C groundwater dating is estimating the A_o of the water at the time of recharge. In other words, ^{14}C groundwater dating requires sorting out the dilution of the live carbon with the dead carbon contributions in the sample. It is also important to bear in mind that in most groundwater systems a radiocarbon age is not an estimate of the actual age of the groundwater, as in plant material, but it is an estimate of the average age of the total dissolved inorganic carbon (TDIC) in the groundwater. When properly executed, the age estimate of the TDIC may be considered a reasonable surrogate for an estimate of absolute age of the groundwater sample.

At this point it is appropriate to elaborate on the meaning of radiocarbon groundwater age. Like the rings in a tree, where the ^{14}C content in successive rings for each year is determined and a composite sample of the tree yields the average age of the tree, the age of groundwater is also a composite number. A pail full of water that issues from a spring or well clearly did not fall as a discrete pail of water at the recharge zone a specific number of years and days ago. Rather, it is the integration of many precipitation events that have undergone further convective and dispersive mixing in the aquifer on its way to the sampling point. In a closed, porous aquifer, such as well-layered sandstone, the age calculated can be considered the mass-weighted mean residence time of the water represented in the sample. In water controlled by flow in fractures, the age is an even more difficult concept because it also incorporates the effect of flow paths having various flow lengths and flow times.

For the previously described reasons and for others that follow, it is not wholly appropriate to use the word "age" in relationship to groundwater ^{14}C. Better language could include mean radiocarbon residence time or estimated mean radiocarbon age. Nonetheless, the term "age" has historically and conventionally been applied to groundwater and remains in common usage.

As described in Chapter 9, groundwater may acquire carbon from several sources. The initial ^{14}C content of groundwater (Figure 11.5) is the result of the contributions of:

1. live carbon, that is, ^{14}C from precipitation and from the biogenic production of CO_2 in the soil zone,
2. dead carbon from the weathering of minerals in the soil zone and the dissolution of minerals in the aquifer,
3. carbon accompanying isotopic exchange, and
4. both live and dead carbon by other processes. In most groundwater systems these additional processes are of limited importance; however, they may be significant locally.

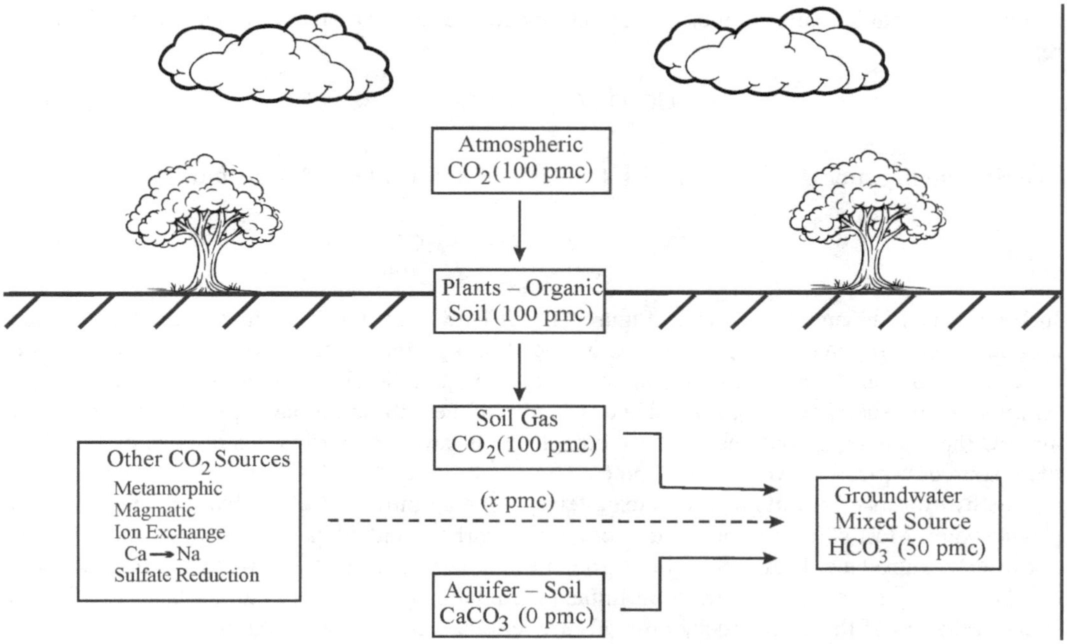

FIGURE 11.5 Diagram illustrating the acquisition of live and dead carbon in groundwater. For illustrative purposes, the atmosphere ^{14}C is considered to be 100 pmc. Most groundwater acquires about 50 percent of its carbon from soil gas and about 50 percent from the dissolution of soil and aquifer minerals. In a limited number of situations atmospheric CO_2 directly entering the groundwater system is the primary source of live carbon. In some situations other CO_2 sources can significantly impact the carbon content of groundwater

11.4.1.1 Live and Dead Carbon

Almost all groundwater ^{14}C comes from CO_2 produced by root respiration and by the decay of organic matter in the soil zone. Contributions of live carbon from the biogenic production of CO_2 in the soil typically had a ^{14}C content of ≈100 pmc prior to the industrial revolution. Since the industrial revolution, the soil zone ^{14}C activity first declined due to atmospheric dilution by dead carbon from the burning of fossil fuels, and then it increased dramatically due to massive contributions of live carbon from atmospheric weapons testing.

Atmospheric CO_2 also contributes directly to groundwater recharge as H_2CO_3 in precipitation. Under most conditions the contribution of live carbon from atmospheric CO_2 may be neglected because the partial pressure of atmospheric CO_2 (log $P_{CO_2} \approx 10^{-3.5}$ atm) is small compared to the P_{CO_2} of most groundwater (log $P_{CO_2} \approx 10^{-2}$ atm). However, in areas of little or no vegetation, particularly in arid regions with sandy soil, atmospheric CO_2 may be the major source of groundwater carbon. Recall that this carbon is considered to have a nominal ^{14}C activity of 100 pmc if it fell with precipitation prior to the industrial revolution. The activity of post-industrial revolution precipitation would be less than 100 pmc if it fell prior to atmospheric weapons testing and substantially greater than 100 pmc if it fell after the advent of atmospheric testing. Regardless of the A_o, the small bicarbonate content, due to the groundwater's low P_{CO_2}, is often an identifying trait in precipitation-dominated groundwater systems. As a rule of thumb, if the measured ^{14}C is 100 pmc or greater, the groundwater age is considered recent. In most cases, if the ^{14}C is > ~ 55 pmc, the water will turn out to be recent.

The most important weathering process in groundwater radiocarbon dating is the dissolution of carbonate minerals, although hydrolysis of silicate minerals may be local contributors. The oxidation of organic matter in the soil (CH_2O) to carbonic acid and the accompanying ^{14}C may be represented as:

$$\underset{(100\text{ pmc})}{CH_2O + O_2} \rightarrow \underset{(100\text{ pmc})}{HCO_3^-} + H^+ \tag{11.12}$$

The H^+ from Equation11.12 can then dissolve soil zone and aquifer carbonate minerals:

$$H^+ + \underset{(0\text{ pmc})}{CaCO_3} = Ca^{2+} + \underset{(0\text{ pmc})}{HCO_3^-} \tag{11.13}$$

In Equations 11.12 and 11.13 and in Figure 11.5, it is evident that in a simple carbonate system approximately 50 percent of the carbon introduced into groundwater is live carbon from the root zone and 50 percent is dead carbon from mineral dissolution. Soluble carbon in soil zone and aquifer minerals is usually considered dead because most of these minerals have existed long enough for any ^{14}C they may have contained to have completely decayed. Contributions of dead carbon from these sources typically have a ^{14}C of 0 pmc.

The hydrolysis of silicate minerals (e.g., feldspars) may introduce either live or dead carbon to groundwater. Most silicate minerals do not contain carbon, and silicate hydrolysis generally does not directly introduce dead carbon into groundwater. The congruent and incongruent dissolution of silicate minerals can, however, result in the precipitation of secondary carbon-bearing minerals in the soil zone. If these secondary minerals are less than 50,000 years old, they will contain live carbon. Subsequent dissolution of secondary carbonate minerals can either dilute or increase the ^{14}C activity groundwater in non-carbonate-bearing aquifers such as granitic terrains.

The general form of hydrolysis reactions and their ^{14}C may be represented by the weathering of a Ca-feldspar:

$$\underset{(\text{Ca-feldspar})}{CaAl_2Si_2O_8} + 6H_2O + \underset{(\text{live soil zone carbon})}{2H_2CO_3} = CaAl^{2+} + 2Al(OH)_3 + 2H_4SiO_4 + \underset{(\text{live carbon})}{2HCO_3^-} \tag{11.14}$$

The live carbon released (Equation 11.14) will either remain in ionic form and contribute to the overall A_o of the groundwater or will precipitate as a carbonate mineral in the soil zone. If weathering reactions such as represented by Equation 11.14 are the principal source of DTIC, the groundwater will have an A_o ^{14}C activity close to 100 pmc or greater than 100 pmc if the groundwater recharged during the post-atmospheric testing era. A low P_{CO_2} content and a more negative than anticipated $\delta^{13}C$ content may sometimes be used to identify such groundwater.

The live bicarbonate represented by Equation 11.14 may precipitate in the soil zone and later dissolve by percolating recharge water. If the bicarbonate precipitated more than about 50,000 years ago, the carbon released by subsequent dissolution reactions may be considered dead. However, the dissolution of soil zone caliche may contribute carbon whose ^{14}C activity is neither completely live nor dead.

11.4.1.2 Carbon Isotopic Exchange

Isotopic exchange processes do not alter the average dissolved carbon content of the aqueous CO_2 and H_2CO_3 or the concentration of the ionic species HCO_3^- and CO_3^{2-}, but they can affect the ^{14}C activity. Exchange occurs in both open and closed aqueous environments. In the unsaturated zone (open system), all of these aqueous species can undergo isotopic exchange with gaseous CO_2. Because enrichment of ^{14}C in isotopic exchange reactions is approximately two times greater than the enrichment of ^{13}C (Bigeleisen and Mayer, 1947; Craig, 1954), ^{13}C enrichment may be used as an analog for ^{14}C enrichment.

Isotopic enrichment of carbon species of interest in groundwater investigations are briefly discussed in Chapter 9, and the shorthand epsilon (ε) notation for several reactions are illustrated in Figure 9.3. For example, the equation for calculating the temperature-dependent ^{13}C enrichment factor for isotopic exchange between CO_2 and HCO_3^- (ε_2) can be represented as (Fontes, 1983):

$$\varepsilon_2=(^{13}CO_{2(g)} - \delta HCO_3^-) = (-9483/T + 23.89)‰ \tag{11.15}$$

where T is the temperature in Kelvin. Other ^{13}C enrichment factors of interest in ^{14}C dating are (Mook et al., 1974; Mook, 1980):

$$\varepsilon_1 = (^{13}CO_{2(g)} - {}^{13}CO_{2(aq)}) = (-373/T + 0.19)‰ \tag{11.16}$$

$$\varepsilon_3 = (^{13}CO_{2(g)} - {}^{13}CO_{2(aq)}) = (-9866/T + 0.78)‰ \tag{11.17}$$

$$\varepsilon_5 = (^{13}CO_{2(g)} - \delta CaCO_3) = (-5380/T + 9.15)‰ \tag{11.18}$$

$$\varepsilon_9 = (\delta CaCO_3 - \delta HCO_3^-) = (-4232/T + 15.1)‰ \tag{11.19}$$

More complex forms of equations for calculating some isotopic enrichment factors are in Appendix D. At isotopic equilibrium ^{14}C exchange between $CO_{2(g)}$ and HCO_3^- can be written as:

$$A_{o(HCO_3^-)} = A_{CO_{2(g)}} - \varepsilon^{14}C\left[CO_{2(g)} - HCO_3^-\right] = A_{CO_{2(g)}} - 0.2\varepsilon_2 \tag{11.20}$$

Activities are expressed in percent rather than permil. It is assumed that the $\varepsilon^{14}C$ % = 0.2ε ^{13}C ‰ (Fontes and Garnier, 1979). Assuming the A_{CO_2} was 100% at 25 °C, the ^{14}C activity of HCO_3^- in isotopic equilibrium with $CO_{2(g)}$ would be:

$$A_{o(HCO_3^-)} = 100\% - (-7.92 \times 2 \times 10^{-3}) = 101.6\% \tag{11.21}$$

Pre-bomb groundwater bicarbonate in isotopic equilibrium with atmospheric CO_2 is slightly enriched in ^{14}C due to isotopic fractionation.

Once the groundwater system becomes closed to soil gas, CO_2 carbon ion exchange can occur between the TDIC and aquifer carbonate minerals. At isotopic equilibrium, ^{14}C exchange between TDIC and solid carbonate can be written as:

$$A_o = \{A_{CO_{2(g)}} - \varepsilon^{14}C\left[CO_{2(g)} - HCO_3^-\right]\} + \{A_{CaCO_3} - \varepsilon^{14}C\left[CaCO_3 - HCO_3^-\right]A_{CaCO_3} - 0.2\varepsilon_9\} \tag{11.22}$$

In Equation 11.22 ε_9 is multiplied by 0.2 rather than 2 because ε_9 is in permil and A_o is in percent. Open system exchange with soil zone CO_2 (Equation 11.20) results in an increase in A_o, whereas as a closed-system exchange with carbonate minerals (Equation 11.22) results in a decrease in A_o because the ^{14}C of carbonate minerals is 0 pmc.

11.4.1.3 Other ^{14}C Factors

Although the TDIC, and hence the initial $A^{14}C$, of most groundwater is controlled in a predictable way, locally or even regionally, the carbon content of a groundwater can be greatly influenced by a host of other processes. Some of the less common yet locally important processes include contributions of both dead and live carbon from endogenic and exogenic CO_2. In general, this carbon has a $A^{14}C$ of 0 pmc. It is difficult to interpret radiocarbon data that has acquired such additional carbon because the quantity of dead carbon dilution cannot be readily measured or estimated. Additional carbon, commonly in the form of CO_2, can be introduced into groundwater from the anaerobic biochemical decomposition of organic material, sulfate reduction, coalification, volcanic activity, methane production, and the metamorphism of carbonate rocks. Methanogenesis is particularly problematic in groundwater systems that encounter organic-rich rocks such as coal, where coal bed CH_4 $^{14}C \sim 0$ pmc. On the other hand, the ^{14}C of landfill methane may be 100 pmc.

Ion exchange reactions such as $Ca^{2+} \rightarrow 2Na^{+}$ can also lead to the dilution of the $A^{14}C$, and it can alter groundwater $\delta^{18}O$. Removal of Ca^{2+} by ion exchange may cause the dissolution of additional dead carbonate minerals in order to maintain saturation with respect to calcite and dolomite.

Important Takeaway – Determination of the ^{14}C age of a groundwater sample requires knowledge or an estimate of its current 14C activity (A_t) and its initial 14C activity (A_o). Once A_o has been determined, the age of the groundwater can be calculated using the law of radioactive decay [$t = \left(\frac{5730}{ln2}\right)\left(ln\frac{A_t}{A_o}\right)$,] where A_t is the measured 14C activity of the groundwater. In most groundwater systems, A_o is about 50 pmc, where 50 percent of the initial carbon is live carbon from soil zone root respiration and the decay of organic matter and 50 percent is dead carbon from the dissolution of soil zone and aquifer carbonate minerals. Soil zones that do not contain primary carbonate minerals may only contribute live carbon to the A_o (100 pmc), but secondary soil zone carbonate minerals may contribute live or dead carbon. Other factors such as isotopic exchange, sulfate reduction, coalification, volcanic activity, methane production, and the metamorphism of carbonate rocks can affect the A_o and post A_o ^{14}C activity. Isotopic exchange is well understood, but other factors are more difficult to quantify.

11.4.2 Models for Estimating A_o

Estimating A_o is the nub of radiocarbon groundwater dating. From the previous discussion it is apparent that numerous factors may influence the A_o of a groundwater and that the A_t may be further diluted by subsequent geologic and biologic processes. Nevertheless, under good conditions and within the limits of laboratory analytical error, groundwater ages may be estimated by using the solute and isotopic chemistries of the groundwater and by applying correction procedures. Approximate groundwater ages may be determined for a set of groundwater samples with more certainty than absolute ages, and approximate ages may be used to evaluate groundwater flow directions and velocities. For this reason, it is sometimes useful to represent the calculated age with a ± value (e.g. 3500 ± 500 yrs) to indicate a level of uncertainty. There is no set procedure for determining uncertainty in calculated ages. Uncertainty can be based on the range of calculated ages using different soil gas $\delta^{13}C$ values, differences between the calculated results using different models, and on the laboratory-reported pmc range.

Models and methods that have been developed to calculate A_o have been summarized by Fontes and Garnier (1979) and more recently by Han and Plummer (2016). The methods include single-sample based models, statistically based models, and geochemical mass-balance models. Single-sample based models are most commonly used. Statistically based models can be used where all of the TDIC ^{14}C and $\delta^{13}C$ data in a groundwater investigation plot as a curved relationship (Gonfiantini and Gian, 2003; Han et al., 2014). Statistically based models assume the ^{14}C content is affected by geochemical processes along the flow path. The geochemical mass-balance approach adjusts one of the single-sample models for the hypothetical or observed chemical evolution along the flow path.

No model has universal applicability, and the application of ^{14}C dating spans the time of great climatic changes associated with the ending of Pleistocene glaciation, when there were significant perturbations in the atmospheric ^{14}C content. The hydrogeochemist should have an understanding of the geologic controls and the hydrogeochemistry of the groundwater system before attempting to estimate the age of a groundwater. The various models should not be applied systematically and indiscriminately to all groundwater-dating problems. Rather, each model is intended to describe a specific geochemical process under specified conditions and should be applied only when the situation being modeled is believed to be appropriate. For some circumstances, such as the contributions of external CO_2 from the metamorphism of carbonates or groundwater affected by coal bed CH_4, age estimates also require knowledge of the water's carbon content before and after the external contributions.

Estimating A_o requires the collection of water samples that are representative of the aquifer at the sampling location, laboratory analysis of A_t and $\delta^{13}C$ of the sample's carbon, determination of the soil and aquifer mineralogy and $\delta^{13}C$ content, estimation of the soil gas $\delta^{13}C$ at the time of recharge, and a determination as to whether the groundwater system is closed or open. In addition to this data, the groundwater sample should be analyzed for pH, major ions, and discharge temperature. The P_{CO_2} and HCO_3^- activity coefficients should be calculated. Determination of the δ^2H and $\delta^{18}O$ of the groundwater is also advised because these analyses can aid in the understanding of the origin and flow path of the groundwater. If young groundwater or possible mixing with young groundwater is anticipated, samples should also be analyzed for ^3H or the $^3He/^3H$ ratio.

Both the ^{14}C and $\delta^{13}C$ of the sample are easily determined by laboratory analysis at a reasonable cost. Other parameters such as the soil zone gas and mineral ^{14}C and $\delta^{13}C$ are usually not readily available, and the laboratory analysis of a single or even a few soil zone gas and mineral samples may not be representative. For example, in a single stratigraphic member of a regional carbonate aquifer at a scale of 10^{-2} to 10^5 m, Muller and Mayo (1986) found an average $\delta^{13}C$ of -1.85 ± 1.5 ‰, but with a variation of 6.28 ‰.

The δ_g content of the soil gas is dependent on the composition of the vegetative cover at the recharge area during the time of recharge. Large variations in $\delta^{13}C$ soil gas CO_2 values are observed between the three terrestrial plant groups (Chapter 9). When estimating the soil gas $\delta^{13}C$, consideration should be given to the possible climate conditions and the vegetation communities that may have been present at the time of recharge. In the absence of local data, good starting points are:

$$\delta^{13}C \text{ carbonate} = 0.0\ ‰$$

$$\delta^{13}C \text{ soil gas} = -18 \text{ to } -25\ ‰$$

$$^{14}C \text{ carbonate} = 0 \text{ pmc}$$

$$^{14}C \text{ soil gas} = 100 \text{ pmc}$$

Because many of the variables used in groundwater dating models must be estimated based on an analysis of the likely plant communities at the time of groundwater recharge and on the geologic recharge conditions, it is advisable to run each model with a range of estimated recharge values, particularly soil gas $\delta^{13}C$. By running several versions of each model, it is possible to understand the sensitivity of the calculated age to input variabilities and to better understand the ± value of the calculated results.

In the literature, mathematical model equations often contain their own symbolic representations, which can lead to confusion. A simplified set of symbolism has been adopted here, and it may differ from symbolism in other published sources.

Single-sample-based models, which include no $\delta^{13}C$ exchange correction, empirical approaches, simple mixing, simple mixing with isotopic exchange, and mixed open- and closed-systems, are mainly used in groundwater investigations. Single-sample modeling does not imply that only one sample should be collected and analyzed but rather that each water sample should be analyzed independently. Most of the models assume that the decrease in ^{14}C from A_o is only the result of radioactive decay. Single-sample models are in the form of mathematical equations that attempt to account for A_o contributions of dead and live carbon from various sources and for the effects of isotopic exchange.

It is important to note that most single-sample modeling focuses on geologic and hydrogeologic conditions in the recharge zone, as this is the area where groundwater typically acquires most, if not all, of its carbon.

If the TDIC is of pure biogenic origin, the measured A_o of a groundwater sample is usually not adjusted prior to application of the law of radioactive decay, and no A_o correction is required. In such systems the TDIC may be in equilibrium with soil zone CO_2 or may be completely transformed into HCO_3^-. No major correction is needed for the following conditions:

1. Open groundwater systems in equilibrium with soil zone CO_2. In an open groundwater system, the TDIC is usually in the form of HCO_3^- and H_2CO_3, whose respective molar concentrations, represented as a and b, respectively, are only controlled by the partial pressure of soil zone $CO_{2(g)}$. In this system the pH is usually low. At isotopic equilibrium the $\delta^{13}C$ and $A^{14}C$ of the TDIC may be represented by the equations (Fontes, 1983; Wigley, 1975; Deines et al., 1974):

$$\delta_{TDIC} = a\left(\delta_{CO_{2(g)}} - \varepsilon_2\right) + b\left(\delta_{CO_{2(g)}} - \varepsilon_9\right) \tag{11.23}$$

   and

$$A_{o(TDIC)} = a\left(A_{CO_{2(g)}} - 0.2\varepsilon_2\right) + b\left(A_{CO_{2(g)}} - 0.2\varepsilon_9\right) \tag{11.24}$$

   For example, at the conditions $T = 25°C$, pH = 7, log $P_{CO_2} = 10^{-2}$ atm, soil gas $\delta^{13}C$ (δ_g) = −25 ‰, and the soil gas ^{14}C (A_g) = 100 pmc, the calculated δ_{TDIC} is −17 ‰ and the calculated $A_{o(TDI)}$ is 101.6 pmc.

   The H_2CO_3 contribution is negligible because the pH is 7. At other pH values, the parameters a and b would have to be fractionally apportioned. This method has been applied to groundwater flow in fractured lava flows (Hufen et al., 1974), and it is applicable to some fractured carbonate aquifers where there is a high production of soil zone CO_2.
2. Complete transformation of soil CO_2 into TDIC. In non-carbonate aquifers, the soil zone CO_2 can be transformed into bicarbonate ions during the weathering of silicate minerals by hydrolysis. Such TDIC would have a $\delta^{13}C$ and A_o equal to the carbon in the soil gas.

The empirical approach (Vogel and Ehhalt, 1963) assumes:

$$A_o = B \tag{11.25}$$

where B is an arbitrary value for pmc such as 85 for non-carbonate soil and 50 for carbonate soil. The model has no theoretical basis but can be useful when scanning a ^{14}C dataset and making a first-order approximation during the early stages of data analysis.

Simple mixing models assume that A_o is the result of the mixing of carbon from two reservoirs, such as the soil zone and carbonate minerals in the soil zone. The two most commonly used mixing models are the Tamers and Pearson. The Tamers model (Tamers, 1967, 1975) was first proposed by Vogel and Ehhalt (1963) and is the starting point for several other models. In the model, half of the bicarbonate comes from soil gas CO_2 and half comes from dissolution of carbonate minerals in the soil zone and aquifer. A_o may be stoichiometrically expressed as:

$$A_o = \frac{mCO_2 + 0.5\,mHCO_3^-}{mCO_2 + mHCO_3^-}\left(A_g\right) \tag{11.26}$$

where m is the molal concentration and A_g is the $A^{14}C$ of the soil gas, usually assumed to be 100 pmc. In the more complex models, which account for isotopic fraction and which are described later, the Tamers formula will be expressed in the form:

$$A_o = \frac{a + 0.5b}{a + b} A_g \tag{11.27}$$

If the $A^{14}C$ of the mineral is not approximately zero, as in the case of Holocene caliche, or soil with recent carbonate precipitates, the expression becomes:

$$A_o = \frac{(a + 0.5b)A_g + (0.5b)A_m}{a + b} \tag{11.28}$$

where A_m is the $A^{14}C$ of the solid mineral phase, a is the concentration of $CO_{2\ (aq)}$, and b is the concentration of HCO_3^-.

The Tamers model does not take into account isotopic exchange, external sources of CO_2, incongruent dissolution of dolomite, or other processes. It assumes that all the CO_2 is from the soil zone and that all bicarbonate is from the dissolution of carbonate minerals.

The simple isotope mixing model, commonly referred to as the Pearson model (Ingerson and Pearson, 1964; Pearson and Hanshaw, 1970; Pearson et al., 1972; Pearson and Swarzenki, 1974), calculates A_o on the basis of the $\delta^{13}C$ content of each species. The Pearson model is based on the fact that:

1. ^{14}C and $\delta^{13}C$ behave chemically and isotopically identically, except that ^{14}C is subject to radioactive decay, and
2. a factor of 0.2 relates their isotopic fractionation factors between $\delta^{13}C$ in permil and ^{14}C in pmc. What this means is the $\delta^{13}C$ can be treated as analogous to non-decaying ^{14}C.

The Pearson model is illustrated in Figure 11.6, where the variables have been assigned the values:

δ_m (mineral carbonate), measured = 0 ‰ (PDB)

A_m (mineral carbonate), estimated = 0 pmc

δ_g (soil gas), estimated = –20 ‰ (PDB)

A_g (soil gas), estimated = 100 pmc

δ_s (water sample), measured = –10 ‰ (PDB)

The mineral carbonate in the soil zone and aquifer have been assigned values of δ_m = 0 ‰ and A_m = 0 pmc, which are typical of the vast majority of old (>50,000 years) carbonate rocks of marine origin.

Because the groundwater in Figure 11.6 acquires carbon from soil gas CO_2 and carbonate minerals, the A_0 of the groundwater must lie somewhere along the mixing line between the two end members labeled A and B. If, as in the example, the δ_s of the water sample is determined by laboratory analysis to be –10 ‰, and it is assumed that this value represent both the initial and final δ_g content of the water (i.e., there is no carbon exchange), then the initial $A^{14}C$ (A_o) content may be projected to 50 pmc. The estimated A_o is then used with the measured $A^{14}C$ (A_t) of the water sample to estimate the groundwater age by means of the law of radioactive decay.

Rather than graphically solve for A_o, the following equation, based on similar triangles, is used:

$$A_0 = \frac{\delta_s - \delta_m}{\delta_g - \delta_m}(A_g - A_m) + A_m \tag{11.29}$$

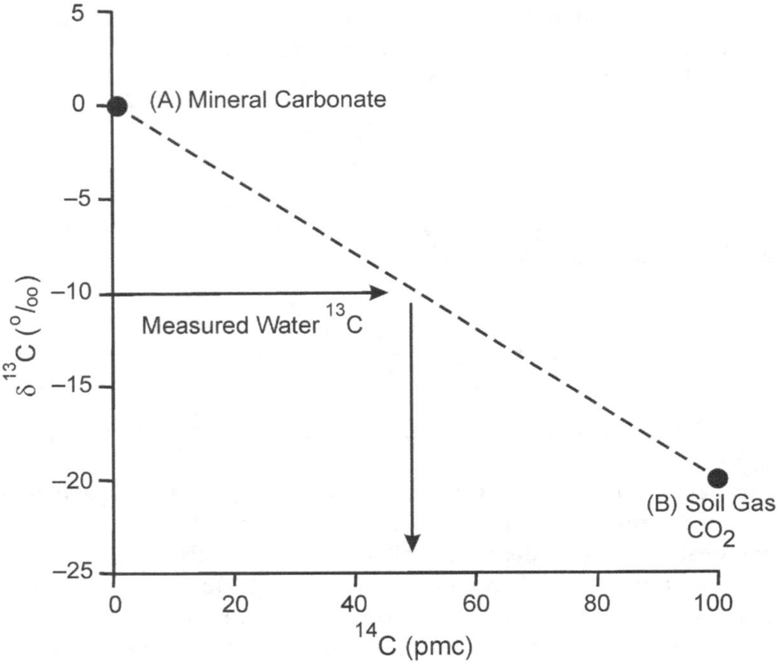

FIGURE 11.6 Graphical representation of the Pearson simple isotope mixing^{14}C dating model with typically assumed $\delta^{13}C$ and ^{14}C values for the mineral carbonate and soil zone $CO_{2\,(g)}$. The measured TDIC $\delta^{13}C$ is −10 ‰ when the assumed values are (A) carbonate mineral $\delta^{13}C = 0$ ‰, and $^{14}C = 0$ pmc, and (B) soil gas $\delta^{13}C = -20$ ‰ and $^{14}C = 100$ pmc. Projecting the measured TDIC $\delta^{13}C$ to the mineral carbonate-soil gas CO_2 mixing line yields a ^{14}C A_o of 50 pmc.

The Pearson model may be applied to groundwater that reacts with soil zone CO_2 and then with carbonate minerals when the system becomes closed. The model will account for a small amount of isotopic exchange because exchange can be considered mixing. However, the model cannot be used when isotopic exchange is extensive.

The Gonfiantini model, a modified version of the Pearson model (Gonfiantini, 1972), may be used for open systems to account for complete isotopic exchange between soil zone CO_2 and bicarbonate ions. The model assumes that the TDIC is HCO_3^- and the pH is around 7. The model may be mathematically represented as:

$$A_o = \frac{\delta_s - \delta_m}{\delta_g - \varepsilon_2 - \delta_m}\left(A_g - A_m\right) + A_m \tag{11.30}$$

The model tends to increase the A_o value when the system is closed to CO_2 to account for the dissolution of carbonate minerals. Under tropical climatic conditions, the denominator can approach zero, and thus A_o will approach infinity. This model is most useful when solute water chemistry data are not available.

The Mook model (Mook, 1967, 1972, 1976, 1980) considers simple mixing, isotopic exchange, and chemical equilibrium for all reactions involving carbon-bearing compounds (gaseous and aqueous CO_2, carbonic acid, bicarbonate, carbonate, and solid carbonates). The idea behind this approach is to calculate the isotopic mass balance at each step in the chemical processes involving carbon (i.e., dissolution in the unsaturated zone, isotopic exchange in the unsaturated zone, and re-equilibrium between H_2CO_3 and HCO_3^- in the saturated zone). In the simplest terms, the Mook model consists of the Tamers model plus a complex corrective term (*K*) for isotopic exchange:

$$A_o = \left[\frac{\overbrace{(a+0.5b)+0.5bA_m+0.5(A_g-A_m)}^{\text{(Tamers formula)}}}{d}\right]\times K \quad (11.31)$$

where *K* is the isotopic exchange corrective term:

$$K = \left[\frac{\delta_s-(a+0.5b)\delta_g-0.5b\delta_{ms}}{0.5(\delta_g-\delta_{ms})-\epsilon_2}\right]/d \quad (11.32)$$

where *d* is the TDIC. The Mook model does not take into account possible isotopic exchange between dissolved carbon in the groundwater and the carbon in the aquifer skeleton. The corrective term is also very sensitive to variations in δ_g, ε_2, and numerical instabilities. Best results are usually obtained in semiarid and tropical regions where the soil gases have more positive δ_g values.

Other mixing models with isotopic exchange models include the Wigley model (Wigley, 1975), the Eichinger model (Eichinger, 1983), and the Evans model (Evans et al., 1978). The Wigley model provides for closed-system carbonate exchange during incongruent dissolution. The Evans model is similar to the Wigley model in that it corrects for incongruent dissolution exchange. The Eichinger model is the Tamers model with a correction for TDIC exchange with carbonate minerals.

The widely used Fontes model (Fontes and Garnier, 1979), often referred to as the global exchange-mixing model, is applicable to closed systems. The model is a modified version of the Tamers model that considers isotope exchange accompanying closed-system mixing of carbon from $CO_{2\,(g)}$ and solid mineral carbonate. Partial isotopic exchange (nonequilibrium exchange) is treated as complete exchange. Like the Mook model, the Fontes model contains the Tamers formula plus a corrective term (K′) for isotopic exchange:

$$A_o = \left(\frac{1-0.5d}{\text{TDIC}}\right)A_g + \left(\frac{0.5d}{\text{TDIC}}\right)A_m + K' \quad (11.33)$$

where K′ is the isotopic exchange corrective term:

$$K' - \left(A_g - 0.2\varepsilon_5 - A_m\right)\left(\frac{\delta_s-(d/0.5d)\delta_m-1-(d/0.5d)]\delta_g}{\delta_g-\varepsilon_5-\delta_m}\right) \quad (11.34)$$

The Fontes and the Pearson models are universally accepted as standard models for ^{14}C age determinations of carbonate rocks. Both models may be programed on spreadsheets such as Excel and are included as part of some commercially available geochemistry software.

Han and Plummer (2013) noted that the Fontes model tends to underestimate ages for groundwater that have undergone carbon exchange. Like the Fontes model, the Han and Plummer model is a modified version of the Tamers model:

$$A_o = T + \left({}^{14}C_x - {}^{14}C_i - 0.2\varepsilon_{2\text{or}9}\right)\times\frac{\delta^{13}C_{\text{TDIC}} - \frac{C_a}{C_t}\delta^{13}C_{\text{ai}} - \frac{C_b}{C_T}\delta^{13}C_i}{\delta^{13}C_x - \delta^{13}C_i - \varepsilon_{2\text{or}9}} \quad (11.35)$$

where T = Tamers, C_x = the concentration of CO_2 or HCO_3^-, C_a is the concentration of $CO_{2\,(aq)}$, C_b is the concentration of HCO_3^-, C_T is the TDIC, and $\delta^{13}C_{ai}$ is the isotopic composition of dissolved soil carbon in equilibrium with soil gas CO_2, and $\delta^{13}C_i$ is the solid carbon isotopic composition.

Mixed open and closed-system models consider changes in the ^{14}C accompany groundwater movement along the flow path as the system transitions from open to closed.

After acquiring carbon in the open recharge zone, the carbon can experience additional evolution along the flow path after the system becomes closed. Gonfiantini (1972) and Salem et al. (1980) developed a model that accounts for both open and closed system evolution. The model, known as the IAEA model, can be represented as:

$$A_o = \frac{^{14}C_{\text{soil gas)}} - 0.2\varepsilon_2}{\delta^{13}C_{\text{soil gas}} - \varepsilon_2} \delta^{13}C_{\text{TDIC}} \tag{11.36}$$

The IAEA model can only be used where unsaturated carbon exchange has reached equilibrium.

Oescheger (Geyh, 2000) developed a model for the addition of dead carbon after the system became closed. The model only considers the case where additional dead carbon is added and does not include carbon exchange:

$$A_o = \frac{\text{TDIC}_i}{\text{TDIC}} \times {}^{14}C_i \tag{11.37}$$

where the subscript i is the initial condition prior to the closed system acquisition of dead carbon. $^{14}C_i$ may be considered to be the Tamers solution. The additional dead carbon may be from magmatic CO_2, oxidation of organic matter, and dissolution of carbonate accompanying Ca–Na ion exchange. Both the IAEA and Oeschger models can be quite useful for specific conditions.

The hydrogeologist is often concerned with the direction and velocity of groundwater flow and may be less concerned with the absolute age of the groundwater. In this case, knowledge of the relative groundwater ages may be sufficient and may often be attained with greater ease, accuracy, and confidence than can absolute ages. Determination of relative ^{14}C age is based on the assumptions that the recharge area was covered with a homogeneous vegetation cover and that the groundwaters of interest underwent similar chemical dissolution and isotopic fractionation processes. The relative age (θT) of the sample $k + 1$ relative to the upgradient sample k may be determined by the equation:

$$\theta T = \frac{-8270 \ln A_s(k+1)}{A_s(k)} \tag{11.38}$$

If the initial carbon content of the groundwater has subsequently been increased by dissolution limestone in the aquifer, the relative age may be calculated (Pearson and Hanshaw, 1970):

$$\theta T = (-8270)\ln\frac{A_s(k+1)}{A_a(k)} - \frac{d(k+1}{d(k)} \tag{11.39}$$

Model selection is extremely important when attempting to date groundwater. There is no correct model for all conditions, and there is no silver bullet for any system. Before selecting a groundwater ^{14}C age model(s), the likely geologic and biologic conditions of the recharge area at the time of recharge, the flow path conditions (i.e., mineralogy, closed vs. open system, groundwater mixing, etc.), potential external carbon inputs, the P_{CO_2} of the groundwater, the HCO_3^- $\delta^{13}C$, and the ^{14}C pmc should be evaluated for possible clues regarding the groundwater system and soil zone–recharge area–aquifer conditions. Comparing the calculated ages with the results of more than one model is usually advisable. The following is a general guideline for model selection. No correction is needed for:

1. non-carbonate rock, high porosity, high P_{CO_2} soil, and low pH systems,
2. non-carbonate open systems with respect to CO_2,
3. carbonate rock, fractured, active soil with high P_{CO_2}, and basic pH,
4. open system with respect to CO_2; non-carbonate rock, non-carbonate soil, and low porosity media (large clay fraction) with slightly acid to basic pH, and
5. systems where there has been a complete transformation of CO_2 into dissolved carbon.

Models suitable for non-carbonate rock, slightly carbonated soil, systems with weak biologic activity, and low porosity media (i.e., intertropical dry to semi-arid zones) and closed systems with no or little isotopic exchange include Tamers (1967), Pearson (Pearson and Hanshaw, 1970), Gonfiantini (1972), Fontes and Garnier (1979), Mook (1972, 1976, 1980), and Han and Plummer (2016).

Aquifers with carbonate soils, low porosity media, high biologic activity and CO_2 (temperate zone), and closed systems with partial isotopic exchange of CO_2 can be evaluated with Pearson (Pearson and Hanshaw, 1970), Fontes and Garnier (1979), Mook (1972, 1976, 1980), Han and Plummer (2016) models. Terrains with carbonate rocks, carbonate soil, low porosity media, high or low biologic activity, and closed systems with partial exchange with solid carbonate can be modeled with Pearson and Hanshaw (1970), Fontes and Garnier (1979) and Han and Plummer (2016) models.

11.4.3 A_o Numerical Instabilities and Model Sensitivity

Models used for calculating A_o become numerically unstable under certain conditions. Numerical instability may lead to great errors in age estimates, and care must be exercised when evaluating the results of numerical calculations to ensure that instabilities are identified. Identification of numerical instabilities is often difficult when computer programs are used for calculations. The problem of identifying instabilities may become acute when only one model or one input variable, such as a single value for soil zone $\delta^{13}C$, is used. Comparison of computational results from two or more models using the same input data is often helpful. Numerical instability can occur when the denominator of an equation approaches zero, when the numerator of certain terms approaches zero, or when input data may have large natural variations.

Except for the Tamers formula, the denominator of the equations or correction factors can approach or even become zero. The denominator approaches zero when the value of the $\delta^{13}C$ of the soil gas CO_2 $\delta^{13}C$ (δ_g) approaches the $\delta^{13}C$ of the carbonate minerals in the soil zone or in the aquifer (δ_m). The models are more stable for temperate climate recharge conditions than they are for semiarid or tropical conditions. Division by zero occurs when:

$$0 = \delta_g - \delta_m \text{ (Pearson model)}$$

$$0 = \delta_g - \varepsilon_2 - \delta m \text{ (Gonfiantini model)}$$

$$0 = (\delta_g - \delta_{ms}) - \varepsilon_2 \text{ (Mook model)}$$

$$0 = (\delta_g - \varepsilon_5 - \delta_m) \text{ (Fontes model)}$$

It is generally assumed that old marine carbonates and hence soil carbonate minerals derived from marine carbonates have a δ_m of approximately 0 ‰ (Fontes, 1983). $\delta^{13}C$ variation in the mineral carbonate can cause appreciable uncertainty in the calculated A_o in models that use the δ_m content of the aquifer.

The sensitivity of groundwater dating models to $\delta^{13}C$ variations in marine carbonate minerals were analyzed by Muller and Mayo (1986). $\delta^{13}C$ variations in soil gas CO_2 are graphically illustrated with the Pearson model in Figures 11.7 and 11.8. It is important to note that each model reacts differently to changes in the variables used in sensitivity analyses and that over-generalizing the results of the Pearson example should be avoided. In the examples, input parameters have been selected to represent intermediate (Figure 11.7) and semiarid (Figure 11.8) recharge climates, typical groundwater TDIC $\delta^{13}C$ contents, and the maximum and the minimum mineral carbonate $\delta^{13}C$ content (δ_m). Although the numerical interactions due to changes in input parameters are complex, some generalizations may be made.

By setting the TDIC $\delta^{13}C$ (δ_s) at −10 ‰, it is apparent that variations in δ_m have a slightly greater overall effect on the predicted intermediate climate A_o than they do on the semiarid climate A_o. The predicted range of A_o varied 17 pmc (49 to 66 pmc) for the intermediate climate (Figure 11.7), whereas the predicted range of A_o only varied 12 pmc (76 to 88 pmc) for the semiarid climate (Figure 11.8). The intermediate climate results are more affected by δ_m variations than are semiarid climate results because the slope of the mixing line is steepest for intermediate climate. What is more significant than the range of A_o values is the fact that average A_o of the intermediate climate is 25 pmc smaller than the semiarid climate.

Changing the TDIC $\delta^{13}C$ content (δ_s) from −10 to −5 ‰ has an even greater impact on predicted A_o values than do variations in the δ_m content (Figures 11.7 and 11.8). When combined with natural variations in the mineral carbonate, potential variations in δ_m can have a profound effect on the calculated value of A_o. For example, in both the intermediate and semiarid climate examples, a TDIC

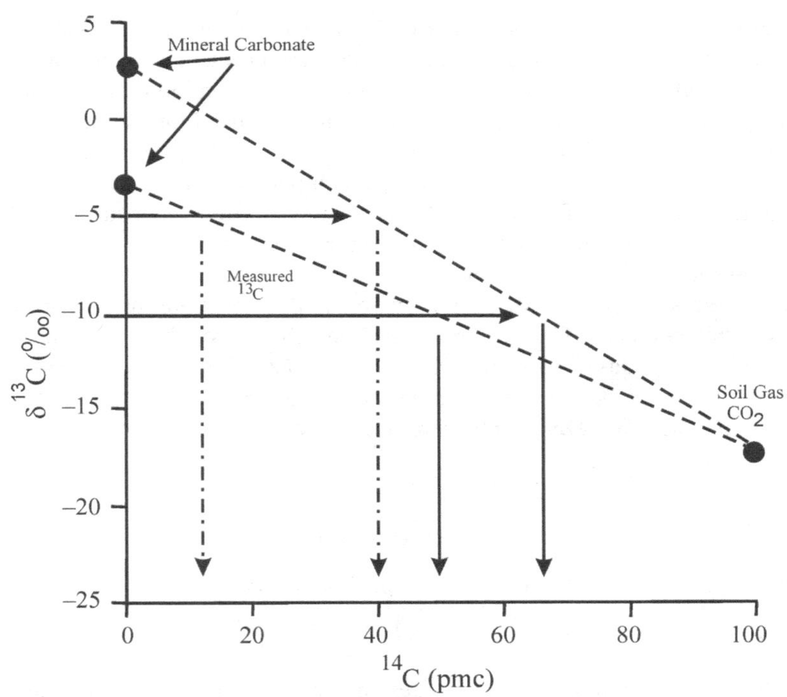

FIGURE 11.7 Graphical representation of the Pearson simple isotope mixing ^{14}C dating model for an intermediate climate with a range of mineral carbonate $\delta^{13}C$ values. The $\delta^{13}C$ values of the mineral carbonate are based on findings of Muller and Mayo (1986). Measured TDIC $\delta^{13}C$ is set at −10 and −5 ‰. Assumed values are carbonate mineral $\delta^{13}C$ = 2.5 and −4 ‰, and the ^{14}C = 0 pmc. Assumed soil gas $\delta^{13}C$ = −17.5 ‰ and ^{14}C = 100 pmc. Projecting the measured TDIC $\delta^{13}C$ of −5 and −10 ‰ to the mineral carbonate-soil gas CO_2 mixing lines yields a variety of ^{14}C A_o ranging from modern to paleo-groundwater recharge.

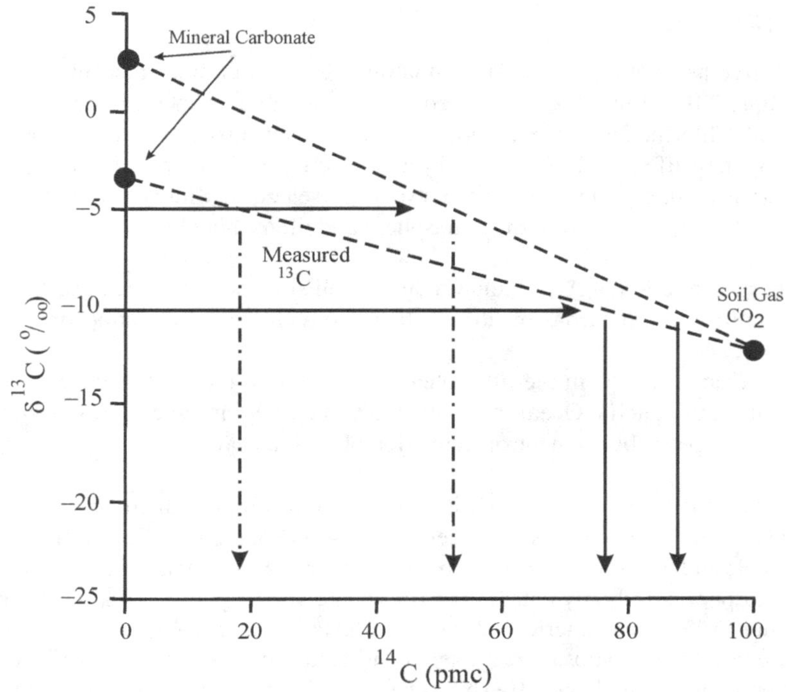

FIGURE 11.8 Graphical representation of the Pearson simple isotope mixing ^{14}C dating model for a semiarid climate with a range of mineral carbonate $\delta^{13}C$ values. The $\delta^{13}C$ values of the mineral carbonate are based on findings of Muller and Mayo (1986). Measured TDIC $\delta^{13}C$ is set at −10 and −5 ‰. Assumed values are carbonate mineral $\delta^{13}C = 2.5$ and −4 ‰, and the $^{14}C = 0$ pmc. Assumed soil gas $\delta^{13}C = -12.5$ ‰ and $^{14}C = 100$ pmc. Projecting the measured TDIC $\delta^{13}C$ of −5 and −10 ‰ to the mineral carbonate-soil gas CO_2 mixing lines yields a variety of ^{14}C A_o ranging from modern to paleo-groundwater recharge.

$\delta^{13}C$ of −10 ‰ yields A_o that is 50 pmc or greater, which suggest modern recharge water. When the TDIC $\delta^{13}C$ is −5 ‰, the A_o is less than 50 pmc, which suggests paleo-groundwater recharge. Fortunately, δ_s can be measured with considerable accuracy. However, unrecognized contributions of carbon from methanogenesis, ion-exchange, metamorphic $CO_{2\,(g)}$, etc., can greatly alter the stable isotopic composition of the TDIC. Post recharge carbon contributions can profoundly affect predicted A_o values and can lead to grossly incorrect estimates of groundwater age.

Figures 11.7 and 11.8 illustrate how sensitive models can be to assumed carbonate and soil gas values, and in the case of the Pearson model how less negative soil gas $\delta^{13}C$ results in progressively larger calculated A_o values and younger ^{14}C age determinations. For many groundwater age estimations, it is apparent that the error values assigned to a calculated radiocarbon age should reflect uncertainties in the δ_m as well as the uncertainty in the laboratory-determined A_s.

Important Takeaway – Evaluating groundwater ages may be performed using singe-sample-based models, statistically based models, and geochemical mass-balance models. Most analyses use a single-sample-based model. Models have been developed for simple mixing of live and dead carbon and models that also incorporate isotopic fractionation. The most commonly used models are the Pearson simple mixing and a modified version of the Fontes and Garner mixing with ion exchange model. These models are included in several groundwater geochemical analysis programs and may easily be used in spreadsheet programs. All of the models rely on the measured ^{14}C and $\delta^{13}C$ of the groundwater and on estimates of other parameters including the ^{14}C and $\delta^{13}C$ of soil gas and aquifer minerals, particularly carbonate minerals. Model results are sensitive to estimated parameters.

11.5 CHLORINE-36

Chlorine-36 is used as a natural tracer for estimating the age of groundwater that recharged 100 ka to 1 Ma (Phillips, 2013). Using ^{36}Cl, it is theoretically possible to date groundwater as old as 2.5 million years old. Chlorine-36 is infrequently used because few groundwaters are older than 100 ka. Chlorine-36 has a half-life of 301,000 ± 4,000 years (Bentley et al., 1986). The chloride ion is one of the most conservative anions in water due to its very high electron affinity and solubility properties. Chlorine-36 is naturally produced in the atmosphere and lithosphere by:

1. atmospheric spallation of Ar, K, and Ca due to collisions with cosmic rays,
2. the slow collision of neutrons (neutron activation) with ^{36}Ar in the atmosphere and upper lithosphere, and
3. neutron activation of ^{35}C in the lithosphere. Atmospheric thermonuclear weapons testing carried out in the Pacific Ocean between 1952 and 1958 introduced a large pulse of ^{36}Cl into the stratosphere by the neutron activation of ^{35}Cl in seawater.

About 30 percent of the near surface ^{36}Cl is of meteoric origin, of which 40 percent is produced in the troposphere and 60 percent is produced in the stratosphere (Bentley et al., 1986; Oeschger et al., 1969). Stable and unstable chloride in the troposphere has a mean residence time of only about one week. Stratospheric ^{36}Cl only enters the troposphere during periods of atmospheric mixing in a manner similar to ^{3}H. Atmospheric ^{36}Cl mixes with stable oceanic Cl, resulting in stable $^{36}Cl/Cl$ ratios near the coast, but the ratios increase near continental centers (Phillips, 2013). The concentration of ^{36}Cl in precipitation and dry fall exhibits strong latitudinal and continental trends. The greatest concentration of ^{36}Cl fall-out occurs at mid-latitudes. Superimposed on the latitudinal effect is the exponential increase in the $^{36}Cl/Cl$ ratio inland from ocean areas.

The process of calculating an absolute ^{36}Cl groundwater age is similar to calculating a ^{14}C age. The law of radioactive decay is applied to the difference between the initial (R_o) and final or measured (R) ^{36}Cl content of the water. As with ^{14}C dating, estimates of the initial ^{36}Cl content must include corrections for the effects of post-recharge isotopic dilution and contributions of ^{36}Cl. Chlorine-36 dating differs from ^{14}C dating in that the long half-life of ^{36}Cl requires an accounting of the contributions from the in situ ^{36}Cl production, whereas such an accounting is seldom required for the relatively short half-lived ^{14}C. Estimating mineral and groundwater ^{36}Cl ages are described in detail by Bentley et al. (1986) and Phillips (2013) and are summarized here.

The measured ^{36}Cl content (R) of an old groundwater is the sum of the initial meteoric contribution (R_o), which has undergone decay since the time of recharge, plus steady-state hypogene (subsurface) contribution (R_{se} = secular equilibrium ratio). R_{se} is the steady-state contribution of ^{36}Cl due to the hypogene neutron flux acting on ^{35}Cl. Because the neutron flux is regulated by U and Th decay, the flux varies greatly from one aquifer to the next according to the aquifer mineralogy. The change of R over time may be described as:

$$R = R_o \exp^{-\lambda}36^{t} + R_{se}(1 - \exp^{-\lambda}36^{t}) \tag{11.40}$$

where R is the measured $^{36}Cl/Cl$ atom ratio of the sample. R_o is the initial $^{36}Cl/Cl$ atom ratio of the sample and is assumed to have a 100 percent meteoric origin. $\text{Exp}^{-\lambda}{}_{36}{}^{t}$ is the radioactive decay of ^{36}Cl over time (t), λ_{36} is the ^{36}Cl decay constant (2.3×10^{-6}). The decay constant assumes that the chloride concentration in the rock and water is less than 1,000 ppm. Age (t), in years, may be estimated using the equation:

$$t = \frac{1}{\lambda_{36}} \ln \frac{R - R_{se}}{R_o - R_{se}} \tag{11.41}$$

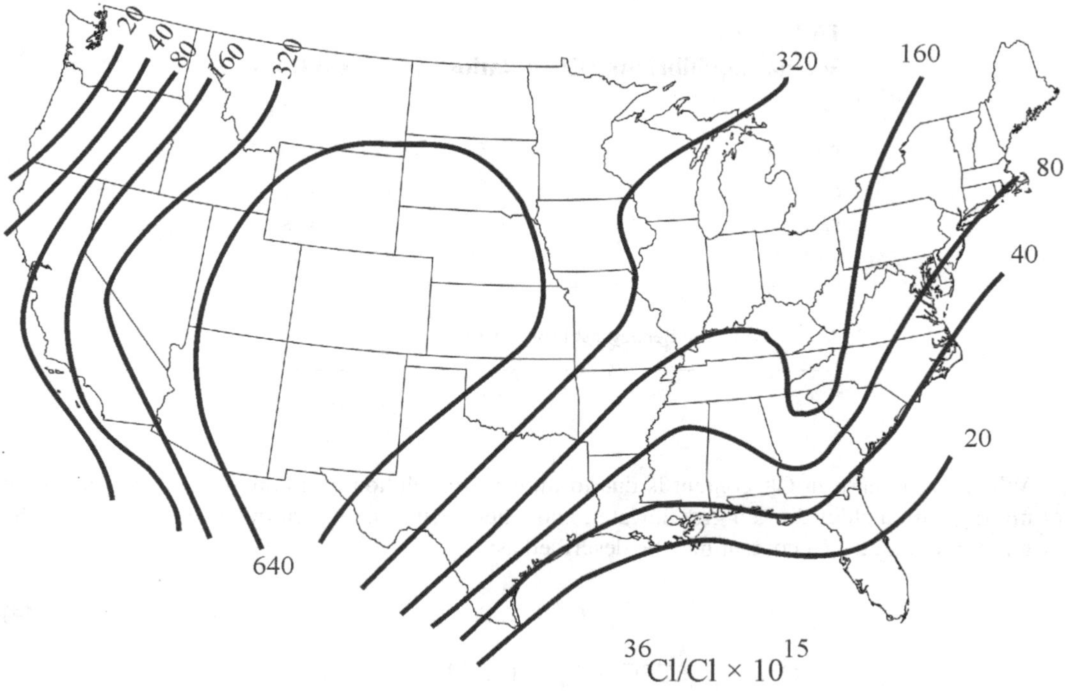

FIGURE 11.9 ^{36}Cl/Cl (×10^{15}) ratios in precipitation and dry fall over the United States. After Bentley et al., 1986.

Approximate values of R_o for pre-bomb precipitation over the United States and southern Canada are have been estimated by various authors (Moysey et al., 2003; Bentley et al., 1986; Phillips, 2000). The estimates are based on analyses of the measured R in groundwater samples. In general, R values are low along coastal areas and peak in the Great Plains (Figure 11.9). R_{se} for various geological environments are shown in Table 11.2. R_{se} may be calculated using the equation:

$$R_{se} = \frac{0.7577\phi_n \sigma 35}{\lambda_{36}(\Sigma N_j \sigma_j + N_{35}^{wW}\sigma_{35} + N_{H20}^{W}\sigma_{H20}} \tag{11.42}$$

where the symbolism is defined in the Chapter notation.

In arid environments and in sedimentary basins, the Cl$^-$ content of groundwater is often elevated due to the effects of evapotranspiration, mineral dissolution (with and without halite beds), diffusion of Cl$^-$ from aquitards, and ion filtration. When the chloride concentration is high and the increase in chloride ions is due to the dissolution of bedded Cl-bearing salt, which is expected to have a $\frac{^{36}\text{Cl}}{\text{Cl}}$ atom ratio of less than 10^{-16}, the absolute ^{36}Cl concentration is used to calculate t rather than the $\frac{^{36}\text{Cl}}{\text{Cl}}$ ratio:

$$t = \frac{1}{\lambda_{36}} \ln \frac{C^{36}}{C_o^{36}} = \frac{1}{\lambda_{36}} \ln \frac{RC}{R_o C_o} \tag{11.43}$$

where C^{36} is the measured ^{36}Cl content (atoms of ^{36}Cl/L), C_o^{36} is the initial ^{36}Cl content (atoms of ^{36}Cl/L), C is the measured Cl$^-$ content (atoms Cl$^-$/L), and C_o is the original Cl$^-$ content (atoms Cl$^-$/L).

TABLE 11.2
Secular Equilibrium ^{36}Cl/Cl Ratios for Typical Rocks

Rock type	^{36}Cl/Cl (R_{se}) × 10^{15}
Granite	30.1
Sandstone	4.68
Shale	12.5
Limestone	10.9

Source: Data from Bentley and Davis, 1982.

When an increase in Cl^- content is due to mineral dissolution of a non-bedded salt, the result of mixing with a chloride-rich groundwater, and when secular equilibrium has been involved, the origin of the mixed ^{36}Cl content may be described as:

$$C^{36} = RC \tag{11.44}$$

$$RC = R_o C_o e^{-\lambda}36^t + R_{se} C_o (1 - e^{-\lambda}36^t) + R_{se}\left(C - C_o\right) \tag{11.45}$$

and the age (t), in years, may be estimated using the equation:

$$t = \frac{1}{\lambda_{36}} \ln \frac{C\left(R - R_{se}\right)}{C_o\left(R_o R_{se}\right)} \tag{11.46}$$

If the increase in Cl^- concentration is the result of ion filtration, Equation 11.41 may be used because ion filtration will not appreciably change the ^{36}Cl/Cl atom ratio.

Unlike ^{3}H and ^{14}C, global fallout of ^{36}Cl was produced by relatively few atmospheric nuclear tests. Only those tests which converted seawater ^{35}Cl, via neutron activation, and which penetrated the troposphere yielded ^{36}Cl contributions to global ^{36}Cl rainout. Because bomb ^{36}Cl was rapidly flushed from the atmosphere as precipitation and dry fall, when found in groundwater it may be considered as a pulse of radioactive recharge. The predicted concentration of ^{36}Cl in precipitation from nuclear testing was as great as 4×10^4 atoms of ^{36}Cl/m^{2-} sec. For comparison, natural chloride fallout has a ^{36}Cl content of about 30 atoms of ^{36}Cl/m^{2-} sec.

11.6 CHLOROFLUOROCARBONS AND SULPHUR HEXAFLUORIDE

Anthropogenic chlorofluorocarbons (CF-11, CFC-12, and CFC-113) and sulfur hexafluoride (SF_6) are excellent trace gases for hydrogeologic studies because they have been released globally at known rates and they dissolve in groundwater following Henry's law (Chambers et al., 2019). They are best used to calculate modern groundwater ages up to about 80 years for piston flow systems (Figure 11.10). For post-1990 groundwater recharge, using a single CFC or SF6 may lead to a non-unique age, but a unique age is possible when multiple CFCs are used. In most groundwater systems the salinity and pressure at the location where the gas enters the unconfined aquifer are well constrained. In shallow groundwater systems the recharge temperature may be represented by the temperature of the water at the sampling point. A recharge temperature uncertainty of ± 2 °C will result in an age uncertainty of <1 year for groundwater that recharged prior to the mid-1970s but can lead

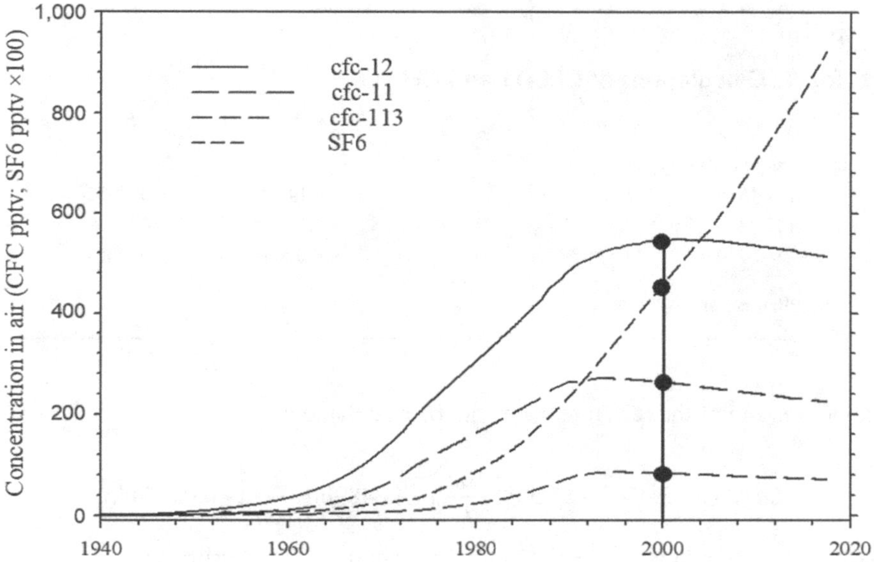

FIGURE 11.10 Atmospheric concentrations of CFCs and SF6 (data from USGS, 2024). Analysis of a single CFC or SF6 may result in a non-unique age, but when multiple analyses, such as those represent by the filled circles in the year 2000, are used, a unique age is possible.

to considerable uncertainty for recharge after 1990 (Cook et al., 2006). A more precise recharge temperature may be obtained by an analysis of other dissolved gases such as argon and nitrogen.

An in-depth discussion of using CFCs is found in IAEA (2006). Dating groundwater using CFCs (Plummer et al., 2006) is summarized here. The concentration (C) of the ith CFC in water is:

$$C_i = K_{H_i P_i} \tag{11.47}$$

where K_{H_i} is Henry's law constant for the ith CFC. The partial pressure (P_i) is:

$$p_i = x_i(P - P_{H_2O}) \tag{11.48}$$

and K_{H_i} can be calculated as:

$$\ln K_H = a_1 + a_2\left(\frac{100}{T}\right) + a_3 \ln\left(\frac{T}{100}\right) + S[b_1 + b_1]\left(\frac{T}{100}\right) + b_3\left(\frac{T}{100}\right)^2 \tag{11.49}$$

where T is in degrees Kelvin and S is salinity in parts per thousand. The salinity correction is not usually needed in groundwater studies but is important for saline water. Constants (a and b) for CFC calculations of K_H at 0–40 °C and 0–40 ‰ salinity are listed in Table 11.3. When the recharge elevation is less than 3,000 m an approximation of the recharge pressure can be calculated by:

$$\ln P = -H / 8{,}300 \tag{11.50}$$

where H is the recharge elevation in meters. In mountainous terrain the recharge elevation can be uncertain. For waters recharged prior to 1987, the calculated age may be only a few years off, for a recharge elevation uncertainty of 1,000 m. The uncertainty is considerably greater for post 1990 groundwater recharge.

TABLE 11.3
Constants for K_H Calculations of CFC-11 and CFC-12

CFS	a_i	a_2	a_3	b_1	b_2	b_3
K_H in mole/L at standard pressure (2023.35 hPa)						
CFC-11	−134.1536	203.2156	57.2320	−0.144490	0.092952	−0.0159977
CFS-12	−122.3246	182.5306	50.5898	−0.145633	0.092509	−0.0156627
CFC-113	−134.2430	203.8980	54.9583	− 0.026320	0.005874	–

Source: Data from Plummer et al., 2006.

The vapor pressure of the recharge water can be calculated as:

$$\ln H_2O = 23.4543 - 67.509\left(\frac{100}{T}\right) - 4.8489\ln\left(\frac{T}{100}\right) - 0.000544S \quad (11.51)$$

The recharge age is calculated by determining the recharge CFC pressure x (pptv):

$$x(\text{pptv}) = \frac{C}{K_H\left(P - P_{H_2O}\right)} \quad (11.52)$$

Equations 11.47, 11.48, 11.50, and 11.52 are used to calculate C, K_H, P and P_{H_2O}, respectively. The value of x is then looked up in a table of atmospheric CFC vs. year.

Annual atmospheric CFS and SF6 values are shown in Figure 11.10.

Important Takeaways – Using the law of radioactive decay groundwater, ^{36}Cl can be used for the unusual case of dating very old groundwater (100 ka to 1 Ma). The measured 36Cl content is the sum of the initial meteoric contribution (R_o), which has undergone decay since the time of recharge, plus steady-state hypogene (subsurface) contribution (R_{se}).

At the other extreme of groundwater ages, both chlorofluorocarbons and sulfur hexafluoride can be used to date piston flow groundwater that contains anthropogenic gas. This dating is based on the application of Henry's law rather than the law of radioactive decay.

Problems

1. You have been hired by a mining company to evaluate a groundwater flow system in a basin and range terrain that contains mixed carbonate and argillaceous bedrock. Basin and range are mountain blocks and down-dropped valleys where the boundary between the blocks and valleys are usually normal faults. The mine site has monitoring and sampling wells that intercept each of the stratigraphic layers and which can be sampled. Of particular interest is Big Spring, which is located slightly downgradient and about 1 mile from the proposed open pit mine. Big Spring has a variable discharge rate of 4 m^3/min in the late spring and about 1 m^3/min in the late summer/early fall. A nearby town owns the Big Spring water right. The concern of the town is what impact will mine dewatering have on the flow of Big Spring. You can instrument the spring discharge and visit the site and collect samples monthly for up to 24 months. Your budget only includes sufficient funds to collect samples for age dating eight times. The starting date of the field work is flexible.

   Provide a summary of geochemical sampling plan to help sort out the flow system(s) that support Big Spring. Groundwater dating, solute chemistry, and δ^2H and δ^{18}O are key elements in the plan.

2. The measured ^{3}H from roof drip samples in an underground longwall coal mine have consistently been non-detect (ND). ^{14}C dating is not possible because of mine methane. The underground workings are located about 1 mile from a dammed lake. The lake water ^{3}H is about 5 TU. After several years of mining, a new longwall panel intercepted a fault that has a steady groundwater flux of about 0.2 m^3/min. The initial tritium in the water was ND. After about six months, the measured tritium was consistently about 5 TU. Possible sources of the fault water include the lake and a shallow groundwater system. The mining company claims it is the shallow aquifer; the lake owners claim it is the lake.

   The owners of the lake have brought a lawsuit against the mining company claiming water rights infringement. You have been hired by an arbitrator to help sort the source of the underground flow.
   a. What will be your testimony, and what additional data would you like to have?
   b. What would be your testimony if the ^{3}H was 5 TU in the springtime and 1 TU in the late summer/early fall?
3. The measured ^{14}C of two groundwater samples were 27.5 and 57.9 pmc. The waters had a discharge temperature of 18 °C. Using the Pearson graphical method, calculate the ages of the waters assuming a) a confined aquifer in carbonate rock, and b) a confined aquifer in pine forest granitic terrain.
4. A carbon sample has a ^{14}C activity of 6.78 dpm/g.
   a. Assuming the sample is from an Egyptian coffin, how old is the coffin? You may neglect any $\delta^{13}C$ correction that should be applied to the laboratory result.
   b. Assuming the sample is from the Ogallala aquifer in the middle of the United States, how old is the water? How old would the water be if the ^{14}C activity were half as great?
5. What are the different considerations you would have to apply when dating unconfined and confined aquifers using ^{14}C and ^{3}H?
6. A small discharge spring issues from the damage zone of a fault. The fault cuts through a complex of leucocratic granite and carbonate rocks. The measured $\delta^{13}C = -13$ ‰, and the $^{14}C = 37$ pmc. It is unclear if the water recharged in the granite or the carbonate or both. Calculating a range of groundwater ages is relatively easy by using one of the available computer codes.

   What differences would you expect in the calculated ages assuming granite vs. carbonate recharge, and how might numeral instabilities affect the results?

Chapter Notation

0.75776	isotopic abundance of ^{35}Cl
°K	temperature in degrees Kelvin
λ_{36}	^{36}Cl decay constant ($2.3 \times 10^{-6}\ a^{-1}$)
δ	delta unit used for reporting isotopic abundances in ‰ difference relative to a standard
$\delta^{13}C$	$\delta^{13}C$ content in permil (‰ relative to the PDB standard)
δ_s	$\delta^{13}C$ content of total dissolved carbon in sample (‰ PDB)
δ_m	$\delta^{13}C$ content of carbonate minerals in soil and aquifer (‰ PDB)
δ_g	$\delta^{13}C$ content of soil gas CO_2 (‰ PDB)
ϕ_n	subsurface neutron flux, which may be calculated by the methods of Morrison and Pine (1955), and Feige and others (1968)
σ_{35}	the neutron activation cross-section of ^{35}C for conversion to ^{36}Cl
ε_2	isotopic enrichment factor between $CO_{2\,(g)}$ and HCO_3^-
ε_8	isotopic enrichment factor between $CO_{2\,(g)}$ and $H_2CO_3{}^{o}$
a	molar concentration of aqueous CO_2
b	molar concentration of HCO_3^-

c	molar concentration of CO_3^{-2}
d	total dissolved inorganic carbon or TDIC ($a + b + c$)
a^{-1}	rock and water chloride concentration (is assumed to be less than 1,000 ppm)
atm	atmosphere
A_x	^{14}C content, also known as activity, of substance × (pmc)
A_o	initial groundwater ^{14}C content (pmc)
A_s	measured groundwater ^{14}C content (pmc)
A_t	^{14}C content of groundwater (pmc) at time 0; in a sample, this is known as A_s
A_g	^{14}C content of the soil gas in the recharge zone (pmc)
A_m	^{14}C content of carbonate minerals in the soil zone and aquifer (pmc)
A_{ms}	^{14}C content of carbonate minerals in the soil zone (pmc)
A_{mm}	^{14}C content of carbonate minerals in the aquifer (pmc) (note: $A_m = A_{ms} + A_{mm}$)
β	beta particle
C	concentration of radioactive isotope at sampling point
C_o	concentration of radioactive isotope at recharge point
dpm g^{-1}	disintegrations per minute per gram
dpm L^{-1}	disintegrations per minute per liter
exp	exponent to base e (e^x)
j	elements U and Th in the rock matrix
Log	logarithm to the base 10
ln	natural logarithm (logarithm to the base e); ln a = 2.303 log a
m	molality
n	neutron
Nw_{35}^{w}	number of ^{35}Cl atoms per kg of saturate porous media.

$$N_{w35}^{w} = 0.7577 \frac{n_e}{\beta} C_{cl} A$$

where

ne = porosity

C_{Cl} = chloride concentration (mol L^{-1})

A = Avogadro's number

β = bulk rock density

$N^{w}_{H20}\sigma_{H20}$	number of water atoms per kg of saturated porous media times the neutron activation cross-section
N	subsurface neutron production
p	proton
pCi L^{-1}	picocuries per liter
P_{CO2}	partial pressure of CO_2 gas
P	atmospheric pressure
pmc	percent modern carbon
Q	flux of groundwater flow (length^3/time)
R	measured $^{36}Cl/Cl$ ratio of the sample
R_o	initial $^{36}Cl/Cl$ ratio of the sample
R_{se}	secular equilibrium ratio; estimate of the steady state hypogene (subsurface) contribution of ^{36}Cl due to U and Th decay
t	turnover time = Q/V
t	mean isotopic residence time (age)
T	lifetime of an isotope
$T½$	half-life of an isotope; same as B

TDIC	total dissolved inorganic carbon
TU	tritium unit; 1 TU = 7.2 dpm/L or 3.2 pCi/L
V	volume of groundwater in storage (lenght^3)
x_o	distance between recharge and discharge point

REFERENCES CITED

Archer, D., et al., 2009, Atmospheric lifetime of fossil fuel carbon dioxide, *Annual Review of Earth and Planetary Sciences*, 37, 117–134.

Beck, J.W., et al., 2001, Extremely large variations of atmospheric ^{14}C concentration during the last glacial Period, *Science*, 292, 2453–2458.

Bentley, H.W., and S.D. Davis, 1982, *Application of AMS to hydrogeology*, in *2nd Annual symposium on acceleration mass spectrometry*, Argonne National Laboratories.

Bentley, H.W., et al., 1986, Chapter 10 – Chlorine-36 in the terrestrial environment, in *Handbook of environmental isotope geochemistry*, Elsevier, 2(B), 427–480.

Bigeleisen, H., and M.G. Mayer, 1947, Calculations of equilibrium constants for isotopic exchange reactions, *Journal Chemical Physics*, 15, 261–267.

Bouchaou, L., 2008, Application of multiple isotopic and geochemical tracers for investigation of recharge, salinization, and residence time of water in the Souss – Massa aquifer, southwest of Morocco, *Journal of Hydrology*, 352(3–4), 267–287.

Braziunas, T.F., et al., 1995. The preindustrial atmospheric $^{14}CO_2$ latitudinal gradient as related to exchanges among atmospheric, oceanic, and terrestrial reservoirs, *Global Biogeochemical Cycles*, 9(4), 565–584.

Chae, J.-S., and G. Kim, 2019, Seasonal and spatial variations of tritium in precipitation in Northeast Asia (Korea) over the last 20 years, *Journal of Hydrology*, 74, 794–800.

Chambers, L.A., et al., 2019, Use and application of CFC-11, CFC-12, CFC-113 and SF6 as environmental tracers of groundwater residence time: A review, *Geoscience Frontiers*, 10(5), 1643–1652.

Cheng, H., et al., 2018, Atmospheric $^{14}C/^{12}C$ changes during the last glacial period from Hulu cave, *Science*, 362(6420), 1293–1297.

Cook, P.G., et al., 2006, Effects and process that can modify apparent DFC age, in *Use of chlorofluorocarbons in hydrology – a guidebook*, International Atomic Energy Agency (IAEA).

Craig, H., 1954, Carbon-13 in plants and the relationship between carbon-13 and carbon-14 variations in nature, *Journal Geology*, 62, 115–149.

Deines, P., et al., 1974, Stable carbon isotope ratio and the existence of a gas phase in the evolution of carbonate groundwaters, *Geochimica et Cosmochimica Acta*, 38, 1147–1164.

Dinauer, A., et al., 2020, Mysteriously high $\Delta^{14}C$ of the glacial atmosphere: influence of ^{14}C production and carbon cycle changes, *Climate of the Past*, 16, 1159–1185.

Eby, M., et al., 2009, Lifetime of anthropogenic climate change: millennial time scales of potential CO_2 and surface temperature perturbations, *Journal Climate*, 22, 2501–2511.

Eichinger, L., 1983, A contribution to the interpretation of ^{14}C groundwater ages considering the example of a partially confined sandstone aquifer, *Radiocarbon*, 25(2), 347–356.

Evans, G.V., et al., 1978, *Some problems in the interpretation of isotope measurements in British aquifers: International symposium on isotope hydrology*, International Atomic Energy Agency (IAEA), Neuherberg/Muenchen, Germany.

Eyrolle, F., et al., 2018, An updated review on tritium in the environment, *Journal of Environmental Radioactivity*, 181, 128–137.

Feige, Y., et al., 1968, Production rates of neutrons in soils due to natural radioactivity, *Journal Geophysical Research*, 73(10), 3135–3142.

Ferronsky, V.I., and V.A., Polyakoc, 1982, *Environmental isotopes in the hydrosphere*, Wiley and Sons.

Florkowski, T., et al., 1988, Influence of the Chernobyl accident on the natural levels of tritium and radiocarbon, *International Journal of Radiation Applications and Instrumentation, Part A. Applied Radiation and Isotopes*, 39(1), 77–79.

Fontes, J.C., 1983, Dating of groundwater in guidebook on nuclear techniques in hydrology, 1983 ed, *International Atomic Energy Agency (IAEA), Technical Report*, 91, 285–317.

Fontes, J.C., and J.M. Garnier, 1979, Determination of the initial ^{14}C activity of the total dissolved carbon: A review of the existing models and a new approach, *Water Resources Research*, 15(2), 399–413.

Fritz, P., and J.C. Fontes, 1980, *Introduction in handbook of environmental isotope geochemistry,* Elsevier, 1(A), 1–19.

Gat, J., et al., 2001, *Environmental isotopes in the hydrogeological cycle – principles and applications*, International Atomic Energy Agency (IAEA) II Atmospheric Water.

Geyh, M.A., 2000, An overview of ^{14}C analysis in the study of groundwater, *Radiocarbon*, 42, 99–114.

Gil-Márquez, J.M., et al., 2020, Groundwater dating tools (^3H, ^3He, ^4He, CFC-12, SF6) coupled with hydrochemistry to evaluate the hydrogeological functioning of complex evaporite-karst settings, *Journal of Hydrology*, 580, 124263.

Gilmore, T., et al., 2021, The $^3H/^3He$ Groundwater age-dating method and applications, *Derbyana*, 42.

Golubenko, K., et al., 2022, Zonal mean distribution of cosmogenic isotope (^7Be, ^{10}Be, ^{14}C, and ^{36}Cl) production in stratosphere and troposphere, *Journal of Geophysical Research: Atmospheres*, 127.

Gonfiantini, R., 1972, Notes on isotope hydrology, internal publication, IAEA, 1972 cited by Fontes, J.Ch., 1983, Dating of groundwater in Guidebook on nuclear techniques in hydrology, 1983 ed, *International Atomic Energy Agency (IAEA) Technical Report*, 91, 285–317.

Gonfiantini, R., and Z. Gian, 2003, Carbon isotope exchange rate of DIC in karst groundwater, *Chemical Geology*, 197, 319–336.

Graven, H., et al., 2017, Compiled records of carbon isotopes in atmospheric CO_2 for historical simulations in CMIP6, *Geoscientific Model Development*, 10, 4405–4417.

Han, L-F., and L.N. Plummer, 2013, Revision of Fontes & Garnier's model for the initial ^{14}C content of dissolved inorganic carbon used in groundwater dating, *Chemical Geology*, 351, 105–114.

Han, L.F., and L.N. Plummer, 2016, A review of single-sample-based models and other approaches for radiocarbon dating of dissolved inorganic carbon in groundwater, *Earth Science Reviews*, 152, 119–142.

Han, L-F., et al., 2014, The curved ^{14}C vs. $\delta^{13}C$ relationship in dissolved inorganic carbon: A useful tool for groundwater age- and geochemical interpretations, *Chemical Geology*, 387.

Harms, P.A., et al., 2016, Distribution of tritium in precipitation and surface water in California, *Journal of Hydrology*, 534, 63–72.

Hufen, T.H., et al., 1974, *Radiocarbon, 13C and tritium in water samples from basaltic aquifers on the island of Oahu-Hawaii in isotope techniques in groundwater hydrology 1974. Proceedings of symposium, Vienna*, International Atomic Energy Agency (IAEA), 2, 111–126.

IAEA, 2006, *Use of chlorofluorocarbons in hydrology – a guidebook*, International Atomic Energy Agency.

IAEA, 2013, *Isotope methods for dating old groundwater*, International Atomic Energy Agency.

IAEA GNIP, 2024, *International atomic energy agency, global network isotopes in precipitation*, www.iaea.org/services/networks/gnip, accessed March 31, 2024.

Ingerson, R.L., and F.J. Pearson Jr, 1964, Estimation of age and rate of motion in groundwater by the Carbon-14 method, in *Recent research in the fields of hydrosphere, atmosphere and nuclear geochemistry*, Maruzen, Tokyo, 263–283.

Kashiwaya, K., et al., 2017, Spatial variations of tritium concentrations in groundwater collected in the southern coastal region of Fukushima, Japan, after the nuclear accident, *Scientific Reports*, 7, 12578.

Lamarche, V.C., Jr, and T.P. Harlan, 1973, Accuracy of tree ring dating of bristlecone pine for calibration of radiocarbon time scale, *Journal Geophysical Research*, 78, 8849–8858.

Levin, I., et al., 2010, Observations and modelling of the global distribution and long-term trend of atmospheric $14CO_2$, *Tellus B – Chemical and Physical Meteorology*, 62(1), 26–46.

Libby, W.F., 1955, *Radiocarbon dating*, 2nd ed, University Chicago Press.

Manning, A.H., et al., 2005, $^3H/^3He$ age data in assessing the susceptibility of wells to contamination, *Groundwate*r, 43, 353–367.

Michael, H.H., and E.K. Ralph, 1970, Correction factors applied to Egyptian radiocarbon dates from the era before Christ, in *Radiocarbon variations and absolute chronology*, Wiley, 109–119.

Michel, R.L., et al., 2018, *Tritium deposition in precipitation in the United States, 1953–2013*, U.S. Geological Survey Science Investigations Report 2018-5086.

Mook, W.G., 1967, *The dissolution-exchange model for dating groundwater with* 14C, in *Interpretation of environmental isotope and hydrochemical data in groundwater hydrology*, International Atomic Energy Agency (IAEA).

Mook, W.G., 1972, On the reconstruction of the initial ^{14}C content of groundwater from the chemical and isotopic composition, in *Proceeding of 8th international conference on radiocarbon dating*, Royal Society New Zealand, 1, 342–352.

Mook, W.G., 1976, The dissolution-exchange model for dating groundwater with 14C, in *Interpretation of environment isotope and hydrochemical data in groundwater hydrology*, International Atomic Energy Agency (IAEA), 213–225.

Mook, W.G., 1980, Carbon-14 in hydrogeological studies, in *Handbook of environmental isotope geochemistry*, Elsevier, 1(A), 49–74.

Mook, W.G., et al., 1974, Carbon isotope fractionation between dissolved bicarbonate and gaseous carbon dioxide, *Earth and Planetary Science Letters*, 22(2), 169–176.

Morrison, P., and J. Pine, 1955, Radiogenic origin of helium isotopes in rocks, *Annals New Your Academy of Science*, 62(3), 69–92.

Moysey, S., et al., 2003, The distribution of meteoric 36Cl/Cl in the United States: A comparison of models: *Hydrogeology Journal*, (11), 615–627.

Muller, A.B., and A.L., Mayo, 1986, ^{13}C variation in limestone on an aquifer-wide scale and its effects on groundwater ^{14}C dating models, *Radiocarbon*, 28(3), 1041–1054.

Nir, A., 1964, On the interpretation of tritium "age" measurements of groundwater, *Journal Geophysical Research*, 69, 2589–2595.

Nir, A., et al., 1966, Natural tritium, *Reviews Geophysics*. 4, 411–456.

Oeschger, R.H., et al., 1969, *The constancy radiation from isotope studies in meteorites and on the Earth*, in *Radiocarbon variations and absolute chronology, 12th Nobel Symposium*, Wiley Interscience, 471–496.

Pearson, F.J., Jr, and B.B. Hanshaw, 1970, Sources of dissolved carbonate species in groundwater and their effects on carbon-14 dating, in *Isotope hydrology 1970*, International Atomic Energy Agency (IAEA), 271–286.

Pearson, F.J., Jr, and W.W. Swarzenki, 1974, Carbon-14 evidence for the origin of arid region groundwater, northeastern Province, Kenya, in *Isotope techniques in groundwater*, International Atomic Energy Agency (IAEA), 95–108.

Pearson, F.J., Jr, et al., 1972, Carbon-14 dates of water from Arkansas Hot Springs, in *Proceeding of eighth international conference on radiocarbon dating*, Royal Society New Zealand, 1, 330–247.

Phillips, F.M., 2000, Chlorine-36, in *Environmental tracers in subsurface hydrology*, Kluwer Academic, Boston, 299–348.

Phillips, F.M., 2013, Chapter 65 – Chlorine-36 dating of old groundwater, in *Isotope methods for dating old groundwater*, International Atomic Energy Agency (IAEA), 125–152.

Plummer, L., et al., 2006, Principles of chlorofluorocarbons dating, in *Use of chlorofluorocarbons in hydrology*, International Atomic Energy Agency (IAEA), 17–30.

Poreda, R.J., et al., 1988, Tritium and helium isotopes as hydrologic tracers in a shallow unconfined aquifer, *Journal of Hydrology*, 103(1–2), 1–9.

Post, V.E.A., et al., 2019, Behaviour of tritium and tritiogenic helium in freshwater lens groundwater systems: Insights from Langeoog Island, Germany, *Geofluids*, 1494326.

Salem, O., et al., 1980, Groundwater flow patterns in the western Libyan Arab Jamahiriya evaluated from isotopic data, in *Hydrology investigation with isotope techniques*, International Atomic Energy Agency (IAEA), 165–179.

Schell, W.R., et al., 1974, World distribution of environmental tritium, in *Physical behavior of radioactive contaminants in the atmosphere*, International Atomic Energy Agency (IAEA), 374–400.

Shapiro, S.D., et al., 1999, Characterizing a sewage plume using the 3H-3He dating technique, *Groundwater*, 37, 861–878.

Solomon, D.K., and E.A. Sudicky, 1991, Tritium and helium-3 isotope ratios for direct estimation of spatial variations in groundwater recharge, *Water Resources Research*, 27(9), 2309–2319.

Solomon, D.K., et al., 1993, A validation of the $^3H/^3He$ method for determining groundwater recharge, *Water Resources Research*, 29(9), 2951–2962.

Solomon, D.K., et al., 1998, Chapter 9 – Dissolved gases in subsurface hydrology, in *Isotope tracers in catchment hydrology*, Elsevier, 291–318.

Solomon, D.K., and P.G. Cook, 2000, 3H and 3He, in *P. Environmental tracers in subsurface hydrology*, Kluwer Academy, 397–424.

Stuiver, M., 1967, Origin and extent of atmospheric 14C variations during the past 10,000 years, in *Radioactive dating and methods of low-level counting*, International Atomic Energy Agency (IAEA), 27–40.

Stuiver, M., and P.D. Quay, 1980, Patterns of atmospheric ^{14}C changes, *Radiocarbon*, 22, 166–176.

Suess, H.E., 1955, Radiocarbon concentration in modern wood, *Science*, 22(3166), 415–417.

Tamers, M.A., 1967, Surface-water infiltration and groundwater movement in arid zones of Venezuela, in *Isotope hydrology 1967*, International Atomic Energy Agency (IAEA), 339–351.

Tamers, M.A., 1975, Validity of radiocarbon dates of groundwater, *Geophysical Surveys,* 2, 217–239.

Tolstikhin, I.N., and I.L. Kamensky, 1969, Determination of groundwater age by theT-^3He method, *Geochemistry International*, 6, 810–811.

USGS, 2024, *Air curves CFS, SF6, 2H excel spreadsheet*, usgs.gov/lab/software/air curv/, accessed April 2024.

van der Plicht, J., 2007, Radiocarbon dating variations in atmospheric ^{14}C, in *Encyclopedia of Quaternary science*, Elsevier, 2923–2931.

Vogel, J.C., and D. Ehhalt, 1963, The use of carbon isotopes in groundwater studies, in *Interpretation of environmental isotope hydrochemical data in groundwater hydrology*, International Atomic Energy Agency (IAEA), 195–212.

Wigley, T.M.L., 1975, Carbon 14 dating of groundwater from closed and open systems, *Water Resources Research*, 11(2), 324–328.

Wu, S., 2017. Variation of atmospheric $14CO_2$ and its spatial distribution, *Journal of Environmental Radioactivity*, 169–170, 116–121.

Part III

Data Manipulation and Geothermometers

12 Project Design, Data Management, and Data Manipulation
Computer and Graphical Methods

12.1 PROJECT DESIGN AND DATA MANAGEMENT

12.1.1 Project Design

The first step in any investigation is to establish the project's purpose and objects. Without a clearly defined purpose and a set of objectives, considerable time and money may be wasted and the results of the investigation may be of little value. This holds true for pure research studies, in house investigations, and in consulting projects.

The purpose is the overall goal of the investigation. This goal may or may not be fully met at the end of the investigation. The objectives are a list of achievable, relevant, and time-bound lines of inquiry that lead to achieving the purpose. The purpose is the *why*, and the objectives are the *what* and *how*. Creating the purpose and objectives starts with developing a set of questions. The questions are then changed into statements, which frequently involve adding action verbs. Once the purpose has been clearly defined, objectives can be developed. The objectives are not methods of investigation, such as the sampling and analyses of well waters for ^{14}C. Methods of study are the actual field, laboratory, and analytical activities that you will perform to meet the objectives. Considering the broad range of solute and isotopic processes outlined in Chapters 1–11, selecting the methods of study is extremely important to the success or failure of the investigation. In hydrogeochemistry, field and laboratory work are expensive in terms of time and money and are often in limited supply. Each of the geochemical processes can be considered as a tool for solving specific hydrogeochemistry and groundwater flow problems, and considerable thought should be given when selecting which tools to select from the tool basket.

The hierarchy of purpose, objectives, and methods can be represented as:

Purpose
- Objective 1
  - Methods
    - e.g., measure water levels in wells; collect and analyze spring waters for major ions and selenium; plot the data as Stiff and trilinear diagrams; calculate SI, etc.
- Object 2
  - Methods
- Objective 3
  - Methods

For example, the owners of a coal power plant located on an arm of the Chesapeake Bay in Virginia have been sued by an environmental organization for polluting the bay with leachate from their ash storage facility. You have been hired by the owners of the power plant to help defend them in federal court. As with most consulting projects, the power plant owners and their lawyers don't know how you should proceed. In many consulting jobs, the owners don't have a clear idea of the big question(s) they need answered. As with all research and most in-house and consulting projects, you will have to develop the purpose statement and the project objectives. For the ash storage facility, the overall questions that the court would like answered is: Does leachate from the ash facility leak into the bay, and does the leachate contain contaminates of concern (COC)? If it does leak COCs in the bay, a secondary question might be: What mitigation measures do the coal plant operators plan on taking? These two questions lead to two different purpose statements and sets of objectives. The purpose statements could be: 1) determine whether ash leachate and COCs are discharging into the bay from the ash storage facility and 2) develop leachate COC mitigation measures.

Objectives for the two different ash facility purposes could be:

Purpose 1: Determine if the ash leachate and COCs are discharging into Chesapeake Bay.

Objectives:

1. Characterize the hydrologic regime of the ash storage facility.
2. Characterize the flux of both the groundwater and the leachate from the facility into the bay.
3. Characterize the solute and possible COC concentrations in the ash leachate.

Purpose 2: Develop mitigation measures for the COCs in the ash leachate that discharges into Chesapeake Bay.

Objectives:

1. Characterize the hydrologic regime of the ash storage facility.
2. Characterize the flux of groundwater and leachate from the facility to the bay.
3. Characterize the solute composition and concentrations of the leachate.
4. Determine which of the leachate solutes are COCs.
5. Develop mitigation measures for the COCs.

The purpose of an investigation could be global in scale or extremely limited. For example, global questions regarding climate change could be: Is climate change actually occurring, or is climate change human caused? The purpose statements for these two questions would require very different sets of objectives. A more limited investigation question could be whether underground mine water is leaking from mine A to mine B.

It is important to keep in mind that a project's purpose and objectives are not set rules that must be followed. They are guidelines that can be modified as data becomes available and new leads need to be followed. The purpose and objectives of most hydrologic investigation are not limited to hydrogeochemistry. In most investigations, both physical hydrogeology and hydrogeochemical methods are used, and they must be integrated in both the project design and the final conclusions. Solutes and isotopes can provide the answers to critical questions in the scheme of a project's purpose and objectives.

12.1.2 Data Management

Once the field and laboratory data have been collected, the data need to be organized for efficient analysis. Although no set and hard rules can be established for evaluating geochemical data, the analysis of the geochemistry of groundwater systems should follow a logical progression. The specific steps in the analysis must be tailored to meet the objectives of the investigation and the data available. The first step in most analyses should be the compilation of appropriate data tables. A group of data tables, which can be easily computerized, are collectively called the database. Both spreadsheets and database programs can be used for data management. The database should include all pertinent geologic and other potentially relevant physical data in addition to geochemical data.

Data should be organized in such a fashion so as to group waters together that have similar geochemistry, have common geologic or hydrogeologic characteristics, have a similar location, or illustrate the chemical evolution of waters. Organization of the data is often an interactive process, which proceeds as additional knowledge is gained. Analysis of raw data often yields a great volume of secondary data that also need to be included in the database.

The database may be built using any one of several commercially available database programs specifically designed for hydrogeology or by using common commercially available spread sheets and database programs. Commercially available hydrogeologic database programs often have active links to geochemical and graphical programs. Most spread sheets are generally more flexible than hydrogeologic database programs and are easily amenable to the exportation and importation of data.

Errors are common in field and laboratory data, and efforts need to be taken to verify the validity of the data and to correct the errors prior to performing further analysis. Errors may be systematic or random and may occur during sampling and laboratory procedures and when reporting and manipulating the data. Systematic errors are usually the easiest to identify because they affect a whole suite of analyses in the same fashion. For example, when using atomic absorption procedures on water with elevated salinity, chemical interference can result in the under detection of Na^+. Another common systematic error is row and column shifting when manipulating spreadsheet data. Random errors can easily go undetected. Such an error is evident in Figure 12.1, where the laboratory report TDS in mg/L is plotted against the calculated TDS in meq/L. The mg/L data were obtained by converting the individual laboratory cation and anions analyses reported as mg/L to meq/L and then summing the results. From the figure, it appears that the laboratory reported TDS of sample R2 is about twice as great as the likely TDS of the sample. In some datasets, random errors can be detected by simply plotting the data as a scatter plot and looking for outliers. A summary of common data collection, data entry, and data analysis errors are listed on Table 12.1.

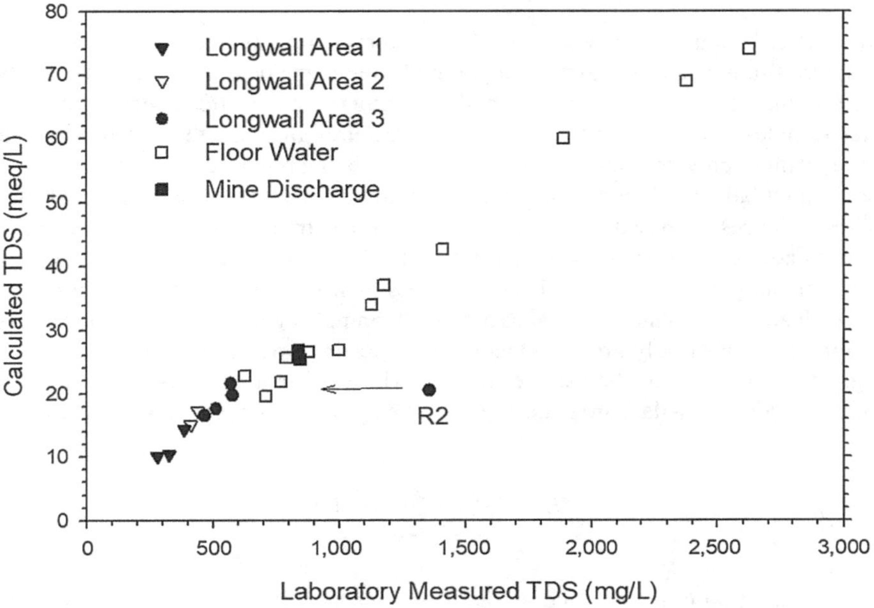

FIGURE 12.1 Scatter plot of laboratory-reported TDS vs. calculated TDS for groundwater in an underground coal mine. All of the data except for R2 fall along the regression line, which defines the relationship between mg/L and meq/L. The laboratory-reported value for R2 is about twice as great as the likely actual value.

TABLE 12.1
Summary of Common Types of Data Errors in Hydrogeochemistry Analysis

Evaluation stage	Common errors
Data collection	Improper labeling of samples Instrument error Instrument drift Improper instrument calibration Writing error
Data entry and data cleaning	Incorrect transfer of data from instrument Values entered in the wrong field Inadvertent deletion of duplication during database handling Outlier and inconsistencies carried over from the instrument Values entered incorrectly Values incorrectly changed during data cleaning Insufficient number of fields (i.e., °C and °F, conductivity ms and μS) Data format Transposition errors Data cleaning errors
Data analysis	Incorrect extraction from database Data incorrectly coded Inadvertent deletion or duplication during analysis Sorting errors (spreadsheets)

The overall performance of a geochemical laboratory can be monitored by submitting samples of known composition in addition to the samples collected in the field. For samples collected as part of some hazardous waste investigations, formal sampling protocols often specify the use of blanks, wash water samples, and spiked samples. Such procedures increase the reliability of laboratory data, but they cannot prevent some erroneous data from being reported.

A cation–anion balance should be performed on all solute data that include major ion analysis. Cation–anion balances are based on the fact that water is electrically neutral, and the total electrical charges of the cations are equal to the total electrical charges on the anions. Laboratories should be able to perform analyses with an error balance of less than 1 percent; however, laboratory data are acceptable with an error balance as great as 5 percent. Samples with very small and very large TDS concentrations are particularly prone to laboratory reporting errors. Occasionally, data with error balances greater than 5% must be used, especially when the data have historical significance and when other time-dependent data are not available. The error balance, E, of a sample is calculated as:

$$E = \frac{\sum_{\text{cations}} - \sum_{\text{anons}}}{\sum_{\text{cations}} + \sum_{\text{anions}}} \times 100 \tag{12.1}$$

where the concentration of cations and anions are in meq/L. Error balances cannot be calculated when the data are in mg/L. Concentrations of minor and trace constituents are usually so small as to have a negligible effect on the error balance calculation. The small concentrations make identifying trace species errors particularly difficult.

12.2 COMPUTER ANALYTICAL METHODS

Computer methods are powerful tools for performing routine thermodynamic calculations, for calculating parameters such as saturation indices, and for analyzing the geochemical evolution of water. Some codes will also perform ^{14}C dating analysis. Free computer codes such as the USGS WATEQF (Plummer et al., 1976) have relieved us from the need to perform the tedious and often error-prone manual thermodynamic calculations, and codes such as the USGS PHREEQC (Parkhurst et al., 1980) can be used to model the chemical evolution of groundwaters under different physical and chemical conditions. Codes such as the USGS NETPATH interpret net geochemical mass-balance reactions between initial and final waters. Free computer software is available from the web pages of the USGS and the IAEA. Several commercially available comprehensive groundwater chemical analysis packages are also available. These programs may ease the pain in analyzing chemical data because they commonly include database management, statistical analysis packages, thermodynamic calculations, chemical evolution analysis, interfaces for USGS and other free software, and tabular and graphical presentations of the data and results. The thermodynamic heart of most commercial programs is WATEQF and PHREEQC. PHREEQC and many of the commercially available packages allow the user to select from several different thermodynamic databases, and it is possible to program additional thermodynamic species in many of the programs. When adding species, it is important to make sure the thermodynamic data are internally consistent. Several USGS computer codes that can be downloaded online are described here.

WATEQF (Plummer et al., 1976) is a USGS chemical speciation code that incorporates temperature, pH, E_h, dissolved oxygen, alkalinity, and the chemical analysis of a water sample as inputs. The code has a thermodynamic database of several hundred species and uses both the Debye-Hückel and Davies methods to calculate activity coefficients. Because the program is open source, it is possible to add thermodynamic data for species not included in the code. The program calculates the distribution of aqueous species, ion activities, and mineral saturation indices and other data. The program code assumes homogeneous aqueous phase equilibria, except for redox species. Equilibrium with respect to mineral solubility is not assumed. The code cannot be used to evaluate flow path geochemical evolution. WATEQF is the underlying thermodynamic code used in most free and commercially available aqueous geochemistry programs including PHREEQC. It may be used as stand-alone program when geochemical evolution modeling is not required. Ball and Nordstrom (1991) have prepared an easy-to-understand user's manual for WATEQF.

As an example, the WATEQF analysis of water from McGill Hot Spring in Nevada is shown in Table 12.2. Key elements of the model output include: 1) a summary of the input data, 2) the calculated distribution of species including molality, activity, and activity coefficients (gamma), 3) calculated P_{CO_2} and mole ratios of selected elements and species, and 4) the ion activity products (IAP) for mineral species, equilibrium constant (K_T) and the saturation indices (IAP/KT and log IAP/KT) for the aquifer temperature. The WATEQF example output (Table 12.2) shows that redox data were not supplied, the waters are supersaturated with respect to calcite and dolomite and undersaturated with respect to gypsum. The calculated $P_{CO_2} = 4.438026 \times 10^{-3}$ and the anions and cations are in good balance.

NETPATH (Plummer et al., 1976, 1994) is a USGS inverse geochemical modeling code that can be used to interpret net geochemical mass-balance reactions between an initial and a final water along a groundwater flow path. NETPATH-WIN is an easy-to-use Microsoft Windows® user interactive version of NETPATH that is available on the USGS web pages (El-Kadi et al., 2011). NETPATH can be used to compute the mixing proportions of two to five initial waters and net geochemical reactions that can account for the observed composition of final water. The program utilizes previously defined chemical and isotopic data for waters in a hydrochemical system. For a set of minerals and (or) gas phases hypothesized to be the reactive phases in the system, NETPATH calculates the mass transfers in every possible combination of the selected phases that can account

TABLE 12.2
Sample WATEQF Output for McGill Hot Spring, Nevada.

McGill Hot Spring

INITIAL SOLUTION

TEMPERATURE = 77.00 DEGREES C PH = 8.100
ANALYTICAL EPMCAT = 5.100 ANALYTICAL EPMAN = 4.936

***** OXIDATION – REDUCTION *****

DISSOLVED OXYGEN = .000 MG/L
EH MEASURED WITH CALOMEL = 99.0000 VOLTS FLAG CORALK PECALC IDAVES
MEASURED EH OF ZOBELL SOLUTION = 99.0000 VOLTS 2 1 0 0
CORRECTED EH = 99.0000 VOLTS
PE COMPUTED FROM CORRECTED EH = 100.000

*** TOTAL CONCENTRATIONS OF INPUT SPECIES ***

SPECIES	TOTAL MOLALITY	LOG TOTAL MOILALITY		TOTAL MG/LITRE
CA	2	1.35E-03	−2.8704	5.40E+01
MG	2	8.64E-04	−3.0634	2.10E+01
NA	1	6.53E-04	−3.1853	1.50E+01
K	1	2.56E-05	−4.592	1.00E+00
CL	−1	1.27E-04	−3.8963	4.50E+00
SO4	−2	2.08E-04	−3.6814	2.00E+01
HCO3	−1	4.39E-03	−2.3572	2.68E+02

*** CONVERGENCE ITERATIONS ***

ITERATION	S1-ANALCO3	S2-SO4 TOT	S3-FTIOT	S4-PTOT	S5-CLTOT
1	6.36E-04	9.95E-05	0.00E+00	0.00E+00	0.00E+00
2	4.17E-05	4.24E-06	0.00E+00	0.00E+00	0.00E+00
3	−1.44E-06	−1.73E-07	0.00E+00	0.00E+00	0.00E+00

****DESCRIPTION OF SOLUTION ****

ANAL.	COMP.	PH	ACTIVITY	H2O = .9999
EPMCAT	5.10	4.44	8.1	PCO2 = 4.438026E−03
EPMAN	4.94	4.31		LOG PCO2 = −2.3528

TEMPERATURE PO2 = 0.0000 +00

EH = ******	PE = 100.00	77.00 DEG C	PCH4 = .000000E+00
PE CALC S = 100.000		CO2 TOT = 4.177312E-03	
PE CALC DOX = 100.000	IONIC STRENGTH	DENSITY = 1.0000	
PE SATO DOX = 100.000	6.41E-03	TDS = 383.5MG/L	
TOT ALK = 4.429E+00 MEQ	CARB ALK = 4.394E+00 MEQ		
ELECT = 1.289E-01 MEQ			

TABLE 12.2 (*Continued*)
Sample WATEQF Output for McGill Hot Spring, Nevada.

IN COMPUTING THE DISTRIBUTION OF SPECIES,
PE = 100.000 EQUIVALENT EH = 6.948VOLTS

DISTRIBUTION OF SPECIES
– ------------------------------------

SPECIES	(charge)	PPM	MOLALITY ACTIVITY	LOG ACT	GAMMA
1 CA	2	4.34E+01	1.08E-03	7.52E-04	–3.124 6.94E-01
2 MG	2	1.84E+01	7.56E-04	5.27E-04	–3.278 6.98E-01
3 NA	1	1.49E+01	6.47E-04	5.90E-04	–3.229 9.11E-01
4 K	1	9.98E-01	2.55E-05	2.32E-05	–4.634 9.09E-01
64 H	1	8.70E-06	8.64E-09	7.94E-09	–8.19.20E-01
5 CL	–1	4.50E+00	1.27E-04	1.15E-04	–3.938 9.09E-01
6 SO4	–2	1.47E+01	1.53E-04	1.06E-04	–3.974 6.92E-01
7 HCO3	–1	2.29E+02	3.76E-03	3.43E-03	–2.465 9.13E-01
18 CO3	–2	2.79E+00	4.66E-05		3.24E-05–4.49
86 H2CO3	0	3.59E+00	5.79E-05		5.80E-05–4.237
27 OH	–1	5.06E-01	2.98E-05	2.71E-05	–4.5689.09E-01
19 MGOH	1	1.95E-01	4.71E-06	4.31E-06	–5.366 9.15E-01
23 MGSO4 AQ	0	3.69E+00	3.06E-05	3.07E-05	–4.513 1.00E+00
22 MGHCO3	1	3.12E+00	3.66E-05	3.33E-05	–4.478 9.10E-01
21 MGCO3 AQ	0	3.05E+00	3.61E-05	3.62E-05	–4.441 1.00E+00
29 CAOH	1	4.30E-02	7.53E-07	6.88E-07	–6.162 9.14E-01
32 CASO4 AQ	0	3.22E+00	2.37E-05	2.37E-05	–4.626 1.00E+00
30 CAHCO3	1	5.27E+00	5.22E-05	4.76E-05	–4.322 9.13E-01
31 CACO3 AQ	0	1.88E+01	1.88E-04	1.87E-04	–3.729 9.93E-01
44 NASO4	–1	5.68E-02	4.77E-07	4.35E-07	–6.361 9.13E-01
43 NAHCO3	0	9.54E-02	1.14E-06	1.14E-06	–5.944 1.00E+00
42 NACO3	–1	3.00E-01	3.62E-06	3.30E-06	–5.481 9.13E-01
94 NACL	0	3.97E-33	6.80E-38	6.81E-38	–37.16 71.00E+00
46 KSO4	–1	5.55E-03	4.11E-08	3.75E-08	–7.426 9.13E-01
63 HSO4	–1	4.14E-05	4.26E-10	3.89E-10	–9.411 9.11E-01

MOLE RATIONS FROM ANALYTICAL MOLAILITY	MOLE RATIONS FROM COMPUTED MOLALITY	LOG ACTIVITY RATIOS
CL/CA = 0.094209	CL/CA = 0.11725	CA/H2 = 13.0761
CL/MG = 0.14695	CL/MG = 0.16798	MG/H2 = 12.922
CL/NA = 0.19454	CL/NA = 0.19611	NA/H1 = 4.8708
CL/K = 4.9632	CL/K = 4.9712	K/H1 = 3.466
CL/ AL =1.2698E+26	CL/AL = 1.2698E+26	K/H1 = 24.3

(*Continued*)

TABLE 12.2 (*Continued*)
Sample WATEQF Output for McGill Hot Spring, Nevada.

CL/FE = 1.2698E+26	CL/FE = 1.2698E+26	FE/H2 = 16.2
CL/SO4 = 0.60965	CL/SO4 = 0.82737	CA/MG = 0.154
CL/HCO3 = 0.60965	CL/HCO3 = 0.033815	NA/K = 1.4048
CA/MG = 1.5598	CL/MG = 1.4326	
NA/K = 25.513	NA/K = 25.349	

PHASE	IAP	KT	LOG IAP	LOG KT	IAP/KT	LOGIAP/ KT
18 ANHYDRIT	7.98E-08	1.41E-05	−7.098	−4.852	5.68E-03	−2.246
22 ARAGONIT	2.43E-08	1.45E-09	−7.614	−8.839	1.68E+01	1.225
151 ARTIN	1.34E-20	2.51E-19	−19.873	−18.601	5.34E-02	−1.273
20 BRUCITE	3.86E-13	4.81E-12	−12.413	−11.317	8.02E-02	−1.096
13 CALCITE	2.43E-08	1.12E-09	7.614	−8.951	2.17E+0	1.337
12 DOLOMITE	4.15E-16	7.64E-19	−15.382	8.117	5.44E+02	2.735
19 GYPSUM	7.98E-08	1.72E-05	−7.098	−4.764	4.64E-03	−2.334
65 HALITE	6.81E-08	4.81E+01	−7.167	1.682	1.42E-09	−8.849
118 HUNTITE	1.21E-31	4.85E-34	−30.918	−33.314	2.49E+02	2.396
39 HYDMAG	43.485	−40.598	−2.887			
11 MAGNESIT	1.71E-08	1.23E-09	−7.768	−8.912	1.39E+01	1.144
67 MIRABI	3.69E-11	8.99E+00	−10.433	0.954	4.11E-12	−11.386
59 NAHCOL	2.02E-06	7.19E-01	−5.694	−0.143	2.81E-06	−5.551
61 NATRON	1.13E-11	2.53E+00	−10.949	0.403	4.45E-12	−11.352
150 NESQUE	1.71E-08	1.44E-06	−7.768	−5.841	1.18E-02	1.927
66 THENAR	3.70E-11	5.74E-01	−10.432	−0.241	6.44E-11	10.191
62 THRNAT	1.13E-11	6.61E-01	−10.948	−0.18	1.71E-11	−10.768
60 TRONA	2.28E-17	1.76E-03	−16.642	−2.754	1.29E-14	−13.888

for the observed changes in the selected chemical and (or) isotopic compositions observed along the flow path. The calculations are useful for interpreting geochemical reactions, mixing proportions, evaporation and/or dilution of waters, and mineral mass transfer in the chemical and isotopic evolution of natural and environmental waters. Rayleigh distillation calculations are applied to each mass-balance model that satisfies the constraints to predict carbon, sulfur, nitrogen, and strontium isotopic compositions at the end point, including radiocarbon dating. DB-WIN is the interactive interface used to enter analytical data into NETPATH-WIN. DB-WIN includes a tab for performing WATEQF calculations. For simple mixing and easy-to-evaluate chemical evolution pathways, NETPATH is often preferred over the more complex PHREEQC. NETPATH is particularly useful for eliminating possible reaction pathways.

PHREEQC is the go-to computer code for performing a wide variety of real-world aqueous geochemical calculations based on ion-association and has capabilities for 1) speciation and saturation-index calculations, 2) reaction-path and advective-transport calculations involving specified irreversible reactions, mixing of solutions, mineral and gas equilibria, surface-complexation reactions, and ion-exchange reactions, and 3) inverse modeling, which finds sets of mineral and gas mole transfers that account for composition differences between waters within specified compositional uncertainties. PHREEQC relieves us of the computational burden and allows us to focus and test numerous likely reaction pathways.

PHREEQC features include the capabilities to

1. use redox couples to distribute redox elements among their valence states in speciation calculations,
2. model ion-exchange and surface-complexation reactions,
3. model reactions with a fixed-pressure, multicomponent gas phase,
4. calculate the mass of water in the aqueous phase during reaction and transport calculations,
5. keep track of the moles of minerals present in the solid phases and automatically determine the thermodynamically stable phase assemblage,
6. simulate advective transport in combination with PHREEQC reaction-modeling capability, and
7. perform inverse modeling calculations that allow for uncertainties in the analytical data.

The user interface has a simplified approach to redox reactions, which includes explicit mole-balance equations for hydrogen and oxygen and input that is modular and is in a completely free format, and uses mineral names and standard chemical symbolism rather than index numbers. The equation solver optimizes a set of equalities subject to both equality and inequality constraints that are used to determine the thermodynamically stable set of phases in equilibrium with a solution. The interface starts with one or more user-defined solutions where the temperature, redox, cations, and anions are entered. The interface contains tabs where the initial solution can be subjected to various reactants and reactions. Calculations or data management operations may be performed in the following order: 1) initial solution or speciation calculations, 2) initial exchange calculations, 3) initial surface calculations, 4) initial gas-phase calculations, 5) batch-reaction calculations, 6) inverse-modeling calculations, 7) advective-transport calculations, 8) advective-dispersive transport calculations, 9) cell batch-reaction calculations, 10) copy operations, 11) dump operations, and 12) delete operations. Numerous examples of PHREEQC problems are available on the internet, and the user's guide to PHREEQC (Parkhurst and Appelo, 1999) is a good starting point. Most commercially available programs that contain PHREEQC also contain several example problems.

12.3 GRAPHICAL METHODS

Despite the great power of computer methods, considerable and often critical insight can be gained by graphically displaying the data. Graphical methods are powerful tools for developing and visualizing groundwater geochemical evolution, for the relationships of groundwater chemistry to the groundwater flow, and for displaying groundwater chemistry. Graphs permit the rapid inspection of complex and large geochemical datasets. Graphs often lead to the identification and correlation of hydrostratigraphic and other water-bearing units, the identification of geochemical evolution pathways and ground-water mixing, and the identification of anomalous data. Graphical representations have the additional advantage of conveying geochemical information to people who do not have scientific or technical training. Geochemical graphs may be drawn by hand, but free and commercially available computer graphic programs allow the easy construction of trial and final graphs. Several of the commercially available aqueous chemistry computer programs will create graphs using the input database. The advantage of using graphing computer programs is the ease with which numerous chemical relationships can be explored and dismissed when they do not provide useful information. It is common to disregard considerably more trial graphs than those actually found to be useful.

Over the years a variety of graphical forms have been developed for displaying and analyzing solute data. Important graphical forms have been introduced by Collins (1923), Durov (1948), Maucha (1949), Schoeller (1935), Stiff (1951), Piper (1944), and others. No attempt is made here to describe all graphical forms or the evolution of all forms. Solute data are usually converted to meq/L for graphical analysis, although other units such as mg/L may be applicable to specific graphical forms and investigations.

In order to compare graphical methods a common dataset (Table 12.3) is used in many of the examples that follow. With reference to Table 12.3, both Antelope Island crystalline and sedimentary systems are located on Antelope Island in the Great Salt Lake (Figure 12.2). The southern Farmington Bay system is located in the non-island lowland southeast of the island, and the Weber delta and the Wasatch Range and thermal systems are located east of the island.

Bar graphs are among the simplest methods to graphically represent solute and isotopic data and can be used to display both composition and concentration. Collins (1923) first used bar diagrams to represent ion concentrations. In the sample bar graph (Figure 12.3) mean cation and anion concentrations from Table 12.1 are graphically compared. In the figure, the height of each bar represents the overall concentration, and the locations of the horizontal divisions within each bar represent the concentrations of individual species. Cations on the left side of each bar and anions on the right are arranged so that chemically related species are opposite each other (e.g., Na^+ is opposite Cl^-, Ca^{2+} is opposite HCO_3^-). Sodium and potassium concentrations are commonly combined. A different pattern or color may be used to represent each species.

In Figure 12.3, the bar heights and distribution of ions indicate that the mineralogically similar Antelope Island and Wasatch Range crystalline rock aquifers and the mineralogically similar sedimentary aquifers have undergone different geochemical evolutions. The somewhat elevated TDS and Na–Cl in the island crystalline aquifer waters were attributed to precipitation of evaporated lake water and the dry fall of west desert dust aerosols. The very high TDS and Na–Cl of Antelope Island and South Farmington Bay sedimentary aquifer waters were attributed to contributions of Great Salt Lake brine.

Pie diagrams are useful for illustrating the distribution of the components of a sample population and the overall sample size. The mean or median concentration of a group of samples is often represented by the pie radius. Pie diagrams may be standalone diagrams or they may be placed on maps, cross-sections, and on other location and spatial diagrams. Pie slices show the percentage that each constituent contributes to the total concentration of the sample. Water chemistry percentages are usually calculated using meq/L concentrations, although mg/L concentrations are sometimes useful. Stand-alone pie diagrams based on data in Table 12.3 are shown in Figure 12.4. In the figure, the cations occupy the top portion of the pie and anions occupy the lower portion. As with bar diagrams, the locations of cations and anions are commonly plotted opposite of their common chemically reacting ion. Pie diagrams are also useful for displaying non-geochemical data that may be important for understanding the distribution of data.

Vector diagrams (Maucha, 1949) represent major ion chemistry, trace elements, and organic compounds as vectors, where the length of each vector represents the concentration. The data in Table 12.3 are plotted as vector diagrams in Figure 12.5. The usual arrangement in vector diagrams is to plot the Na^+ and Cl^- pair on the vertical vector and the Ca^{2+} and HCO_3^- and the Mg^{2+} and SO_4^{2-} pairs on the horizontal vectors. Vector diagrams are not widely used.

12.3.1 Shape Diagrams

Several types of shape diagrams, also known as pattern diagrams, have been developed to represent ion chemistry. All of the diagrams represent solute contents and types of water as recognizable shapes and concentrations by size. Shape diagrams are useful for displaying the spatial distributions of solute chemistries on maps, cross-sections, and well logs. The major limitation of shape diagrams is that a separate diagram is usually constructed for each water sample, although average

TABLE 12.3
Mean Solute Chemistry of Groundwater in the Vicinity of Antelope Island, Great Salt Lake, Utah

Type	Number samples	°C	pH	Ca^{2+}	Mg^{2+}	meq/L Na^+	K^+	HCO_3^-	SO_4^{2-}	Cl^-	Total	mg/L SiO_2	error %
Island crystalline	12	14.0	8.2	2.53	1.42	4.06	0.07	3.19	1.52	3.19	15.98	14.4	1.13
Island sedimentary	8	13.6	8.2	4.60	2.21	10.40	0.42	4.60	1.97	10.99	45.06	16.0	0.38
S. Farmington Bay	14	18.6	7.9	1.81	1.88	18.65	0.14	4.65	0.65	17.65	55.43	28.6	0.85
Weber delta	17	14.8	7.6	2.81	1.04	0.99	0.06	3.77	0.27	0.50	9.44	18.2	3.81
Wasatch Range	4	13.3	7.7	0.97	0.45	0.52	0.30	1.42	0.35	0.20	4.94		0.00
Wasatch Fault thermal	5	53.8	7.1	26.30	4.59	158.54	7.97	3.94	7.57	183.27	392.19	25.8	0.67

Type	log P_{CO2}	calcite	dolomite	gypsum	Log halite	quartz
Island crystalline	−3.06	0.32	0.64	−1.99	−6.55	0.45
Island sedimentary	−2.99	0.75	1.23	−1.77	−5.89	0.54
S. Farmington Bay	−2.63	0.16	0.31	−3.22	5.11	0.75
Weber delta	−2.33	0.07	0.07	−2.86	−7.98	0.63
Wasatch Range	−2.91	−0.83	−0.83	−3.84	−8.69	
Wasatch Fault thermal	−1.72	0.64	0.64	−1.21	−3.58	0.23

Source: After Mayo and Klauk, 1991.

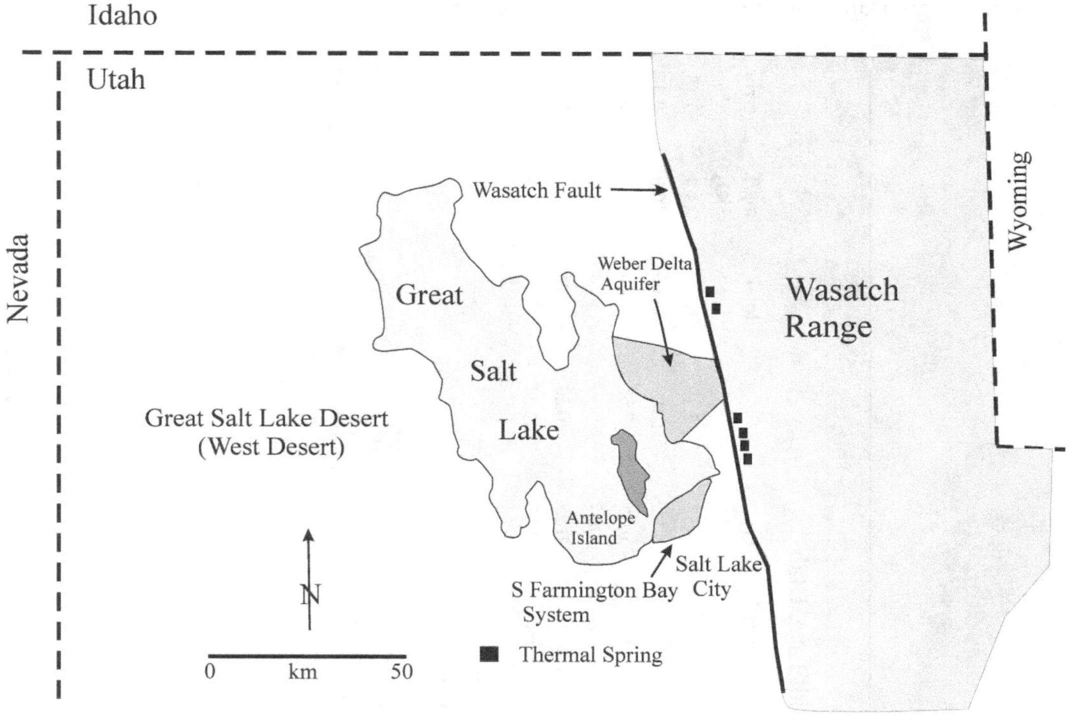

FIGURE 12.2a Map of Antelope Island, the Great Salt Lake and the Wasatch Range, Utah (modified after Mayo and Klauk, 1991).

FIGURE 12.2b Photo showing the abrupt change in topographic relief (2,225 m) at the boundary between the West Desert (elevation ~ 1280 m) and the Wasatch Range (elevation ~3,500 m). The view looks across the Great Salt Lake and the Weber delta from Antelope Island to the Wasatch Range. After Cressler (2007).

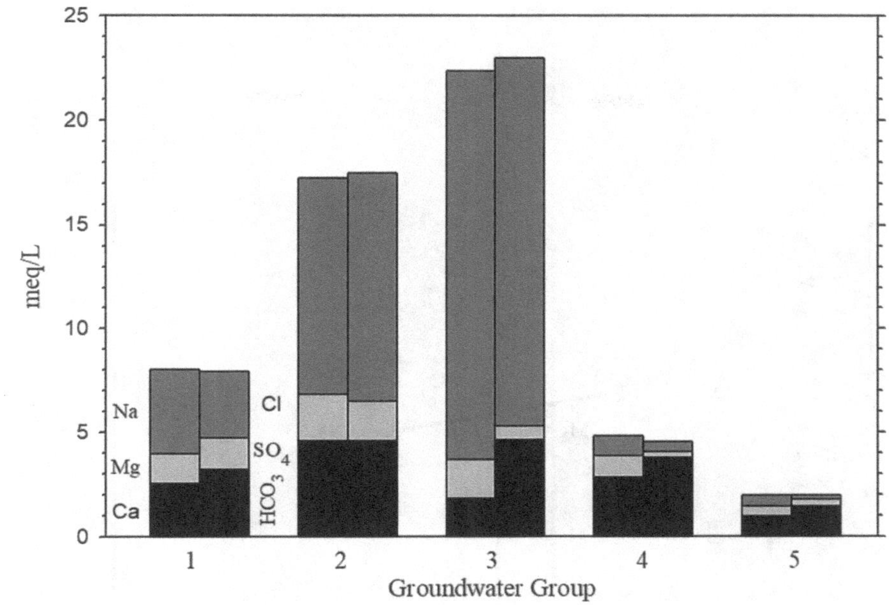

FIGURE 12.3 Bar graphs representing the major ion concentrations of groundwaters in the vicinity of Antelope Island, Great Salt Lake, Utah. The groundwater groups are: 1) Antelope Island crystalline bedrock, 2) Antelope Island sediment bedrock, 3) South Farmington Bay sediments, 4) Weber Delta sediments, and 5) Wasatch Range bedrock. Data are from Table 12.3.

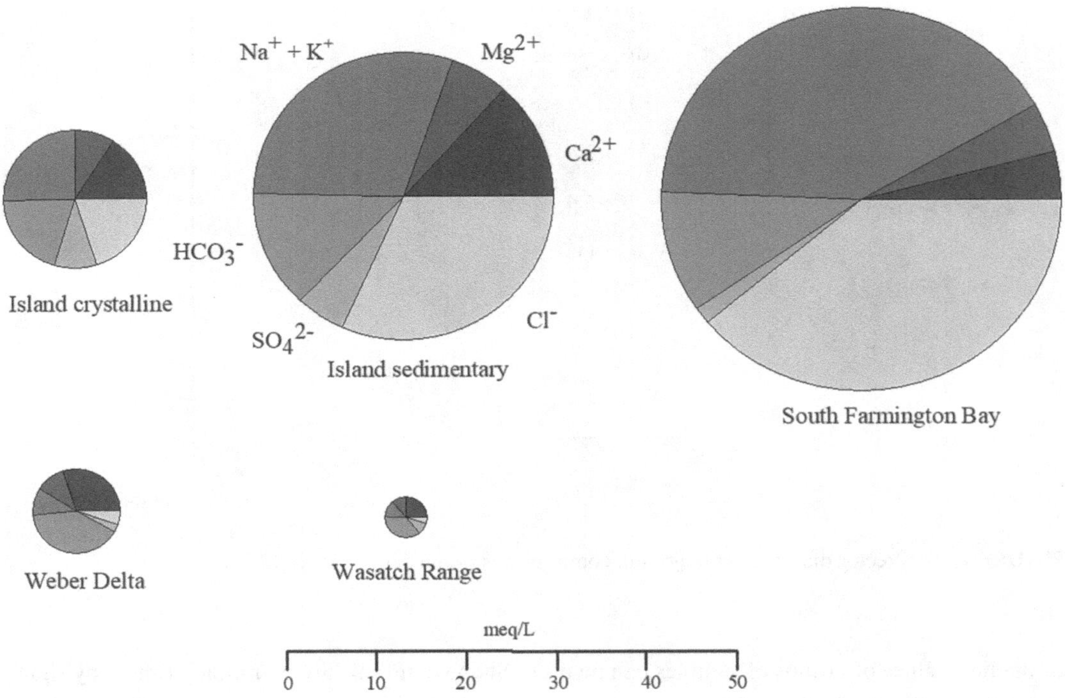

FIGURE 12.4 Pie diagrams of the major ion chemistry of the data in Table 12.3. Most of the waters have unusually high Na^+ and Cl^- concentrations relative to the concentrations of other major ions.

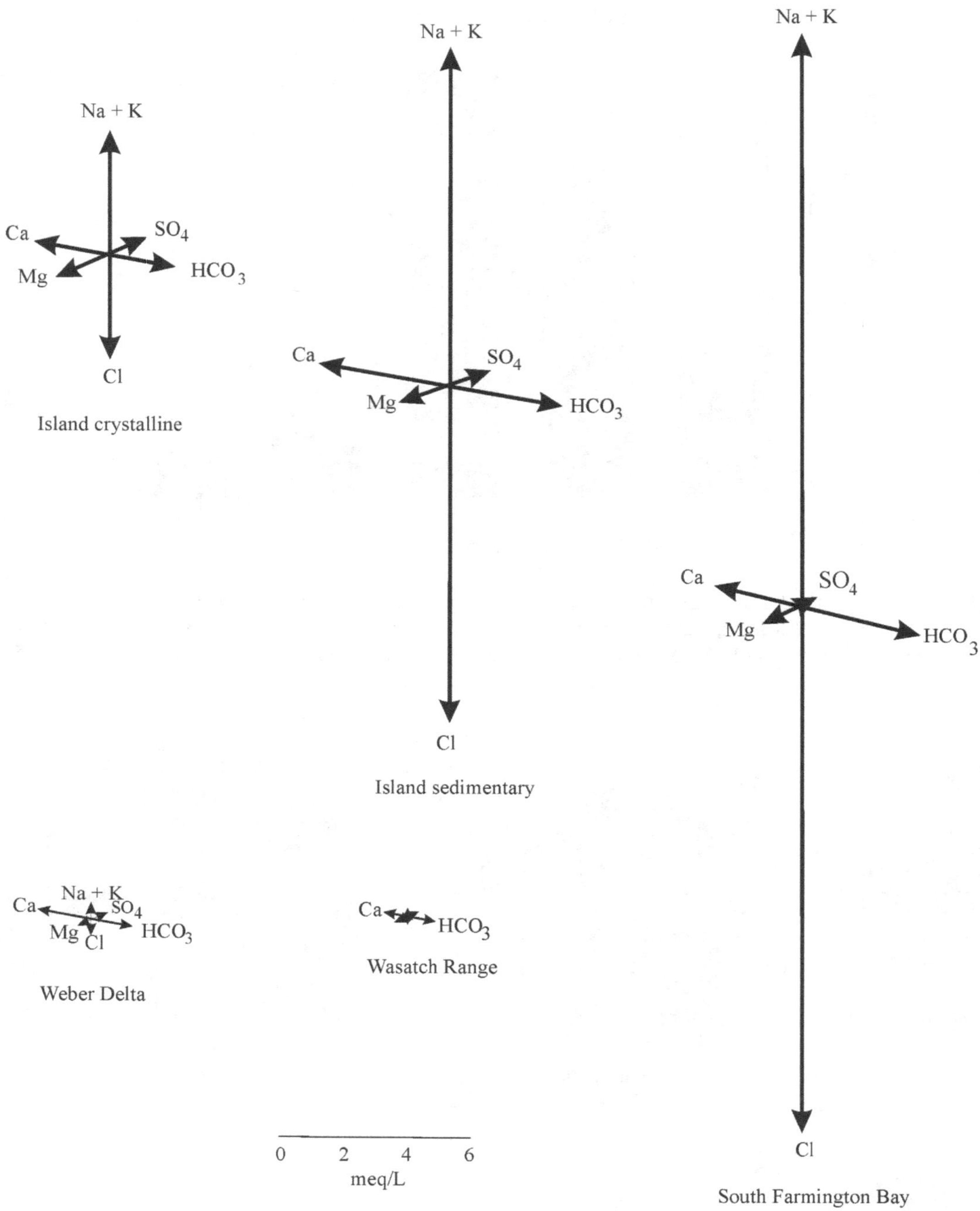

FIGURE 12.5 Vector diagrams of major ion compositions of the data in Table 12.3.

or median values of groups of samples can be used. Shape diagrams are commonly drawn by showing concentrations on an independent concentration scale, and the axes of each individual diagram are usually not labeled. Common shape diagrams include four corner diagrams (Colby et al., 1956), radial diagrams (Lloyd and Heathcote, 1985), and trilinear diagrams (Stiff, 1951).

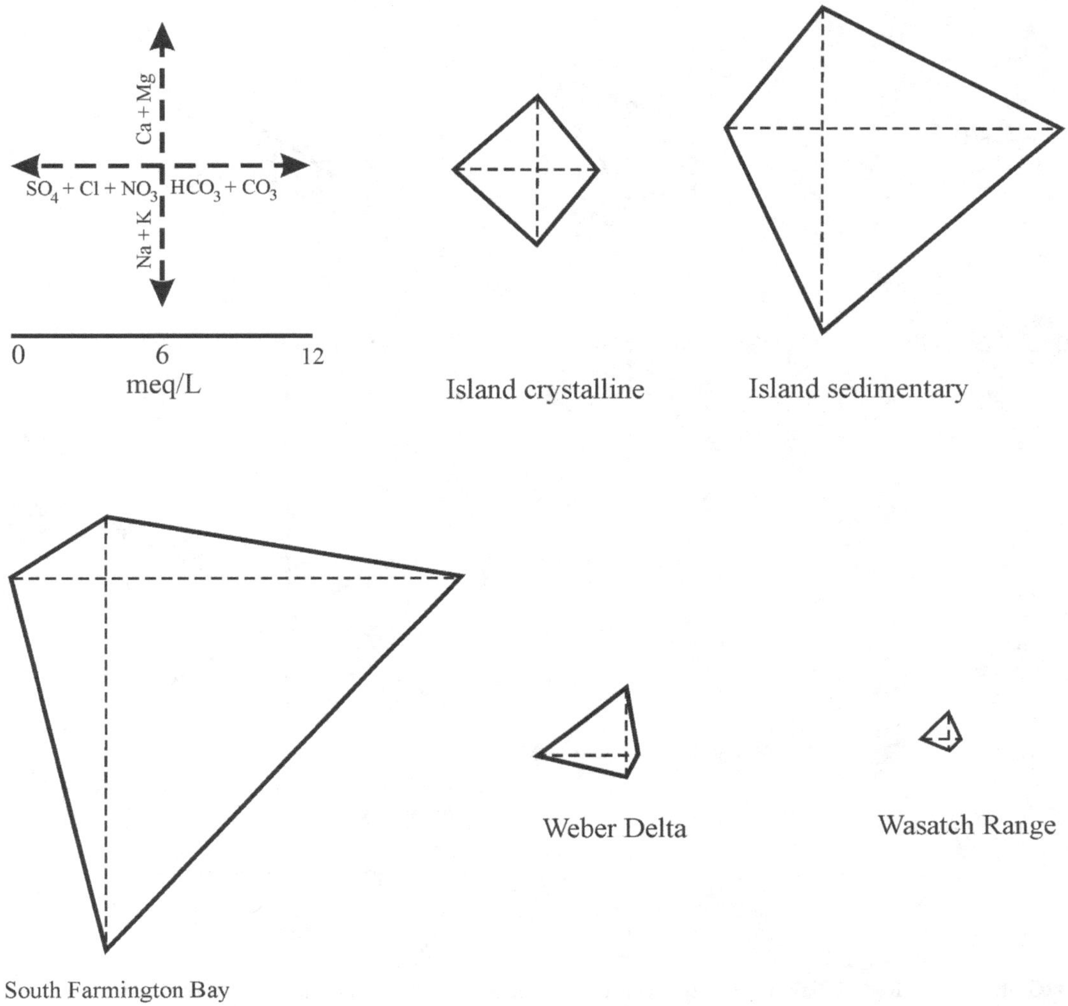

FIGURE 12.6 Four corner diagrams of the data in Table 12.3. Cations are on the vertical axis and anions are on the horizontal axis.

Four corner diagrams are diagrams that plot the concentrations of the combined species $Ca^{2+} + Mg^{2+}$, $HCO_3^- + CO_3^{2-}$, $Na^+ + K^+$, and $SO_4^{2-} + Cl^- + NO_3^-$ on the four rectangular coordinates. The data in Table 12.3 are shown as four corner diagrams in Figure 12.6. As with most diagrams, solute concentrations are in meq/L.

These kite shaped diagrams have received little application because they combine numerous ionic species on each axis, and some of the combinations have limited geochemical significance. Four corner diagrams can also be used to display trace elements and organic compounds. To meet the needs of a particular investigation, any arrangement of species is possible.

Radial diagrams are another shape diagram that can be used on maps and cross sections. The data in Table 12.3 are plotted as radial pattern diagrams on Figure 12.7. The diagram consists of six arms, each of which represents a different ionic species. The arms radiate from the origin at 60° intervals. Cations in the upper half of the diagram and anions in the lower half are located so as to place chemically related species opposite each other. The length of each arm is determined by the ion concentration. Radial diagrams can be used to display major ions, trace elements, and organic compounds. The diagrams are not commonly used.

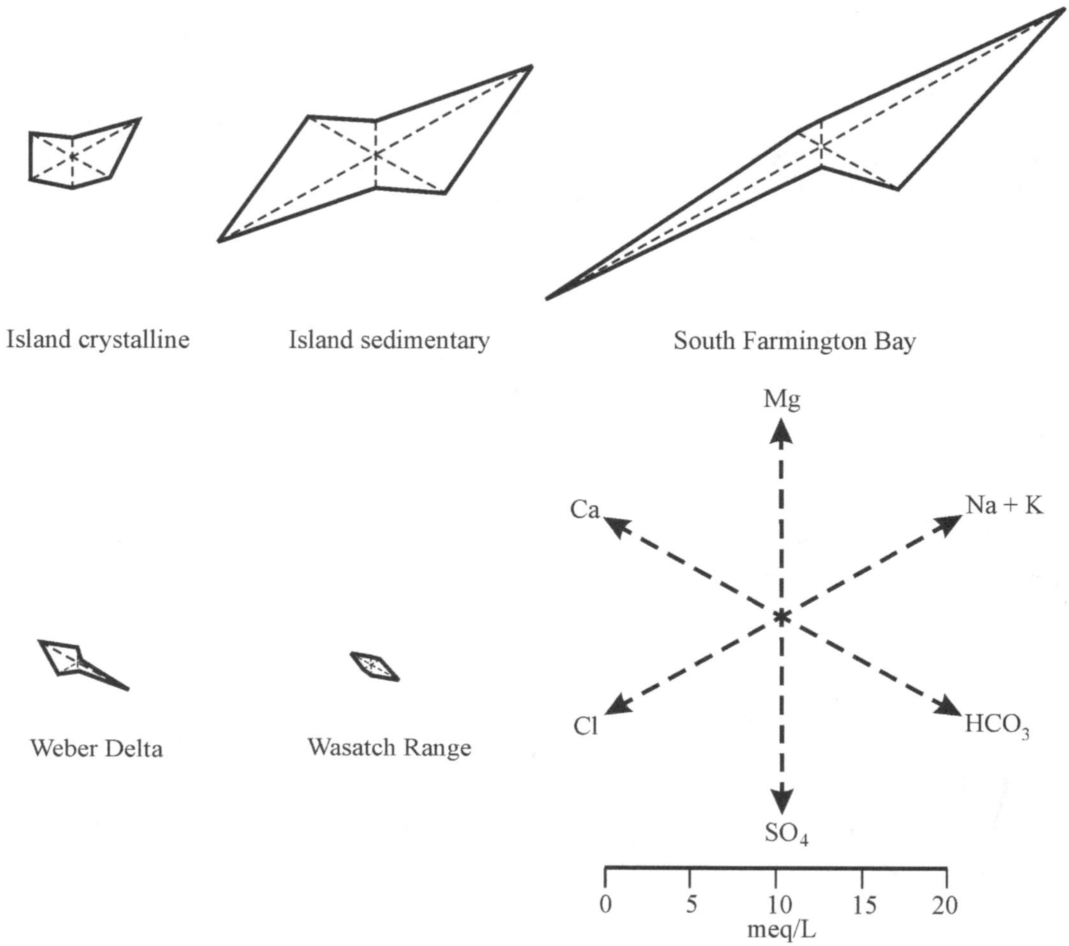

FIGURE 12.7 Radial diagrams of the data in Table 12.3. Cations are on the top half of the diagram, and anions are on the bottom half.

Stiff diagrams (Stiff, 1951) are the most commonly used pattern diagram and are one of the more useful graphical forms for representing major ion chemistry. The shape of the diagram indicates solute composition, and the size of the diagram indicates solute concentration. Stiff diagrams are the easiest pattern diagram to construct and visually interpret because the axes are horizontal and are parallel to each other. The diagrams are often placed on maps, cross-sections, stratigraphic sections, well logs, etc. The diagrams use 3 or 4 parallel horizontal axis which extend on both sides of a vertical zero (0) concentration axis. Cation concentrations are plotted on the left side, and anion concentrations are plotted on the right side of the diagram. The locations of cations and anions correspond to commonly related chemical species. The top three horizontal axis are typically reserved for the seven major ions and the optional lower axis, if used, is for CO_3^{2-} and Fe(total). Additional axes can be added, and axes can be labeled as needed. It is often useful to combine HCO_3^- and CO_3^{2-}. Na–Cl are commonly plotted on the top axis, Ca–HCO_3 on the second axis, and Mg–SO_4 on the bottom axis. Concentrations are in meq/L. The axes used for plotting purposes are usually omitted from completed diagrams, and concentrations are usually displayed on a separate diagram.

Stiff diagram shapes can be a great aid in the preliminary interpretation of groundwater chemical evolution and for displaying groundwater chemistry to non-technical people and groups. Being

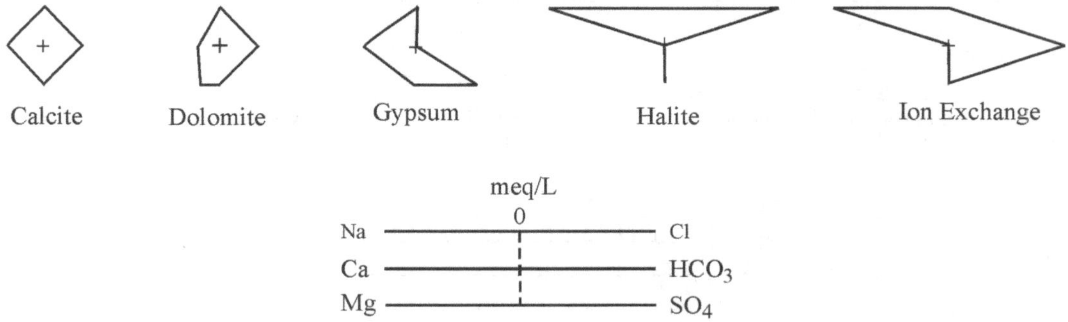

FIGURE 12.8 Typical shapes of Stiff diagrams of water that have encountered selected minerals and Na-ion exchange. The + in each diagram represent the intersection of the Ca–HCO_3 axis with the 0 meq/L concentration. Diagrams are not scaled to mineral saturation index. If the halite diagram were scaled to the gypsum diagram the halite diagram would not fit on the page.

familiar with the typical shape for groundwater with respect to common rock forming minerals is the starting point for analyzing Stiff diagrams. Typical shapes are illustrated in Figure 12.8. In the figure, the sizes of the diagrams are not scaled to mineral saturation. If mineral saturation were displayed, the size of the halite diagram would dwarf the other diagrams. Carbonate mineral dissolution results in two distinct shapes. Calcite has a rhombus shape with essentially no Cl^-, Na^{2+}, SO_4^{2-}, and Mg^{2+}, or dolomite adds Mg^{2+} while reducing Ca^{2+} by an equal amount. It is important to keep in mind that calcite and dolomite are end members of Ca–Mg–HCO_3^- mineralogy and that almost all aquifer carbonate minerals are not pure calcite or dolomite but rather are a Ca_x–Mg_y–HCO_3^- mineral. Gypsum and anhydrite dissolution have equal amounts of Ca^{2+} and SO_4^{2-}. Halite dissolution results in an inverted triangle shape, and ion exchange starts with the calcite or dolomite shape and substitutes Na^{2+} for Ca^{2+} and Mg^{2+}.

Stiff diagrams seldom have the shape of a single mineral. Instead, the shapes represent the encounter of water with several minerals along the flow path. As an example, the data in Table 12.3 are plotted as Stiff diagrams in Figure 12.9. A cursory examination of the diagrams reveals several clues about the history of the groundwaters and their potential chemical evolution, starting with the Wasatch Range, which is a high elevation mountain range downwind of the Great Salt Lake (Figure 12.2). The water has a composition and concentration of water that has been in contact with carbonate minerals, plus a source of Na^+, Cl^-, and SO_4^{2-}. The water is undersaturated with respect to both calcite and dolomite (Table 12.3), which is consistent with a mixed bedrock lithology of granites, metamorphic, and carbonate rocks. The SO_4^{2-} could be from gypsum in the carbonate rocks. However, precipitation is a probable source when considered in light of the Na^+ and Cl^- concentrations that are from the evaporation of the nearly halite saturated Great Salt Lake water and dry and wet fall from west desert dust. If it were available, $\delta^{34}S$ data might help sort out the sulfate source. The Weber delta, located between the Wasatch Range and the Great Salt Lake, has water that is similar to Wasatch Range water but with greater concentrations of all ions. Weber delta sediments are derived from the erosion of the Wasatch Range with minor contributions of Pleistocene Lake Bonneville sediments. Weber delta water is consistent with carbonate mineral dissolution in that the water is saturated with respect to calcite and nearly saturated with respect to dolomite. The South Farmington Bay aquifer is down gradient of the Weber delta and contains significant amounts of clay-rich sediments from the Pleistocene Lake Bonneville. Despite the fact that the Farmington Bay sediments contain clay, ion exchange is not anticipated because the concentration of Na^+ and Cl^- are similar. The elevated Na^+ and Cl^- concentrations may be the result of the dissolution of small amounts of halite that is common in the aquifer sediments.

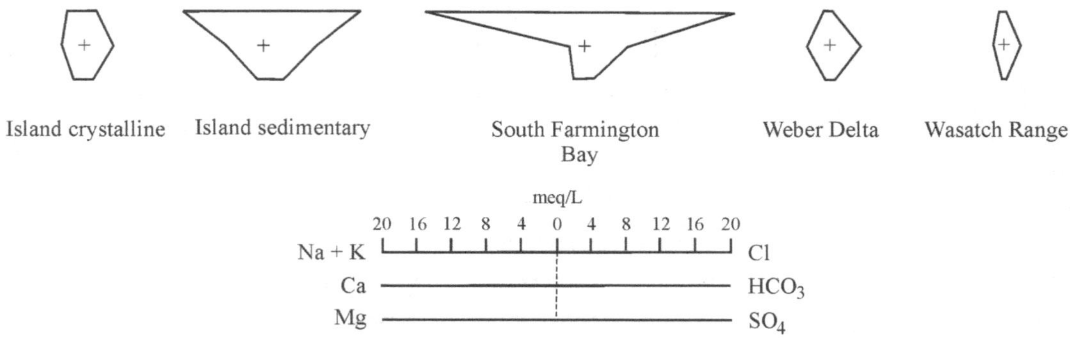

FIGURE 12.9 Stiff diagrams of the data in Table 12.3. All of the groundwaters are from Antelope Island in Utah's Great Salt Lake and the surrounding environment.

Antelope Island contains both crystalline and near-shore sedimentary aquifers. The lithology of the crystalline aquifer is similar to the Wasatch Range. The Na^+ and Cl^- concentrations are similar to the South Farmington Bay waters, suggesting in situ dissolution of halite. The calcite and dolomite supersaturation and the elevated Ca^{2+}, HCO_3^-, and Mg^{2+} suggests carbonate mineral dissolution and may indicate that mineral precipitation has not kept up with dissolution, possibly due to the ionic strength effect. The island's crystalline aquifer Stiff diagram has a similar shape as the Wasatch Range, except the ion concentration is greater and the water is supersaturated with respect to both calcite and dolomite. The Ca^{2+} and Mg^{2+} concentrations are consistent with carbonate mineral dissolution, but the Na^+ and Cl^- concentrations require an external source. The most likely Na^+ and Cl^- source is evaporated Great Salt Lake water and dry fall of west desert dust.

Trilinear diagrams are among the most powerful graphical tools available. Over the years, a variety of trilinear forms have been developed and applied to groundwater systems. Trilinear diagrams have the advantage over Stiff diagrams and many other graphical forms in that more than one water analysis can be represented on a single diagram. Plotting multiple analyses on a single diagram allows for the identification of chemically similar or chemically different groundwater types and for the visualization of chemical mixing and evolution. Most trilinear forms contain two or more pyramid shapes on which cations and anion concentrations are separated. Solute concentrations, in meq/L, are plotted as a percentage of the total concentration of either cations or anions or as a percentage of the total ion concentration. Two trilinear forms, the Durov and Piper, are described here.

The Durov diagram (Durov, 1948) is a plot where the percentage of major ions, in meq/L, are plotted on two separate triangles positioned at 90° (Figure 12.10). Anions are on the upper triangle and cations on the sidewise triangle. The data in each triangle are projected to a common rectangular grid. The plot commonly contains two additional rectangular grids where pH vs. Cl and TDS or conductivity vs. Na + K is plotted. The scales of the auxiliary rectangular grids are determined by the pH and TDS of the waters. The triangles and auxiliary rectangular plots share the scales for Cl and Na + K. The typical configuration of axis labels is shown in Figure 12.10, but it is sometimes useful to label the axes differently to better visualize water groups. In the figure, the plotting location of island sedimentary system waters from Table 12.3 are shown. Durov plots are not commonly used due to their complexity and the popularity of Piper diagrams.

Hill (1940) made the first attempt to construct a diagram containing the trilinear form. This diagram was refined by Piper (1944) and is widely used. It is a common practice to refer to the Piper diagram as a trilinear diagram. Care should be exercised when making such a reference because other trilinear forms are still in use and have application-specific advantages. It is preferable to refer to the Piper form as either the Piper diagram or as a trilinear diagram in the manner of Piper. In the original Piper diagram, both solute concentrations and contents were represented. Modern

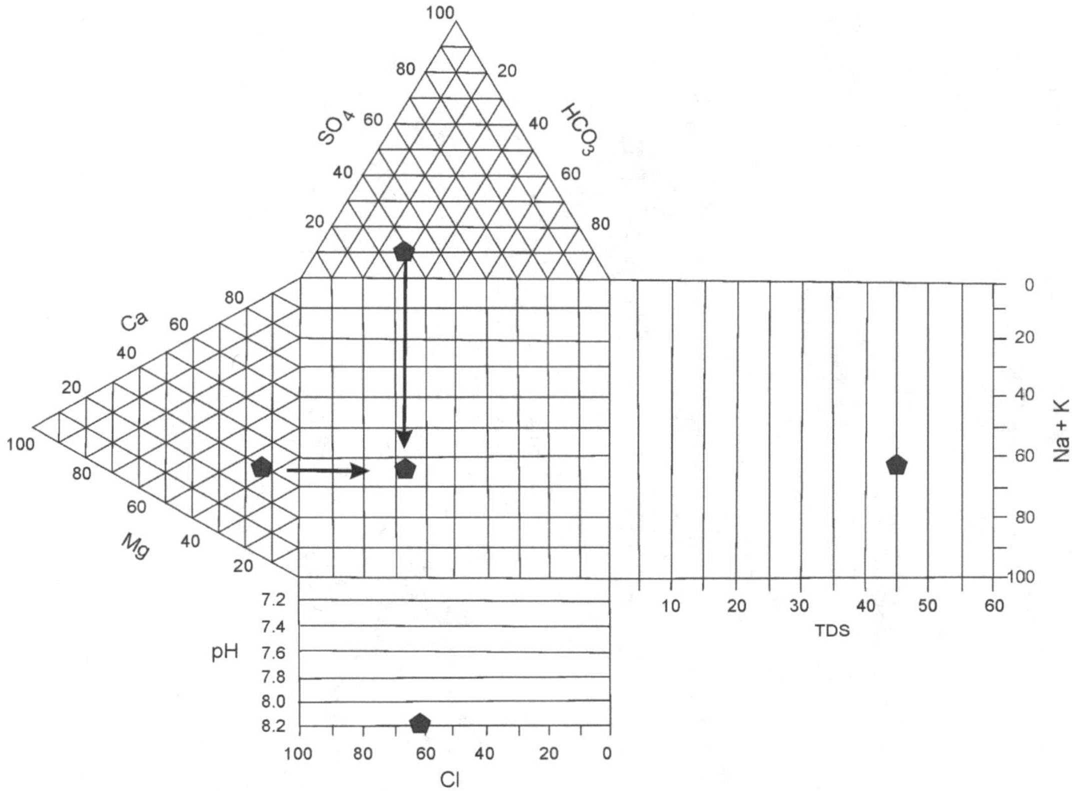

FIGURE 12.10 Construction of a Durov diagram using Antelope Island sedimentary aquifer data from Table 12.3.

versions usually omit concentrations. Concentration was illustrated by the size of each element in the diagram. The Piper diagram consists of two smaller triangles for cations and anions and a larger diamond-shaped central diagram on which the small triangle data are projected. The cations are plotted on the lower left triangle and the anions are plotted on the lower right triangle. The data in each triangle is based on either the percentage of each species relative to the total concentration (meq/L) of either the cations and anions or the total sum of the cations and anions (Figure 12.11). When the percentages are based on the sum of the cations and anions individually, the scale of the triangular plots is 0 to 50, and it is 0 to 100 when the percentages are based on the sum of the cations and anions.

Piper diagrams are usually constructed using major ion chemistry, but Piper, Stiff, and other diagrams can be modified to accommodate minor water chemistry by substituting low concentration cations, anions, and non-ionic species for high concentration species.

The procedure for constructing the Piper diagram in Figure 12.11 is:

1. Convert major ion solute data to meq/L.
2. Calculate the percentage of each individual cation and anion relative to the sum of cations and anions. For illustrative purposes, the mean solute concentrations of Island crystalline water (Table 12.3) have been used. Based on the sum of the cations and anions, the percentage concentration of the major ions are cations (Ca^{2+} = 15.8, Mg^{2+} = 8.9, $Na^+ + K^+$ = 25.8) and anions ($HCO_3^- = 20.0$, $SO_4^{2-} = 9.5$, $Cl^- = 20.0$).

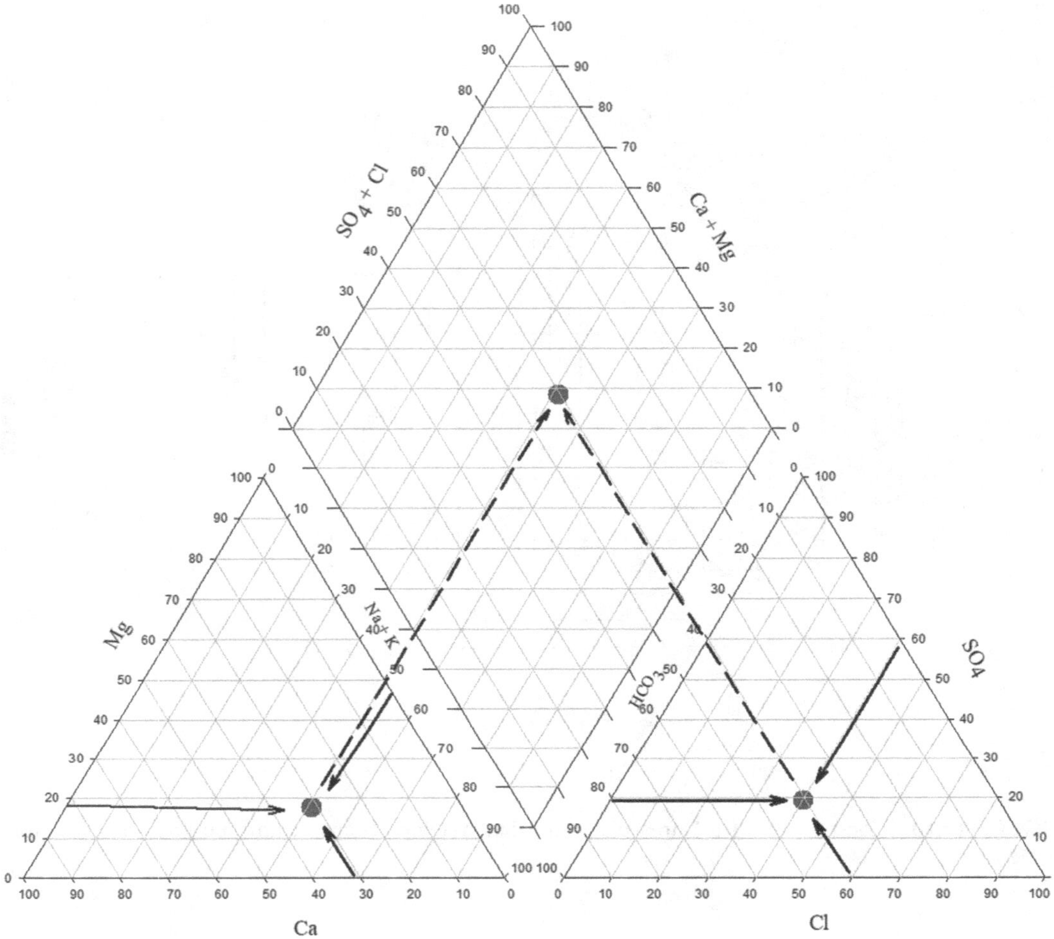

FIGURE 12.11 Construction of a Piper diagram. The triangle on the left represents the cations, the small triangle on the right represents the anions, and the large diamond in the center represents the composite solute concentration. The plotted location of cations and anions are then projected to their intersection on the diamond plot. The individual solute concentrations are plotted as a percentage of their meq/L concentrations of the cations and anions, respectively. In some Piper diagrams, the small triangular fields are represented as 0 to 50 percent.

3. Plotting locations of cations (in the lower left triangle) and of anions (in the lower right triangle) are the intersections of the percentage of each cation or anion projected parallel to the appropriate axis.
4. The plotting location in the large diamond is the intersection of the projection of the cation location in the triangle parallel to the $SO_4^{2-} + Cl^-$ axis, and the projection of the anion location in the triangle parallel to the $Ca^{2+} + Mg^{2+}$ axis.
5. Concentrations can, if desired, be represented by using open circles as plotting symbols. The radius of the circle can then be used to represent the total solute concentration.

Piper plots are particularly useful for distinguishing the chemistries of different aquifers and for showing the flow path chemical evolution. For example, the evolution of water types is illustrated in Figure 12.12, and the chemical evolution along flow paths is shown in Figure 12.13.

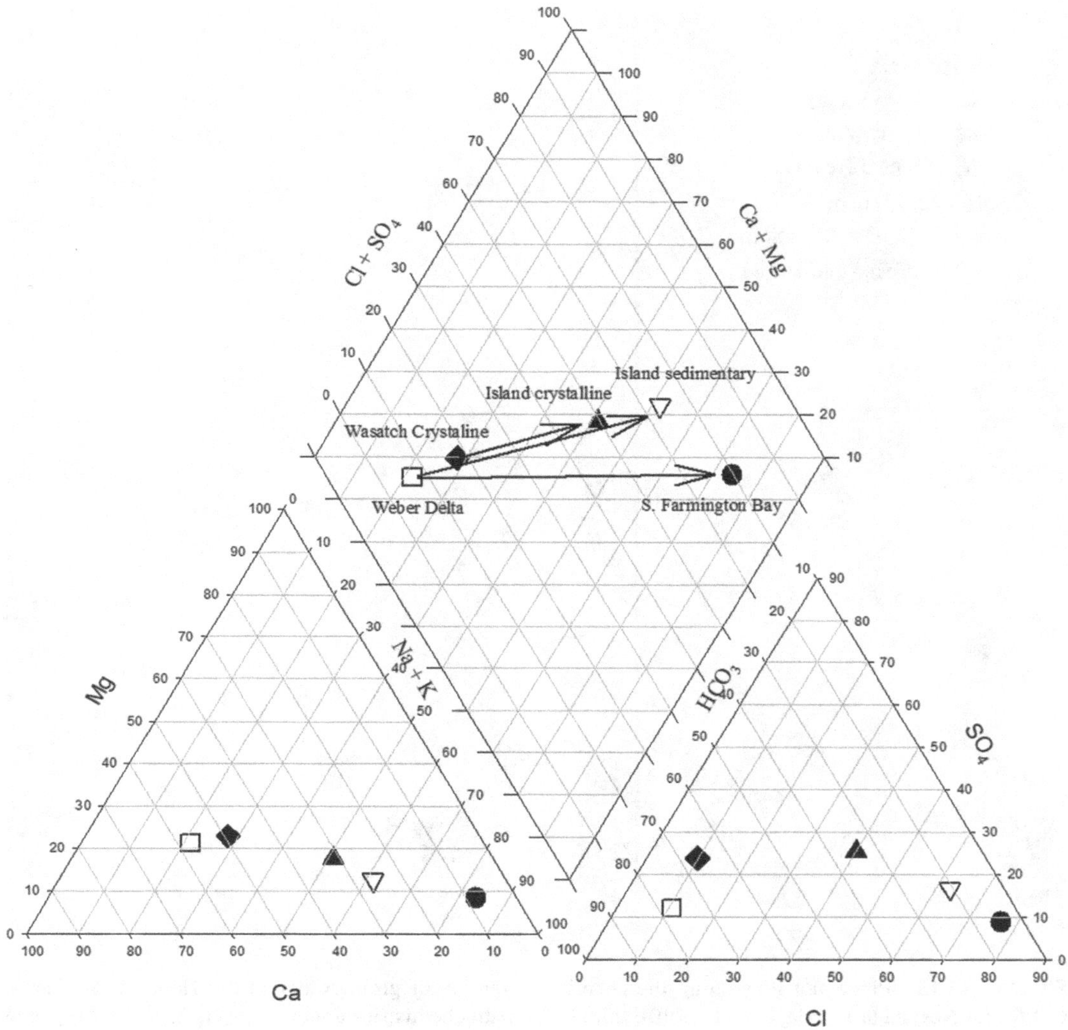

FIGURE 12.12 Piper plot of the data in Table 12.3 with arrows showing potential chemical evolution. Weber Delta type water can evolve from a Ca^{2+}–Mg^{2+} rich water to a Na^{+}–K^{+} rich water in the Antelope Island sedimentary and Southern Farmington Bay aquifers. Wasatch Range crystalline aquifer $Ca^{2+} - HCO_3^-$ type waters can evolve to island crystalline aquifer mixed ion-type waters.

12.3.2 Hydrogeochemical Facies

Back (1960, 1966) defined the concept of hydrochemical facies on the basis of plotting locations on the Piper diagram (Figure 12.14). Hydrochemical facies are a description of water chemistry in terms of the dominant ions. The descriptions are useful for describing the geochemical evolution of water and may be applied to maps, cross-sections, fence diagrams, etc. The facies are determined by their dominate cations and anions, which reflects the bedrock lithology at the recharge area, the lithology along the groundwater flow path, and groundwater mixing. Cation facies include calcium, magnesium, and sodium and potassium type water, and anion facies include bicarbonate–carbonate, chloride, and sulfate type water. The total water type is defined by the major cations and anions.

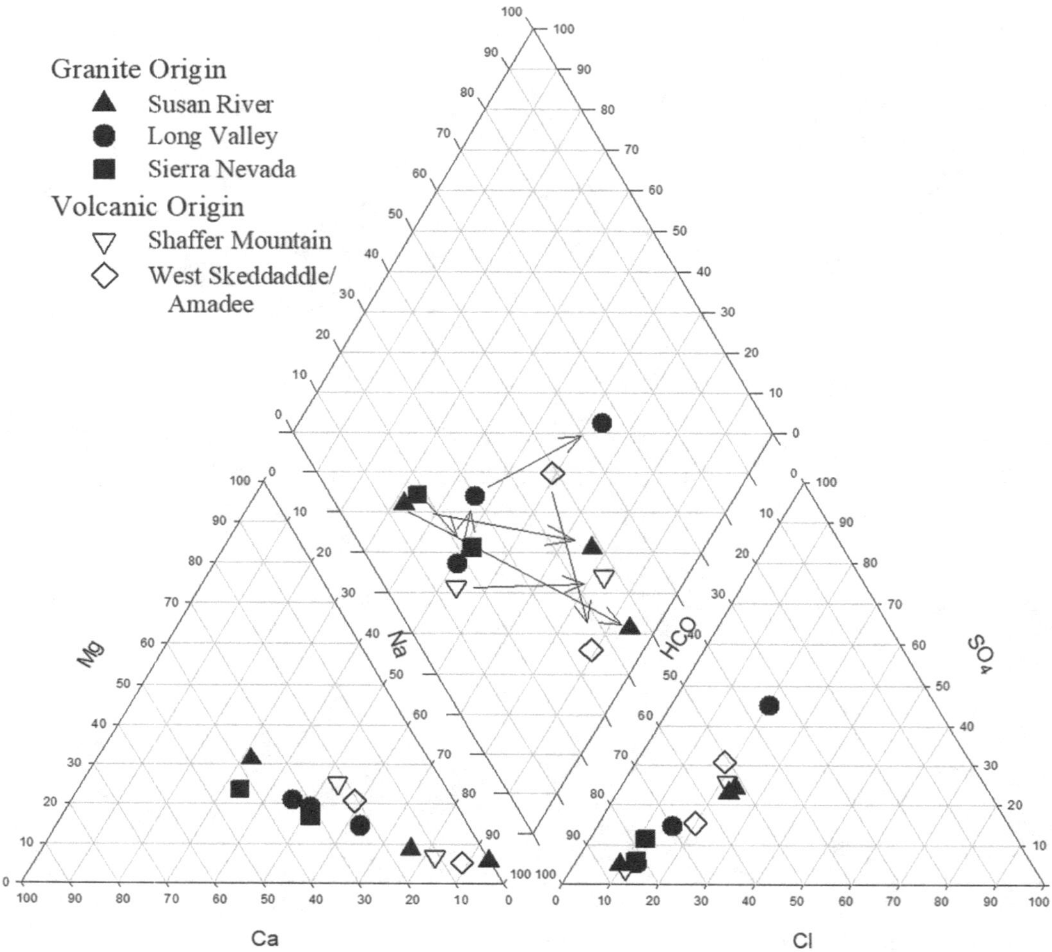

FIGURE 12.13 Piper plot illustrating the chemical evolution of groundwater in the Honey Lake Basin, California-Nevada (after Mayo et al., 2010). Initial flow path chemistries are Ca^{2+}-HCO_3^- or Na^+-HCO_3^- type water. Down gradient, most of the waters interact with clay-rich playa sediments and evolve into Na^+–mixed anion–HCO_3^-, and Na^+–Cl^- type water.

Lloyd and Heathcote (1985) described the significance of many common hydrochemical facies. A modified summary of their description is:

1. HCO_3^- and Ca^{2+} are the dominant ions: Often indicates recharge or flow through limestone, sandstone, and recharge in some granitic rocks when the TDS is very low.
2. HCO_3^- is the dominant anion with Ca^{2+} and Mg^{2+} as the dominant cations: Often indicates dolomite dissolution; where Ca^{2+} and Na^+ are the dominant cations, partial ion exchange is possible.
3. HCO_3^- and Na^+ are the dominant ions: Often indicates ion exchange. In sodium-rich environments where excess CO_2 produces HCO_3^-, other processes may occur (Winograd and Farlekas, 1974).
4. SO_4^- is the dominant anion or mixed anions with Ca^{2+} as the dominant cation: When the TDS is elevated, this often indicates recharge in gypsum or anhydrite terrains; when the

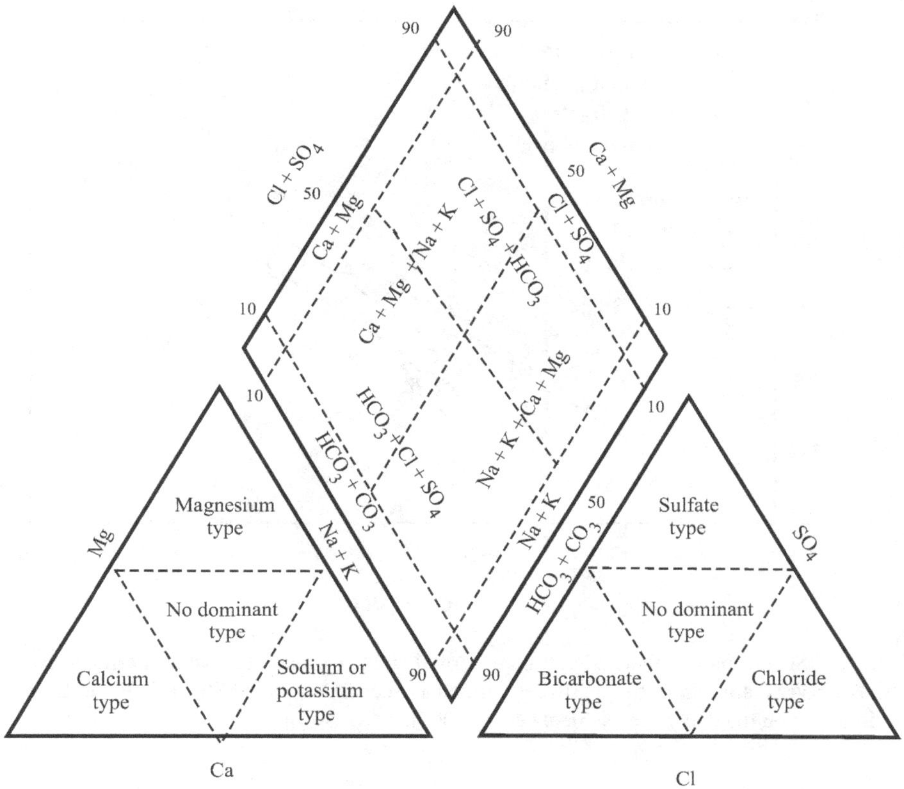

FIGURE 12.14 Plotting locations of hydrogeochemical facies. The facies are determined by their dominate cations and anions, which reflect the bedrock lithology at the recharge area, the lithology along the groundwater flow path, and groundwater mixing.

TDS is low, this may indicate recharge in lavas. It may also represent mixed water or simple dissolution.

5. No dominant ions: Usually indicates mixing.
6. SO_4^- is the dominant anion or mixed anions with Na^+ as the dominant cation: This water seldom naturally occurs and indicates probable mixing.
7. Cl^- and Ca^{2+} are the dominant ions: This water seldom naturally occurs and may indicate cement contamination in a well or reverse ion exchange in Mg^{2+} – Cl^- waters.
8. Cl^- is the dominant cation and no dominant cations: This water may result from reverse ion exchange in Na^+–Cl^- waters.
9. Cl^- and Na^+ are the dominant ions: May indicate dissolution of halite, recharge from excessive evaporation or evapotranspiration, seawater intrusion, connate water, or deep sedimentary water. The diagram does not provide much help in determining sources of Na^+–Cl^- waters.

12.3.3 Scatter Plots

Scatter plots are a convenient way to explore linear, log-linear, and log-log relationships between various solute, other geochemical parameters (i.e., temperature, pH, conductivity), isotopic, geologic, and topographic data. Scatter plots are particularly useful for displaying and analyzing large datasets and for visualizing data trends. Each datum in an x–y scatter plot is represented with

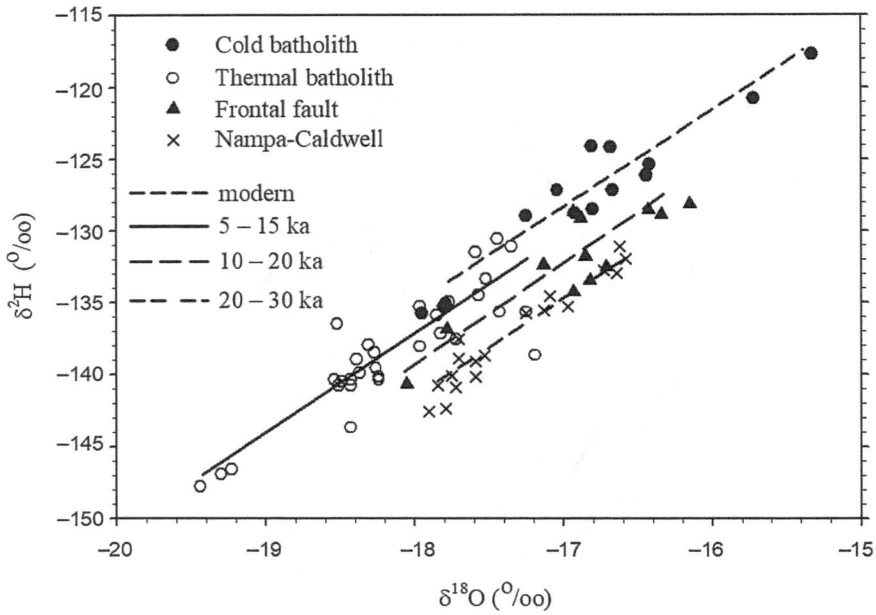

FIGURE 12.15 Scatter plot of thermal and non-thermal stable isotopic data from the Idaho Batholith and adjacent Snake River Plain, Idaho. Statically determined age dependent Local Meteoric Water Lines (LMWL) were superimposed on the data (after Schlegel et al., 2009).

a symbol. A single symbol can be assigned to an individual datum or to each member in a data group. For example, in a multilayered aquifer, all waters in horizon 1 could be represented as an open box (□) and all horizon 2 waters could be represented as a filled box (■). Scatter plots can be constructed by using any one of a number of commercially available computer programs including spreadsheets and scientific graphing packages. Some spreadsheet and graphical programs will also analyze the data using a variety of statistical methods and will automatically plot statistical trend lines. Forms of x–y plots, such as δ^2H vs. $\delta^{18}O$, are commonly plotted relative to established x- and y-axes. In most plots the axes are left to the creativity of the researcher.

Two examples of scatter plots include Schlegel et al. (2009), who used scatter plots and statistically generated trend line equations to identified local meteoric water lines (LMWL) for the thermal and non-thermal groundwater systems in the Idaho Batholith and adjacent Snake River Plain, Idaho (Figure 12.15). Elsewhere, as part of a preliminary data analysis of 282 groundwater samples in the Cretaceous Bohemian Basin, Czechia, the concentrations of Ca^{2+} vs. HCO_3^- and Ca^{2+} vs. SO_4^{2-} indicates that neither simple limestone or gypsum dissolution can be responsible for the Ca^{2+} in most of the waters and that considerably more data analysis is needed (Figure 12.16).

12.3.4 Tie Line Diagrams

Binary and tertiary tie line diagrams are used to illustrate theoretical relationships and to analyze data trends. The plots can employ single or multiple tie lines. In some diagrams the tie lines are isothermal or isobaric lines that span two or more phase regions, where each end of the tie line represents an end member. For example, an idealized binary liquid–vapor pressure diagram is a theoretical tie line plot (Figure 12.17). The mixture consists of two liquids, A and B. The diagram shows the mole fraction of liquid A in both the liquid and vapor phase as a function of pressure. The total vapor pressure of the binary system is calculated using Raoult's law, and Dalton's law is used

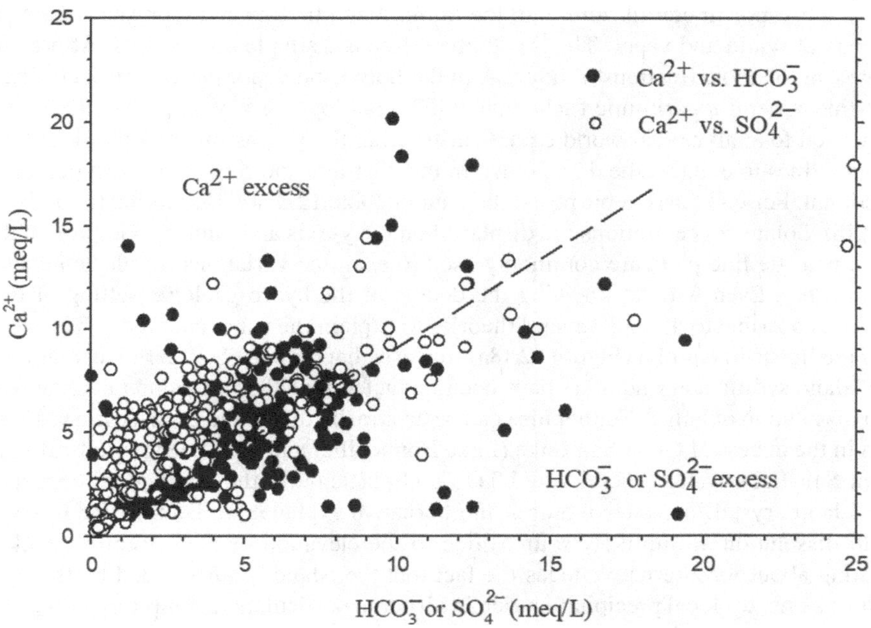

FIGURE 12.16 Scatter plot of Ca^{2+} vs. HCO_3^- and Ca^{2+} vs. SO_4^{2-} of groundwater in the Cretaceous Bohemian Basin, Czechia (unpublished data). The dashed line represents calcium vs. anion equilibrium. Most of the waters have a Ca^{2+} excess vs. HCO_3^- and many of the waters have excess SO_4^{2-}, both of which suggest the water is not derived from simple limestone or gypsum dissolution.

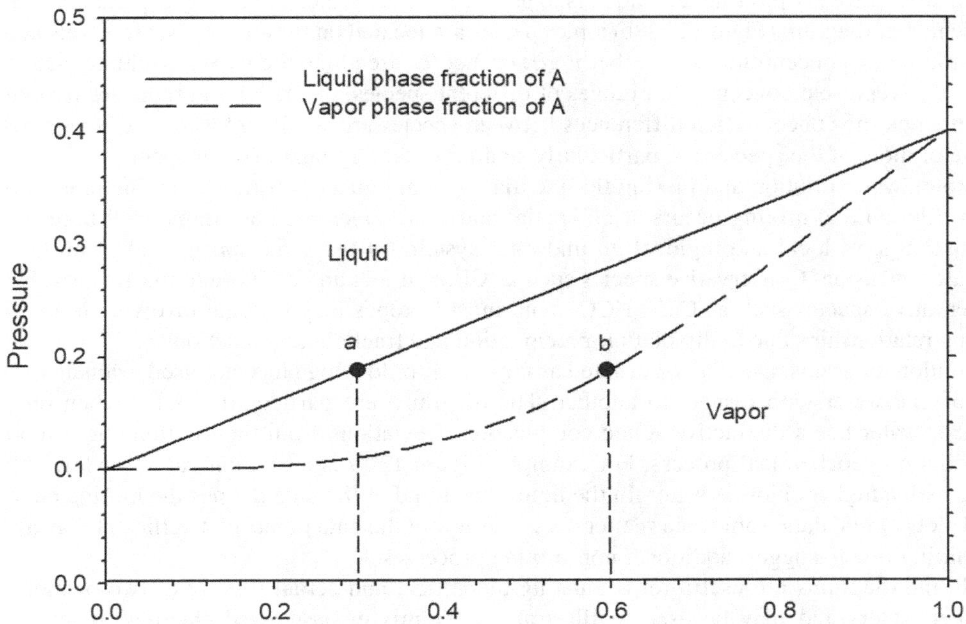

FIGURE 12.17 Idealized liquid–vapor pressure tie line plot of a binary liquid solution showing the mole fraction of both the liquid and vapor of volatile liquid A. The mole fractions are partial pressure dependent. At pressure 0.2, *a* and *b* represent the mole fraction of liquid A.

to calculate the vapor in equilibrium with the liquid. Each tie line end represents the equilibrium compositions of liquid and vapor. The liquid phase line is a simple mixing line. At pressure 0.2, *a* and *b* represent the mole fractions of liquid A in the liquid and vapor phases, respectively.

Linear-linear, semi-logarithmic (Schoeller, 1962), and log–log x–y and x–y–z tie line plots are frequently used to analyze real world data. Constructing the graphs involves plotting the data and then drawing lines to connect the data points. In the diagrams plotting the underlying data as symbols is optional. For comparison purposed the data in Table 12.3 are used in the following example (Figure12.18). Solute concentrations are displayed on the y-axis and ionic species are on the x-axis.

Linear-linear tie line plots are commonly used to examine variations in selected chemical and other parameters. Even without knowing the details of the hydro-geologic setting of the data in Table 12.3 it is possible to develop several theories to explain the water chemistry. The elevated Na^+ and Cl^- in the linear-linear plot (Figure 12.18a) suggests that both the Southern Farmington Bay and Antelope Island sedimentary aquifers have been impacted by either intrusion of saline water or by the in situ dissolution of halite. Neither idea can be arbitrarily dismissed because the sediments were laid down in the ancestral Great Salt Lake (Lake Bonneville) and have been periodically inundated with Great Salt Lake water. The elevated TDS and particularly the Na^+ and Cl-concentration of Antelope Island crystalline water requires an alternative explanation. Both saline water intrusion and in situ dissolution are unlikely sources due to the elevation of the crystalline rock massive. When casting about for alternative ideas the fact that the island is surrounded by the hypersaline Great Salt Lake makes local precipitation rich in airborne particulates a tempting source. The large-scale differences between the ion concentrations in the water groups compress the concentrations of most anions and cations. An alternative tie line plotting technique for plotting data containing one or more outlier concentrations or data with very large concentration differences is to use a break in the concentration axis (Figure 12.18c). Placing the data break between 5.1 and 5.2 meq/L permits an easier examination of the low concentration ions.

An alternative x-y plotting technique is to use logarithmic scales to visually separate the data. In the Schoeller diagram (Figure 12.18b) ionic species are located on the – axis at set intervals and logarithmic solute concentrations, in either mg/L or meq/L, are along the y-axis. Multiple y-axes may be used to represent concentration ranges of different species. Schoeller diagrams are particularly useful when the concentration differences between species are small and when the concentrations of one or more of the species are particularly smaller or larger than the other species.

Groundwater dilution and mixing tie-line diagrams are an *x–y* form of a tie line plot. Groundwater dilution and mixing occurs at all spatial and time scales such as simple well bore mixing, the discharge of local and regional groundwater systems at the same spring, leaky aquifers, and seawater intrusion. Conservative species such as Cl^-, and δ^2H and $\delta^{18}O$ often mix linearly, but non conservative species such as Ca^{2+}, HCO_3^-, and most isotopes may become involved in non-linear mixing relationships due to dissolution-precipitation and fractionation reactions.

Dilution diagrams, usually constructed as log–linear or log–log plots, are used evaluate the dilution of one water with respect to another. The diagrams are particularly useful when one end-member water has a distinctive solute composition. Deviations from the dilution curve represent non-dilution geochemical process. For example, Figure 12.19 is a dilution curve for low chloride water with a higher chloride water. In the figure, the trend of the data defines the mixing curve. As in all plots of real data, some data scatter occurs. A few of the data points plot sufficiently away from the mixing line to suggest additional non-mixing processes.

Mixing diagrams are useful for evaluating theoretical and actual mixing of two or more end-member waters and may be used to illustrate water mixing-associated chemical reactions and chemical evolution. Conservative constituents are usually used as end-members, although non-conservative parameters are sometimes justified on the basis of the envisioned geochemical reactions and chemical evolution. Mixing diagrams may have linear, log–linear, or log–log scales and straight or curved mixing lines. Straight lines are usually drawn between conservative end-member species and non-evolving non-conservative species. Curved lines often represent evolving chemistry of

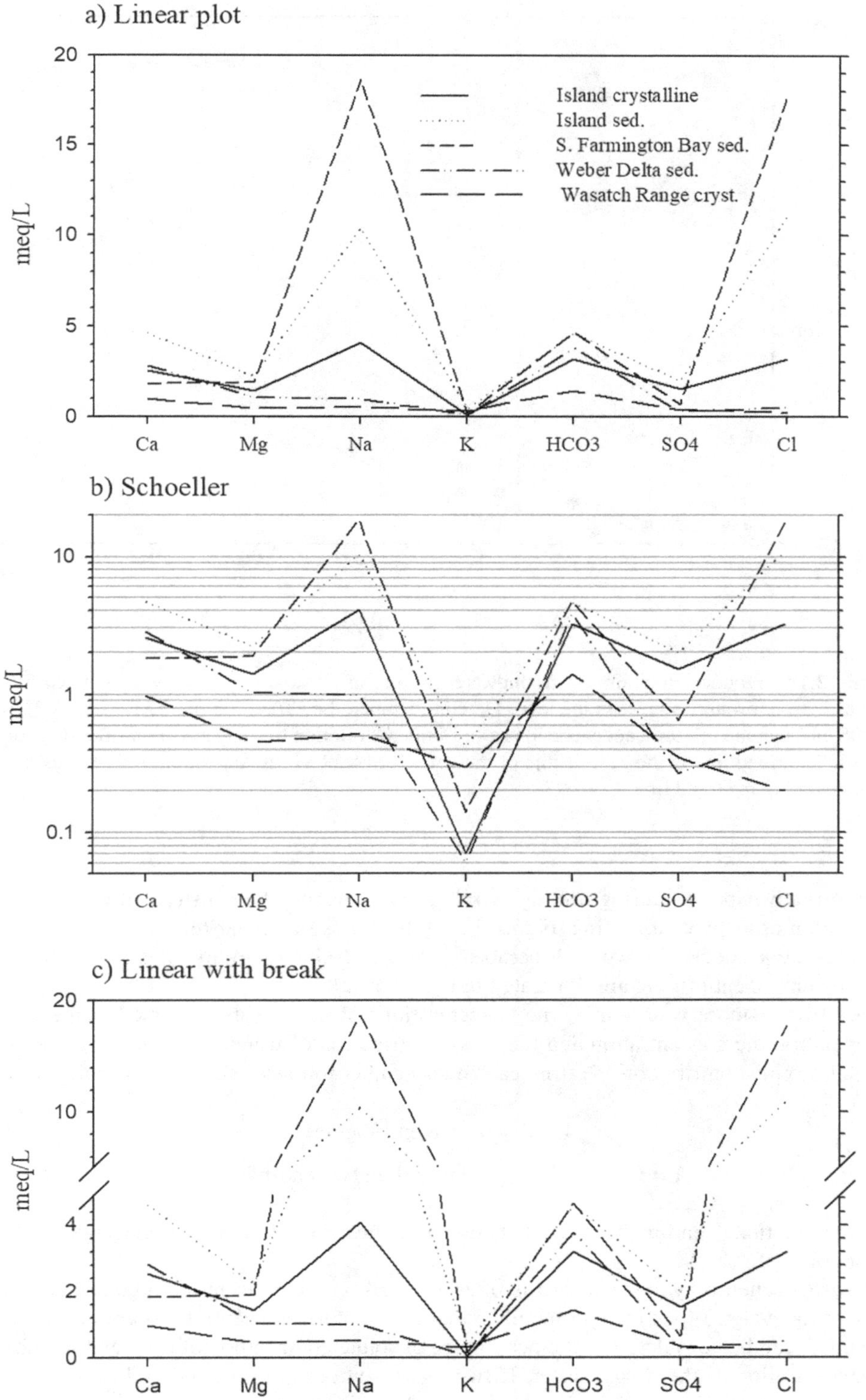

FIGURE 12.18 Comparison of linear (a), Schoeller (b), and linear with x–y break (c) tie line plots of the data in Table 12.3. Showing the underlying data as symbols is optional in tie line plots. Here the differences in the underlying data are represented by changes in the slope breaks of the tie lines.

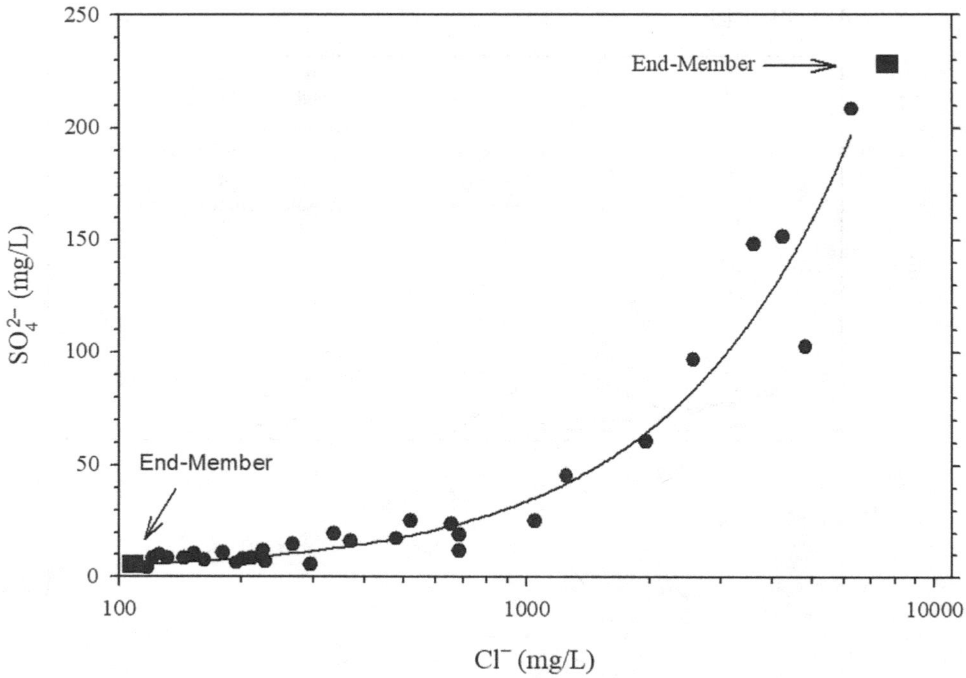

FIGURE 12.19 Dilution diagram for groundwater mixing of Cl^- and SO_4^{2-} rich waters. Each filled circle represents a groundwater sample. In the low TDS end-member, the Cl/SO_4 ratio is 15.5/1 and is 32.8/1 in the high TDS end-member. Significant excursions away from the dilution line suggests non-dilution geochemical processes. The trend line is the best fit line of the data. It may be necessary to remove obvious outlier data before calculating the trend line.

non-conservative species during mixing. Mixing diagrams may be constructed with the x-axis as concentration or as percent mixing. Figure 12.20 illustrates the linear mixing of conservative and non-conservative species in two-end-member systems. Mixing diagrams illustrating ion exchange and three component mixing are illustrated in Figure 12.21.

When two or more waters mix, the concentrations of conservative species in the final water will depend on the concentration and the mass contribution of water from each end member. The fractional mixing contribution (f) from each end-member follows a simple mass balance equation:

$$1 = f_1 + f_2\text{: two end-members}$$

$$1 = f_1 + f_2 + f_3 \ldots \text{ for more than two end-members.}$$

where the fractional end-member contribution is a function of both concentration and percent contribution.

In hydrogeochemistry, our concern is often focused on the concentration contribution and not the flux of the water. When the end-member and final solution concentrations are known or can be estimated, f_x can be considered as the percentage of solute contribution of each end-member to the final concentration of the mixed water. If two water sources are involved and concentrations are known, the concentration contribution from the end-members can be written as:

$$Cm = C_1 f + C_2 (1 - f) \tag{12.2}$$

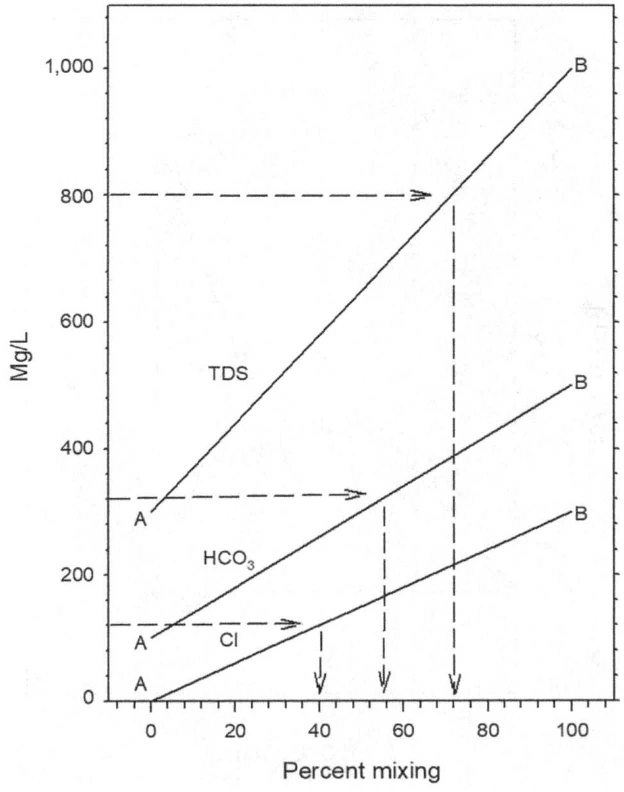

FIGURE 12.20 Two-water mixing diagram illustrating three mixed systems TDS, HCO_3^-, and Cl^- having end-member concentrations of 300 and 1,000, 100 and 500, and 0 and 100 mg/L, respectively. If the mixed waters have TDS, HCO_3^-, and Cl^- concentrations of 800, 325, and 120 mg/L, the mixing would be 40, 60, and 72 percent, respectively.

where
C_m = concentration in the mixed water
C_1 – concentration in end-member 1
C_2 = concentration in end-member 2
f = proportion of end-member 1
$1 - f$ = proportion of end-member 2

If the species are conservative and are not stable isotopes, Equation 12.2 can be rearranged as:

$$f = \frac{C_m - C_2}{C_1 - C_2} \tag{12.3}$$

Example 12.1 Mixing fraction calculation

Problem: Pumping well SW-67 is open to two water-bearing horizons. Calculate the contribution from each horizon when the well is at rest and when pumping.

In the San Luis Valley, Colorado, closed basin, the TDS in a production well fluctuates greatly between resting and pumping episodes (Mayo, 2010). The well screen is open to water-bearing horizons 2 and 3, which have significantly different water quality. The beginning and end of test water quality in pumping well SW-67 and observation wells are listed in Table E12.1.

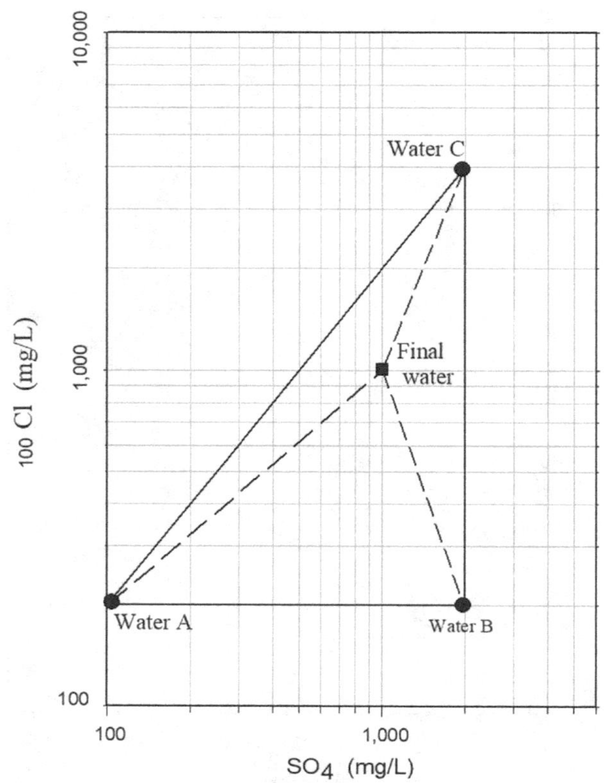

FIGURE 12.21 When three water are mixed, the concentration of the final water will fall in the triangle defined by the mixing lines of end-member waters A–B, A–C, and B–C. If the water sample plots on or very near one of the mixing lines, the water is only a mixture of the two end-members.

Solution: Using Equation 12.3, the calculated fractional contribution of SO_4^{2-} in horizon 2 to the mixed SW-67 water at the beginning end of the pumping test is:

$$f = \frac{70.3 - 11.7}{88.9 - 11.7} = 0.76 \tag{E12.1.1}$$

$$f = \frac{20.1 - 11.5}{67.0 - 11.5} = 0.2 \tag{E12.1.2}$$

and the calculated contribution of Cl^- for beginning and end of the test is:

$$f = \frac{32.6 - 5.6}{39.6 - 5.6} = 0.8 \tag{E12.1.3}$$

$$f = \frac{10.0 - 5.11}{29.6 - 5.11} = 0.2 \tag{E12.1.4}$$

The calculated values are consistent with plotting locations on the $Cl^- - SO_4^{2-}$ mixing diagram for the well waters (Figure E12.1). All of the data plot along or very near the mixing line between observation wells 3B and well 2A. The calculated values in Equations E12.1.1 to E12.1.4 and the mixing line plot demonstrate that, at rest, the pumping well bore is dominated by horizon 2 water, and by horizon 3 water when the well is pumping. This change is due to the fact that at rest the

TABLE E12.1
Beginning and End of Water Quality Test, San Luis Valley Closed Basin

Well	Horizon	Screened Interval (m)	Time	‰				TDS		pH	meq/L						
				$\delta^{34}S$	$\delta^{13}C$	$\delta^{2}H$	$\delta^{18}O$	mg/L	meq/L		Ca^{2+}	Mg^{2+}	Na^{+}	K^{+}	HCO_3^-	SO_4^{2-}	Cl^-
SW-67	2&3	16.5–33.2	B	7.6	-6.0	−83	−9.9	10,600	339.3	8.1	7.04	4.24	155.0	1.16	69.00	70.30	32.6
			E	12.9	−6.0	−96	−13.0	3,530	116.0	8.5	2.24	2.36	52.5	0.23	28.54	20.30	10.0
2A			B	7.8	−7.0	−88	−11.0	12,800	401.8	7.4	8.00	14.00	173.0	1.39	76.87	88.90	39.6
			E	14.5	−6.6	−102	−15.0	10,400	342.3	8.7	4.24	3.56	162.0	0.60	75.30	67.00	29.6
3B			B	7.8	−9.2	−73	−7.9	2,100	68.6	7.6	2.28	2.13	28.6	.043	17.83	11.70	5.6
			E	11.3	−5.0	−78	−9.7	2,170	71.5	8.5	0.70	0.70	34.4	0.14	19.27	11.50	5.1

B = beginning of test, E = end of test

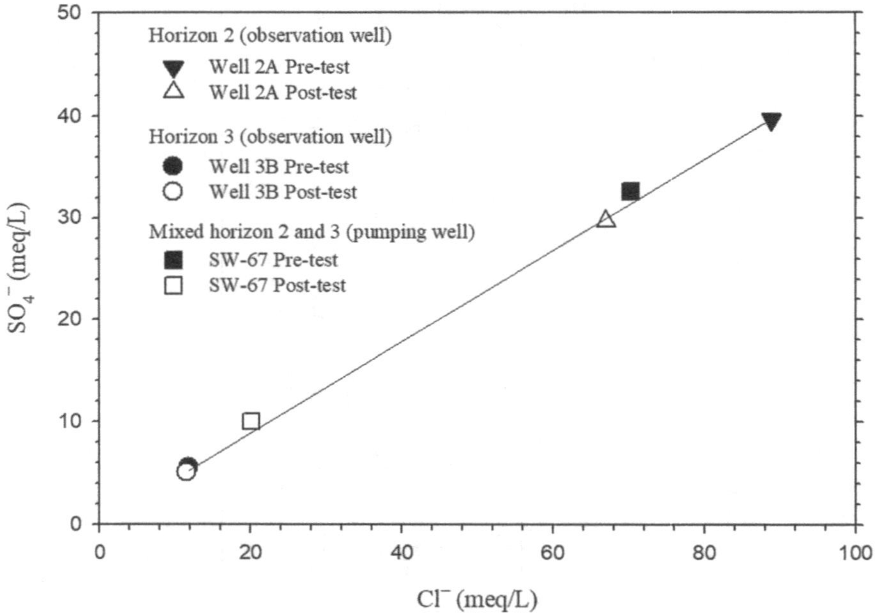

FIGURE E12.1 Mixing line of Cl^- and SO_4^{2-} concentrations in pre- and post-pumping test discharge water in a shallow San Luis Valley, Colorado, aquifer (data from Mayo, 2010). The pre-test pumping well bore mixed-water chemistry was dominated by horizon 2 water, and by horizon 3 water after the test. This change is due to the fact that when the well is at rest the hydraulic head in horizon 2 is greater than horizon 3, and the transmissivity of horizon 3 is greater than horizon 2.

hydraulic head in horizon 2 is greater than horizon 3 and the transmissivity of horizon 3 is greater than horizon 2.

Bjerrum plots are a common way to display the speciation concentrations of acids as a function of pH. The acid may be monoprotic such as ammonium, diprotic such as phosphatic, or triprotic such as carbonate. pH is plotted on the x-axis and concentration is plotted on the y-axis. Concentration may be mole fraction or activity. Because of the wide range of calculated concentrations, concentration data are usually plotted on a logarithmic scale. Examples of phosphate, borate, sulfide, and ammonia Bjerrum plots are shown in Figure 12.22.

Example E12.2 Construction Bjerrum plot for the $CO_2(H_2CO_3)-HCO_3^- - CO_3^{2-}$ System

Problem: Construct a Bjerrum plot for the dissolved CO_2 system.

Solution: Bjerrum plots are constructed by preparing a spreadsheet of activity fraction vs. pH for each reaction product using temperature-dependent equilibrium constants, K_1, K_2, etc. The calculated spreadsheet data are then plotted as tie lines with pH on the x-axis and activity on the y-axis. For the dissolved inorganic carbonate system (CO_2- H_2CO_3-HCO_3^--CO_3^{2-}), reactions are:

$$DIC = [CO_2] + [HCO_3^-] + [CO_3^{2-}] \quad (E12.2.1)$$

$$CO_2 + H_2O \rightleftharpoons H_2CO_3 \quad (E12.2.2)$$

$$H_2CO_3^- \rightleftharpoons H^+ + HCO_3^- \quad (E12.2.3)$$

$$HCO_3^- \rightleftharpoons H + CO_3^{2-} \quad (E12.2.4)$$

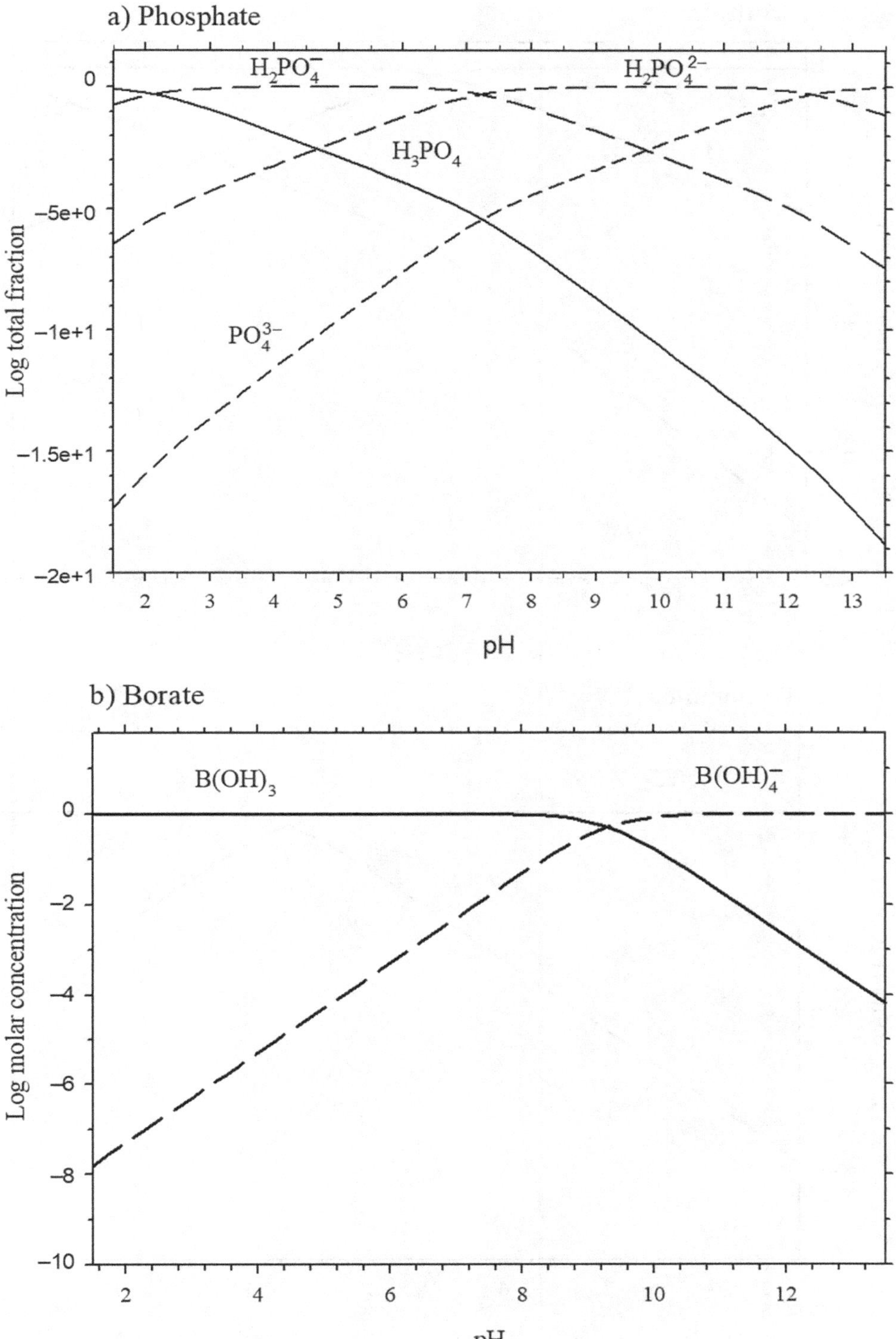

FIGURE 12.22 Bjerrum plots for the acids of (a) phosphate, (b) borate, (c) sulfide, and (d) ammonia at 25 °C.

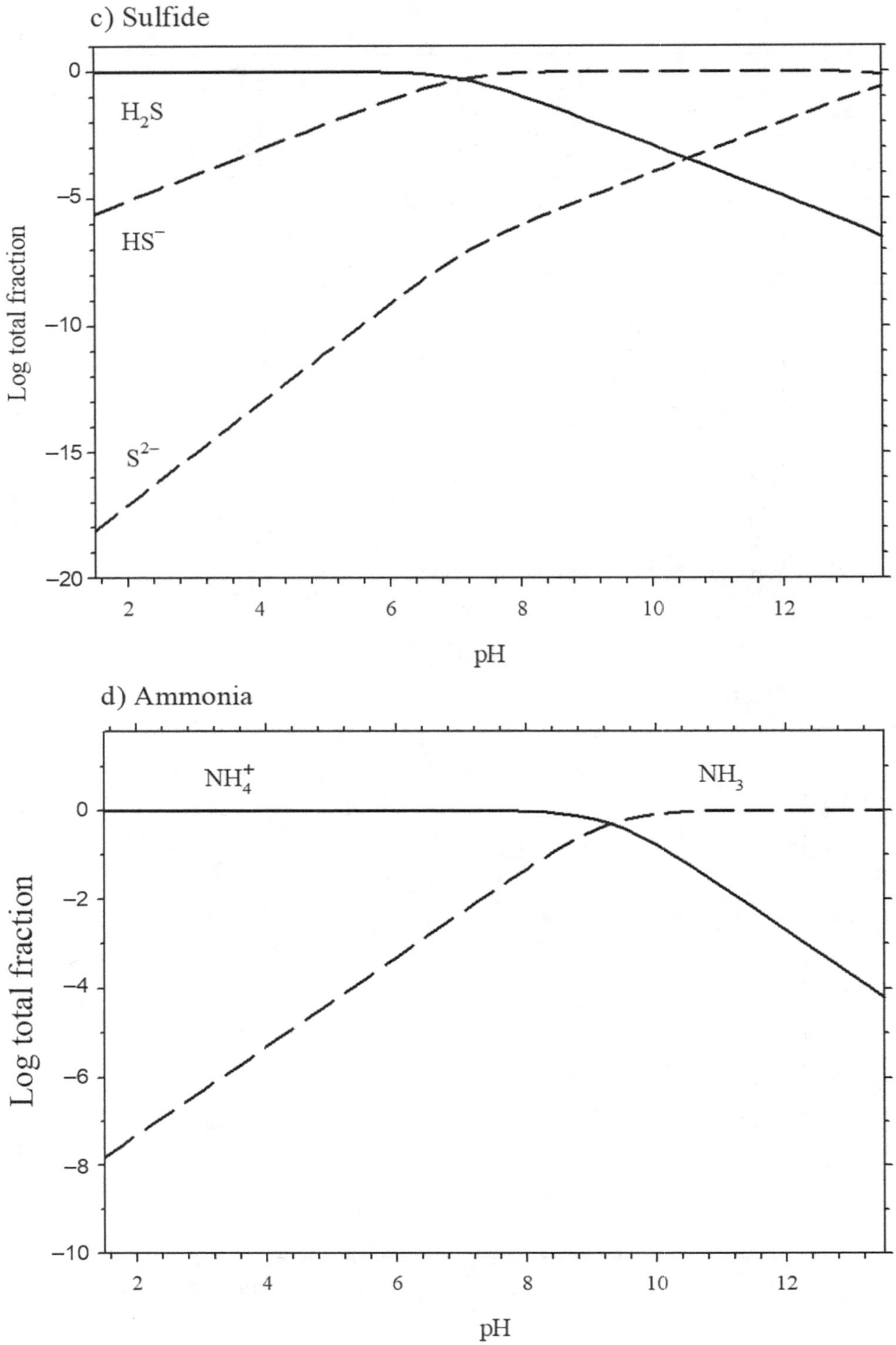

FIGURE 12.22 (Continued)

and the equilibrium constants at 25 °C are:

$$K_1 = \frac{a_{H^+}\ a_{HCO_3^-}}{a_{H_2CO_3}} = 4.467x10^{-7} \quad \text{(E12.2.5)}$$

$$K_2 = \frac{a_{H^+}\ a_{CO_3^{2-}}}{a_{HCO_3^-}} = 4.68 - 10^{-11} \quad \text{(E12.2.6)}$$

Using the equilibrium constants, the activity fraction of each species is calculated using:

$$[H_2CO_3] = \frac{[H^+]^2}{[H^+]^2 + K_1[H^+] + K_1K_2} \times DIC \quad \text{(E12.2.7)}$$

$$[HCO_3^-] = \frac{K_1[H^+]}{[H^+]^2 + K_1[H^+] + K_1K_2} \times DIC \quad \text{(E12.2.8)}$$

$$[CO_3^{2-}] = \frac{K_1K_2}{[H^+]^2 + K_1[H^+] + K_1K_2} \times DIC \quad \text{(E12.2.9)}$$

DIC can be set at any value. In the example, the DIC is 100 percent carbon. Using equations E12.2.7 to E12.2.9, the calculated carbon fractions, as a function of various pH values, are listed in Table E12.2.1 and are plotted in Figure E12.2.1.

12.4 CONTOUR MAPS

Contour maps are an excellent way to analyze and display the spatial relationships of groundwater solute and isotopic chemistry. The maps may be hand drawn but are easily constructed using either free or commercially available software. Examples of contour maps include: 1) shallow groundwater chromium concentrations associated with leakage from natural gas compressor station unlined evaporation ponds in Hinkley, California (Figure 12.23), 2) the ^{14}C concentrations in the Honey Lake closed basin, California-Nevada (Figure 12.24), and 3) the extent of methanogenesis in the San Luis Valley, Colorado, upper confined aquifer base on the ^{13}C concentrations (Figure 12.25). In Figure 12.23 two types of contours are shown. The open-ended water level contours demonstrate the direction of groundwater flow direction and the closed chromium concentration contours illustrate the extent of 0.01 and 0.05 mg/L chromium concentrations. Groundwater flow direction was critical in establishing the source of the far field chromium contamination, and the 0.05 mg/L contour line helped established the area where chromium exceeded the drinking water standard. At Honey Lake the ^{14}C contours complemented the calculated point source ^{14}C ages by illustrating the likely spatial distribution of paleo-groundwater (Figure 12.24). Many wells in the agricultural San Luis Valley exsolve CO_2 and methane gas and have elevated TDS (Figure 12.25). These factors suggested possible methanogenesis. The $\delta^{13}C$ was contoured map was constructed to determine the spatial extent of methanogenesis in the upper confined aquifer and by extension the location of a buried ancestral closed basin and paleo-swamp.

12.5 EQUILIBRIUM DIAGRAMS

Equilibrium diagrams are a convenient way to graphically display and evaluate equilibrium reactions. Multiple water analyses can be evaluated simultaneously on the same diagram. The diagrams represent equilibrium with respect to two or more master variables that are plotted on x-, y-, and sometimes z-axes. A master variable can be as simple as temperature or pH and can be as complex as the log of the ratio of the chemical activities of two or more constituents. In one form or another, all equilibrium diagrams incorporate temperature. Most research and textbook examples of

TABLE E12.2
Calculated Activity Fractions for the CO_2 System as a Function of pH

	Activity fraction (log)				Activity fraction (log)		
pH	H_2CO_3	HCO_3^-	CO_3^{2-}	pH	H_2CO_3	HCO_3^-	CO_3^{2-}
1				7.25	−0.9518	−0.0518	−3.1318
1.25				7.5	−1.1803	−0.0303	−2.8603
1.5				7.75	−1.4181	−0.0181	−2.5981
1.75				8.0	−1.6616	−0.0116	−2.3416
2	−1.94E-05			8.25	−1.909	−8.99E-03	−2.089
2.25	−3.45E-05			8.5	−2.1594	−9.40E-03	−1.8394
2.5	−6.13E-05	−3.8501		8.75	−2.413	−0.013	−1.593
2.75	−1.09E-04	−3.6001		9.0	−2.6708	−0.0208	−1.3508
3	−1.94E-04	−3.3502		9.25	−2.9352	−0.0352	−1.1152
3.25	−3.45E-04	−3.1003		9.5	−3.2102	−0.0602	−0.8902
3.5	−6.13E-04	−2.8506		9.75	−3.5015	−0.1015	−0.6815
3.75	−1.09E-03	−2.6011		10	−3.8167	−0.1667	−0.4967
4	−1.94E-03	−2.3519		10.25		−0.2629	−0.3429
4.25	−3.44E-03	−2.1034		10.5		−0.3943	−0.2243
4.5	−6.09E-03	−1.8561		10.75		−0.5599	−0.1399
4.75	−0.0108	−1.6108		11.0		−0.7541	−0.0841
5	−0.019	−1.369		11.25		−0.9693	−0.0493
5.25	−0.0332	−1.1332		11.5		−1.1984	−0.0284
5.5	−0.0574	−0.9074		11.75		−1.4362	−0.0162
5.75	−0.0973	−0.6973		12.0		−1.6792	−9.19E-03
6	−0.1604	−0.5104		12.25		−1.9252	−5.19E-03
6.25	−0.2539	−0.3539		12.5		−2.1729	−2.93E-03
6.5	−0.3825	−0.2325	−4.0625	12.75		−2.4216	−1.65E-03
6.75	−0.5456	−0.1456	−3.7256	13.0		−2.6709	−9.28E-04
7	−0.7379	−0.0879	−3.4179	13.25		−2.9205	−5.22E-04

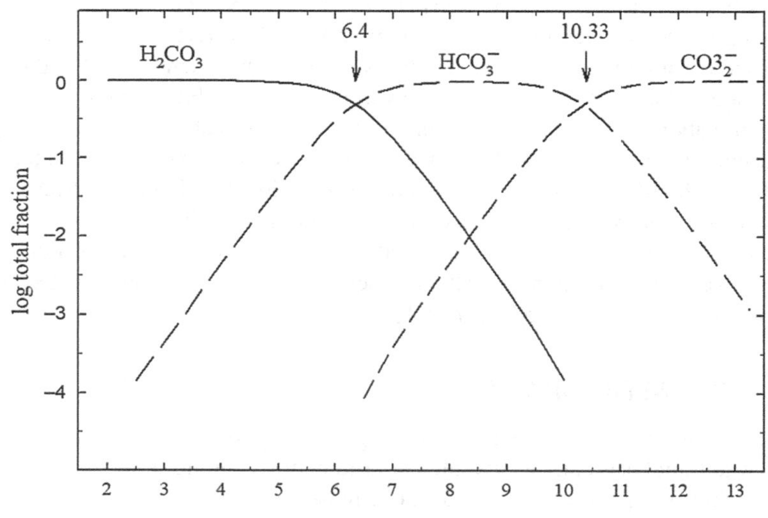

FIGURE E12.2 Bjerrum plot of the $CO_2(H_2CO_3)-HCO_3^- - CO_3^{2-}$ system. Most groundwaters have a pH between the break-over points pH 6.4 and 10.33. Between pH 6.4 and 10.33, the system is dominated by HCO_3^-. The plot was constructed using the data calculated using Equations E12.2.7 to E12.2.9 and as displayed in Table E12.2.

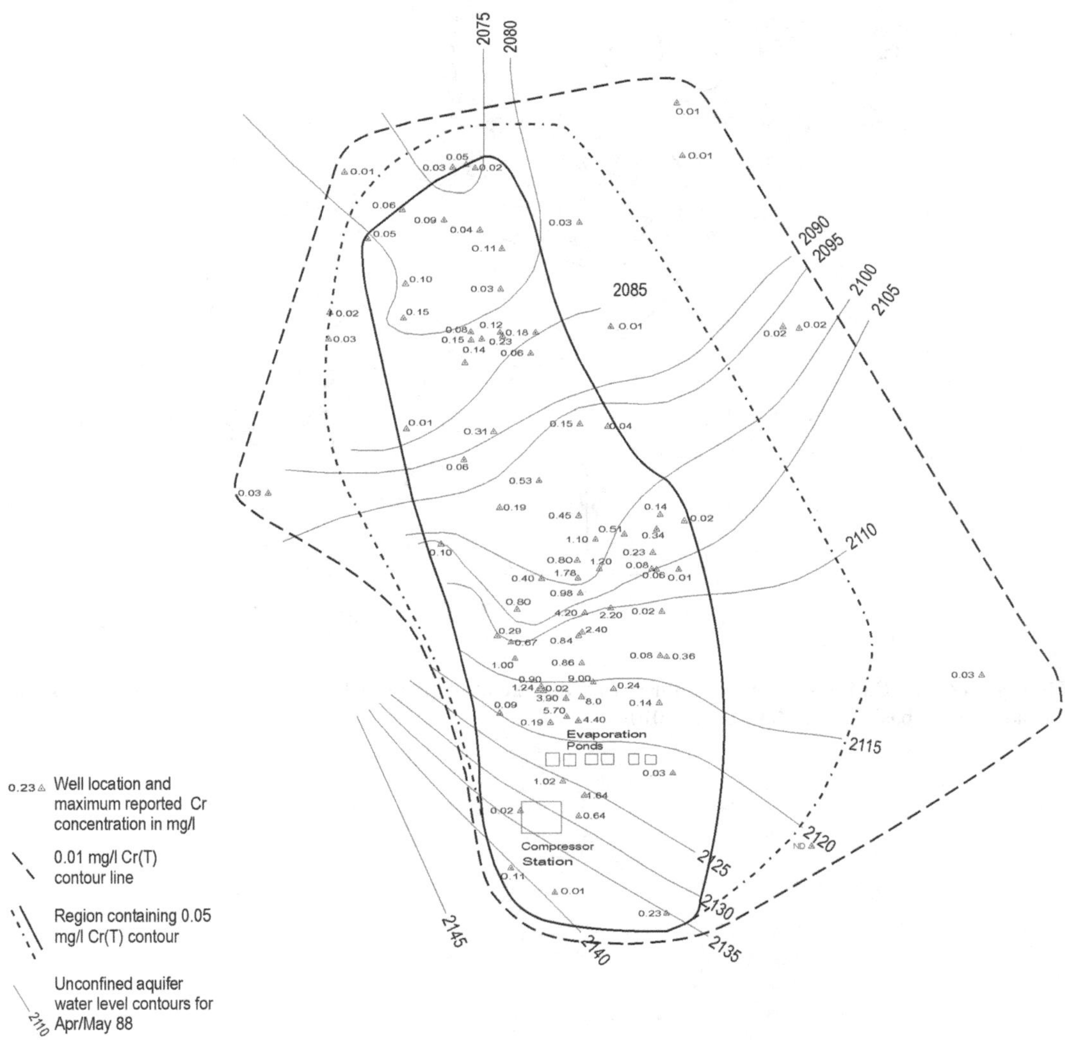

FIGURE 12.23 Contour map of groundwater chromium contamination due to leakage from unlined cooling water evaporation ponds at a natural gas compressor station located near Hinkley, California (unpublished data). Both open and closed contours were used to demonstrate groundwater flow direction and the extent of chromium concentration in the major court case, known as the Erin Brockovich case after the movie. The drinking water standard is 0.05 mg/L and concentrations as in excess of 1 mg/L were measured in many wells.

equilibrium diagrams are constructed for the standard conditions of 25 °C and 1 atm pressure. Few groundwater systems exist at this standard temperature; therefore, if temperature is not one of the master variables, the diagram must be constructed for the calculated equilibrium at the temperature of the groundwater system under consideration.

Despite the fact that an endless number and variety of equilibrium relationships can be evaluated with equilibrium diagrams, most diagrams belong to one of four types:

1. E_h-pH or pE-pH,
2. solubility – congruent dissolution,
3. solubility – incongruent dissolution, and
4. partial pressure and fugacity.

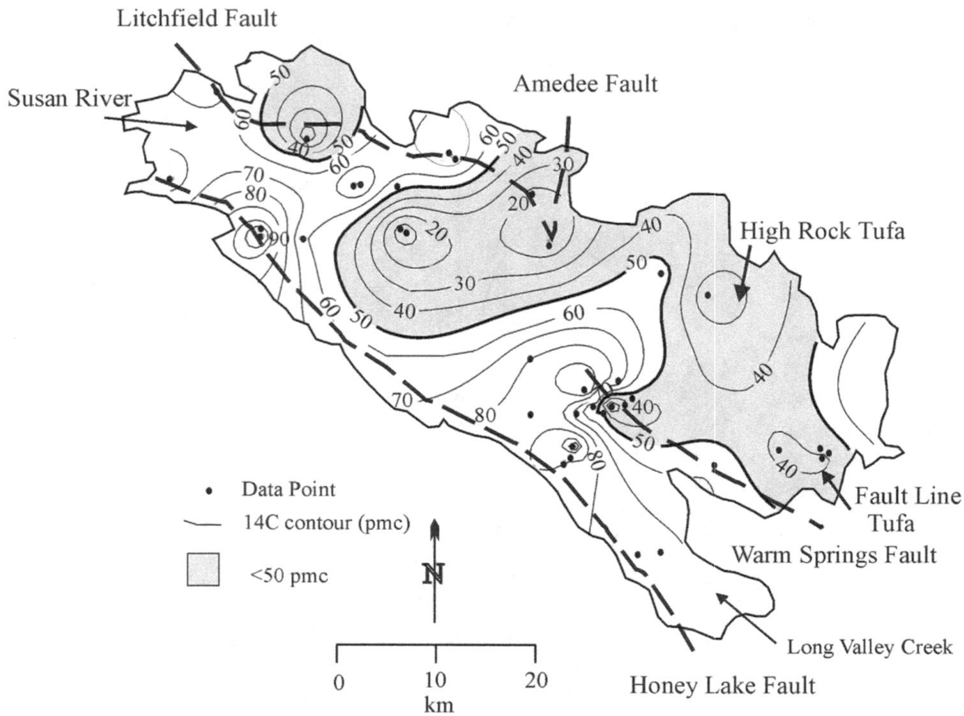

FIGURE 12.24 Carbon-14 contours illustrating the extent of likely paleo-groundwater in the California-Nevada, closed basin (after Mayo et al., 2010).

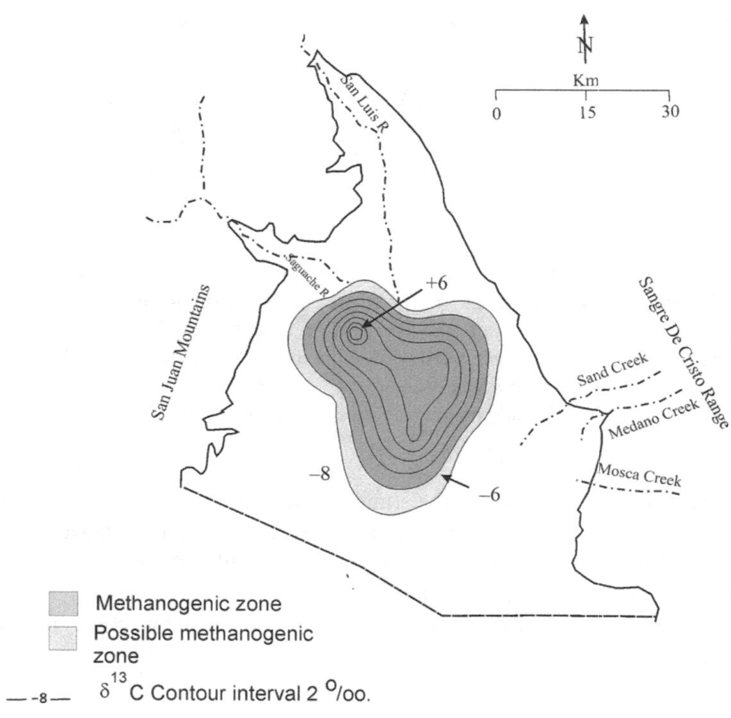

FIGURE 12.25 Contours of the ^{13}C content indicating the location of buried paleo-swamp deposit in the San Luis Valley, Colorado, closed basin (after Mayo, 2020). $\delta^{13}C$ values greater than −6 ‰ suggest methanogenesis.

Construction of equilibrium diagrams often requires the computation of K_{eq} and other values using Gibbs free energy data. Gibbs free energy data for individual species are found in numerous reference sources. Data compilations of species of interest to hydrogeology have been published by Brookins (1988), Drever (1988), Garrels and Christ (1965), Krasukopf (1979), Lindsay (1979), Tardy and Garrels (1974), Woods and Garrels (1987), and others. A table of selected values is included in Appendix C. When using Gibbs free energy values from multiple sources it is important to ensure that the data are internally consistent. A first order check on consistency can be accomplished by cross-comparing values for several critical species contained in each source.

In the literature both kilocalories/mole (kcal/mole) and kilojoules/mole (kJ/mole) are used as reporting units for thermodynamic data. The conversion between these two standard units at 25 °C and 1 atm pressure is 1 kcal = 4.184 kJ. The data in Appendix C is in kJ/mole, and both kcal and kJ have been used in example problems.

12.5.1 E_H-pH Diagram

E_h-pH or, in some literature, pE-pH diagrams are used to describe mineral stability relationships for soluble and insoluble species whose equilibrium concentrations are both oxidation–reduction and pH dependent. The diagram form, introduced by Pourbaix (1949), has E_h plotted on the y-axis and pH plotted on the x-axis. E_h-pH diagrams show how the stability of various solid and aqueous species changes with shifts in pH and the electron potential of a solution.

Since the early 1980s many authors have chosen to represent electrical potential as unit-less pE rather than as E_h, which has units of volts. The main practical difference between the two diagram forms is the scale of the vertical axis, which vary according to the relationship:

$$E_{h_{volts}} = \frac{2.304RT}{F} pE \tag{12.4}$$

where:
E_h = the redox potential of the solution in volts.
R = gas constant
1.98717 cal/mole-°K
8.314×10^{-3} kJ/mole-°K
T = temperature (°K)
F = Faraday constant
9.65×10^4 cal/mole
23.06 kcal/volt
96.42 kJ/volt-mole

The methods and equations used for constructing both E_h-pH and pE-pH diagrams are similar. Only the construction of a E_h-pH diagram is illustrated here. Before attempting to construct a E_h-pH diagram, a review of the concepts of pE and E_h (Chapter 2) is recommended. Aqueous chemical reactions involving the transfer of both protons (i.e., hydrogen ions) and electrons have the general form:

$$\alpha[A] + b[B] + w[\mathrm{H_2O}] = c[C] + d[D] + m[H^+] + n \tag{12.5}$$

where,
a, b, w, c, d, m = moles of corresponding substances,
[] = chemical activity of the product or reactant (activity of pure solids and liquids ≈ 1; gas activity = partial pressure in atm), and
n = number of electrons released.

Some redox reactions are written such that n, [C], [D], and [H^+] are on the left side of the equation and [A] and [B] are on the right side of the equation. If the reaction is independent of E_h or pH, n or m will be 0. The Nerst equation, which integrates the two master variables E_h (or pE) and pH into a single equation, can be used to evaluate chemical reactions such as Equation 12.4:

$$E_h = E^o - \frac{RT}{nF} \ln \frac{[\text{oxidized state or products}]}{[\text{reduced state or reactants}]} \tag{12.6}$$

where E^o, the standard potential of the redox reaction, and E_h are in volts. At 25 °C, Equation12.6 becomes:

$$E_h = E^o - \frac{0.0592}{nF} \ln \frac{[\text{oxidized state or products}]}{[\text{reduced state or reactants}]} \tag{12.7}$$

E^o is related to Gibbs free energy as follows:

$$E^o = -\frac{\Delta G^o_R}{nF} \tag{12.8}$$

where ΔG^o_R is the Gibbs free energy of the reaction. Substitution of Equation12.5 into Equation12.6 yields:

$$E_h = E^o - \frac{RT}{nF} \ln \frac{[C^c][D^d][H^+]m}{[A^a][B^b]} \tag{12.9}$$

Equation12.9 assumes the activity of water is unity (1). Converting to the base 10 logs, Equation12.9 can be written as the Nerst equation:

$$E_h = E^o - 2.303\frac{RT}{nF} \log \frac{[C^c][D^d][H^+]m}{[A^a][B^b]} \tag{12.10}$$

At 25 °C and 1 atmosphere, equation 12.9 is:

$$E_h = E^o - \frac{0.592}{n} \log \frac{[C^c][D^d][H^+]m}{[A^a][B^b]} \tag{12.11}$$

Construction of E_h-pH diagrams is a two-step process. The first step is to define the stability limits of water for the desired temperature, and the second step is to define the equilibrium relationships for the selected species at the desired temperature. Equilibrium relationships define the boundaries between species at the selected temperature. The general procedure for the construction of E_h-pH diagrams follows:

Step 1

Determine and plot the stability field of water for the desired temperature. The stability field of water is defined by $O_{2(g)} - H_2O$ and $H_{2(g)}$ boundaries,

Step 2

a Write balanced equations for possible chemical reactions.
b. Look up or calculate the E^o values for each reaction.
c Calculate E_h-pH relationships for each reaction at the desired temperature, using the appropriate form of the Nerst equation (Equations 12.9 or 12.10).
d. Plot the boundary relationship developed by Step 2c in a manner similar to plotting the stability of water.

An area represents the stability of each substance, a line represents equilibrium between two species, and equilibrium between three species is represented by a point, known as a triple point. Four species cannot coexist at equilibrium at the same location in a two-dimensional E_h-pH diagram. Equilibrium lines representing E_h-independent reactions are vertical, and pH-independent reactions are horizontal. Equilibrium lines representing all other reactions slope.

Example 12.3 E_h-pH diagram for the Fe-H2O system.

Problem: Construct a E_h-pH diagram for the dissolution of iron. Assume that the water is in contact with the atmosphere, has a temperature of 25 °C, and the total Fe concentration is 10^{-6} molar.
Solution:

Step 1 – Determine the stability field of water

The formation of water may be represented by the reduction of oxygen:

$$2H_2O = O2_{(g)} + 4H^+ + 4e \qquad \text{(E12.3.1)}$$

By convention, half-cell reactions are always written as reductions. When Equation1E2.3.1 is considered in terms of Equation 12.5:

$$a[A] + b[B] + w[H_2O] = c[C] + d[D] + m[H^+] + n$$

the activity of species A [A] must be placed on the right side of the equation and species C, D and *n* [C], [D], and *n* are placed on the left side. In the reaction in Equation E12.3.1 it is apparent that the stability of water is both pH and E_h dependent. Above certain E_h–pH limits water will be oxidized to gaseous $O_{2(g)}$ and H^+:

$$O_2 + 4H^+ + 4e = 2H_2O \qquad \text{(E12.3.2)}$$

and below certain limits water will be reduced to $H_{2(g)}$:

$$2H^+ + 2e = H_{2(g)} \qquad \text{(E12.3.3)}$$

The stability of water may be calculated by substituting Equations E12.3.2 and E12.3.3 into the Nerst equation (Equation 12.9 or 12.10). The activity of O_2 and H_2 may be set at 1. Above partial pressures of 1 atm, the gases would escape into the atmosphere in open systems. Although only technically correct at extremely low solute concentrations, the activity of water is also considered unity. At 25 °C the oxidization of water may be represented by substituting Equation E12.3.2 into Equation 12.10:

$$Eh = E^o - \frac{0.0592}{4}\text{pH}^4 \qquad \text{(E12.3.4)}$$

$$= E^o - 0.592\,\text{pH} \qquad \text{(E12.3.5)}$$

Values of E^o can be calculated using ΔG_R^o:

$$\Delta G_R^o = \Delta G_p^o - \Delta G_r^o$$

The procedure for calculating E^o using Gibbs free energy values follows. Using the ΔG_R^o values listed in Table E12.3.1, calculate the ΔG_R^o of the

$$G_R^o = c\Delta G_C^o + d\Delta G_D^o - a\Delta G_A^o - b\Delta G_B^o \quad \text{(E12.3.6a)}$$

$$= 2\Delta G_{H_2O(l)} - \Delta G_{O_2} - 4\Delta G_{H^+}$$

$$= 2(-237.18 \text{ kJ/mole}) - 1\times 0 - 4\times 0 \quad \text{(E12.3.6c)}$$

$$= -474.36 \text{ kJ/mole}$$

Next, calculate E^o by substituting Equation E12.3.6c into Equation 12.7:

$$E^o = -\frac{\Delta G_R^o}{nF}$$

$$E^o = -\frac{-474.37 \text{ kJ/mole}}{(4)96.42 \text{ kJ/volt} - \text{mole}} \quad \text{(E12.3.7a)}$$

$$= 1.230 \text{ volt} \quad \text{(E12.3.7b)}$$

By substituting this value into Equation E12.3.5:

$$E_h = 1.230 - 0.0592\,\text{pH} \quad \text{(E12.3.8)}$$

By a similar calculation it can also be shown that:

$$pE = 20.78 - \text{pH} \quad \text{(E12.3.9)}$$

where the pressure of O_2 is 1 atm. For partial pressures of O_2 less than 1 atm, Equations E12.3.8 and E12.3.9 may be written as:

$$E_h = 1.23 - 0.0592\text{pH} - \frac{\log P_{O_2}}{4} \quad \text{(E12.3.10)}$$

and

$$pE = 20.78\ \text{pH} - \frac{\log P_{O_2}}{4} \quad \text{(E12.3.11)}$$

The reduction of water to $H_{\_2(g)}$ may be calculated in a similar fashion. At 25 °C:

$$E_h = E^o - \frac{0.0592}{2}\log \text{pH}^2 \quad \text{(E12.3.12)}$$

Because E^o for the hydrogen cell half reaction is 0, equation (E12.3.12) may be simplified to:

$$E_h = 0.000 - 0.0592 \log \text{pH} \quad \text{(E12.3.13a)}$$

$$= -0.0592 \text{ pH} \quad \text{(E12.3.13b)}$$

It can also be shown that:

$$pE = 0 - \text{pH} \quad \text{(E12.3.14)}$$

where P_{O_2} is 1 atm. For partial pressures of H_2 other than 1 atm, Equations E12.3.13a and E12.3.14, respectively may be written as:

$$E_h = -0.0592 \text{ pH} \frac{\log P_{H_2}}{2} \quad \text{(E12.3.15)}$$

$$\text{pE} = 0 - \text{pH} \frac{\log P_{H_2}}{2} \quad \text{(E12.3.16)}$$

When the correspondingly appropriate boundary equations are plotted on a E_h-pH or pE –pH diagram, they plot as straight lines that define the limits of the stability of water. All geochemical reactions of interest in hydrogeology will plot inside the water stability field. The stability of water as defined by Equations E12.3.8 and E12.3.13b is shown in Figure E12.3.1.

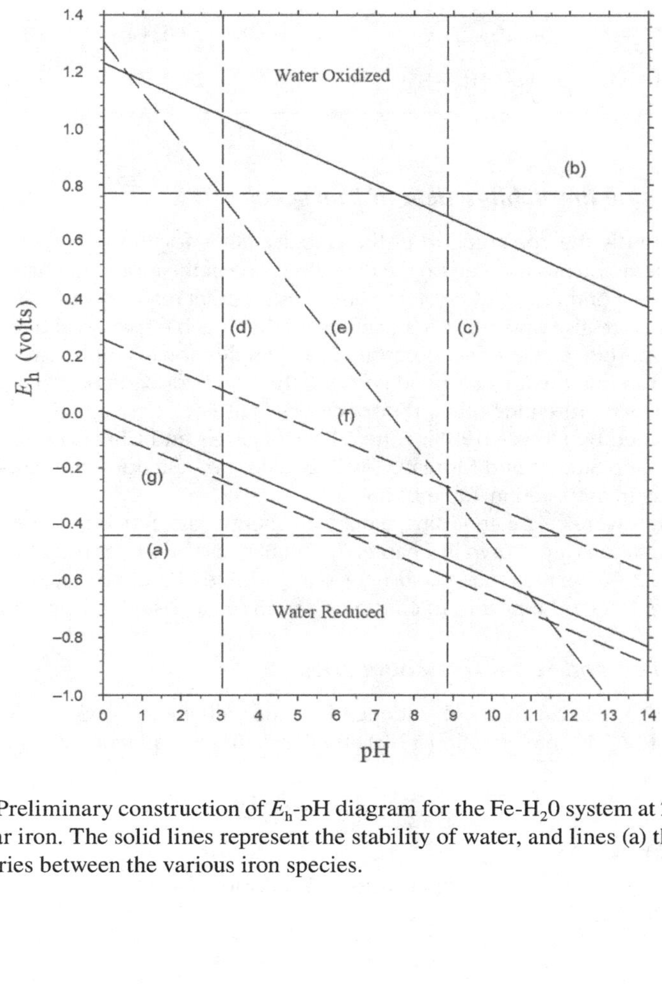

FIGURE E12.3.1 Preliminary construction of E_h-pH diagram for the Fe-H_20 system at 25 °C and a total iron content of 10^{-6} molar iron. The solid lines represent the stability of water, and lines (a) through (g) define the equilibrium boundaries between the various iron species.

TABLE 12.3.1
ΔG_R^o Values of Water.

Species	kJ/mole
H_2O	−237.18
O_2	0
H^+	0

TABLE 12.3.2
K_{eq} and E° Values for Various Iron Reactions.

Reaction	K_{eq}	E° (volts)	
$Fe^{2+} + e = Fe$		−0.44	(E12.3.17)
$Fe^{3+} + e = Fe^{2+}$		0.77	(E12.3.18)
$Fe(OH)_2 = Fe^{2+} + 2OH^-$	−16.3		(E12.3.19)
$Fe(OH)_3 = Fe^{3+} + 3OH$	38.9		(E12.3.20)
$Fe(OH)_3 + 3H^+ + e = Fe^{2+} + 3H_2O$		0.948	(E12.3.21)
$Fe(OH)_2 + 2H^+ + 2e = Fe + 2H_2O$		−0.064	(E12.3.22)
$Fe(OH)_3 + H^+ + e = Fe(OH)_2 + H_2O$		0.257	(E12.3.23)

Step 2 – Determine the stability field of iron species

The second step is the construction of the stability fields for the iron species. Selecting the appropriate species to be represented on the diagram is one of the more difficult tasks in constructing E_h-pH diagrams. For the Fe–H_2O system numerous different reactions are listed in Appendix E, and other reactions are also possible. Clearly not all of the possible species and reactions can be of hydrogeologic importance in every circumstance. To simplify the example, we will only consider the oxidation of iron into the ions Fe^{2+} and Fe^{3+} and the iron hydroxide precipitates $Fe(OH)_2^o$ and $Fe(OH)_3^o$. Diagrams representing other iron species such as FeO, hematite (Fe_2O_3), and magnetite (Fe_3O_4) are illustrated by Drever (1988), Fetter (1993), Garrels and Christ (1965), Richardson and McSween (1989), and Stumm and Morgan (1981). Equilibrium reactions of interest and K_{eq} and E^o values for our system are listed in Table E12.3.2.

K_{eq} and E^o values were calculated using Gibbs free energy data, found in Appendix C. Examples of K_{eq} and E^o calculations are shown in Chapter 2. Calculated ΔG_R^o values used for this example are listed in Table E12.3.3. The next step is to use the data in Table E12.3.3 to calculate the iron species boundaries. Other selected equilibrium constants are 25 °C are listed in Appendix E.

Calculate Fe – Fe^{2+} and Fe^{2+} – Fe^{3+} Boundaries

Reactions E12.3.17 and E12.3.18 are pH independent and will plot as horizontal lines. By substituting E12.3.17 and E12.3.18 into Equation 12.10 and E12.3.18 into Equation 12.10, we obtain:

$$E_h = E^o - 0.0592 \log \frac{Fe^{2+}}{Fe} \quad \text{(E12.3.24a)}$$

$$E_h = -0.44(\text{volts}) - 0.0592 \log \frac{Fe^{2+}}{Fe}$$

TABLE E12.3.3
Calculated ΔG_R^o Values for Various Iron Species at 25 °C

	ΔG_R^o (kJ/mole)
Fe	0.0
Fe^{2+}	−78.9
Fe^{3+}	4.6
$Fe(OH)_{2(s)}$	−486.6
$Fe(OH)_{3(s)}$	−699.0
H^+	0.0
H_2O	−237.2
OH	−157.3

and

$$E_h = E^o - 0.0592\log\frac{Fe^{3+}}{Fe^{2+}}$$
$$= 77(\text{volts}) - 0.0592\log\frac{Fe^{3+}}{Fe^{2+}} \quad \text{(E12.3.25b)}$$

At equilibrium, changes in total iron concentration will not affect equilibrium ratios of charged iron species, because $[Fe^{3+}] = [Fe^{2+}]$ in the reaction in Equation E12.3.24b and $[Fe^{2+}] = [Fe]$ in the reaction in Equation E12.3.25b. Because log 1 = 0, Equations E12.3.24b and E12.3.25b may be simplified to:

$$E_h = -0.44 \text{ volts} \quad \text{(E12.3.26)}$$

and

$$E_h = 0.77 \text{ volts} \quad \text{(E12.3.27)}$$

Plotting Equation E12.3.26 as dashed line (a) and Equation E12.3.27 as dashed line (b) on Figure E12.3.1, define the equilibrium boundary conditions for Fe–Fe^{2+} and Fe^{2+}–Fe^{3+}, respectively. Note that these equations define equilibrium boundary conditions that lie within and beyond the stability limits of water.

Calculate $Fe(OH)_2$ – Fe^{2+} and $Fe(OH)_3$ – Fe^{3+} **Boundaries**

The reactions in Equations E12.3.18 and E12.3.19 are electron independent and will plot as vertical lines. These reactions can be expressed as:

$$\log K = \log Fe^{2+} + 2\log[OH^-] \quad \text{(E12.3.28)}$$

and

$$\log K = \log\,[Fe^{3+}] + 3\log[OH] \quad \text{(E12.3.29)}$$

Because E_h-pH diagrams use pH as a variable, OH^- must be expressed in terms of pH:

$$\log\left(OH^-\right) = pH + \log K_w \quad (E12.3.30)$$

where Kw is the equilibrium constant for water. Substituting Equation E12.3.30 into Equation E12.3.28 yields:

$$\log K = \log [Fe^{2+}] + 2pH + 2\log k_w \quad (E12.3.31a)$$

which can be solved for pH as:

$$\begin{aligned}
2pH &= \frac{\log K}{\log[Fe^{2+}] + 2\log K_w} \\
&= \log K - \log\left[Fe^{2+}\right] - 2\log K_w \\
pH &= \frac{\log K - \log Fe^{2+}}{2} - K_w \\
pH &= \frac{-16.3 - \log 10^{-6}}{2} - 14 \\
&= \frac{-16.3 + 6}{2} + 14 \\
pH &= 8.85
\end{aligned} \quad (E12.3.31b)$$

In a similar fashion, substituting Equation E12.3.30 into Equation E12.3.29 yields:

$$pH = \frac{\log K - \log Fe^{3+}}{3} - K_w \quad (E12.3.32a)$$

$$\begin{aligned}
pH &= \frac{-38.8 - \log 10^{-6}}{3} + 14 \\
pH &= 3.07
\end{aligned} \quad (E12.3.32b)$$

The boundaries for $Fe(OH)_2$–$Fe^{2+}$and $Fe(OH)_3$–Fe^{3+} are plotted on Figure E12.3.1 as dashed vertical lines (c) and (d) which intercept the pH axis at 8.85 and 3.07, respectively. As in the case of previously defined boundary equations, these boundaries are also defined beyond the stability limits of water.

Calculate $Fe(OH)_3$ – $Fe(OH)_2$, $Fe(OH)_2$ – Fe, and $Fe(OH)_3$ – Fe^{2+} Boundaries

Boundaries for $Fe(OH)_3$–Fe^{2+}, $Fe(OH)_3$–$Fe(OH)_2$, and $Fe(OH)_2$–Fe reactions are both pH and E_h dependent. These reactions are described by Equations E12.3.21, E12.3.22, and E12.3.23, respectively. Calculation of E_h-pH relationships for each reaction follows:

$$\log\left(OH^-\right) = pH + \log K_v \quad (E12.3.30)$$

Substituting Equation E12.3.21 into the Nerst equation (Equation 12.10) yields:

$$E_h = E^o - 0.0592 \log \frac{\left[H^+\right]^3}{Fe^{2+}}$$

$$= 0.948 - \left(0.0592\left(3\log\left[H^+\right]\right)\right) - \left(-0.0592 \log 10^{-6}\right) \quad \text{(E12.3.33a)}$$

$$= (0.948 + 0.355) - 0.178 \text{ pH}$$

$$E_h = 1.303 - 0.178 \text{ pH} \quad \text{(E12.3.33b)}$$

where $[Fe(OH)_3]$ and $[H_2O]$ are = 1:

$$Fe(OH)_2 + 2H^+ + 2e = Fe + 2H_2O \text{ Boundary}$$

The E_h-pH equilibrium boundary for Equation E12.3.22 can be calculated in a similar fashion:

$$E_h = E^o - \frac{0.0592}{2}\log\frac{\left(H^+\right)^2}{1} \quad \text{(E12.3.34a)}$$

$$E_h = -0.064 - 0.0592 \text{ pH} \quad \text{(E12.3.34b)}$$

where $[Fe(OH)_3]$, [Fe] and $[H_2O]$ are 1.

$$Fe(OH)_3 + H^+ + e = Fe(OH) + H_2O \text{ Boundary}$$

The equilibrium boundary for Equation E12.3.23 can be determined as follows:

$$E_h = E^o - \frac{0.0592}{2}\log\frac{\left(H^+\right)}{1} \quad \text{(E12.3.35a)}$$

$$= 0.257 - 0.0592 \text{ pH} \quad \text{(E12.3.35b)}$$

where E^o and E_h are in volts. Plotting E12.3.33b as dashed line (e), E12.3.34b as dashed line (f), and E12.3.35b as dashed line (g) on Figure E12.3.1 define the equilibrium boundary conditions for $Fe(OH)_3$–$Fe(OH)_2$, $Fe(OH)_2$–Fe, and $Fe(OH)_3$–Fe^{2+}, respectively.

It is apparent from the overlapping boundary lines that each segment of each line cannot be valid in the system and that completion of the E_h-pH diagram requires the elimination of unneeded line segments. The completed diagram is illustrated in Figure E12.3.2. The intersection of lines (b), (d) and (e) define the triple point for Fe^{2+}–Fe^{3+}–$Fe(OH)_3$ and the intersection of lines (c), (e) and (f) define the triple point for Fe^{2+}–$Fe(OH)_2$–$Fe(OH)_3$. Line (g) only exists beyond the stability limits of water and can be completely eliminated. The location of line (g) illustrates why native iron does not occur in nature and why raw iron exposed to moisture and the atmosphere immediately begins to rust.

Example E12.3 demonstrates how tedious and error prone the construction of E_h-pH diagrams can be. Fortunately, some commercially available groundwater geochemical programs can be used to construct the diagrams.

12.5.2 Congruent Solubility Diagrams

Congruent dissolution diagrams graphically display equilibrium relationships for mineral species that dissolve congruently. The diagrams can be used to illustrate which minerals are stable in a solution and the saturation condition of a solution relative to one or more mineral species.

Concentrations are represented by chemical activities because dissolved species exhibit nonideal behavior in dilute solutions. The activity of a specific element or ion is represented on the y-axis, and a different variable such as pH or the activity of another ion is represented on the x-axis. A line represents equilibrium between two substances. The areas between the lines do not represent stability fields in the same sense that they do in E_h–pH diagrams. If a chemical analysis plots on an equilibrium line, the water sample is in equilibrium with the two species represented by the line. On the other hand, if a chemical analysis plots off the line, the water is either undersaturated or supersaturated with respect to the species being evaluated.

The construction of a quartz solubility diagram is described in Example E12.4.

Example E12.4 Solubility diagram for dissolution of quartz.

Problem: Construct a solubility diagram for silica species in equilibrium with the dissolution of quartz. Assume a water temperature of 25 °C.

Solution: The solubility of quartz and amorphous silica can be characterized by the following equilibria (Seward, 1974; Stumm and Morgan, 1981), where log *K* values at 25 °C are listed

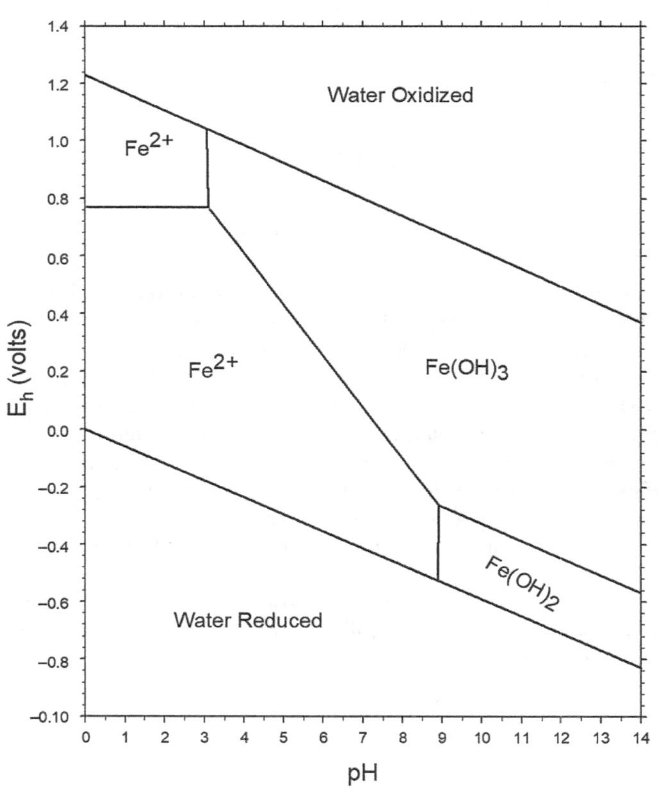

FIGURE E12.3.2 Completed E_h-pH diagram for the Fe–H_2O system after the removal of overlapping lines.

TABLE E12.4.1
Equilibrium Constants for Various Silica Species

Reaction	log *K*	
$SiO_2\left(\text{quartz}\right) + 2H_2O = H_4SiO_{4(aq)}$	−3.7	(E12.4.1)
$SiO_2\left(\text{amorphous}\right) + 2H_2O = H_4SiO_{4(aq)}$	2.7	(E12.4.2)
$H_4SiO_{4(aq)} = H_3SiO_4^- + H^+$	−9.83	(E12.4.3)
$H_3SiO_4^- = H_2SiO_4^{2-} + H^+$	12.57	(E12.4.4)
$H_3SiO_4^- = H_2SiO_4^{2-} + H^+$	12.57	(E12.4.5)

in Table E12.4.1. By rearranging Equation E12.4.1 (Table E12.4.1) in in terms of the equilibrium constant for quartz, we have:

$$K = \frac{[H_2O]^2 + \left[H_4SiO_{4(aq)}\right]}{[SIO_{2(quartz)}]} \quad \text{(E12.4.5a)}$$

Because the activities of H_2O and SiO_2 = 1, Equation E12.4.5a can be written as:

$$-3.7 = \log[H_4SiO_{4(aq)}] \quad \text{(E12.4.5b)}$$

Solving Equation E12.4.2 in a similar fashion, we have:

$$K = \frac{[H_2O]^2 + \left[H_4SiO_{4(aq)}\right]}{[SIO_{2(amorphous)}]} \quad \text{(E12.4.6a)}$$

and

$$-2.7 = \log[H_4SiO_{4(aq)}] \quad \text{(E12.4.6b)}$$

The next step is to use the equations representing the dissociation of silicic acid in terms of pH. Similar to carbonic acid, silicic acid is polyprotic and subsequently has progressive dissociation constants. The first dissociation constant is defined as K_1 and the second constant as K_2, K_1, K_2, and K_3 are commonly used as acid dissociation constants identifiers and are not specific to a particular acid. Assuming $[H_2O]$, $[SiO_{2\ (qtz)}]$, and $[SiO_{2\ (amorphous)}]$ = 1, the data in Equation E12.4.3 may be expressed as:

$$K_1 = \frac{[H_3SiO_4^-] + [H^+]}{[H_4SiO_{4(aq)}]} \quad \text{(E12.4.7)}$$

By substituting the appropriate equilibrium constants and rearranging the equation elements, an equation can be developed that expresses the relationship between pH and log $[H_3SiO_4^-]$:

$$-9.82 = \frac{\log[H_3SiO_4^-] + \log[H^+]}{\log[H_4SiO_{4(aq)}]} \quad \text{(E12.4.8a)}$$

$$\begin{aligned}
-9.82 &= \log[H_3SiO_4^-] + \log[H^+] - \log[H_4SiO_{4(aq)}] \\
-9.82 &= \log[H_3SiO_4^-] + \log[H^+] - (-3.7) \\
-9.82 - 3.7 + pH &= \log[H_3SiO_4^-] \\
-13.52 + pH &= \log[H_3SiO_4^-]
\end{aligned} \quad \text{(E12.4.8b)}$$

In a similar fashion, an equation can be developed for the relationship between pH and log $[H_2SiO_4^{2-}]$:

$$K_2 = \frac{[H_2SiO_4^{2-}] + [H^+]}{[H_3SiO_{4]}} \quad \text{(E12.4.9)}$$

and

$$= 12.57 = \frac{[H_2SiO_4^{2-}] + [H^+]}{[H_3SiO_{4]}} \quad \text{(E12.4.10a)}$$

$$-12.57 = \log\,[H_2SiO_4^{2-}] + \log\left[H^+\right] - \log[H_3SiO_4^{2-}]$$

by substituting Equation E12.4.8b into Equation E12.4.10a and solving:

$$\begin{aligned}
-12.57 &= \log\,[H_2SiO_4^{2-}] + \log\left[H^+\right] - (-13.52) + \log\left[H^+\right] \\
-26.09 + 2pH &= \log[H_2SiO_4^{2-}]
\end{aligned} \quad \text{(E12.4.10b)}$$

The total dissolved silica content (SiO_2) is the sum of the charged and uncharged species:

$$SiO_2 = \left[H_4SiO_{4\,(aq)}\right] + [H_3SiO_4] + [H_2SiO_4^{2-}] \quad \text{(E12.4.11)}$$

If activity coefficients are neglected Equation E12.4.1 may be represented as:

$$SiO_2 = H_4SiO_{4(aq)}\left(1 + \frac{K_1}{\left[H^+\right]} + \frac{K_1K_2}{\left(H^+\right)^2}\right) \quad \text{(E12.4.12)}$$

The silica solubility diagram (Figure E12.4.1) has been constructed using Equations E12.4.5b, E12.4.6b, E12.4.8b, and E12.4.10b. The dissolution of both quartz (Equation E12.4.5b) and amorphous silica (Equation E 12.4.6b) are pH independent and plot as straight lines (a) and (b), respectively. Above a pH of about 9, $H_4SiO_{4(aq)}$ begins to dissociate according to Equation E12.4.3. The $H_4SiO_{4(aq)}$–$H_3SiO_4^-$ equilibrium boundary, defined by Equation E12.4.8b, plots as line (c). In a similar fashion, line (d), derived from Equation E12.4.10b, represents the $H_3SiO_4^-$–$H_2SiO_4^{2-}$ equilibrium boundary. Lines representing the total silica and total amorphous silica activities above pH ~ 9 are approximate and are not rigorously defined.

12.5.3 Incongruent Solubility Diagrams

Incongruent dissolution fundamentally differs from congruent dissolution in that one or more of the products of incongruent dissolution is a solid phase. This difference makes the form of incongruent stability diagrams similar to E_h-pH diagrams. The stability of each solid or amorphous phase and the neutral and ionic species is represented by an area, equilibrium between two species is represented by a line, and equilibrium between three species is represented by a triple point. Many

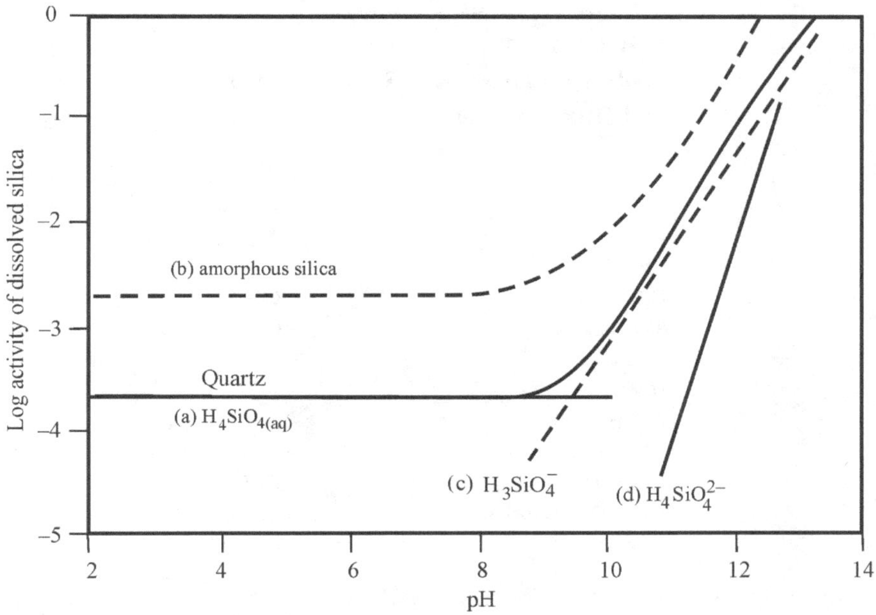

FIGURE E12.4.1 Quartz solubility diagram at 25 °C. Amorphous silica and H_4SiO_4 are pH independent and plot as horizontal lines. Lines (c) and (d) represent $H_4SiO_{4(aq)}$–$H_3SiO_4^-$ and $H_3SiO_4^-$–$H_2SiO_4^{2-}$ equilibrium.

incongruent solubility diagrams are referred to as activity diagrams because chemical activities are represented on both x- and y-axes.

The construction of an activity diagram for the incongruent dissolution of microcline (K-feldspar) and muscovite (K-mica) is shown in Example E12.5.

Example 12.5 Stability diagram for the incongruent solution weathering of a granitic rock.

Problem: Construct a stability diagram for the incongruent weathering of a granitic rock containing microcline (K-feldspar) and muscovite (K-mica). Assume an aquifer temperature of 25 °C.

Solution: This diagram will predict the solid phases (i.e., solid reaction products) that will be in equilibrium at various ion concentrations. Weathering of microcline ($KAlSi_3O_8$) and muscovite ($KAl_3Si_2O_{10}(OH)_2$) may be represented by:

$$\underset{\text{(microcline)}}{3KAlSi_3O_8} + 12H_2O + 2H^+ = \underset{\text{(muscovite)}}{KAl_3Si_3O_{10}(OH)_2} + 6H_4SiO_4 + 2K^+ \quad \text{(E12.5.1)}$$

$$\underset{\text{(microcline)}}{2KAlSi_3O_8} + 9H_2O + 2H^+ = \underset{\text{(kaolinite)}}{Al_2Si_2O_5(OH)_4} + 4H_4SiO_4 + 2K^+ \quad \text{(E12.5.2)}$$

$$\underset{\text{(muscovite)}}{2KAl_3Si_3O_{10}(OH)_2} + 3H_2O + 2H^+ = \underset{\text{(kaolinite)}}{3Al_2Si_2O_5(OH)_4} + 2K^+ \quad \text{(E12.5.3)}$$

$$\underset{\text{(muscovite)}}{KAl_3Si_3O_{10}(OH)_2} + 9H_2O + H^+ = \underset{\text{(gibbsite)}}{3Al(OH)_3} + 3H_4SiO_4 + K^+ \quad \text{(E12.5.4)}$$

$$\underset{\text{(kaolinite)}}{Al_2Si_2O_5(OH)_4} + 5H_2O = \underset{\text{(gibbsite)}}{2Al(OH)_3} + 2H_4SiO_4 \quad \text{(E12.5.5)}$$

TABLE E12.5.1
Gibbs Free Energies of Selected Silica and Other Species

	ΔG_R^o (kJ/mole)
H^+	0.00
K^+	67.70
H_2O	−56.69
SiO_2 (amorphous)	−203.33
SiO2 (quartz)	−204.75
H_4SiO_4	−312.66
$Al(OH)_3$ (gibbsite)	−276.10
$Al_2Si_2O_5(OH)_4$ (kaolinite)	−906.84
$KAl_3Si_3O_{10}(OH)_2$ (muscovite)	−1,337.45
$KAlSi_3O_8$ (microcline)	−894.90

Because the activities of solid phases in Equations E12.5.1 through E12.5.5 are unity, equations can be represented in terms of ion concentrations of K^+, H^+, and H_4SiO_4. Aluminum is considered to be conserved in the solid species. In real waters, minor amounts of aluminum will go into solution. However, the dissolution of aluminosilicate minerals generally does not substantially alter the dissolved aluminum content of most natural waters. At aquifer temperatures above approximately 200 °C the dissolved silica concentration will be controlled by the solubility of quartz, which means that at elevated aquifer temperatures dissolved silica can also be considered a conserved species (Henley et al., 1984). At typical aquifer temperatures the dissolution of aluminosilicate minerals such as microcline and muscovite will be affected by the activity of dissolved silica.

Construction of the stability diagram requires using equations that define equilibrium in terms of the chemical activities of K^+, H^+, and H_4SiO_4. Gibbs free energies, taken from Appendix C, were used to calculate equilibrium constants (Table E12.5.1).

Let us use the weathering of microcline to muscovite (Equation E12.5.1) to demonstrate the calculations used for the construction of an incongruent dissolution diagram:

$$\underset{\text{(microcline)}}{3KAlSi_3O_8} + 12H_2O + 2H^+ = \underset{\text{(muscovite)}}{KAl_3Si_3O_{10}(OH)_2} + 6H_4SiO_4 + 2K^+$$

Using Equation E12.5.6a and the Gibbs free energy data from Table E12.5.1, the Gibbs free energy of the reaction can be calculated:

$$\Delta G_R^o = \Delta G_p^o - \Delta G_r^o \tag{E12.5.6a}$$

$$((-1337.45) + 6(-312.66) + 2(-67.70)) - (3(-894.9) + 12(-56.678))$$

$$= 3348.81 + 3364.84 \tag{E12.5.6c}$$

$$= 16.03 \text{kcal/mole}$$

Because the ΔG_R^o is positive, the reaction spontaneously proceeds from right to left. Substituting Equation E12.5.6c into Equation 2.36:

$$\log K = -0.733\Delta G_R^o = (-0.733)(16.03)\text{kcal/mole} \quad \text{(Equation 2.36)} \tag{E12.5.7a}$$

$$= -11.75 \text{kcal/mole}$$

$$K = 10^{-11.75} \tag{E12.5.7b}$$

Applying the law of mass action:

$$K = \frac{[KAl_3Si_3O_{10}(OH)_2]\,[H_4SiO_4]^6[K^+]^2}{[KAlSi_3O_8]^3[H_2O]^{12}[H^+]^2} \quad \text{(E12.5.8)}$$

Because the activity of solid species and water are considered unity, Equation E12.5.8 can be simplified to:

$$K = \frac{[H_4SiO_4]^6[K^+]^2}{[H^+]^2} \quad \text{(E12.5.9)}$$

By substituting Equation E12.5.7b into Equation E12.5.9:

$$10^{-11.75} = \frac{[K^+]^2}{[H^+]^2}[H_4SiO_4]^6 \quad \text{(E12.5.10)}$$

Taking the logarithm of Equation E12.5.10 and simplifying:

$$-11.75 = 6\log H_4SiO_4 + 2\log K^+ - 2\log[H^+] \quad \text{(E12.5.11a)}$$

$$-11.75 = 6\log[H_4SiO_4] + 2\log\frac{[K^+]}{[H^+]}$$

$$-5.87 = 3\log[H_4SiO_4] + \log\frac{[K^+]}{[H^+]}$$

and

$$\log\frac{[K^+]}{[H^+]} = 3\log[H_4SiO_4] - 5.87 \quad \text{(E12.5.11b)}$$

Equilibrium boundary conditions, relating log $[K^+]/[H^+]$ to log $[H_4SiO_4]$, for microcline, muscovite, kaolinite, gibbsite, and amorphous silica are calculated in a similar fashion.

Microcline–muscovite

$$\underset{\text{(microcline)}}{3KAlSi_3O_8} + 12H_2O + 2H^+ = \underset{\text{(muscovite)}}{KAl_3Si_3O_{10}(OH)_2} + 6H_4SiO_4 + 2K^+$$

$$\Delta G_R^o = 16.03 \text{ kcal/mole} \quad \text{(E12.5.12a)}$$

$$K = 10^{-11.75}$$

$$\log\frac{[K^+]}{[H^+]} = -3\log[H_4SiO_4] - 5.87 \quad \text{(E12.5.12b)}$$

Microcline–kaolinite

$$\underset{\text{(microcline)}}{2KAlSi_3O_8} + 9H_2O + 2H^+ = \underset{\text{(kaolinite)}}{Al_2Si_2O_5(OH)_4} + 4H_4SiO_4 + 2K^+$$

$$\Delta G_R^o = 7.103 \text{ kcal/mole} \quad \text{(E12.5.13a)}$$

$$K = 10^{-5.21}$$

$$\log\frac{[K^+]}{[H^+]} = -2\log[H_4SiO_4] - 2.60 \qquad \text{(E12.5.13b)}$$

Muscovite–kaolinite

$$\underset{\text{(muscovite)}}{2KAl_3Si_3O_{10}(OH)_2} + 3H_2O + 2H^+ = \underset{\text{(Kaolinite)}}{3Al_2Si_2O_5(OH)_4} + 2K^+$$

$$\Delta G_R^o = -10.96 \text{ kcal/mole} \qquad \text{(E12.5.14a)}$$

$$K = 10^{8.03}$$

$$\log\frac{[K^+]}{[H^+]} = 4.02 \qquad \text{(E12.5.14b)}$$

Muscovite–gibbsite

$$\underset{\text{(muscovite)}}{KAl_3Si_3O_{10}(OH)_2} + 9H_2O + H^+ = \underset{\text{(gibbsite)}}{3Al(OH)_3} + 3H_4SiO_4 + K^+$$

$$\Delta G_R^o = -13.65 \text{ kcal/mole} \qquad \text{(E12.5.15a)}$$

$$K = 10^{-10.01}$$

$$\log\frac{[K^+]}{[H^+]} = -3\log[H_4SiO_4] - 10.01 \qquad \text{(E12.5.15b)}$$

Kaolinite–gibbsite

$$\underset{\text{(Kaolinite)}}{Al_2Si_2O_5(OH)_4} + 5H_2O = \underset{\text{(gibbsite)}}{2Al(OH)_3} + 2H_4SiO_4$$

$$\Delta G_R^o = 12.76 \text{ kcal/mole} \qquad \text{(E12.5.16a)}$$

$$K = 10^{-9.3}$$

$$\log[H_4SiO_4] = -4.68 \qquad \text{(E12.5.16b)}$$

Amorphous silica

$$SiO_2\,(\text{amorphous}) + 2H_2O = H_4SiO_4$$

$$\Delta G_R^o = 4.04 \text{ kcal/mole} \qquad \text{(E12.5.17a)}$$

$$K = 10^{-2.96}$$

$$\log[H_4SiO_4] = -2.96 \qquad \text{(E12.5.17b)}$$

On Figure E12.5.1, plotting Equation E12.5.12b as line (a), Equation E12.5.13b as line (b), Equation E12.5.14b as line (c), Equation E12.5.15b as line (d), Equation E12.5.16b as line (e), and Equation E12.5.17b as line (f) defines the equilibrium boundary conditions for microcline–muscovite, microcline–kaolinite, muscovite–kaolinite, muscovite–gibbsite, kaolinite–gibbsite, and the solubility of amorphous silica, respectively. The diagram is completed by removing overlapping line segments (Figure E12.5.2).

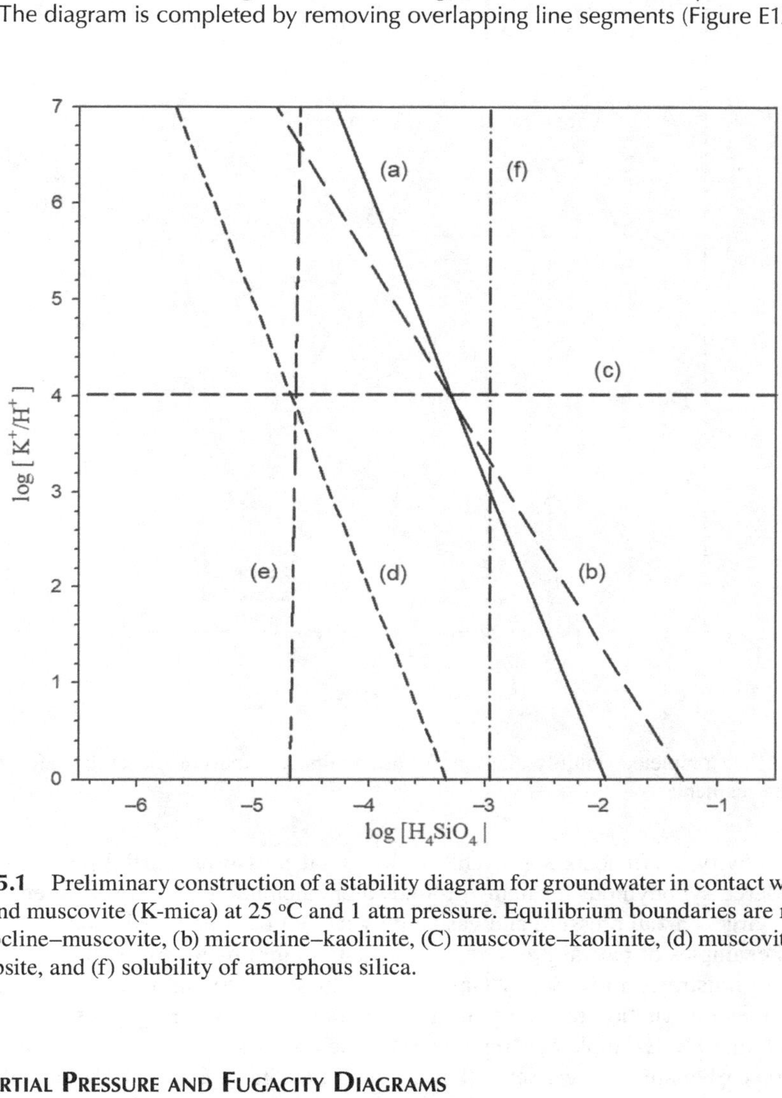

FIGURE E12.5.1 Preliminary construction of a stability diagram for groundwater in contact with microcline (K-feldspar) and muscovite (K-mica) at 25 °C and 1 atm pressure. Equilibrium boundaries are represented by lines (a) microcline–muscovite, (b) microcline–kaolinite, (C) muscovite–kaolinite, (d) muscovite–gibbsite, (e) kaolinite–gibbsite, and (f) solubility of amorphous silica.

12.5.4 Partial Pressure and Fugacity Diagrams

All oxide, carbonate, and sulfide minerals are in equilibrium with gaseous O_2, CO_2, and S_2, respectively. Non-mineral species such as water and soil gases are also in equilibrium with their representative gases. Gas–gas, gas–liquid, and gas–mineral stability relationships may be represented on partial pressure and fugacity diagrams. Partial pressure diagrams represent stability in terms of the partial pressure of one or more gases. Fugacity diagrams represent stability in terms of the chemical activity of one or more gases. In a pure gas, fugacity equals the chemical activity. Recall from Equation 3.21 that fugacity (f) equals the partial pressure (P) times a dimensionless gas activity coefficient (y).

$$f_{\mathrm{i}} = y_i P_{\mathrm{i}}$$

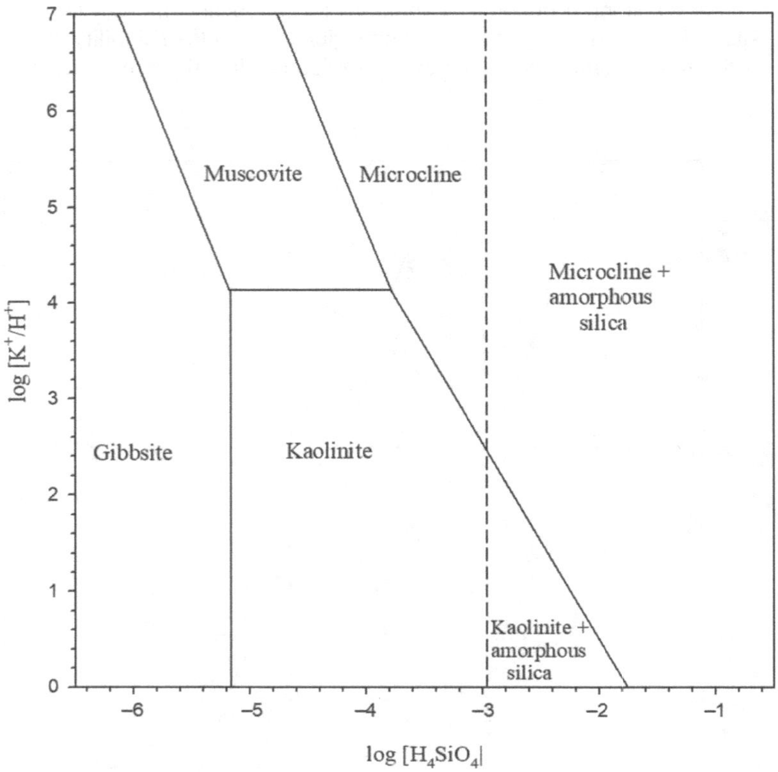

FIGURE E12.5.2 Completed stability diagram for microcline and muscovite stability after the removal of overlapping line segments.

Because gas activity coefficients approach 1 at low total pressures, partial pressure and fugacity may be considered synonymous for many groundwater calculations. However, when total gas pressures become large, partial pressure and gas activity cannot be considered equal.

Numerous examples of partial pressure and fugacity diagrams for oxide, carbonate, and sulfide gas systems are illustrated in works by Garrels and Christ (1965) and Faure (1991). Example 12.6 demonstrates the construction of a diagram for the stability of water with respect to the fugacity of the gases O_2 and H_2. Example 12.7 demonstrates the construction of a stability diagram for iron oxides and iron carbonate with respect to the fugacity of the gases O_2 and CO_2. In both examples, it is assumed that partial pressure and fugacity are equivalent.

Example E12.6 Partial pressure (or fugacity) diagram for the stability of water.

Problem: The stability of water is the most important geochemical parameter in hydrogeochemistry. All geochemical reactions of hydrogeologic interest must occur within the stability range of waters. In Example E12.3 we saw how water becomes unstable beyond pH and E_h limits. Our task here is to construct a diagram for water with respect to the gas phases O_2 and H_2. Assume a temperature of 25 °C.

Solution: From the equation representing the dissociation of water:

$$2H_2O_{(l)} = 2H_{2(g)} + O_{2(g)} \tag{E12.6.1}$$

it is evident that the stability of water is affected by the partial pressure of both $H_{2(g)}$ and $O_{2(g)}$. The equilibrium expression for reaction E12.6.1 is:

$$K = \frac{(P_{H_2})^2(P_{O_2})}{(H_2O)^2} \quad \text{(E12.6.2)}$$

Because $[H_2O] = 1$:

$$K = (P_{H_2})^2(P_{O_2}) \quad \text{(E12.6.3)}$$

The Gibbs free energy relationship of reaction E12.6.1 may be represented as:

$$\Delta G_R^o = \Delta G_p^o - \Delta G_r^o$$

Substituting numerical values from Appendix C and solving:

$$\begin{aligned} \Delta G_R^o &= 0 + 0 - 2(-56.687) \\ &= 113.37 \text{ kcal/mole} \end{aligned} \quad \text{(E12.6.4a)}$$

Substituting Equation E12.4.4b into Equation 2.36

$$\begin{aligned} \log K &= -0.733\Delta G_R^o \\ &= (-0.733)(113.37) \\ &= -83.10 \end{aligned} \quad \text{(E12.6.4)}$$

and

$$K = 10 - 83.1$$

substituting this into Equation E12.6.3:

$$10^{-83.1} = (P_{H_2})^2(P_{O_2}) \quad \text{(E16.4.4d)}$$

From the very small equilibrium constant, it is obvious that the dissociation of water into the gases H_2 and O_2 is negligible. At the surface of the earth, the partial pressure of H_2 and O_2 in water must remain less than 1 atmosphere, or else gas bubbles will form and the dissociated gas will escape into the atmosphere. The stability of water may therefore be evaluated with respect to 1 atmosphere of pressure.

From Equation E12.6.4d $P_{H_2} = 10^{-41.55}$atm. When P_{O_2} and $P_{H_2} = 1$ atm, $P_{O_2} = 10^{-83.1}$atm and the stability of water is:

$$P_{O_2} = [O_2] = 10^{-83.1} \quad \text{(E12.6.5a)}$$

By taking logarithms:

$$\log P_{O_2} = \log [O_2] = -83.1 \quad \text{(E12.6.5b)}$$

In a similar fashion it can be shown that:

$$P_{H_2} = [H_2] = 10^{-41.55} \quad \text{(E12.5.6a)}$$

and

$$\log P_{H_2} = \log [H_2] = -41.55 \tag{E12.4.6b}$$

From Equations E12.6.5b and E12.6.4b, the two endpoints for the stability of water can be plotted as a function of the partial pressure of the gases O_2 and H_2. The plot is shown on Figure E12.6. The line connecting the two endpoints defines the stability of water when the total pressure is 1 atm.

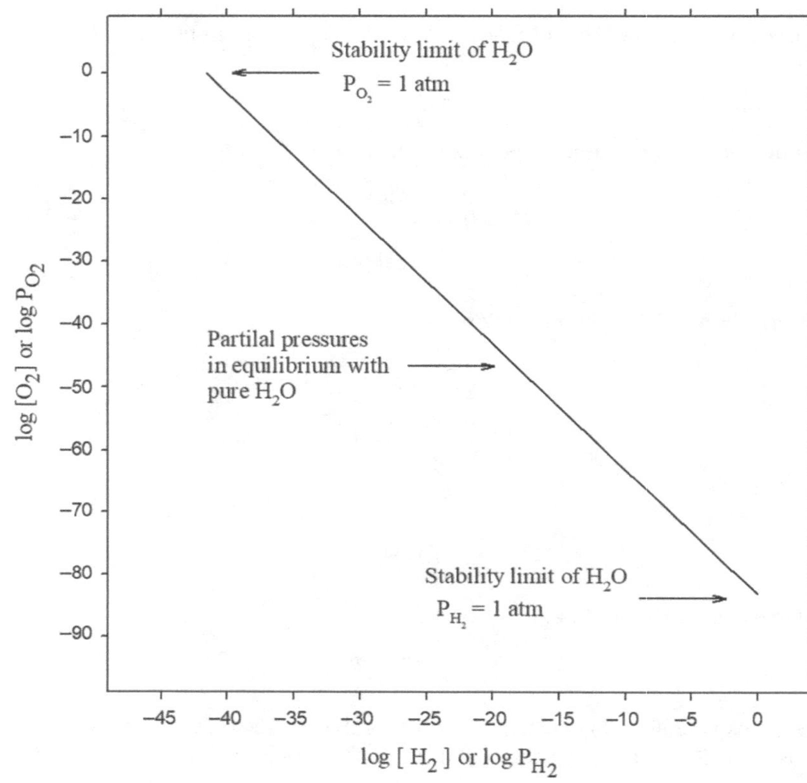

FIGURE E12.6 Construction of the partial pressure diagram for water in terms of the partial pressures of the gases O_2 and H_2.

Example E12.7 Iron carbonate partial pressure diagram relative to O_2 and CO_2 fugacity.

Problem: Construct a partial pressure (fugacity) diagram for iron oxides and iron carbonates in terms of the fugacity of O_2 and CO_2. The common iron oxide and iron carbonate minerals which may be stable in contact with water are hematite (Fe_2O_3), magnetite (Fe_3O_4), and siderite ($FeCO_3$). Assume an aquifer temperature of 25 °C.

Solution: Gibbs free energies, taken from Appendix C, were used to calculate equilibrium constants (Table E12.7).

TABLE E12.7
Gibbs Free Energy for Selected Iron Species

ΔG_R^o	kJ/mole
O_2	0.00
$CO_{2(g)}$	94.25
$Fe_2O_{3(s)}$ (hematite)	−177.60
$Fe_3O_{4(s)}$ (magnetite)	−242.69
$FeCO_{3(s)}$ (siderite)	−160.30

Possible reactions, calculated ΔG_R^o values, and log activities are shown in the following. ΔG_R^o values and log activities were calculated in the same fashion as described in Examples E12.1 through E12.4.

Magnetite–Hematite

$$\underset{\text{(magnetite)}}{2Fe_3O_{4(s)}} + O_{2(g)} = \underset{\text{(hematite)}}{3Fe_2O_{3(s)}} \tag{E12.7.1}$$

$$\Delta G_R^o = -47.42 \text{ kcal/mole}$$

$$K = 10^{-1034.76}$$

Magnetite–Siderite

$$\underset{\text{(magnetite)}}{Fe_3O_{4(s)}} + 3CO_{2(g)} = \underset{\text{(siderite)}}{3FeCO_{3(s)}} + O_{2(g)} \tag{E12.7.3}$$

$$\Delta G_R^o = 44.55 \text{ kcal/mole} \tag{E12.7.4a}$$

$$K = 10^{-32.66}$$

$$\log [O_2] = 6\log [CO_2] - 65.31 \tag{E12.7.4b}$$

Hematite–Siderite

$$\underset{\text{(hematite)}}{Fe_2O_{3(s)}} + 2CO_{2(g)} = \underset{\text{(siderite)}}{2FeCO_{3(s)}} + O_{2(g)} \tag{E12.7.5}$$

$$\Delta G_R^o = 46.08 \text{ kcal/mole} \tag{E12.7.6a}$$

$$K = 10^{-33.78}$$

$$\log [O_2] = 4\log [CO_2] - 67.55 \tag{E12.7.6b}$$

The equilibrium conditions as defined by Equations E12.7.2b, E12.7.4b, and E12.7.6b are plotted as lines (a), (b) and (c), respectively, on Figure E12.7.1. The diagram is completed in Figure E12.7.2 by removing overlapping line segments and adding the stability limits of water as defined by Equations E12.6.b and E12.6.6b.

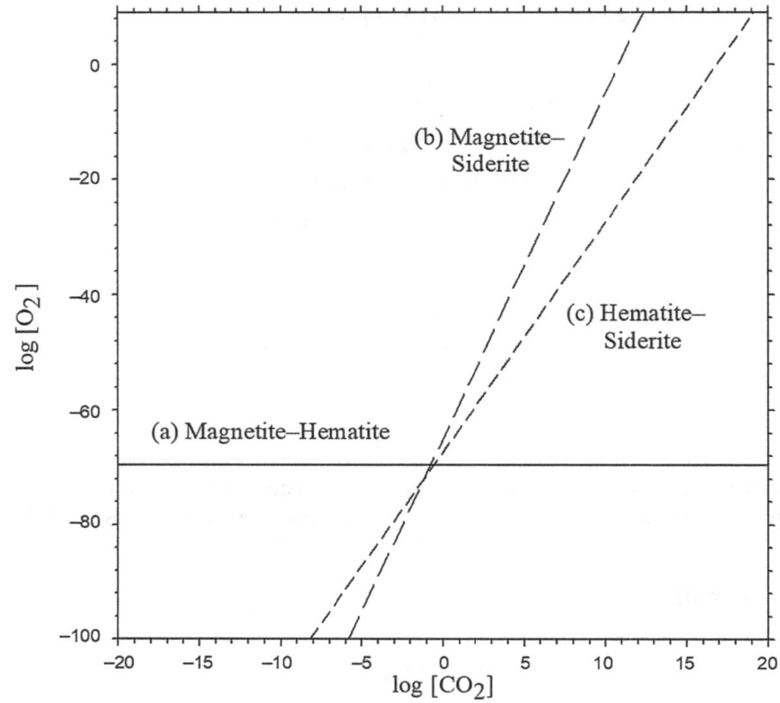

FIGURE E12.7.1 Preliminary construction of a partial pressure (fugacity) diagram for iron oxides and iron carbonates relative to the fugacity of O_2 and CO_2. Equilibrium between magnetite–hematite, magnetite–siderite, and hematite–siderite are represented by line segments (a), (b), and (c), respectively.

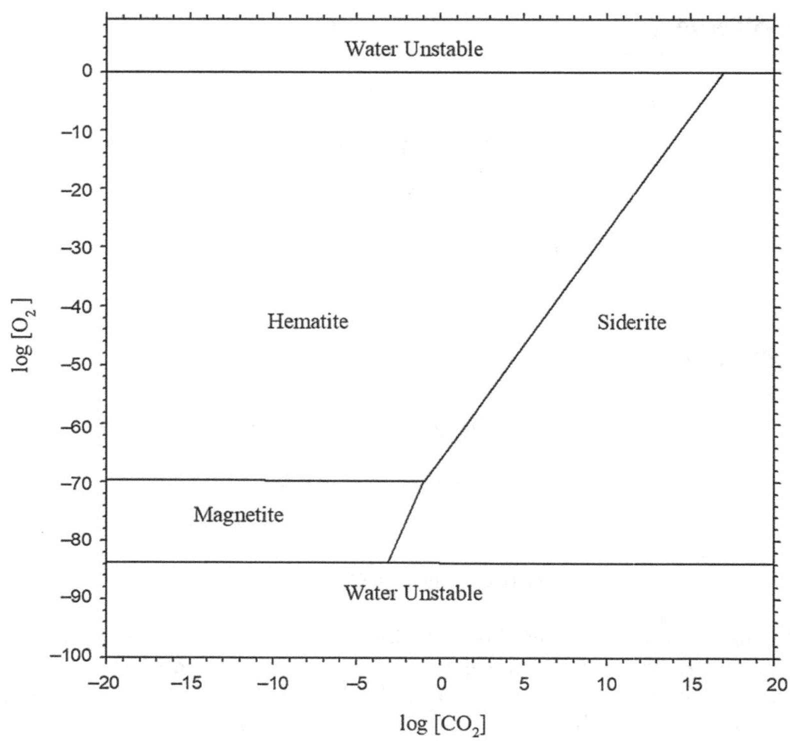

FIGURE E12.7.2 Final partial pressure diagram of iron oxides and iron carbonates relative to the fugacity of O_2 and CO_2 completed by removing overlapping line segments and adding lines representing the stability of water.

12.5.5 Problems

Prior to the construction of a water treatment facility, the discharge from a coal mine resulted in iron staining in a nearby creek. The solute, gas, and isotopic compositions of coal mine waters collected from inside the mine, where older sealed mine workings discharge into the main mine drainage, are listed in the following table. The coal is interbedded with un-metamorphosed sandstone. Seals are closures that isolate older mine tunnels, usually with block walls, from active mine workings. Use these data for the following problems.

Sample Location	date	(gpm)	°C	pH	cond µmhos /cm	TDS	Ca (mg/L)	Mg (mg/L)	Na (mg/L)	K (mg/L)	HCO_3 (mg/L)	Cl (mg/L)	SO_4 (mg/L)	Fe^{3+} (mg/L)
11th Seal	2/18/13	4.3	15.7	7.28	900	499	73.3	24.4	90.2	9.9	538.9	8	35	0.48
	1/15/14	4.6	15.7	7.62	913	525	75.9	25.8	88.5	9.8	536.5	8	32	0.44
	average		15.7	7.45	907	512	74.6	25.1	89.3	9.9	537.7	8	34	0.46
17th Seal	12/18/13	~100	12.5	7.48	928	536	64.4	42.5	64.8	11.2	388.9	12	124	2.53
	1/15/14	~100	12.6	7.33	921	542	65.4	43.2	62.6	10.6	421.9	12	120	2.15
	average		12.6	7.41	925	539	64.9	42.8	63.7	10.9	405.4	12	122	2.34

		Ca (meq/L)	Mg (meq/L)	Na (meq/L)	K (meq/L)	HCO_3 (meq/L)	Cl (meq/L)	SO_4 (meq/L)	Fe^{+3} (meq/L)	sum cat	sum an
11th Seal	12/18/13	3.66	2.00	3.92	0.25	8.83	0.23	0.73	0.02	9.84	9.79
1/15/14	3.79	2.12	3.85	0.25	8.79	0.23	0.67	0.01	10.01	9.69	
	average	3.72	2.06	3.89	0.25	8.81	0.23	0.70	0.02	9.92	9.74
17th Seal	12/18/13	3.21	3.49	2.82	0.29	6.37	0.34	2.58	0.09	9.81	9.29
1/15/14	3.26	3.55	2.72	0.27	6.91	0.34	2.50	0.07	9.81	9.75	
	average	3.24	3.52	2.77	0.28	6.64	0.34	2.54	0.08	9.81	9.52

	Field DO (mg/L)	Lab pH
17th Seal		
1/15/2014	2.44	3.80
3/7/2014	6.28	7.18
3/7/2014	7.35	7.82
3/11/2014	5.77	8.30
11th Seal		
1/14/2014	4.03	4.20

Log Saturation Indices (SI)

	Calcite	Dolomite	Gypsum	Halite
11th Seal	0.43	0.56	−2.1	−7.7
17th Seal	0.19	0.36	1.63	−1.7

1. Confirm that the SIs listed in the table are correct; you may do the calculations manually or use WATEQF or PHREEQC.
2. Use the major ion data in the table to construct bar, Stiff, and trilinear (Piper) diagrams.
3. What is the hydrogeochemical facies of the mine waters?

4. Construct a tie-line plot of the major ions using both meq/L and mg/L. What is the most appropriate type of tie-line for the data (linear–linear, semi-log etc.)? Discuss the suitability of using meq/L vs. mg/L in the diagrams.
5. Construct a E_h-pH or pE-pH diagram for the Fe^{2+}, Fe^{3+}, goethite system. Place the mine water in the diagram. Assume the pE of the water samples is 2.
6. Based on an analysis of your graphic illustrations and the data in the table, summarize the likely geochemical evolution of the mine waters and the chemical reactions.

Chapter notation

[]	effective molar concentration or chemical activity
Δ	change
°C	temperature in degrees Centigrade
°K	temperature in degrees Kelvin
°	uncharged species (e.g., used as a superscript)
a, b, c, d	moles of species *a, b, c,* and *d*, respectively

subscripts

(aq)	aqueous
(g)	gas
(l)	liquid
(s)	solid

a, b, c, d	moles of species a, b, c, and d, respectively
atm	atmosphere (unit of pressure)
cal	calorie (1 cal = 4.184 J)
cal/mole	calories per mole
e^-	electron
E_h	redox potential of a solution in volts.
E^o	standard potential of a redox reaction in volts at STP
ΔG_R	Gibbs free energy of a reaction at temperature T
ΔG_R	Gibbs free energy of formation of the reaction at (STP)
ΔG°_P	Gibbs free energy of formation of the products at (STP)
ΔG°_r	Gibbs free energy of formation of the reactants at (STP)
J	joule (J = 0.239 cal)
K	thermodynamic equilibrium constant
K_{eq}	thermodynamic equilibrium constant
Kcal/mole	kilocalories per mole
KJ/mole	kilojoules per mole
K_w	equilibrium constant for water
log	logarithm to the base 10
ln	natural logarithm (logarithm to the base e); ln a = 2.303 log a
M	molality
mole	one gram formula weight of a substance
n	number of electrons in a balanced half reaction
P	pressure
Px	partial pressure of x
pE	base-10 log of activity of electrons/L (dimensionless)
pE^o	standard base-10 log activity of electrons/L at STP
pH	negative log of H^+ concentration

R universal gas constant
8.3143 J/mol – °K
1.98717 cal/mol – °K
STP standard conditions (25 °C or 298 °K, 1 atm pressure)

REFERENCES CITED

Back, W., 1960, Origin of hydrochemical facies in groundwater in the Atlantic Coastal Plain, *Proceeding International Geological Congress*, I, 87–95.

Back, W., 1966, *Hydrochemical facies and groundwater flow patterns in northern part of Atlantic Coastal Plain*, U.S. Geological Survey Professional Paper, 498-A.

Ball, W., and D.K. Nordstrom, 1991, *User's manual for WATEQF, with revised thermodynamic data base and text cases for calculating speciation of major, trace, and redox elements in natural waters*, U.S. Geological Survey Open-File Report 91-183.

Brookins, D.G., 1988, *Eh-pH diagrams for geochemistry*, Springer.

Colby, B.R., et al., 1956, *Sedimentation and chemical quality of surface waters in the Wind River Basin, Wyoming*, U.S. Geological Survey Water-Supply Paper 1373.

Collins, W.D., 1923, Graphic representation of analyses, *Industrial and Engineering Chemistry*, 15, 394.

Cressler, A.M., 2007, *U.S. Geological Survey*, www.usgs.gov/media/images/great-salt-lake-and-wasatch-range-antelope-island-davis-county-ut

Drever, J.I., 1988, *The geochemistry of natural waters*, 2nd ed, Prentice-Hall.

Durov, S.A., 1948, Natural waters and graphic representation of their composition, *Akademiya Nauk SSSR Doklady*, 59, 87–90.

El-Kadi, A.I., et al., 2011, NETPATH-WIN: An interactive user version of the mass-balance model, NETPATH, *Ground Water*, 49(4), 593–599.

Faure, G., 1991, *Principles and applications of geochemistry*, Macmillan Publishing Company.

Fetter, C.W., 1993, *Contaminant hydrogeology*, Macmillan.

Garrels, R.M., and C.L. Christ, 1965, *Solutions, minerals and equilibria*, Freeman, Cooper and Company.

Henley, R.A., et al., 1984, Fluid-mineral equilibria in hydrothermal systems. *Society of Economic Geologists*, 1, 267.

Hill, R.A., 1940, Geochemical patterns in Coachella Valley, California, *American Geophysical Union Transactions*, 21, 46–49.

Krasukopf, K.B., 1979, *Introduction to geochemistry*, McGraw-Hill.

Lindsay, W.L., 1979, *Chemical equilibria in soils*, Wiley.

Lloyd, J.W., and J.A., Heathcote, 1985, *Natural inorganic hydrochemistry in relation to groundwater – an introduction*, Clarendon Press.

Maucha, R., 1949, The graphical symbolization of the chemical composition of natural waters, *Hidrologie*, 12, 117–118.

Mayo, A.L., 2010, Ambient well bore and aquifer mixing, pumping stress, and water quality in long-screened wells: What is sampled and what is not? *Hydrogeology Journal*, 18, 823–837.

Mayo, A.L., 2020, Mapping paleo-lacustrine environments and long-term climatic trends using groundwater chemistry, *Hydrogeology Journal*, 28, 987–1002.

Mayo, A.L., R. Henderson, and D. Tingey, 2010, Chemical evolution of shallow playa groundwater in response to post-pluvial isostatic rebound, Honey Lake Basin, California-Nevada, USA, *Hydrogeology Journal*, 18, 725–747.

Mayo, A.L., and R.H. Klauk, 1991, Contributions to the solute and isotopic groundwater chemistry, Antelope Island, Great Salt Lake, Utah, *Journal of Hydrology*, 127, 307–335.

Parkhurst, D.L., and C.A.J. Appelo, 1999, *User's guide to PHREEQC (version 2)*, U.S. Geological Survey Water Resources Investigation Report 99-4259, 312.

Parkhurst, D.L., et al., 1980, *PHREEQE – A computer program for geochemical calculations*, U.S. Geological Survey Water-Resources Investigations, 80–96.

Piper, A.M., 1944, A graphical procedure in the geochemical interpretation of water analyses, *American Geophysical Union Transactions*, 25, 914–923.

Plummer, L.N., et al., 1976, *WATEQF – a FORTAN IV version of WATEQ, a computer program for calculating chemical equilibrium in natural waters*, U.S. Geological Survey Water Resource Investigation 76-13.

Plummer, L.N., et al., 1994, An interactive code (NETPATH) for modeling net geochemical reactions along flow path, Version 2.), U.S. Geological Survey Water-Resources Investigations. Report 94-4169.

Pourbaix, M., 1949, *Thermodynamics of dilute aqueous solutions*, Edward Arnold and Co, London.

Richardson, S.M., and H.Y. McSween, 1989, *Geochemistry – pathways and processes*, Prentice Hall.

Schlegel, M.E., A.L. Mayo, S.T. Nelson, and D. Tingey, 2009, Paleo-climate of the Boise area, Idaho from the last glacial maximum to the present based on groundwater δ^2H and $\delta^{18}O$ compositions, *Quaternary Research*, 71, 172–180.

Schoeller, H., 1935, Utilit, de la notion des exchanges de bases pour la comparison des eaux souterraines, *Société Géologie Comptes Rendus Sommaire et Bulletin*, 5, 651–657.

Schoeller, H., 1962, *Les Eaux Souterraines. Hydrologie dynamique et chimique, Recherche, Exploitation et Évaluation des Ressources*, Les Eaux Souterraines.

Seward, T.M., 1974, Determination of the first ionization constant of silicic acid from quartz solubility in buffer solutions, *Geochimica et Cosmochimica Acta*, 38, 1651–1664.

Stiff, H.A., Jr, 1951, The interpretation of chemical water analysis by means of patterns, *Journal of Petroleum Technology*, 3(10), 15–17.

Stumm, W., and J.J. Morgan, 1981, *Aquatic chemistry*, Wiley and Sons.

Tardy, Y., and R.M. Garrels, 1974, A method of estimating Gibbs energy of formation of layer silicates, *Geochimica et Cosmochimica Acta*, 38, 1101–1116.

Winograd, I.J., and G.M. Farlekas, 1974, Problems in ^{14}C dating of waters from aquifers of deltaic origin in Isotope hydrology, *International Atomic Energy Agency*, 69–93.

Woods, T.L., and R.M. Garrels, 1987, *Thermodynamic values at temperature for natural inorganic materials: A summary*, Oxford University Press.

13 Geothermometers

13.1 INTRODUCTION

Chemical and isotopic geothermometers use solute and isotopic compositions to estimate in situ aquifer temperatures. Aquifer temperature estimates provide insight into groundwater flow pathways and circulation depths. They are used during the analysis of geothermal resources and are useful for estimating subsurface thermal conditions for reconnaissance investigations where deep geothermal systems are not economically reachable by drilling. The thermometers include water and solutes, steam and gas, and isotopes. Water, solutes, steam, and gas thermometers are collectively known as chemical geothermometers. Geothermometers have been developed on the basis of theoretical considerations of solute and isotopic equilibria and by analyzing empirical data. The thermometers are based on the idea that groundwater solute and isotopic equilibrium reached at the aquifer temperature does not re-equilibrate as the water flows from the geothermal reservoir to the sampling point.

Geothermal water may cool during the ascent to the surface (conductive) and by boiling and steam loss (adiabatic) due to the rapid reduction in the hydrostatic head. Conductive cooling does not alter the solute or isotopic composition, but it may result in changes to the water's mineral saturation, which can affect the solute composition and induce additional chemical reactions. Geothermometry assumes that there is no change in the composition due to conductive cooling. Boiling is accompanied by steam loss and a subsequent increase in the solute concentrations of the residual solution.

Groundwater is heated by close proximity to magmatic bodies and at depth by the geothermal gradient. The geothermal gradient, which is the increase in crustal temperature with increasing depth, is primarily due to radiogenic decay in the mantle and upper crust. Away from tectonic plate boundaries and in stable tectonic areas, the gradient is typically about 25 to 30 °C/km. At tectonic boundaries mantle heat flow can be appreciably greater. In areas of high heat flow, such as the tectonically active portions of the western United States, it can be as great as 40 to 50 °C/km. The geothermal gradient means that, at depth, all groundwater is thermal and only shallow groundwater systems (i.e., < ~ 0.5 km deep) have aquifer temperatures approaching mean annual air temperature. The temperature of shallowly circulating groundwater is usually the mean annual air temperature. All deep groundwater undergoes temperature-regulated thermodynamic chemical evolution, and some isotopes undergo isotopic exchange. In some locations thermal groundwater escapes rapidly to the surface and discharges at thermal springs; elsewhere, the ascent to the surface is slow and the discharge temperature is closer to mean annual air temperature. Hot gas and vapor may escape to the surface in the absence of a discharging liquid. Such discharge locations are known as fumaroles. The rapid ascent of thermal liquids is usually associated with the upward flow along fault damage zones. Elevated geothermal gradients in areas such as the Salton Sea in California, where the East Pacific Rise transitions to the San Andreas Fault, and in Iceland, located on the Mid-Atlantic Ridge, make relatively shallow subsurface geothermal water particularly suitable for large-scale geothermal energy development.

In addition to aiding in geothermal resource investigations, knowledge of the approximate aquifer temperature can be used to estimate maximum groundwater circulation depth using the equation:

$$D = (T - A)/G \tag{13.1}$$

DOI: 10.1201/9781003646129-16

where D is the maximum groundwater circulation depth, T is the calculated or estimated maximum groundwater temperature, A is the mean annual air temperature, and G is the geothermal gradient. All units must be consistent. For example, if the calculated or estimated groundwater temperature is 250 °C, the mean annual air temperature is 11.2 °C, and the geothermal gradient is 60 °C/km, the groundwater circulation depth would be:

$$D = \frac{250 - 11.2}{\frac{60^\circ \mathrm{C}}{\mathrm{km}}} = 3.98 \mathrm{~km}$$

Similar calculations can be made using the regional geothermal gradient and the measured discharge temperature at a thermal spring or well. For example, in the Idaho Batholith Schlegel et al. (2009) and Mayo et al. (2014) calculated groundwater circulation depths up to 6.8 km (average 5 km) for ~150 °C thermal spring discharges. The waters had an average ^{14}C age of ~20,000 ybp.

Early attempts to develop aqueous geothermometers include work by White (1965, 1968) and Ellis and Mahon (1967). Geothermometer development accelerated during the 1970s as part of the first big U.S. push for domestic non-petroleum-based energy in response to the oil embargos. Since then, many geothermometers have been developed, and existing thermometers have undergone refinement. Arnórsson (2000) and Cioni and Marini (2020) discuss in detail the development, limitations of usefulness of many geothermometers. Some geothermometers including silica are written as polynomial equations (Table 13.1). Most non-silica quantitative geothermometers are written in the form:

$$\mathrm{T}\ (^\circ\mathrm{C}) = \frac{A}{\log k + B} - 373.15 \tag{13.2}$$

where A and B are equation specific constants, and K is the thermodynamic equilibrium constant for the reaction. For example, in the Na–K thermometer the parameter K is the solute concentrations of Na^+ and K^+ (i.e., log $\frac{Na^+}{K^+}$) with concentrations in either moles/L or mg/L. The parameter K should not be confused with the element K.

Important Takeaway – Groundwater is heated by close proximity to magmatic bodies and, at depth, by the geothermal gradient. Thermally driven rock–water interactions can alter the solute and isotopic composition of the heated water. Alkali elements, silica, and isotopic geothermometers can be used to calculate in the in situ aquifer temperature of heated groundwater. Knowledge of aquifer temperatures can be used to assist in reconnaissance investigations of geothermal resources and for calculating groundwater circulation depths.

13.2 QUALITATIVE GEOTHERMOMETERS

Qualitative geothermometers are based on physical observations at discharge locations, anomalous solute concentrations, compositions of escaping gases, and anomalous isotopic data (Williams et al., 2008). Sinter (silica) deposition by thermal groundwater is a reliable indicator that the subsurface temperature exceeds 180 °C (White, 1970) because silicate minerals precipitate rapidly at this temperature. Travertine ($CaCO_3$) deposits may provide qualitative information regarding subsurface aquifer temperatures, but many factors may result in travertine deposition. In non-carbonate terrain, where all groundwater CO_2 is obtained in the recharge zone, travertine deposits indicate cool aquifer temperatures. Lower than anticipated Ca^+ and HCO_3^- concentrations may indicate elevated subsurface temperatures due to temperature-induced calcite precipitation.

TABLE 13.1
Silica Mineral Geothermometers Based on Silica Mineral Solubility

No steam loss		
Quartz	$T\,(^\circ C) = \dfrac{1309}{5.19 - \log C} - 373.15$	(13.3)
Chalcedony	$T\,(^\circ C) = \dfrac{1032}{4.69 - \log C} - 373.15$	(13.4)
α-cristobalite	$T\,(^\circ C) = \dfrac{1000}{4.78 - \log C} - 373.15$	(13.5)
Opal	$T\,(^\circ C) = \dfrac{781}{4.51 - \log C} - 373.15$	(13.6)
Amorphous silica	$T\,(^\circ C) = \dfrac{731}{4.52 - \log C} - 373.15$	(13.7)
Maximum steam loss		
Quartz	$T\,(^\circ C) = \dfrac{1522}{5.75 - \log C} - 373.15$	(13.8)
Chalcedony	$T\,(^\circ C) = \dfrac{1227}{5.20 - \log C} - 373.15$	(13.9)
α-cristobalite	$T\,(^\circ C) = \dfrac{1195}{5.29 - \log C} - 373.15$	(13.10)
Opal	$T\,(^\circ C) = \dfrac{976}{5.02 - \log C} - 373.15$	(13.11)
Amorphous silica	$T\,(^\circ C) = \dfrac{926}{5.03 - \log C} - 373.15$	(13.12)

Log *C* is the SiO_2 concentration in ppm.
Sources: Data from Fournier, 1977. Additional silica geothermometers have been developed and reviewed by Gunnarsson and Arnórsson (2000), Fournier and Potter, 1982; Verma, 2000, 2003, 2014; Verma and Betzler, 2013, and others.

Anomalous concentrations of volatile constituents are often associated with elevated aquifer temperatures. Enrichment of volatile constituents or their derivatives (i.e., B, NH_4^+, Hg, and H_2S) in shallow groundwater may suggest deep aquifer boiling (Tonani, 1970). During boiling, volatile components are partitioned into the steam phase and may escape upward into a cooler overlying aquifer or to the surface. High rations of Cl^-/F^- and Cl^- / SO_4^{2-} in warm groundwaters may indicate high subsurface temperatures (Mahon, 1970; Fournier and Truesdell, 1970). Brondi et al. (1973) suggest that elevated Li^+ in cool groundwater may indicate past elevated aquifer temperature. Soil gas and sediments with elevated mercury (Klusman and Landress, 1979) and helium (Roberts, 1975) have been found to be associated with deep thermal systems. Collaborating geologic and/or hydrogeologic evidence should be established when attempting to use these constituents as qualitative indicators of elevated subsurface temperatures because other factors may be responsible.

The stable isotopic compositions of groundwater may also provide qualitative information regarding existing and paleo-subsurface temperatures. As discussed in Chapter 8, positive $\delta^{18}O$ shifts relative to the GMWL have been observed in many thermal waters (Figure 13.1). Such shifts are often on the order of +3 to +5 ‰. Low $\delta^{18}O$ in aluminosilicate minerals, particularly feldspar, has also been used as an indicator of paleo-geothermal systems in granitoid rocks (Taylor, 1977).

Important Takeaway – Qualitative solute and isotopic geothermometers can be used to indicate existing or paleo-geothermal groundwater conditions. Qualitative thermometers can include sinter and travertine deposits, aqueous volatile constituents, soil and gas Hg and He, high rations of Cl^-/F^- *and* Cl^- / SO_4^{2-}*, and stable* $\delta^{18}O$ *shifts in groundwater relative to the GMWL.*

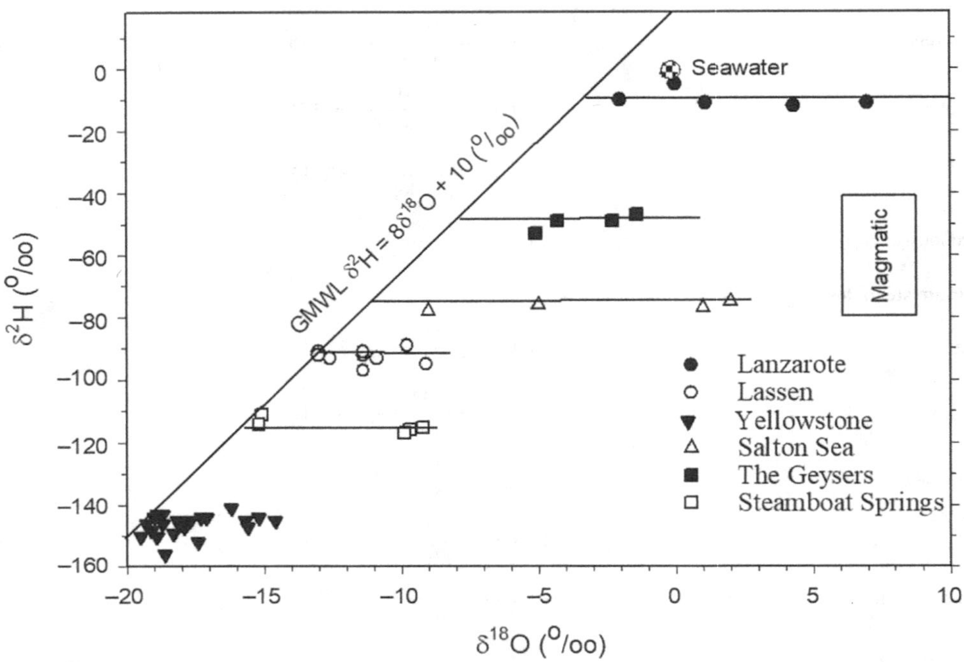

FIGURE 13.1 Positive $\delta^{18}O$ shifts in worldwide geothermal waters. As shift of 5 ‰ or more is is usually indicative of geothermal heating. Data sources: Lanzarote (Arana and Pnichi, 1974), Lassen (Craig, 1963), Yellowstone (Bergfeld et al., 2014), Salton Sea (Craig, 1966), the Geysers (Lowenstern et al., 1999), and Steamboat Springs (Craig, 1963).

13.3 QUANTITATIVE SOLUTE GEOTHERMOMETERS

Quantitative geothermometers have been developed using the solute and isotopic content of thermal groundwaters and their associated gases (Ellis and Mahon, 1977; Fournier, 1977; Truesdell and Hulston, 1980; White, 1970). Geothermometers may be based on univariate reactions such as the dissolution of silica or may be more global in scope, involving a large number of dissolved and gaseous components. Univariate solubility thermometers are highly sensitive to secondary process such as the dilution with other waters and to vapor loss or condensation. Thermometers are sensitive to partial equilibration, dilution, steam loss, etc. The most commonly used geothermometers are based on the solubility of silica, alkali element exchange, and isotopic fractionation.

The assumptions on which quantitative geothermometers are based include (Fournier, 1977):

1. Temperature dependent water and rock reactions fix the amount of dissolved indicator constituents at the groundwater circulation depth.
2. There is an adequate supply of all constituents involved in the temperature-dependent reaction. In other words, the mineral supply is not a limiting factor.
3. Water–rock equilibration occurs at the reservoir temperature.

4. There is little or no re-equilibration or change in the composition at lower temperatures during water's ascent from the reservoir to the surface.
5. Hot water coming from deep in the groundwater system does not become diluted by mixing with cooler shallow groundwater.

Some of these assumptions may or may not be valid for the water investigated. The selection and use of one or more geothermometers is further complicated by the fact that temperature-dependent equilibrium means that some independent knowledge of the subsurface temperature and geologic regime is needed to select the appropriate geothermometer(s). For example, silica-based geothermometry requires an abundant supply of quartz, chalcedony, or other silica oxide mineral. Alkali element geothermometers require an abundant supply of aluminosilicate minerals such as K-feldspar, Ca-Na feldspar, amphibole, and an understanding of the subsurface mineralogy.

Fournier and Truesdell (1974) suggest the following guidelines when distinguishing between conductive and adiabatic geothermometers for boiling and non-boiling springs.

Boiling springs

1. Small flow rate: assume mostly conductive cooling. Apply chemical indicators assuming little or no steam loss.
2. Large flow rate: assume adiabatic cooling. Apply chemical indicators assuming maximum steam loss.

Non-boiling springs

1. Small flow rate: will not have clear-cut interpretation. May indicate that the water has never been very hot, the water is a mixture of water from sources with different temperatures, or that hot water cooled during conductive flow.
2. Large flow rate: assume no conductive cooling. Test to see if geothermometers, particularly Na–K–Ca, suggest chemical equilibrium at the aquifer discharge temperature (± 25 °C). If a higher temperature is indicated, treat as mixed water.

A small flow rate is arbitrarily set at <200 L/min for a single isolated spring discharge and <20 l/min for a single spring in a group of springs. These guidelines should not be considered as set rules for predicting subsurface conditions.

13.3.1 Solubility Geothermometers

Only silica dissolution has been extensively used as a solubility thermometer. The use of silica was proposed by Fournier and Rowe (1966). Silica is useful because it is abundant in most thermal systems, and silica concentrations are not influenced by the common ion effect, complex formation, ion exchange, and the loss of volatile gases. Silica-based thermometers for systems with and without steam loss are shown in Table 13.1. Correlation of the silica content of hot water with the quartz equilibration temperature is illustrated in Figure 13.2. If the water cooled adiabatically before sampling (i.e., by boiling), use curve A. Use curve B if the water cooled mainly by conduction. The quartz geothermometer works best in the temperature range of 150–225 °C. At higher temperatures, silica may precipitate during the ascent of the water. Other silica thermometers work best for lower aquifer temperatures. There is no correct silica-temperature curve, and a graph comparing multiple solubility vs. temperature geothermometers can be constructed using the silica equations from various researchers.

In theory, solubility geothermometers could be developed for many aquifer minerals. However, the applicability of such thermometers is limited because the concentrations of most ions are affected by factors such as ion exchange, the common ion and ionic strength effects, and the order

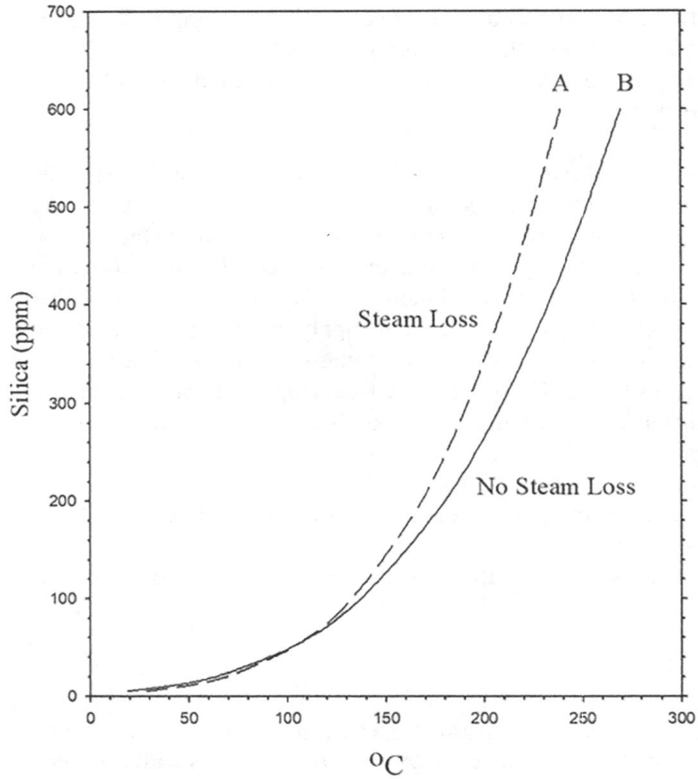

FIGURE 13.2 Solubility of quartz in equilibrium with saturated steam (B), and silica in residual water after steam loss after adiabatic cooling to 100 °C (A). The equilibrium curves are based on Equations 13.3 and 13.8.

of encounter. For example, an anhydrite thermometer was used to estimate hot spring temperature based on Ca^{2+} and SO_4^{2-} concentrations (Saki and Matsubaya, 1974). In practice, this thermometer is only applicable to the unusual water when the Ca^{2+} content is not influenced by the dissolution or precipitation of carbonate minerals.

13.3.2 Exchange Geothermometers

Exchange geothermometers are based on temperature-dependent exchange and alteration reactions between thermal groundwater and aquifer minerals where the ratios of dissolved constituents are temperature dependent. The most common thermometers are based on equilibrium exchange of alkali ions with feldspar and amphibole minerals. Commonly used thermometers are Na–K and Na–K–Ca. Most thermometers were empirically calibrated using down-hole temperatures of well waters collected at depth.

13.3.2.1 Na–K Geothermometer

The original Na–K thermometers (Orville, 1963; Hemley, 1967) were developed for the reaction:

$$K^+ + \text{Na-feldspar} = \text{K-feldspar} + Na^+ \tag{13.13}$$

and for reactions where K^+ exchanges for Na^+ in an aquifer containing muscovite and quartz. Such reactions are common in igneous rocks and in immature sediments. Cioni and Marini (2020) provide a comprehensive review of the mineralogy and ion concentrations effecting Na–K and other thermodynamic-based geothermometers. There are numerous versions of the Na–K thermometer (Table 13.2) because various authors have developed empirical calibrations specific to the Na/K ratio of the thermal water under investigation. There is no universally correct Na–K thermometer, and all are plausible in the temperature range of 100 to 350 °C (Cioni and Marrini, 2020). It is usually best to analyze water using several Na–K thermometers. Na–K thermometers should not be used with reservoir temperatures <100 °C. A calculated high Na–K temperature and a calculated low Na–K–Ca temperature suggest a low aquifer temperature.

13.3.2.2 Na–K–Ca Geothermometer

The empirical Na–K–Ca geothermometer of Fournier and Truesdell (1973) is one of the most commonly used alkali thermometers. The advantage of the Na–K–Ca geothermometer relative to quartz and Na–K thermometers is that does not give high or misleading temperatures for cooler groundwater systems. The Na–K–Ca geothermometer is based on the idea that, except in alkali lake environments and where the dissolution of alkali-bearing carbonate minerals are a factor, the Na^+ and K^+ concentrations are controlled by reactions involving silicate minerals. On the other hand, the absolute concentration of Ca^{2+} is generally controlled by the temperature, pH, and P_{CO_2}-dependent solubility of calcium carbonate minerals. Ca^{2+} does enter into silicate exchange reactions in competition with Na^+ and K^+. In most natural waters, the concentration of K^+ is considerably less than the concentration of Na+. Therefore, during exchange reactions, the change in Ca^{2+} concentration will affect the K^+ concentration more than it will affect the Na^+ concentration.

Application of the Na–K–Ca geothermometer assumes the following conditions:

1. The concentration of Na^+ greatly exceeds the concentration of K^+.
2. Excess silica is present in the solution.
3. Aluminum is conserved in the solid phase.
4. Reactant phase hydrogen ions are balanced with product phase hydrogen ions (i.e., there is no change in the hydrogen ion concentration during the reaction; no pH change).

The Na–K–Ca thermometer is based on the relationship:

$$T = \left[\frac{1647}{\log\left(\frac{Na}{K}\right) + \beta\left(\log\sqrt{\left(\frac{Ca}{Na}\right)} + 2.06\right) + 2.47}\right] - 273.15 \qquad (13.30)$$

which uses two different β values (1/3 and 4/3). The ion concentrations are in mg/L. The β of 1/3 is used when the reservoir temperature is >100 °C and 4/3 is used in the less common case where the temperature is <100 °C. Benjamin et al. (1983) reanalyzed the Fournier and Truesdell data and suggest the following correction for water temperatures >100 °C:

$$T = \left[\frac{1647}{\log\left(\frac{Na}{K}\right) + 0.055(\log\sqrt{\left(\frac{Ca}{Na}\right)} + 1.69}\right] - 273.15 \qquad (13.31)$$

where T is °C. The Na–K–Ca thermometer is not reliable for acidic sulfate rich water containing little chloride. The thermometer can provide reliable results for CO_2- and Ca^{2+}-rich water provided

TABLE 13.2
Na–K Geothermometers

Geothermometer Na/K (mg/kg)	Calibration range (°C)	Comments	Source	
$(^\circ C) = \frac{504.1}{\log \frac{Na}{K} - 0.0594} - 373.15$	60–700		16	(13.14)
$T\,(^\circ C) = \frac{801.5}{\log \frac{Na}{K} + 0.5214} - 373.15$	170–300		4	(13.15)
$T\,(^\circ C) = \frac{640.5}{\log \frac{Na}{K} + 0.2411} - 373.15$	130–375	Sedimentary basin	9	(13.16)
$T\,(^\circ C) = \frac{777}{\log \frac{Na}{K} + 0.47} - 373.15$	4–340		6	(13.17)
$T\,(^\circ C) = \frac{855.6}{\log \frac{Na}{K} + 0.6267} - 373.15$	>125		14	(13.18)
$T\,(^\circ C) = \frac{1217}{\log \frac{Na}{K} + 1.252} - 373.15$	80–340		5	(13.19)
$T\,(^\circ C) = \frac{933}{\log \frac{Na}{K} + 0.762} - 373.15$	Volcanic Iceland		1	(13.20)
$T\,(^\circ C) = \frac{1319}{\log \frac{Na}{K} + 1.468} - 373.15$	25–250	Volcanic Iceland	2	(13.21)
$T\,(^\circ C) = \frac{1390}{\log \frac{Na}{K} + 1.52} - 373.15$	250–350		7	(13.22)
$T\,(^\circ C) = \frac{1178}{\log \frac{Na}{K} + 1.239} - 373.15$	Volcanic		11	(13.23)
$T\,(^\circ C) = \frac{1180}{\log \frac{Na}{K} + 1.08} - 373.15$	30–200	Sedimentary basin	8	(13.24)
$T\,(^\circ C) = \frac{1170}{\log \frac{Na}{K} + 1.42} - 373.15$	50–150	Granitic rocks	10	(13.25)
$T\,(^\circ C) = \frac{1289}{\log \frac{Na}{K} + 1.384} - 373.15$	80–350	(Arnórsson, 1983a)	15	(13.26)
$T\,(^\circ C) = \frac{2.0446}{\log \frac{Na}{K} + 126.01} - 373.15$	91–516		3	(13.27)

TABLE 13.2 (*Continued*)
Na–K Geothermometers

Geothermometer Na/K (mg/kg)	Calibration range (°C)	Comments	Source	
$T\,(^\circ C) = \dfrac{876.3}{\log\dfrac{Na}{K} + 0.6469} - 373.15$	30–350		12	(13.28)
$T\,(^\circ C) = \dfrac{315}{\log\dfrac{Na}{K} - 9.467} - 373.15$	120–270	Sedimentary basin	13	(13.29)

Sources:

1 Arnórsson et al., 1983a
2 Arnórsson et al., 1983b
3 Bayram, 2001
4 Ellis, 1970
5 Fournier, 1979a
6 Fournier and Truesdell, 1973
7 Giggenbach, 1988
8 Kharaka and Mariner, 1989
9 Mercado, 1970
10 Michard, 1990
11 Nieva and Nieva, 1987
12 Santoyo and Díaz-González, 2010
13 Simmons, 2013
14 Truesdell, 1976
15 Verma and Santoyo, 1997
16 White, 1968

calcium carbonate was not deposited after the water moved from the high temperature reservoir. If carbonate minerals were deposited prior to sampling, the geothermometer may result in calculated temperatures that are too high. The thermometer is not reliable for CO_2-rich water with aquifer temperatures less than 75 °C (Paces, 1975), and the thermometer may give too low of temperature estimates if low Ca^{2+} thermal water encounters calcium carbonate minerals during its ascent.

The Na–K–Ca geothermometer yields anomalously high temperate estimates for water rich in Mg^{2+}. The Mg correction is best used for waters with relatively low temperatures. Thermal groundwater may encounter numerous Mg-bearing minerals including amphibole, biotite, chlorite, dolomite, glauconitic, montmorillonite, olivine, pyroxene, serpentine, and vermiculite. Fournier and Potter (1979) proposed a Mg temperature correction for the Na–K–Ca geothermometer. The correction is based on a calculated *R* value:

$$R - \left[Mg^{2+}/\left(Mg^{2+} + Ca^{2+} + K^{+}\right)\right] \times 100 \qquad (13.32)$$

where the concentrations are reacting equivalents. For *R* values between 5 and 50, the Mg temperature correction (Δt_{mg}) is used for estimating Na–K–Ca temperatures > 70 °C:

$$\Delta t_{mg} = a_1 - b_1 R + c_1 \log R - d_1\left(\text{Log } R^2\right)/T - e_1\left(\text{Log } R^2\right)/T^2 + f_1\left(\text{Log } R^3\right)/T^2 \qquad (13.33)$$

where T = the calculated Na–K–Ca temperature (°C), a_1 = 10.66, b_1 = 4.7415, c_1 = 325.867, d_1 = 1.0321 × 10^5, e_1 = 1.9683 × 10^7, and f_1 = 1.605 × 10^7. The calculated correction value is subtracted from the Na–K–Ca calculated temperature. For *R* values <5, the Δt_{mg} is calculated as:

$$\Delta t_{mg} = -a_2 + b_2 \log R + c_2\left(\log R\right)^2 - d_2\left(\log R\right)^2/T - e_2 \log R/T \qquad (13.34)$$

where T = the calculated Na–K–Ca temperature, a_2 = 1.02995, b_2 = 59.97116, c_2 = 145.049, d_2 = 36711.6, e_2 = 1.67516 × 10^7.

Over the years, numerous less frequently used alkali-exchange geothermometers have been developed. Giggenbach (1988) described two K–Mg geothermometers based on equilibrium with chlinochlore (a chlorite mineral) and muscovite:

$$T = \frac{4410}{14.0 - \log \frac{K^2_{\text{chlinochlore}}}{Mg_{\text{chlinochlore}}}} - 273.15 \quad (13.35)$$

$$T = \frac{4410}{9.3 - \log \frac{K^2_{\text{muscovitee}}}{Mg_{\text{mscovite}}}} - 273.15 \quad (13.36)$$

where the concentrations are mg/kg. K–Mg geothermometers were also developed by Fournier (1991):

$$T = \frac{2680}{9.4 - \log \frac{K^2}{Mg}} - 273.15 \quad (13.37)$$

$$T = \frac{6460}{18.10 - \log \frac{K^2}{Mg}} - 273.15 \quad (13.38)$$

Equations 10.37 and 13.38 are used for $T < 162.5$ and > 163.5 °C, respectively. Concentrations are mg/kg. Nieva and Nieva (1987) proposed a Na–K–Mg geothermometer:

$$T = \frac{11140}{18.3 + 6\log\left(\frac{Na}{K}\right) + \log\left(\frac{Mg}{Na^2}\right)} - 273.15 \quad (13.39)$$

where concentrations are in mole/kg. A Na–K thermometer can be derived from Equation 13.30:

$$T = \frac{6428}{118.143 \log \frac{K^2}{Mg}} - 273.15 \quad (13.40)$$

Cioni and Marini (2020) list 16 Na–Mg thermometers developed from the K–Mg thermometer of Giggenbach (1988).

Another class of less frequently used geothermometers involves major-trace element ratios. Kharaka and Mariner (1989) proposed a Mg–Li thermometer for use in sedimentary basins with subsurface temperatures in the range of 30 to 70 °C:

$$T = \frac{2200}{\log\left(\frac{\sqrt{Mg}}{Li}\right) + 5.47} - 273 \quad (13.41)$$

where Mg^{2+} and Li^+ are in mg/kg and T is °C. Fouillac and Michard (1981) developed two Na–Li geothermometers:

$$T = \frac{1000}{\log\left(\frac{\text{Na}}{\text{Li}}\right) + 0.38} - 273 \quad (13.42)$$

$$T = \frac{1195}{\log\left(\frac{\text{Na}}{\text{Li}}\right) - 0.13} - 273 \quad (13.43)$$

where concentrations are moles. These thermometers are based on the exchange of H^+ for Li^+ on lithium-rich clay where:

$$\text{Clay} - \text{Li} + \text{H}^+ = \text{Clay} - \text{H}^+ + \text{Li}^+$$

The exchange may also be with zeolites. Equation 13.42 is for Cl concentrations < 0.3 M and Equation 13.43 is for Cl concentrations > 0.3 M. Sanjuan et al. (2014) proposed a Na–Li thermometer for seawater basalt interactions:

$$T = \frac{920}{\log\left(\frac{\text{Na}}{\text{Li}}\right) - 1/105} - 273 \quad (13.44)$$

where concentrations are moles.

13.3.3 Giggenbach Na–K–Mg Ternary Diagram

Giggenbach (1988) noted that the successful application of an alkali geothermometer requires a determination of the applicability of the geothermometer to the groundwater being investigated. Exchange geothermometers assume that exchange equilibrium has been attained; however, the common exchange geothermometers do not incorporate a mechanism to evaluate the contribution of ions from dissolution vs. exchange reactions. In an attempt to overcome this limitation, he developed a ternary diagram for Na–K–Mg solutions, which allows the evaluation of the ion contributions from both dissolution and exchange reactions. The diagram combines the slow equilibrating Na–K geothermometer and the fast-equilibrating K–Mg geothermometer.

The diagram consists of two parts. The underlying diagram has calibration curves that define the calibrated–partially equilibrated water boundary and the partially equilibrated–immature water boundary (Figure 13.3a). In the diagram sets of Na–K and K–Mg lines radiate from the Mg and Na vertices. Each radiating line represents different Na–K and K–Mg ratios, where each value on each line represents the same temperature. Where the two lines cross, both Na–K and K–Mg are in equilibrium, and the full equilibration curve can be drawn connecting the cross-over points. The location of the partial equilibrium–immature water curve is arbitrary. The water's chemistry is then plotted as the percentage of Na^+, K^+, and Mg^{2+} on the underlying calibration plot using the corrected percentage contributions (fraction) of Na^+, K_+ and Mg^{2+} (Figure 13.3b). Prior to calculating the percentage contributions of each ion, the concentrations are adjusted as follows: $Na^+/1000$, $K^+/100$,

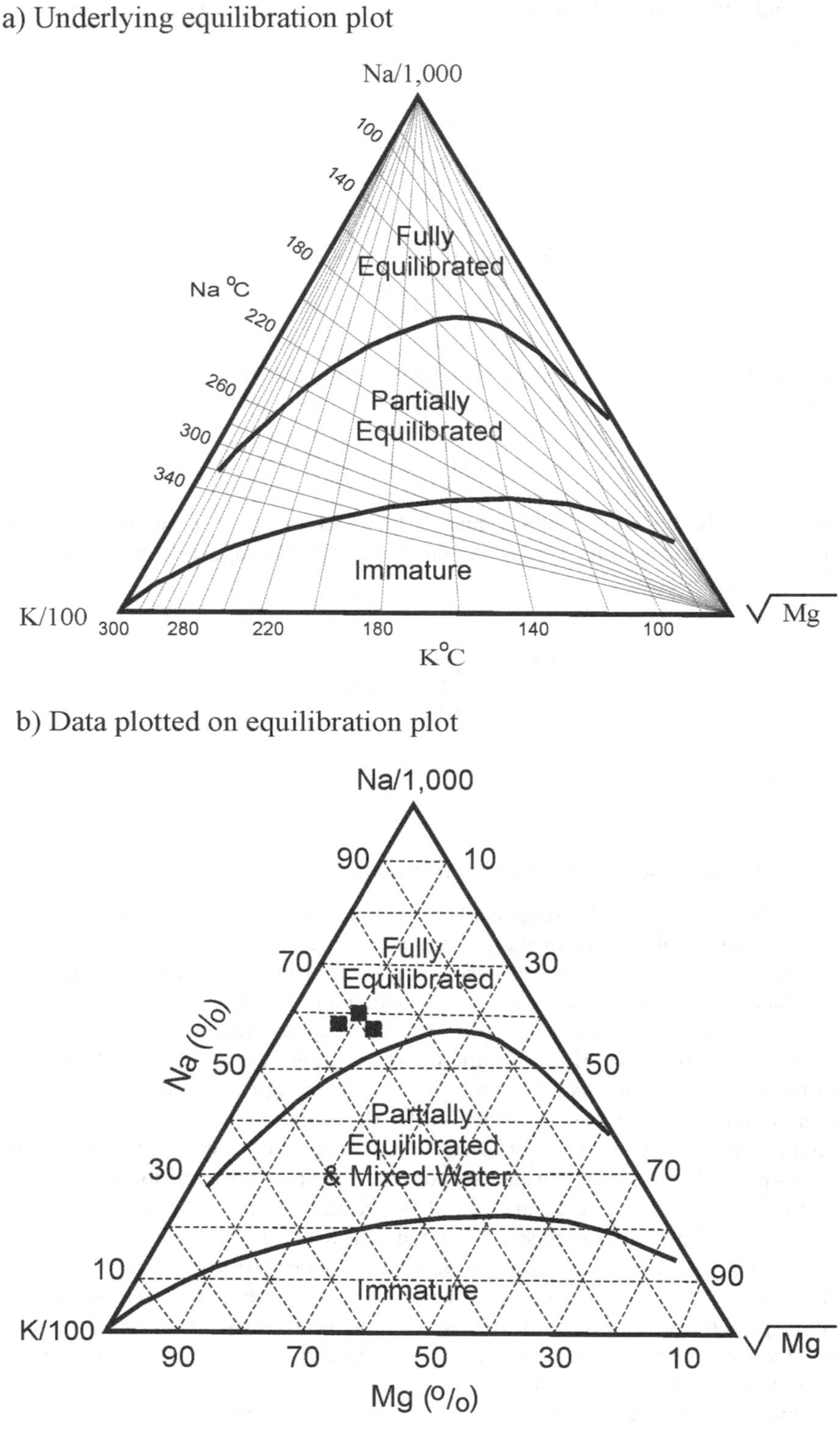

FIGURE 13.3 Giggenbach Na–K–Mg ternary diagram for determining the extent of alkali element exchange equilibrium. The diagram consists of two parts: (a) plotting the Na–K–Mg equilibration curve and (b) plotting the fraction concentration of each ion.

and $\sqrt{Mg^{2+}}$. The Giggenbach ternary diagram has been widely used in geothermal investigations, and several commercially available geochemistry codes can be used to easily construct the tertiary diagram.

The Giggenbach geothermometer is applicable 1) to rising hydrothermal solutions where rock–water interactions are initially diffuse due to magmatic vapors and metasomatism associated with the conversion of $CO_{2(g)}$ to calcite and to acid clays in contact with aluminum silicate minerals and 2) where K-metasomatism accompanies thermal groundwater upwelling. Metasomatism is the chemical alteration of existing rock with hydrothermal fluids and vapors without a change in the volume of the rock.

Important Takeaway – Solute geothermometers include both solubility and exchange thermometers. Most solute thermometers are univariate silica based involving silica equilibria for minerals such as quartz. Most exchange thermometers are alkali element exchange reactions using Na, K, Ca, and Mg. The Na–K–Ca with or without Mg correction is the most commonly used alkali thermometer. There is no universally correct or useful geothermometer, and not all groundwaters are suitable for geothermometry. The Giggenbach ternary diagram has been widely adopted to evaluate the equilibrium vs. non-equilibrium of the alkali ions in solution.

13.3.4 Mixing Geothermometer Models

The cooling of ascending hot waters may occur in several ways. Cooling may be adiabatic (i.e., steam loss due to boiling), conductive (i.e., heat loss to wall rock in the aquifer adjusted for the geothermal gradient), or caused by mixing with shallow cooler water. Thermal springs with discharge rates >100 to 200 L/min are assumed to have little heat loss per unit mass of water to aquifer wall rock during upward movement (Fournier and Truesdell, 1974). For non-mixed, non-boiling systems, the discharge temperature of large flow rate springs may be assumed to be close to the maximum aquifer temperature. Both the solute composition and discharge temperature of thermal waters that have mixed with cooler, shallow groundwater will not be in equilibrium with the deeper subsurface conditions. It is not always readily apparent if mixing has occurred. In addition to the Giggenbach graphical method, there are several qualitative tests that may indicate mixing. Seasonal temporal variations in the discharge rate are often a good indicator of mixed groundwater issuing from intermediate and regional groundwater systems. Mixing is also indicated when there are variations in the discharge temperature or in the content of conservative constituents such as Cl^- and B in several thermal springs located in the same area.

Fournier and Truesdell (1974) developed two mixing models based on silica-enthalpy relationships. Both models require knowledge of the temperature and silica content of the cold water before mixing (Fournier, 1979b). Model 1 assumes boiling of the thermal water before it mixes with the cold water; however, the steam and hot water must rise together. Model 2 assumes that some steam escaped from the hot water prior to mixing with the cold water. The first step in constructing the models is to plot curves depicting quartz solubility and quartz solubility with maximum steam loss (Figure 13.4). Then plot the location of cold and hot water silica-enthalpy locations as A and B, respectively. Extend line A–B, a mixing line, to the quartz solubility curve C. C is the silica-enthalpy location for the hot water component. The original hot water temperature is determined using steam tables (Keenan et al., 1969, 1992). Steam table are extensive and are not included herein. For systems where there is steam loss, plot two points: (A) the cold water and (D) the residual hot water. Point E is the location of the residual water before steam separation (i.e., 100 °C). Draw a horizontal line from E to the maximum steam loss curve (F), then down vertically to the quartz solubility curve (G), where the silica temperature can be determined with the steam tables. The original silica-enthalpy is at G. The curves in Figure 13.4 cannot be used for boiling water because

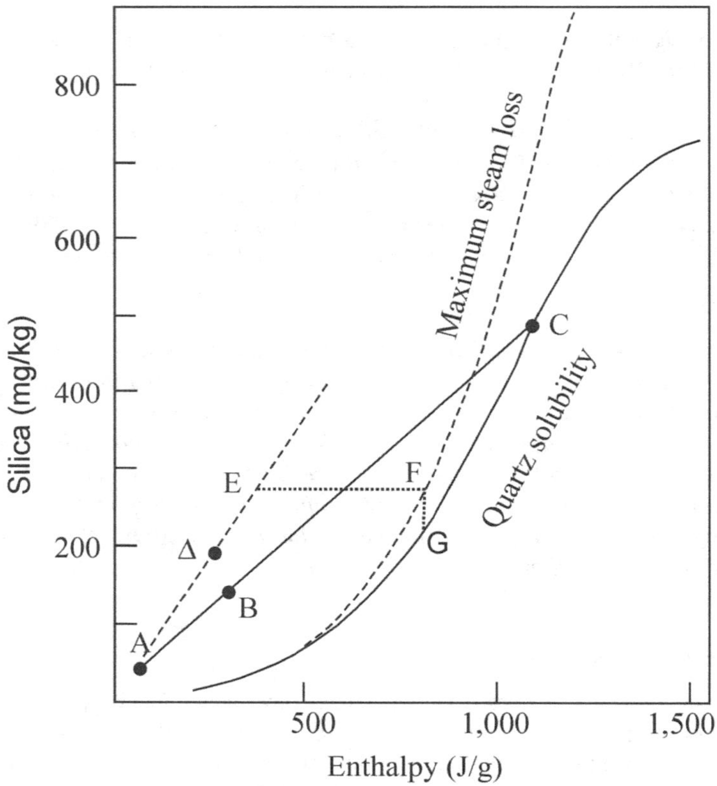

FIGURE 13.4 Dissolved silica-enthalpy curves for calculating the hot-water component of cold–hot mixed thermal groundwater (modified after Truesdell and Fournier, 1977)

heat is lost to the escaping steam. A chloride-enthalpy diagram method has been developed by Truesdell and Fournier (1975) for mixed hot and cold water boiling springs with an initial hot water component of >200 °C.

13.4 GAS GEOTHERMOMETERS

Thermodynamic equilibrium is the basis of gas geothermometers. Ellis (1957) was the first to suggest that steam could be used to estimate aquifer temperatures. Since then numerous gas geothermometers have been developed using liquid and gas from wells and fumaroles (Richet et al., 1977). The three types of gas geothermometers are based on:

1. gas–gas equilibria,
2. gas–mineral equilibria, and
3. mineral-gas with residual gas such as CH_4, H_2 and H_2S.

Arnórsson and Gunnlaugsson (1985) developed five fumarole steam geothermometers for Iceland volcanic geothermal systems (Table 13.3). The thermometers are based on the idea that fumarole gas is regulated by temperature-dependent gas–gas and gas–mineral equilibria. They found that the concentrations of CO_2, H_2S, and H_2 are in equilibria with mineral buffers. Since 1985, there has been a proliferation of gas geothermometers. As part the development of a new and complex methodology for developing gas geothermometers, Acevedo-Aniscasio et al. (2021) compiled a list

TABLE 13.3
Selected Fumarole Gas Thermometers Based on Gas-Gas and Gas-Mineral Equilibria

Thermometer	Concentration	Source	
$T = 44.1 + 269.25(CO_2) - 76.88\ (CO_2)^2 + 9.52\ (CO_2)^3$	mmole/kg	2	(13.45)
$T = 246.7 + 44.8\ H_2S$	mmole/kg	2	(13.46)
$T = 277.2 + 20.99\ H_2$	mmole/kg	2	(13.47)
$T = 341.7 + 29.7 \log \frac{CO_2}{H_2}$	mmole/kg	2	(13.48)
$T = 304.1 + 39.48 \log \frac{H_2S}{H_2}$	mmole/kg	2	(13.49)
$T = \frac{4625}{10.4 \log \frac{CH_4}{CO_2}} - 373.15$	mmole/kg	3	(13.50)
$T = 5.7835 \log\left(X_{H_2S}\right)^2 + \frac{1}{6}\log\left(X_{CO_2}\right) + 662.814$	mole fraction	4	(13.51)
$T = 77.6 + 66.152 \log(H_2S) + 4.811[\log(H_2S)]^2$	mmole/kg	1	(13.52)

Temperature (T) is °C.
Sources:
1 Arnórsson et al., 1998
2 Arnórsson and Gunnlaugsson, 1985
3 Giggenbach, 1991
4 Nehring and D'Amore, 1984

of 43 gas geothermometers and a list of open-source software for computing geothermal reservoir temperatures. Selected gas geothermometers are listed in Table 13.3, and open-source gas geothermometer software are listed in Table 13.4.

13.5 ISOTOPIC GEOTHERMOMETERS

Isotopic geothermometers utilize low atomic mass elements that exhibit valence changes and undergo large isotopic fractionation. The most commonly used elements are hydrogen, carbon, oxygen, and sulfur, which are abundant as gases and solutes in most geothermal groundwaters. In geothermal systems, isotopic fractionation is greatest for the lighter element such as H, C, O, and S. Isotopic exchange occurs between gases and the steam phase, minerals and the gas phase, water and a solute, and a solute and a solute. As discussed in Chapter 7, isotopic fractionation is commonly expressed as the alpha notation (α), where the fractionation between two compounds is:

$$1000\ \ln_{AB} = \delta_A - \delta_B - 10^3 \ln\alpha \qquad (13.53)$$

where δ_A and δ_B are the light/heavy isotopic ratios of the compound of interest. The application of isotope geothermometers has been reviewed by Truesdell and Hulston (1980), Arnórsson (2000), and others. The only commonly used qualitative isotopic geothermometer is the positive $\delta^{18}O$ shift relative to the GMWL. Most isotopic geothermometers are quantitative, involving isotopic exchange in the systems $CO_2 - CH_4$, $H_2 - H_2O$, $SO_4^{2-} - H_2O$, and $SO_4^{2-} - H_2S$ (Table 13.5). Less frequently used geothermometers have been developed for several other molecule combinations.

TABLE 13.4
Open-Source Software for Computing Gas Geothermometer

Program	Language	Thermometers	°C	Concentration	Type	Source
CO2B	Fortran	1	140–370	% mol	vapor	6
Spreadsheet	Excel	5	100–350	% mol	Liquid & vapor	5
GasGeo	Visual Basic	21	100–350	mmol/kg; mole fraction; mmole/mole	Liquid & vapor	4
EQUILGAS	Visual Fortran	3	100–350	% vol	Liquid	2
GasGeoPlus	Fortran	21	100–350	mmol/kg; mole fraction; mmole/mole	Liquid & vapor	3
SYSGASCHEM	Perl	21	100–350	mmol/kg; mole fraction; mmole/mole	Liquid & vapor	2
GasGeol	Java	8	170–374	Mmol/mol	Liquid & vapor	4

Sources:
1 Acevedo-Anicasio et al., 2021
2 Barragán et al., 2016
3 García-Mandujano, 2019
4 Pandarinath et al., 2011

The idea of isotopic geothermometry is quite simple. Isotopic fractionation between two species is temperature dependent. The isotopic equilibrium temperature of a species of interest can be determined by plotting the calculated fractionation (i.e., 10^3 lnα) for the water on a temperature-fractionation graph constructed for the geothermometer of interest. The graph can be constructed using fractionation equations such as those in Table 13.5. Similar to other types of geothermometry, the application of isotopic geothermometers is not always straightforward. Some isotopic reactions equilibrate so rapidly that they only reflect the temperature of the water or gas at the time of collection. Equilibration of other reactions only occurs at aquifer temperatures in the range of 200 to 300 °C. Isotopic compositions of escaping fluids and gases can also be affected by mixing of warm thermal water with cold water near the land surface and by boiling and steam.

13.5.1 Carbon Geothermometers

The carbon dioxide–methane geothermometer assumes that the two gases are in isotopic equilibrium according to the exchange reactions:

$$^{12}CO_2 + ^{13}CH_4 = ^{13}CO_2 + ^{12}CH_4$$

$$C^{16}O_2 + H_2{}^{18}O = C^{16}O^{18}O + H_2{}^{16}O$$

The carbon dioxide–water geothermometers are theoretically based. The thermometers are applicable to warm and very hot systems and have been applied to several geothermal systems worldwide. Another carbon geothermometer is based on carbon dioxide–dissolved bicarbonate. It is based on the reaction of CO_2 in water and the speciation products of the reaction. $CO_{2(g)}$ forms $CO_{2(aq)}$, which remains uncombined with water, and H_2CO_3, which is in equilibrium with HCO_3^- and CO_3^{2-}. Equilibria between the two species is temperature and pH dependent.

TABLE 13.5
Selected Isotopic Fraction Factors ($10^3\ln\alpha$) for Use in Isotopic Geothermometers

Geothermometer	°C Range	
$CO_2 - CH_4$		
$^{12}CO_2 + ^{13}CH_4 = ^{13}CO_2 + ^{12}CH_4$		
$10^3 \ln = \frac{15.25\times10^3}{T} + \frac{2.432\times10^6}{T^2} - 9.56$	100–400	(13.54)
$CH_4 - H_2O$		
$C^1H_3{}^2H + ^1H_2O = ^1H^2HO + C^1H_4$		
$10^3 \ln = -\frac{49.137\times10^3}{T} + \frac{40.828\times10^6}{T^2} + 26.70$	200–600	(13.55)
$CO_2 - H_20$		
$C^{16}O_2 + H_2{}^{18}O = C^{16}O^{18}O + H_2{}^{16}O$		
$10^3 \ln = -\frac{7.849\times10^3}{T} + \frac{2.941\times10^6}{T^2} - 887.720$	CO_2 & steam	(13.56)
$10^3 \ln = -\frac{18.29\times10^3}{T} + \frac{7.626\times10^6}{T^2} + 19.6$	CO_2 & liquid	(13.57)
$H_2 - H_2O$		
$^1H^2H + ^1H_2O = ^1H_2 + ^1H^2HO$		
$10^3 \ln = \frac{396.8\times10^3}{T} + \frac{11.76\times10^6}{T^2} - 217.30$	100–400	(13.58)
$SO_4^{2-} - H_2O$		
$S^{16}O_4 + H_2{}^{18}O = S^{16}O_3{}^{18}O + H_2{}^{16}O$		
$10^3 \ln = \frac{604\times10^6}{T^2} - 2.6$	100–400	(13.59)
$10^3 \ln = \frac{2.68\times10^6}{T^2} - 7.35$	low pH fluid only	(13.60)
$SO_4 - H_2S$		
$^{32}SO_4 + H_2{}^{34}S = ^{34}SO_4 + H_2{}^{32}S$		
$10^3 \ln = \frac{3.25\times10^6}{T^2} - 5.6$		

Temperatures are in °K.
Source: Data are from compilations of Arnórsson (2000), Horita (2001), and Zeebe (2010)

Isotopic exchange accompanies each step of the reaction, but the rate controlling step is controlled by the isotopic exchange between:

$$CO_{2(g)} \text{ and } CO_{2(aq)}$$

At elevated aquifer temperatures, carbon isotopic exchange becomes frozen and resists further exchange as the water rises to the surface. The carbon dioxide–dissolution bicarbonate geothermometer has not been widely used.

13.5.2 Hydrogen Geothermometers

Geothermal fluids can contain five hydrogen bearing compounds (H_2O, H_2S, H_2, CH_4, H_2 NH_3). Two hydrogen isotope geothermometers, H_2–H_2O and H_2–CH_4, have been developed. The equation for the H_2–H_2O reaction is theoretically based (Richet et al., 1977):

$$^{1}H^{2}H + ^{1}H_2O = ^{1}H_2 + ^{1}H^{2}HO$$

13.5.3 Oxygen Geothermometers

Two common oxygen isotope geothermometers are $SO_4^{2-} - H_2O$ and CO_2–H_2O. The temperature-dependent equilibrium fractionations for these reactions are shown in Table 13.5. The sulfate water geothermometer is based on the oxygen isotope exchange between SO_4^{2-} and H_2O. The CO_2–H_2O thermometer is based on the oxygen exchange between the two species. CO_2 is probably in the form of a HCO_3^- complex, which can only be in the liquid phase.

13.5.4 Sulfur Geothermometers

The multiple sulfur valence states and the large isotopic fraction between oxidized and reduce sulfur species make sulfur isotopes particularly appealing as potential geothermometers. Because sulfur isotopic exchange is very fast in acid conditions and slow in alkaline conditions, the pH of the system should be measured or estimated prior to applying sulfur isotopic geothermometers.

Important Takeaway – Isotopic geothermometers are based on temperature-dependent fractionation between the light isotopes in oxygen, hydrogen, and carbon and sulfur isotopes. A limited number of isotopes have been developed, and they are not frequently used.

13.5.5 Problems

1. The chemical analyses of two geothermal waters in Turkey (Guelec, 2003) are presented here. All solute data are in mg/L.

Sample	°C	pH	Ca	Mg	Na	K	HCO_3	SO_4	Cl	SiO_2
1	171	7.4	2,920	71	6,640	1,280	84	150	33,600	94
2	92	6.7	79	10	1,750	158	1,294	537	18,260	2198

   a. Calculate the error balances and characterize the water types.
   b. Estimate the aquifer temperature using quartz, quartz with maximum steam loss, chalcedony, Na/Li, and Na–K–Ca geothermometers.
   c. Which thermometer(s) yielded the best results, and why?

2. The chemical analyses of several geothermal waters in China (Wang et al., 2018) are presented here. All concentrations are in mg/L.

Sample	°C	pH	Ca	Mg	Na	K	HCO_3	SO_4	Cl	SiO_2	Li
1	95.1	8.0	187	1.5	775	65	67	106	1 735	95	1.55
2	81.4	7.4	935	11	1,771	162	72	214	5721	80	2.92

   a. Calculate the error balances and characterize the water types.
   b. Calculate the aquifer temperature using quartz (no steam loss), quartz, chalcedony, Na–K, Na–K–Ca, K–Mg (Giggenbach), and Na–Li thermometers.
   c. Compare and discuss the results.

REFERENCES CITED

Acevedo-Anicasio, A., et al., 2021, GaSGeoT: A computer program for an effective use of newly improved gas geothermometers in predicting reliable geothermal reservoir temperatures, *Geothermal Energy*, 9(1).

Arana, V., and C. Pnichi, 1974, Isotopic compositions of steam sample from Lanzarote, Canary Island, *Geothermics*, 3(4), 142–145,

Arnórsson, S., 2000, *Isotopic and chemical techniques in geothermal exploration, development and use*, International Atomic Energy Agency (IAEA).

Arnórsson, S., and E. Gunnlaugsson, 1985, New gas geothermometers for geothermal exploration calibration and application, *Geochimica et Cosmochimica Acta*, 49(6) 1207–1325.

Arnórsson, S., et al., 1983a, The chemistry of geothermal waters in Iceland. III. Chemical geothermometry in geothermal investigations, *Geochimica et Cosmochimica Acta*, 47, 567–577.

Arnórsson, S., et al., 1983b, The chemistry of geothermal waters in Iceland II. Mineral equilibria and independent variables controlling water compositions, *Geochimica et Cosmochimica Acta*, 47, 547–566.

Arnórsson, S., et al., 1998, *Gas chemistry of the Krafla geothermal field, Iceland*, International Symposium Water-Rock Interaction, Auckland, New Zealand, 613–616.

Barragán, R.M., et al., 2016, EQUILGAS: Program to estimate temperatures and in situ two-phase conditions in geothermal reservoirs using three combined FT-HSH gas equilibria models, *Computers and Geosciences*, 88, 1–8.

Bayram, A.F., 2001, Application of an artificial neural network model to a Na-K geothermometer, *Journal Volcanology and Geothermal Research*, 112(1–4), 75–81.

Benjamin, T., et al., 1983, Thermodynamic parameters and experimental data for the Na-K-Ca geothermometer, *Journal Volcanology and Geothermal Research*, 15(1–3), 167–186.

Bergfeld, D., et al., 2014, *Gas and isotope chemistry of thermal features in Yellowstone National Park, Wyoming*, U.S. Scientific Investigations Report 2011-5012.

Brondi, M. et al., 1973, Lithium as a pathfinder element in the large scale hydrogeochemical exploration for hydrothermal systems, *Geothermics*, 2(3–4), 142–153.

Cioni, R., and L. Marini, 2020, *A thermodynamic approach to water geothermometer*, Springer.

Craig, H., 1963, The isotopic geochemistry of water and carbon in geothermal areas, in *Nuclear geology on geothermal areas, Spoleto, 1963*, Consiglio Nazionale delle Ricerche, 17–53.

Craig, H., 1966, Isotopic compositions and origin of the Red Sea and Salton Sea brines, *Science*, 154, 1544–1547.

Ellis, A.J., 1957, Chemical equilibrium in magmatic gases, *American Journal Science*, 255, 416–431.

Ellis, A.J., 1970, Quantitative interpretation of chemical characteristics of hydrothermal systems, *Geothermics*, 2, 516–528.

Ellis, A.J., and W.A.J. Mahon, 1967. Natural hydrothermal systems and experimental hot water/rock interactions (part II), *Geochimica et Cosmochimica Acta*, 31, 519–538.

Ellis, A.J., and W.A.J. Mahon, 1977, *Chemistry and geothermal systems*, Academic Press.

Fouillac, C., and G. Michard, 1981, Sodium/lithium ratio in water applied to geothermometry of geothermal reservoirs, *Geothermics*, 10(1), 55–70.

Fournier, R.O., 1977, Chemical geothermometers and mixing models for geothermal systems, *Geothermics*, 5, 41–50.

Fournier, R.O., 1979a, Geochemical and hydrologic considerations and the use of enthalpy-chloride diagrams in the prediction of underground conditions in hot-spring systems, *Journal Volcanology and Geothermal Research*, 5(1–2), 1–16.

Fournier, R.O., 1979b, A revised equation for the Na/K geothermometer, *Geothermal Resources Council Transactions*, 5, 1–16.

Fournier, R.O., 1991, *Water geothermometers applied to geothermal energy, Applications of geochemistry in geothermal reservoir development*, UNITAR-UNDP, 37–69.

Fournier, R.O., and R.W. Potter II, 1979, Magnesium correction to the Na-K-Ca chemical geothermometer, *Geochimica et Cosmochimica Acta*, 43, 1543–1550.

Fournier, R.O., and R.W. Potter II, 1982, A revised and expanded silica (quartz) geothermometer, *Bulletin Geothermal Resources Council*, 11(10), 3–12.

Fournier, R.O., and J.J. Rowe, 1966, Estimation of underground temperatures from the silica content of water from hot springs and wet-steam wells, *American Journal Science*, 264(9), 685–697.

Fournier, R.O., and A.H. Truesdell, 1970, Chemical indicators of subsurface temperature applied to hot spring waters of Yellowstone National Park, Wyoming, USA, *Geothermics*, 2, 529–535.

Fournier, R.O., and A.H. Truesdell, 1973, An empirical Na-K-Ca geothermometer for natural waters, *Geochimica et Cosmochimica Acta*, 37, 1255–1275.

Fournier, R.O., and A.H. Truesdell, 1974, Geochemical indicators of subsurface temperature: Part 2, estimation of temperature and fraction of hot water mixed with cold water, *USGS Journal Research*, 2(3).

Fournier, R.O., et al., 1977, *Geochemical indicators of subsurface temperature; Part I, Basic assumptions*, USGS Open-File Report 74-1033.

García-Mandujano, 2019, EO. SYSGASCHEM: Information system based on web technologies for the processing of geochemical databases and the determination of the temperature of geothermal systems, *Universidad Politécnica del Estado de Morelos*, 137.

Giggenbach, W.F., 1988, Geothermal solute equilibria. Derivation of Na-K-Mg-Ca geoindicators, *Geochimica et Cosmochimica Acta*, 52, 2749–2765.

Giggenbach, W.F., 1991, Chemical techniques in geothermal exploration, in *Applications of geochemistry in geothermal reservoir development: series of technical guides on the use of geothermal energy*, F. D'Amore (ed.), UNITAR/UNDP Centre Press, 119–142.

Guelec, N., 2003, *Applications of geothermometry*, Stanford University, pangea.stanford.edu/ERE/pdf/IGAstandard/ISS/2003Turkey/nlgung.pdf

Gunnarsson, I., and S. Arnórsson, 2000, Amorphous silica solubility and the thermodynamic properties of H_4SiO_4 in the range of 0° to 350 °C at Psa, *Geochimica et Cosmochimica Acta*, 64, 2295–2307.

Hemley, J.J., 1967, *Aqueous Na/K ratios in the system $K2O$-Na_2-O-Al_2-SiO_2-H_2O*, Geological Society America, Abstract 94–95.

Horita, J., 2001, Carbon isotope exchange in the system CO_2-CH_4 at elevated temperatures, *Geochimica et Cosmochimica Acta*, 65(12), 1907–1919.

Keenan, J.H., et al., 1969, *Steam tables*, Wiley.

Keenan, J.H., et al., 1992, *Steam tables*, 2nd ed, Kreiger Publications.

Kharaka, Y.K., and R.H. Mariner, 1989, Chemical geothermometers and their application to formation waters from sedimentary basins, in *Thermal history of sedimentary basins,* N.D. Naeser and T.H. McCollon (eds.), Springer, 99–117.

Klusman, R.W., and R.A. Landress, 1979, Mercury in soils of the Long Valley, California, geothermal system, *Journal Volcanology and Geothermal Research*, 5(1–2), 49–65.

Lowenstern, J., et al., 1999, *Gas and isotope geochemistry of 81 steam samples from wells in the Geysers geothermal field, Sonoma and Lake Counties, California*, U.S. Geological Survey Open-File Report 99-304.

Mahon, W.A.J., 1970, Chemistry in the exploration and exploitation of hydrothermal systems, *Geothermics*, 2, 1310–1322.

Mayo, A.L., S. Himes, and D. Tingey, 2014, Self-organizing thermal springs in crystalline rock: Theoretical and laboratory approach to the hydrothermal systems in the middle Fork of the Boise River, Idaho, USA, *Hydrogeology Journal*, 22(1), 25–45.

Mercado, S., 1970, High activity hydrothermal zones detected by Na/K, Cerro Prieto, Mexico, *Geothermics*, 2(2), 1367–1376

Michard, G., 1990, Behaviour of major elements and some trace elements (Li, Rb, Cs, Sr, Fe, Mn, W, F) in deep hot waters from granitic areas, *Chemical Geology*, 89, 117–134.

Nehring, N., and F. D'Amore, 1984, Gas chemistry and thermometry of the Cerro Prieto, Mexico, geothermal field, *Geothermic*, 13, 75–89.

Nieva, D., and R. Nieva, 1987, Chapter 7 Developments in geothermal energy in Mexico-part twelve. A cationic geothermometer for prospecting of geothermal resources, *Heat Recovery Systems*, 7(3), 243–256.

Orville, P.M., 1963, Alkali ion exchange between vapor and feldspar phases, *American Journal Science*, 261, 201–237.

Paces, T., 1975, A systematic deviation from Na-K-Ca geothermometer below 75 °C and above 10^{-4} atm PCO_2, *Geochimica et Cosmochimica Acta*, 39, 541–544.

Pandarinath, K., et al., 2011, *GasGeo-software to estimate the reservoir temperatures of geothermal systems using gas geothermometers*, Proceedings XXI National Geochemistry Congress, Actas INAGEQ, 17.

Richet, P., et al., 1977, A review of hydrogen, carbon, nitrogen, oxygen, sulphur and chlorine stable isotope enrichment among gaseous molecules, *Annual Reviews Earth Planetary Science*, 5, 65–110.

Roberts, A.A., et al., 1975, Helium survey, a possible technique for locating geothermal reservoirs, *Geophysical Research Letters*, 2, 209–210.

Saki, H., and O. Matsubaya, 1974, Isotopic geochemistry of the thermal waters of Japan and its bearing on the Kuroko ore solutions, *Economic Geology*, 69(6), 974–991.

Sanjuan, B., et al., 2014, Use of two new Na/Li geothermometric relationships for geothermal fluids in volcanic environments, *Chemical Geology*, 398, 60–81.

Santoyo, E., and L. Díaz-González, 2010, *A new improved proposal of the Na/K geothermometer to estimate deep equilibrium temperatures and their uncertainties in geothermal systems*, Proceedings World Geothermal Congress 2010, Bali, Indonesia.

Schlegel, M.E., et al., 2009, Paleo-climate of the Boise area, Idaho from the last glacial maximum to the present based on groundwater δ^2H and $\delta^{18}O$ compositions, *Quaternary Research*, 71, 72–80.

Simmons, S.F., 2013, Fluid-mineral equilibria in Great Basin geothermal systems: Implications for chemical geothermometry, in *Proceedings of the 38th workshop on geothermal reservoir engineering*, Stanford University, SGP-TR-198.

Taylor, H.P. Jr, 1977, Water/rock interactions and the origin of H_2O in granitic batholiths, *Journal Geological Society (London)*, 133, 509–558.

Tonani, F., 1970, Geochemical methods for exploration for geothermal energy, *Geothermics* Special Issue (2), 492–515.

Truesdell, A.H., 1976, Summary of section III-Geochemical techniques in exploration, in *Proceedings of 2nd UN symposium on the development and use of geothermal resources*, U.S. Government Printing Office 1, 53–79.

Truesdell, A.H., and R.O. Fournier, 1975, Calculation of deep temperatures in geothermal systems from the boiling spring waters of mixed origin, in *Proceedings of 2nd UN symposium development and use of geothermal resources*, U.S. Government Printing Office, 1, 837–844.

Truesdell, A.H., and R.O. Fournier, 1977, Procedure for estimating the temperature of a hot water component in a mixed water using a plot of dissolved silica vs. enthalpy, *USGS Journal Research*, 4, 49–52.

Truesdell, A.H., and J.R. Hulston, 1980, Chapter 5 – Isotopic evidence on environments in geothermal systems, in *Handbook of environmental isotope geochemistry, the terrestrial environment, A*, Elsevier, 179–226.

Verma, M.P., 2000, Chemical thermodynamics of silica: A critique on its geothermometer, *Geothermics*, 29, 323–346.

Verma, M.P., 2003, Steam tables for pure water as an ActiveX component in visual basic 6.0, *Computers & Geoscience*, 29, 1155–1163.

Verma, M.P., 2014, Geothermometry in exploration and exploitation of geothermal system, in *Geothermal and ocean energy,* J.N. Govil (ed.), Studium Press LLC, 7.

Verma, M.P., and K. Betzler, 2013 Chemical geothermometry in evaluation of geothermal reservoir characteristics: a thermodynamic point of view, *Acta INAGEQ*, 19, 155–161.

Verma, M.P., and E. Santoyo, 1997, New improved equations for Na/K, Na/Li and SiO_2 geothermometers by outlier detection and rejection, *Journal Volcanology and Geothermal Research*, 79, 9–23.

Wang, X., et al., 2018, Hydrogeochemical characteristics and geothermometry applications of thermal waters in coastal Xinzhou and Shenzao Geothermal Fields, Guangdong, China, *Hindawi Geofluids*, 8715080.

White, D.E., 1965, Chapter 4 – Saline waters of sedimentary rocks, in *Fluids in subsurface environments*, American Association Petroleum Geologists, 342–366.

White, D.E., 1968, Environments of generation of some base-metal ore deposits, *Economic Geology*, 63, 301–335.

White, D.E., 1970, Geochemistry applied to the discovery, evaluation, and exploration of geothermal resources, *Geothermics (Special Issue)*, 2, 58–78.

Williams, C.F., et al., 2008, *A review of methods applied by the U.S. Geological Survey in the assessment of identified geothermal resources*, USGS Open-File Report 2008-12.

Zeebe, R.E., 2010, A new value for the stable oxygen isotope fractionation between dissolved sulfate ion and water, *Geochimica et Cosmochimica Acta*, 74, 818–828.

Appendix

Periodic Table of the Elements

1	2	3	4	5	6	7	8	9	10	11	12	13	14	15	16	17	18
1 **H** Hydrogen 1.008																	2 **He** Helium 4.003
3 **Li** Lithium 6.941	4 **Be** Beryllium 9.012											5 **B** Boron 10.811	6 **C** Carbon 12.011	7 **N** Nitrogen 14.007	8 **O** Oxygen 15.999	9 **F** Fluorine 18.998	10 **Ne** Neon 20.180
11 **Na** Sodium 22.990	12 **Mg** Magnesium 24.305											13 **Al** Aluminum 26.982	14 **Si** Silicon 28.086	15 **P** Phosphorus 30.974	16 **S** Sulfur 32.066	17 **Cl** Chlorine 35.453	18 **Ar** Argon 39.948
19 **K** Potassium 39.098	20 **Ca** Calcium 40.078	21 **Sc** Scandium 44.956	22 **Ti** Titanium 47.867	23 **V** Vanadium 50.942	24 **Cr** Chromium 51.996	25 **Mn** Manganese 54.938	26 **Fe** Iron 55.845	27 **Co** Cobalt 58.933	28 **Ni** Nickel 58.693	29 **Cu** Copper 63.546	30 **Zn** Zinc 65.38	31 **Ga** Gallium 69.723	32 **Ge** Germanium 72.631	33 **As** Arsenic 74.922	34 **Se** Selenium 78.971	35 **Br** Bromine 79.904	36 **Kr** Krypton 84.798
37 **Rb** Rubidium 84.468	38 **Sr** Strontium 87.62	39 **Y** Yttrium 88.906	40 **Zr** Zirconium 91.224	41 **Nb** Niobium 92.906	42 **Mo** Molybdenum 95.95	43 **Tc** Technetium 98.907	44 **Ru** Ruthenium 101.07	45 **Rh** Rhodium 102.906	46 **Pd** Palladium 106.42	47 **Ag** Silver 107.868	48 **Cd** Cadmium 112.414	49 **In** Indium 114.818	50 **Sn** Tin 118.711	51 **Sb** Antimony 121.760	52 **Te** Tellurium 127.6	53 **I** Iodine 126.904	54 **Xe** Xenon 131.249
55 **Cs** Cesium 132.905	56 **Ba** Barium 137.328	57-71 Lanthanides	72 **Hf** Hafnium 178.49	73 **Ta** Tantalum 180.948	74 **W** Tungsten 183.84	75 **Re** Rhenium 186.207	76 **Os** Osmium 190.23	77 **Ir** Iridium 192.217	78 **Pt** Platinum 195.085	79 **Au** Gold 196.967	80 **Hg** Mercury 200.592	81 **Tl** Thallium 204.383	82 **Pb** Lead 207.2	83 **Bi** Bismuth 208.980	84 **Po** Polonium [208.982]	85 **At** Astatine 209.987	86 **Rn** Radon 222.018
87 **Fr** Francium 223.020	88 **Ra** Radium 226.025	89-103 Actinides	104 **Rf** Rutherfordium [261]	105 **Db** Dubnium [262]	106 **Sg** Seaborgium [266]	107 **Bh** Bohrium [264]	108 **Hs** Hassium [269]	109 **Mt** Meitnerium [268]	110 **Ds** Darmstadtium [269]	111 **Rg** Roentgenium [272]	112 **Cn** Copernicium [277]	113 **Nh** Nihonium unknown	114 **Fl** Flerovium [289]	115 **Mc** Moscovium unknown	116 **Lv** Livermorium [298]	117 **Ts** Tennessine unknown	118 **Og** Oganesson unknown

57 **La** Lanthanum 138.905	58 **Ce** Cerium 140.116	59 **Pr** Praseodymium 140.908	60 **Nd** Neodymium 144.243	61 **Pm** Promethium 144.913	62 **Sm** Samarium 150.36	63 **Eu** Europium 151.964	64 **Gd** Gadolinium 157.25	65 **Tb** Terbium 158.925	66 **Dy** Dysprosium 162.500	67 **Ho** Holmium 164.930	68 **Er** Erbium 167.259	69 **Tm** Thulium 168.934	70 **Yb** Ytterbium 173.055	71 **Lu** Lutetium 174.967
89 **Ac** Actinium 227.028	90 **Th** Thorium 232.038	91 **Pa** Protactinium 231.036	92 **U** Uranium 238.029	93 **Np** Neptunium 237.048	94 **Pu** Plutonium 244.064	95 **Am** Americium 243.061	96 **Cm** Curium 247.070	97 **Bk** Berkelium 247.070	98 **Cf** Californium 251.080	99 **Es** Einsteinium [254]	100 **Fm** Fermium 257.095	101 **Md** Mendelevium 258.1	102 **No** Nobelium 259.101	103 **Lr** Lawrencium [262]

Appendix A Conversion Factors

Mg/L × F_1 = meq/L, mg/L × F_2 = mmole/L		
Element or Species	F_1	F_2
Aluminum (Al^{+3})	0.11119	0.03715
Ammonium (NH^{4+})	0.05544	0.05544
Barium (Ba^{+2})	0.01456	0.00728
Beryllium (Be^{+3})	0.33288	0.11096
Bicarbonate (HCO_3^{2-})	0.01639	0.01639
Boron (B)		0.09250
Bromide (Br^-)	0.01251	0.01251
Cadmium (Cd^{2+})	0.01779	0.00890
Calcium (Ca^{+2})	0.04990	0.02495
Carbonate (CO_3^{2-})	0.03333	0.01666
Chloride (Cl^-)	0.02821	0.02821
Chromium (Cr)		0.01923
Cobalt (Co^{+2})	0.03394	0.01697
Copper (Cu^{+2})	0.03148	0.01574
Fluoride (F^-)	0.05264	0.05264
Germanium (Ge)		0.01378
Gallium (Ga)		0.01434
Gold (Au)		0.00511
Hydrogen (H^+)	0.99209	0.99209
Hydroxide (OH^-)	0.05880	0.05880
Iodide (I^-)	0.00788	0.00788
Iron (Fe^{+2})	0.03581	0.01791
Iron (Fe^{+3})	0.05372	0.01791
Lead (Pb)		0.00483
Lithium (Li^+)	0.14411	0.14411
Magnesium (Mg^{+2})	0.08226	0.04114
Manganese (Mn^{+2})	0.03640	0.01820
Molybdenum (Mo)		0.01042
Nickel (Ni)		0.01703
Nitrate (NO_2^-)	0.01613	0.01613
Nitrite (NO_3^-).	0.02174	0.02174
Phosphate (PO_4^{3-})	0.03159	0.01053
Phosphate (HPO_4^{2-})	0.02084	0.01042
Phosphate ($H_2PO_4^-$)	0.01031	0.01031
Potassium (K^+)	0.02557	0.02557
Rubidium (Rb^+)	0.01170	0.01170
Silica (SiO_2)		0.01664
Silver (Ag)		0.00927
Sodium (Na^+)	0.04350	0.04350
Strontium (Sr^{+2})	0.02283	0.01141
Sulfate (SO_4^{2-})	0.02082	0.01041

(Continued)

(Continued)

Mg/L × F_1 = meq/L, mg/L × F_2 = mmole/L		
Element or Species	F_1	F_2
Sulfide (S^{2-})	0.06238	0.03119
Titanium (Ti)		0.02088
Uranium (U)		0.00420
Zinc (Zn^{+2})	0.03060	0.01530

Appendix B Stability Constants (β) for the Formation of Complexes and Solids from Metals and Ligands (L).

	OH^-		CO_3^{2-}		SO_4^{2-}		Cl^-		F^-		SIO_3^{2-}	
H^+			HL	10.33	HL	1.99			HL	3.2	HL	13.1
	HL	14.00	H_2L	16.68							H_2L	23.0
			$H_2L_{(g)}$	181.4							H_2L_2	26.6
											H_4L_4	55.9
											H_6L_6	78.2
Al^{3+}	AlL	9.0							AlL	7.0		
	AlL_2	18.7							AlL_2	12.6		
	AlL_3	27.0							AlL_3	16.7		
	AlL_4	33.0							AlL_4	19.1		
	Al_3L_4	42.1										
	$AlL_{3(s)}$	33.5										
Ca^{2+}	CaL	1.15	CaL	3.2	CaL	2.31			CaL	1.1	CaL	4.2
			CaHL	11.59	$CaL_{2(s)}$	4.62			$CaL_{2(s)}$	10.4	CaHL	14.1
			$CaL^{(s)}$	8.22							CaH_2L_2	29.9
Fe^{2+}	FeL	4.50	$Fel_{(s)}$	10.7	FeL	2.2						
	FeL_2	7.40										
	FeL_3	11.00										
	$FeL_{2(s)}$	15.10										
Fe^{3+}	FeL	11.80			FeL	4.0	FeL	1.5	FeL	6.0		
	FeL_2	22.30			FeL_2	5.4	FeL_2	2.1	FeL_2	10.6		
	FeL_4	34.40							FeL_3	13.7		
	Fe_2L_2	25.00										
	$FeL_{3(s)}$	42.70										
K^+					KL							
Mg^{2+}	MgL	2.56	MgL	3.4	MgL	2.36			MgL	1.8		
	Mg_4L_4	16.28	MgHL	11.49					$MgL_{2(s)}$	8.2		
	$MgL_{2(s)}$	11.16	$MgL_{(s)}$	4.54								
Na^+			NaL	1.27	NaL	1.06						
			NaHL	10.08		0.96						

Note: Constants are Base 10.
Source: Data from Morel, 1983; Sillin and Martell, 1964; Smith and Martell, 1976

Appendix C Thermodynamic Data for Common Aqueous Species at 98.15 °K and 105 Pa. The Data Have Not Been Checked for Internal Consistency

Species or compound	ΔG_f° (kJ/mol)	ΔH_f° (kJ/mol)	S° (J/K·mol)	Source
Al (aluminum)				
Al metal	0	0	28.3	9
Al^{3+}	−489.4	−538.4	−332	8
$Al(OH)^{2+}$	−696.54	−778	−204	3
$Al(OH)_2^+$	−901.7	−1000	−16	3
$Al(OH)_3$	−1100.6	−1230	108	3
$Al(OH)_4^-$	−1305.8	−1487	160	3
$Al(OH)_{3\ gibbsite}$	−1154.91	−1293.11	68.45	7
$AlO(OH)_{boehmite}$	−915.8	−990.4	48.4	4
$AlO(OH)_{diaspore}$	920.9	999.5	35.3	10
$Al_2Si_2O_5(OH)_{4\ kaolinite}$	−3797.5	−4119	200.4	8
$Al_2Si_2O_5(OH)_{4\ halloysite}$	−3780.7	−4101.5	203	8
$Al_2Si_4O_{10}(OH)_{2\ prophylite}$	−5266.2	−5640	239.4	8
As (arsenic)				
As	0	0	35.7	9
AsO_2^-	−350	−429	40.6	10
AsO_4^{3-}	−648.4	−888.1	−162.8	10
$As_2O_{3\ arsenolite}$	−576	−657	107.4	9
$As_2O_{3\ clauderite}$	−575.6	−654.8	113.3	4
$As_2S_{2\ realgar}$	−140.6	−142.7	127	9
$As_2S_{3\ orpiment}$	−168.4	−169	163.6	9
Ba (barium)				
Ba^{2+}	−555.4	−532.5	8.4	1
$BaCl_2$	−810.4	−858.6	123.7	10
BaF_2	−1156.8	−1207.1	96.4	10
$BaCO_{3\ witherite}$	−1132.21	−1210.85	112.1	1
$BaSO_{4\ barite}$	−1362.2	−1473.2	132.2	10
$Ba(NO_3)_{2\ nitrobarite}$	−796.6	−992.1	213.8	9
C (carbon)				
$C_{graphite}$	0	0	5.74	10
$HCN_{(aq)}$	119.7	107.1	–	10

(Continued)

Species or compound	ΔG$_f$° (kJ/mol)	ΔH$_f$° (kJ/mol)	S° (J/K·mol)	Source
CN–	172.4	150.6	94.1	6
$CH_{4\,(g)}$	–50.72	–74.8	186.26	8, 10
$CH_{4\,(aq)}$	–34.3	–89	83.7	10
$HCOOH_{(l)}$	–361.4	–424.7	129	10
$CH_3COOH_{\,(aq)}$	–396.6	–485.8	159.8	10
CH_3COO^{-}–	–369.3	–486	86.6	10
$CO_{\,(g)}$	–137.17	–110.53	197.66	8, 10
$CO_{2\,(g)}$	–394.37	–393.51	213.79	6, 10
$CO_{2\,aq)}$	386	–413.26	119.36	10
$H_2CO_{3\,aq)}$	–623.08	–699.65	187.4	10
HCO_3^{-}–	–586.8	–689.9	98.4	8
CO_3^{2-}	–527.9	–677.1	–56.9	8, 10
Ca (calcium)				
$Ca_{\,metal}$	–	0	41.3	9
Ca^{2+}	–552.58	–542.83	–53.1	10
$CaCO_{3\,(aq)}$	–1081.4	–1220	–110	9
$CaSO_{4\,(aq)}$	–1298.1	–1452.1	–33.1	9
$CaO_{\,lime}$	–603.5	–635.1	38.2	9
$Ca(OH)_{2\,portlandite}$	–898.49	–986.09	83.4	10
$CaF_{2\,fluorite}$	–1176.9	–1229.3	68.9	9
$CaCO_{3\,calcite}$	–1128.79	–1206.92	92.9	10
$CaCO_{3\,aragonite}$	–1127.74	–1207.13	88.7	10
$CaMg(CO_3)_{2\,dolomite}$	–2163.4	–2326.3	155.2	10
$CaSO_{4\,anhydrite}$	–1321.79	–1434.11	105.7	10
$CaSO_4{\cdot}2H_2O_{\,gypsum}$	–1797.28	–2022.63	194.1	10
$Ca_5(PO_4)_3OH_{\,hydroxyapatite}$	–6338.3	–6721.6	390.4	9
$Ca_5(PO_4)_3F_{\,fluorapatite}$	–6508.1	–6872.2	387.9	9
$CaAl_2Si_2O_{8\,anorthite}$	–4002.2	–4227.8	199.3	10
$CaAl_2Si_4O_{12}{\cdot}4H_2O_{\,laumontite}$	–6682	–7233.6	485.8	5
$CaMgSi_2O_{6\,diopside}$	–3036.6	–3210.8	143.1	9
$Ca_2Mg_5Si_8O_{22}(OH)_{2\,tremolite}$	–11626.9	–12355.1	548.9	9
Cl (chlorine)				
$Cl_{2\,(g)}$	0	0	223.1	9
Cl^-	–131.2	–167.2	56.5	10
$HCl_{\,(g)}$	–95.3	–92.31	186.9	9
F (fluorine)				
F_2	0	0	202.8	9
F^{-}–	–281.5	–335.35	–13.8	8
$HF_{\,(aq)}$	–296.8	–320.1	88.7	10
$HF_{\,(g)}$	–275.4	–273.3	173.78	9
Fe (iron)				
$Fe_{\,metal}$	0	0	27.3	9
Fe^{2+}	–85.35	–87.4	–110	7
Fe^{3+}	–11.05	–45.56	–290.4	7
$FeCl^{2+}$	–143.9	–180.3	–113	10

(Continued)

(Continued)

Species or compound	$\Delta G_f°$ (kJ/mol)	$\Delta H_f°$ (kJ/mol)	S° (J/K·mol)	Source
$FeSO_{4\,(aq)}$	−823.4	−998.3	−117.7	10
FeO	−251.2	−272	59.8	9
$Fe_2O_{3\ hematite}$	−742.2	−824.2	87.4	10
$Fe_3O_{4\ magnetite}$	−1015.4	−1118.4	146.4	10
$Fe(OH)_{2\,(s)}$	−486.5	−569	88	9
$FeOOH_{\ goethite}$	−488.7	−559.4	60.2	7
$Fe(OH)_{3\ ferrihydrite}$	−696.5	−832	196.7	10
$FeCl_{3\ molysite}$	−333.8	−399.2	142.3	9
$FeCO_{3\ siderite}$	−666.67	−740.57	92.9	10
$Fe_2SiO_{4\ fayalite}$	−1379	−1479.9	145.2	10
$FeS_{2\ pyrite}$	−166.9	−178.2	52.9	10
$FeS_{2\ marcasite}$	−158.4	−169.4	53.9	9
$FeS_{\ pyrrhotite}$	−100.4	−100	60.3	9
$FeAsS_{\ arsenopyrite}$	−50	−42	121	9
$FeSO_4{\cdot}7H_2O_{\ melanterite}$	−2509.6	−3014.4	409.2	10
H (hydrogen)				
$H_{2\,(g)}$	0	0	130.57	10
$H_2O_{\ (g)}$	−228.6	−241.8	188.7	9
$H_2O_{\ (l)}$	−237.14	−285.83	69.95	10
K (potassium)				
$K_{\ metal}$	0	0	64.7	9
K^+	−282.5	−252.17	1010	7
$KCl_{\ sylvite}$	−408.6	−436.5	82.6	9
$K_2SO_{4\ arcanite}$	−1319.7	−1437.7	175.6	9
$KNO_{3\ niter}$	−394.5	−494.5	133.1	9
$KAlSi_3O_{8\ microcline}$	−3742.9	−3681.1	214.2	10
$KAl_3Si_3O_{10}(OH)_{2\ muscovite}$	−5608.4	−5984.4	305.3	10
$KMg_3AlSi_3O_{10}(OH)_{2\ phlogopite}$	–	–	319.7	10
Mg (magnesium				
$Mg_{\ metal}$	0	0	32.7	9
Mg^{2+}	−455.4	−467	−137	8
$Mg(OH)^+$	−626.7	–	–	1
$MgSO_{4\,(aq)}$	−1212.2	−1356	−7.1	10
$MgO_{\ periclase}$	−5694	−601.6	26.9	10
$Mg(OH)_{2\ brucite}$	−833.51	−924.54	63.18	10
$MgCO_{3\ magnesite}$	−1012.1	−1095.8	65.7	10
$MgCO_3{\cdot}3H_2O_{\ nesquehonite}$	−1723.7	−1977.3	195.6	9
$MgSO_4{\cdot}7H_2O_{\ epsomite}$	−2871.2	−3388.7	372	9
$Mg_2SiO_{4\ forsterite}$	−2056.7	−2175.7	95.2	5
$MgSiO_{3\ clinoenstatite}$	−1460.9	−1547.8	67.9	10
$MgSiO_{3\ enstatite}$	−1459.9	−1546.8	67.8	5
$Mg_3Si_2O_5(OH)_{4\ chrysotile}$	−4034	−4361.7	221.2	10
$Mg_3Si_4O_{10}(OH)_{2\ talc}$	−5536	−5915.9	260.8	9
$Mg_5Al_2Si_3O_{10}(OH)_{8\ chlorite}$	−8207.8	−8857.4	465.3	5
$Mg_4Si_6O_{15}(OH)_2{\cdot}6H_2O_{\ sepiolite}$	−9251.6	−10116.9	613.4	5
N (nitrogen)				

(Continued)

Species or compound	$\Delta G_f°$ (kJ/mol)	$\Delta H_f°$ (kJ/mol)	S° (J/K·mol)	Source
$N2_{(g)}$	0	0	191.61	10
NO_2^{--}	−32.2	−104.6	123	10
NO_3^{--}	−111.5	−207.4	146.9	10
$NH_{3\,(g)}$	−16.45	−46.11	192.5	10
$NH_{3\,(aq)}$	−26.5	−80.29	111.3	10
NH_4^+	−79.31	−133.51	113.4	10
$NO_{(g)}$	86.6	90.2	210.8	10
$NO_{2\,(g)}$	51.2	33.1	240.1	9
$N_2O_{(g)}$	104.2	82	219.8	10
$NH_{3\,(g)}$	−16.4	−45.9	192.8	9
$(NH_4)_2SO_{4\ mascagnite}$	−901.7	−1180.8	220.1	9
Na (sodium)				
Na_{metal}	0	0	51.3	9
Na^+	−262.5	−240.34	58.45	8
$NaCO_3^-$	−792.8	−935.9	−49.8	10
$NaHCO_{3\,(aq)}$	−849.7	−943.9	113.8	10
$NaSO_4^-$	−1010.6	−1144.7	108.8	10
$NaCl_{halite}$	−384.2	−411.3	72.1	8
$NaF_{villiaumite}$	−546.3	−576.6	51.3	9
$Na_2CO_3{\cdot}10H_2O_{natron}$	−3427.7	−4081.3	562.7	10
$NaHCO_{3\ nahcolite}$	−851	−950.8	102.7	10
$NaHCO_3{\cdot}Na_2CO_3{\cdot}2H_2O_{trona}$	−2383.4	−2684.9	301.2	10
$Na_2SO_{4\ thenardite}$	−1269.8	−1387.8	149.6	9
$Na_2SO_4{\cdot}10H_2O_{mirabilite}$	−3646.4	−4327.1	592	9
$NaSi_7O_{13}(OH)_{3\ magadiite}$	−6651.9	−241.83	188.73	3
$NaAlSi_3O_{8\ albite}$	−3711.7	−3935.1	207.4	10
$NaAlSi_2O_6{\cdot}H_2O_{analcite}$	−3082.6	−3300.8	234.4	9
O (oxygen)				
$O_{2\,(g)}$	–	0	205.2	9
$O_{2\,(aq)}$	16.4	−11.7	110.9	10
OH^-	−157.2	−230	−10.8	10
S (sulfur)				
S	0	0	31.8	9
$H_2S_{(g)}$	−33.5	−20.6	205.8	9
$H_2S_{(aq)}$	−27.8	−39.7	121	9
HS^-	12.8	−17.6	62.8	10
$S^{2-}{}_{(g)}$	85.8	33	−15	9
$SO_{2(g)}$	−300.2	−296.8	248.1	10
HSO_4^-	−755.91	−887.34	131.8	10
SO_3^{2-}	−486.6	−635.6	29	9
SO_4^{2-}	−744.6	−909.3	20	9
Si (silicon)				
Si_{metal}	0	0	18.8	9
$H_4SiO_{4\,(aq)}$	−1308	−1460	180	9

(Continued)

(Continued)

Species or compound	$\Delta G_f°$ (kJ/mol)	$\Delta H_f°$ (kJ/mol)	S° (J/K·mol)	Source
$SiO_{2\ alpha\ quartz}$	−856.64	−910.94	41.84	10
$SiO_{2\ amorph}$	−849.28	−902.07	46.9	3
$SiO_{2\ cristobalite}$	−854.5	−908.3	43.4	9
$SiO_{2\ tridymite}$	−853.8	−907.5	43.9	9
Sr (strontium)				
Sr_{metal}	0	0	55.4	9
Sr^{2+}	−559.4	−545.8	−33	9
SrO	−561.9	−592	54.4	10
$SrCO_{3\ strontianite}$	−1144.77	−1225.7	97.2	2
$SrSO_{4\ celestite}$	−1340.9	−1453.1	117	10

Sources:

[1] Busenberg and Plummer, 1986
[2] Busenberg et al., 1984
[3] Drever, 1988
[4] Haas et al., 1980
[5] Helgeson et al., 1978
[6] Krauskopf and Bird, 1995
[7] Nordstrom et al., 1984
[8] Robie and Hemingway, 1995
[9] Robie and Hemingway et al., 1978
[10] Reed, 2020

Appendix D Isotopic Fractionation Factors of Interest to Hydrogeology

$$10^3 \ln\alpha = a(10^6 / T_K^{\ 2}) + b(10^3 / T_K) + c$$

Reaction	a	b	c	$10^3\ln\alpha$	Range (°C)
^2H					
Water–vapor	24.844	−76.248	52.612	76 @ 25 °C	0–100
Water–ice				20 @ 25 °C	0
Ice–vapor	24.844	−76.248	71.912	127 @ 0 °C	0
Water vapor – H_2	0	467.6	−303.9	949 @ 100 °C	80–200
Water – H_2O				1325 @ 25 °C	
$CH_4 - H_2$	25	346	−223	1219 @ 25 °C	0–700
Water vapor – CH_4	−7.69	6.1	88.4	22 @ 25 °C	10–250
Water – CH_4				816 @ 25 °C	0–100
^{18}O					
Water–vapor	1.137	−0.4156	−2.0667	9.1 @ 25 °C	0–100
Water–ice				3.1 @ 25 °C	0
Ice–vapor	1.137	−0.4156	1.0333	14.7 @ 25 °C	0
$CO_{2(g)}$ – water	−0.0206	17.9942	−19.97	40 @ 25 °C	0–100
Calcite–water	2.78	0	−2.89	28.4 @ 25 °C	0–500
$CO_{2(g)}$ – calcite	−1.8034	10.611	−2.7798	12.5 @ 25 °C	0–600
SO_4 – water	3.25	0	−5.1	31.5 @ 25 °C	0–500
Anhydrite–water	3.88	0	−2.9	25 @ 25 °C	100–575
Gypsum–water	0	2.3	−3.7	4 @ 25 °C	20–60
$SiO_{2(amor)}$ – water	3.52	0	−4.35	29.4 @ 50 °C	34–93
Quartz–water	1.9189	8.582	−18.977	26 @ 50 °C	0–100
^{13}C					
$CO_{2(g)} - CO_{2(aq)}$	0	0.3793	−0.19	1.1 @ 25 °C	0–35
$HCO_3 - CO_{2(g)}$	0	9.552	−24.1	7.9 @ 25 °C	5–125
$HCO_3 - CO_{2(aq)}$	0	9.866	−24.12	9.0 @ 25 °C	5–125
$CO_3^{\ 2} - HCO_3^-$	0	−0.867	2.52	−0.4 @25 °C	5–125
$CO_3^{\ 2} - CO_{2(g)}$	0.87	0	−3.4	6.4 @ 25 °C	0–100
$CO_3^{\ 2} - CO_{2(g)}$	0	9.137	−22.73	7.6 @ 25 °C	0–100

(*Continued*)

(Continued)

$$10^3 \ln\alpha = a(10^6 / T_K{}^2) + b(10^3 / T_K) + c$$

Calcite – HCO_3^-	0	–4.232	15.1	0.9 @ 25 °C	0–35
$CO_{2(g)}$ – calcite	–2.988	7.6663	–2.4612	–10.4 @ 25 °C	0–100

Source: Data from Clark and Fritz, 1997

Appendix E Selected Equilibrium Constants at 25 °C.

Reaction	log K
$Ca^{+2} + H_2O = CaOH^+ + H^+$	−12.780
$CaCO_{3(calcite)} = CO_3^2 + Ca^{+2}$	−8.480
$CaCO_{3(aragonite)} = CO_3^2 + Ca^{+2}$	−8.336
$CaMgCO_{3(dolomite)} = Ca^{+2} + Mg^{+2} + 2CO_{3-}^{\ 2}$	−17.09
$CaSO_4 : 2H_2O_{(gypsum)} = Ca^{+2} + SO_{4-}^{\ 2} + 2H_2O$	−4.580
$CaSO_{4(anhydrite)} = Ca^{+2} + SO_4^2$	−4.360
$Ca_5(PO_4)_3OH_{(hydroxyapatite)} + 4H^+ = H_2O + 3PO_4^2 + 5Ca^+$	−3.421
$CaF_{2\ (fluorite)} = Ca^{+2} + 2F$	−10.600
$FeCO_{3(siderite)} = Fe^{+2} + CO_3^2$	−10.89
$MnCO_{3\ (rhodochrosite)} = Mn^{+2} + CO_3^2$	−11.130
$SrCO_{3(strontianite)} = Sr^{+2} + CO_3^2$	−9.271
$BaCO_{3(witherite)} = Ba^{+2} + CO_{3-}^{\ 2}$	−8.562
$SrSO_{4(celestite)} = Sr^{+2} + SO_{4-}^{\ 2}$	−6.630
$BaSO_{4(barite)} = Ba^{+2} + SO_{4-}^{\ 2}$	−9.970
$SiO_{2(a)} + 2H_2O = H_4SiO_4$	−2.710
$SiO_{2(chalcedony)} + 2H_2O = H_4SiO_4$	−3.550
$SiO_{2(quartz)} + 2H_2O = H_4SiO_4$	−3.980
$Al(OH)_{3(gibbsite)} + 3H^+ = Al^+3 + 3H_2O$	8.110
$Al(OH)_{3(a)} + 3H^+ = Al^{+3} + 3H_2O$	10.800
$Al_2Si2O_5(OH)_{4(kaolinite)} + 6H^+ = H_2O + 2H_4SiO_4 + 2Al^{+3}$	7.435
$NaAlSi_3O_{8(albite)} + 8H_2O = Na+ + Al(OH)_4^- + 3H_4SiO_4$	_18.002
$CaAl_2Si_2O_{8(anorthite)} + 8H_2O = Ca^{+2} + 2Al(OH)_4^- + 2H_4SiO_4$	−19.714
$AlSi_3O_{8(K\text{-}feldspar)} + 8H_2O = K^+ + H_4SiO_4 + 3H_4SiO_4$	−20.573
$KAl_3Si_3O_{10}(OH)_2 + 10\ H^+ = K+ + 3Al^{+3} + 3H_4SiO_4$	12.703
$Mg_5Al_2Si_3O_{10}(OH)_{8(chlorite)} + 16H^+ = 5Mg^{+2} + 2Al^{+3} + 3H_4SiO_4 + 6H_2O$	68.38
$Ca_{0.165}Al_{2.33}Si_{3.67}O10(OH)_{2(Ca\text{-}montmorrllonite)} + 12H_2O =$ $0.165Ca^{+2} + 2.33Al(OH)_4^- + 3.67H_4SiO_4 + 2H^+$	−5.027

(Continued)

(Continued)

Reaction	log K
$Mg_3Si_4O_{10}(OH)_{2(talc)} + 4H_2O + 6H^+ = 3Mg^{+2} + 4H_4SiO_4$	21.399
$K_{0.6}Mg_{0.25}Al_{2.3}Si_{3.5}O_{10}(OH)_{2(illite)} + 11.2H_2O =$ $0.6K^+ + 0.25Mg^{+2} + 2.3Al(OH)_4^- + 3.5H_4SiO_4 + 1.2H^+$	–40.267
$Mg_3Si_2O_5(OH)_{4)(chrysotle)} + 6H^+ = H_2O + 2H_4SiO_4 + 3Mg^{+2}$	32.200
$Fe_2O_3 + 6H^+_{(hematite)} = 2Fe^{+3} + 3\ H_2O$	–4.008
$FeOOH_{(goethite)} + 3H^+ = Fe^{+3} + 2H_2O$	–1.000
$Fe_3O_{4(magnetite)} + 8H^+ = 2Fe3^+ + Fe^{2+} + 4H_2O$	3.737
$Fe(OH)_{3(a)} + 3H^+ + e = Fe^{+3} + 3H_2O$	4.891
$FeS_{2(pyrite)} + 2H^+ + 2e^- = Fe^{+2} + 2HS^-$	–18.479
$FeOH_{3(ferrihydrite)} + 3H^+ = 2Fe^{3+}H_2O$	–1.0
$FeS_{(ppt)} + H^+ = Fe^{+2} + HS^-$	–3.915
$FeS_{(mackinawite)} + H^+ = Fe^{+2} + HS^-$	–4.648
$Fe^{+2} + H_2O = FeOH^+ + H^+$	–9.500
$Fe^{+2} + Cl- = FeCl^+$	0.140
$Fe^{+2} + CO_3^{\,2} = FeCO_3$	4.380
$Fe_{+2} + HCO_3^- = FeHCO_3^+$	2.0
$Fe^{+2} + SO_4^{\,2} = FeSO_4$	2.250
$Fe^{+2} + HSO_4^- = FeHSO_4^+$	1.08
$Fe^{+2} + 2HS^- = Fe(HS)_2$	8.95
$Fe^{+2} + 3HS^- = Fe(HS)_3^-$	10.987
$Fe^{+2} + F^- = FeF^+$	1.000
$Fe^{+2} = Fe^{+3} + e$	–13.020
$Fe^{+3} + H_2O = FeOH^{+2} + H^+$	–2.19
$Fe_{+3} + 2H_2O = Fe(OH)_2^+ + 2H^+$	–5.67
$Fe^{+3} + 3H_2O = Fe(OH)_3 + 3H^+$	–12.56
$Fe^{+3} + 4H_2O = Fe(OH)_4^- + 4H+$	–21.6
$Fe^{+3} + Cl^- = FeCl^{+2}$	1.48
$Fe^{+3} + 2Cl^- = FeCl^{2+}$	2.13
$Fe^{+3} + SO_4^{\,2} = FeSO_4^+$	4.04
$Fe^{+3} + HSO_4^- = FeHSO_4^{+2}$	2.48
$Fe^{+3} + 2SO_4^{\,2} = Fe(SO_4)_2^-$	5.38
$S + 2H^+ + 2e = H_2S$	4.882

(Continued)

Reaction	log K
$MnO_{2(pyrolusite)} + 4H^{+} + 2e = Mn^{+2} + 2H_2O$	41.380
$Mn_3O_{4(hausmannite)} + 8H^{+} + 2e = 3Mn^{+2} + 4H_2O$	61.030
$MnOOH_{(manganite)} + 3H^{+} + e = Mn^{+2} + 2H_2O$	25.340
$Mn(OH)_{2(pyrochroite)} + 2H^{+} = Mn^{+2} + 2H_2O$	15.200
$NaCl_{(halite)} = Na^{+} + Cl^{-}$	1.582
$KAl_3(SO_4)_2(OH)_{6(alunite)} + 6H^{+} = K^{+} + 3Al^{+3} + 2SO_4^{2} + 6H_2O$	−1.400
$KFe^{3}(SO_4)_2(OH)_{6(jarosite)} + 6H^{+} = 3Fe^{+3} + 6H_2O + K^{+} + 2SO_4^{2}$	−9.210
$Zn(OH)_2 + 2H^{+} = Zn^{+2} + 2H_2O$	11.50
$ZnCO_{3(smithsonite)} = Zn^{+2} + CO_3^{2}$	−10.000
$ZnS_{(sphalerite)} + H^{+} = Zn^{+2} + HS^{-}$	−11.618
$Zn_2SiO_{4(willemite)} + 4H^{+} = 2Zn^{+2} + H_4SiO_4$	15.33
$Cd(OH)2 + 2H^{+} = Cd^{+2} + 2H_2O$	13.650
$CdCO_{3(octavite)} = Cd^{+2} + CO_3^{2}$	−12.1
$PbCO3_{(cerrusite)} = Pb^{+2} + CO_3^{2}$	−13.13
$PbSO_{4(anglesite)} = Pb^{+2} + SO_4^{2}$	−7.79
$PbS_{(galena)} + H+ = Pb^{+2} + HS^{-}$	−12.78

Note: The Values are from WATEQ4F
Source: Ball and Nordstrom, 1991 and PHREEQC (Parkhurst and Appelo, 2013)

APPENDIX REFERENCES

Ball, J.W., and D.K. Nordstrom, 1991, *User's manual for WATEQ4F, with revised thermodynamic data base and text cases for calculating speciation of major, trace, and redox, elements in natural waters*, U.S. Geological Survey Open-File Report 91-183.

Busenberg, E., and L.N. Plummer, 1986, The solubility of $BaCO_3$(cr) (witherite) in CO_2-H_2O solutions between 0 and 90°C, evaluation of the association constants of $BaHCO_3^{+}$(aq) and $BaCO_3$(aq) between 5 and 80°C, and a preliminary evaluation of the thermodynamic properties of Ba^{2+}(aq), *Geochimica et Cosmochimica Acta*, 50(10), 2225–2233.

Busenberg, E., L.N. Plummer, and V.B. Parker, 1984, The solubility of strontianite ($SrCO_3$) in CO_2-H_2O solutions between 2 and 91°C, the association constants of $SrHCO^{+3}$(aq) and $SrCOO_3$(aq) between 5 and 80°C, and an evaluation of the thermodynamic properties of Sr^{2+}(aq) and $SrCO_3$(cr) at 25°C and 1 atm total pressure, *Geochimica et Cosmochimica Acta*, 48(10), 2021–2035.

Clark, I.D., and P. Fritz, 1997, *Environmental isotopes in hydrology*, CRC Press.

Drever, J.I., 1988, *The geochemistry of natural waters*, 2nd ed, Prentice Hall.

Haas, J.L., G.R. Robinson Jr, and B.S. Hemingway, 1980, *Thermodynamic tabulations for selected phases in the system CaO-Al2O$_3$-SiO$_2$-H$_2$O*, U.S. Geological Survey Open File Report 80-908.

Helgeson, H.C., J. Delany, W. Nesbitt, and D.K. Bird, 1978, Summary and critique of the thermodynamic properties of rock forming minerals, *American Journal Science*, 278-A, 1–229.

Krauskopf, K.B., and D.K. Bird, 1995, *Introduction to geochemistry*, 3rd ed, McGraw Hill.

Morel, F.M.M., 1983, *Principles of aquatic chemistry*, Wiley and Sons.

Nordstrom, D.K., D.D. Valentine, J.W. Ball, N. Plummer, and B.F. Jones, 1984, *Partial compilation and revision of basic data in the WATEQ programs*, U.S. Geological Survey Water-Resources Investigation Report 84-4186.

Parkhurst, D.L., and C.A.J. Appelo, 2013, *PHREEQC*, U.S. Geological Survey Techniques.

Reed, J., 2020, Digitizing the NBS tables of chemical thermodynamic properties: Selected values for inorganic and C1 and C2 organic substances in SI Units, *Journal of Research of the National Institute of Standards and Technology*, 125.

Robie, R.A, and B.S. Hemingway 1995, Thermodynamic properties of minerals and related substances at 298.15 K and 1 bar (10^5 pascals) pressure and at higher temperatures, *U.S. Geological Survey Bulletin,* 2131.

Robie, R.A, et al., 1978, Thermodynamic properties of minerals and related substances at 298.15 K and 1 bar (108 pascals) pressure and at higher temperatures, *U.S. Geological Survey Bulletin* 1452, 456 p.

Sillin, L.G., and A.E. Martell, 1964, *Stability constants of metal-ion complexes*, Chemical Society, London, Special Publication, 17.

Smith, R.M., and A.E. Martell, 1976, *Critical stability constants*, Springer.

Index

A

absolute abundance ratio, 183, 193
acid
  acid-base reactions, 18
  Boric, 19
  Bronsted-Lowry, 18
  carbonic, 19
  carboxylic, 117
  humic, 19, 238
  Lewis, 18
  phosphoric, 19
  silicic, 19, 154
  speciation, 146
  strong, 19
  weak, 22
acid mine drainage (AMD), 146
activity
  chemical, 51
  coefficients, 51
  Debye-Hückel, Davies, Truesdell-Jones, Pitzer, 55
  isotopic (groundwater dating), 273
adiabatic and conductive cooling, 375
adsorption - desorption, 32
  double electrical layer, 32
aliphatic hydrocarbons, 114
alkali earth metals, 165
alkali metals, 164
aluminum, 39
aluminosilicate, 135, 150, 163
ammonia, 23
Anderson, et al. v. Pacific Gas and Electric, 167
anion exchange capacity (AEC), 35
Antelope Island, Utah, 215, 320
antimony, 173
aromatic hydrocarbons, 115
Arrhenius equation, 96
arsenic, 172
atmospheric pressure, 5, 76
Avogadro's number, 6

B

barium, 165
benzene ring, 114
beryllium, 165
Bjerrum plot, 342
boron, 23
Bowens reaction series, 151
bromide, 138
Bronsted-Lowry acid, 18
Boyles law, 78

C

C3, C4, and CAM plants, 232
CAS (carbonate associated sulfur), 247
calcite, 123, 128
  calcite system, 130
caliche, 135
canyon diablo ion meteorite (CDT), 196
carbon
  aquifer and soil zone processes, 237
  carbon budget, 119, 230
  carbon cycle, 232
  carbonic acid, 21
  CO_2, 235
  diagenesis and lithification, 238
  earths reservoirs/budget, 113, 230
  exchange processes, 234, 286
  fractionation (^{13}C), 191
  fractionation (atmospheric, oceanic and biologic), 231
  groundwater sources, 234
  humic, 120
  inorganic sources, 239
  isotope reporting standards, 195
  magmatic, 240
  metamorphic, 241
  methanogenesis, 238
  organic, 117, 231
  soil zone, 237
  speciation, 14
  total dissolve inorganic carbon, 272
carbon-associated sulfur (CAS), 247
carbonate system, 24, 123
  equilibrium constants, 127
  minerals, 52
carbonatite (REE), 240
Carbon-14 (^{14}C), 280
  Activity (A_o and A_t), 284
  age dating empirical approach, 289
  age dating model numerical instabilities and sensitivity, 295
  age dating, 283
  A_o estimating models (Fontes, Gonfiantini, Han and Plummer, IAEA, Mook, Pearson, mixing, Tamer), 288
  atmospheric testing, 281
  isotopic exchange (ε), 286
  live and dead carbon, 280, 285
  percent modern carbon (PMC), 196
  reporting standards, 281
  Suess effect, 281
carbon stable isotope, 229
  Aquifer and soil zone carbon, 237
  Carbonate rocks, 237
  CO_2 (groundwater), 235
  connate water, 242
  diagenesis and lithification, 238
  exchange processes, 234, 286
  fractionation (atmospheric, oceanic and biologic), 231
  magmatic, 240
  metamorphic, 240
  methanogenesis, 238
  photosynthetic plants (C3, C4, CAM), 232
  reporting standard, 195
  sulfate reduction, 242

carbonate system, 123
  CO_2 and pH control, 125
carbonic acid, 21
carboxyl group, 117
cation-anion balance, 314
cation exchange capacity (CEC), 35
cesium, 165
Charles law, 78
chelation, 38
chemical activity, 51
  activity coefficient, 51, 315
  calculate activity, 54
  effective molar concentration, 51
  ionic strength, 35, 53
  partial pressure, 51
chemical bonding, 9
  anion, 10
  cation, 10
  covalent bond, 10
  electron transfer, 10
  hydrogen bond, 10
  ionic bond, 10
  organic chemicals, **114**
  water molecule, 11
chemical reactions (inorganic), 17
  acid-base, 18
  adsorption and desorption, 32
  chemical substitution, 32
  congruent and incongruent dissolution, 18, 50, 73, 151, 242, 358
    electrostatic adsorption, 32
    hydrolysis, 22
    hydronium ion (H_30^+), 18
    ion hydration, ion pairs, ion complex, 36
    mineral dissolution, 18, 50, 104
    reduction-oxidation (Redox), 25
chemical reporting units
  chemical equivalence, 7
  formality (f), molality (m), molarity (M), 6
  geochemical, 5
  gram formula weight (GFW), 6
  laboratory units, 5
  mole fraction, 7
  mole, 6
  parts per million (ppm), 6
  SI system, 3
  temperature (°C, °F, °K), 3
China (Sichuan basin), 263
chloride/chlorine, 138, 298
  chloride bromide ratio, 139
  chloride mass balance, 141
  chlorine-36 dating, 298
  isotope reporting standard, 196
  terrestrial abundance, 139
chlorofluorocarbon (CFC), 95, 300
chromium
  diagram, 166
  Hinkley, California, 167
circulation depth (groundwater), 375
closed and open system, 129
coal, 239
CO_2, 22, 86, 98, 113, 125, 133, 235
  atmospheric, 77, 235
  CO_2 – bicarbonate, carbonate, 98, 133
  CO_2 system, 86
  methanogenesis, 135
  pH control, 125
cobalt, 166
colloids, 32
common ion effect, 15, 72, 133
complex ions, 40
  speciation, 41
  trace metals, 42
computer methods, 315
  NETPATH, PHREEQC, WATEQF, 315
congruent and incongruent dissolution, 18, 50, 73, 151, 242, 358
connate water, 222, 242
constants
  equilibrium constant (K), 20
  Henry's law, 82
  ionization, 19
continental effect
  ^{18}O and ^{2}H Tritium, 274, 345
contour map, 345
conversion factors (mg/L-meq/L), 8
coordination chemistry, 37
  Chelation, 38
  Ligand, 37
copper, 166
Cretaceous Bohemian Basin, Czech Republic, 334
crude oil, 116
crustal composition, 124

D

Daltons law, 79
data management, 312
dating (groundwater), 272
Davies activity coefficient, 55, 315
DDT, 111
Debye- Hückel activity coefficient, 54, 315
delta unit (δ), 184
density, 16, 76
  temperature and salinity, 12
diagenesis and lithification, 238
dielectric constant, 10
dipole moment, 11
dimethylsulfide (DMS), 142
dissolved inorganic elements, 125
dissolved organic carbon (DOC), 108
dissolved organic matter (DOM), 108
DNAP and LNAPPL, 109
Dolomite, 123, 128
double layer, 32
dual-isotope approach, 184

E

East Pacific Rise, 275
effective molar concentration, 50
E_h - pE, 66
  antimony, 174
  arsenic, 173
  chromium, 166
  cobalt, 168
  copper, 169
  iron, 149

- lead, 171
- manganese, 150
- mercury, 170
- nitrogen, 158
- sulfur, 145
- zinc, 172

Elbe River, Czech Republic, 263
electrical conductivity, 14
electrical double layer (Gouy layer), 32
electrolytes, 14
elements
- dissolved concentrations, 125
- trace (aqueous), 163

energy
- energy units (calorie, Joule, erg), 59
- Gibbs free energy, 60
- kinetic and potential energy, 59

Enthalpy (H), 59
entropy, 60
equilibrium
- activity, 50
- chemical, 49
- conjugate bases, 20
- constants (Keq), 20, 50, 52, 125
- definition, 50
- dynamic and static equilibrium, 50
- effective molar concentration, 51
- law of mass action, 49, 186
- pH, 19
- reactants and products, 49
- water, 19

equilibrium diagram, 345
- antimony, 174
- arsenic, 173
- chromium, 166
- cobalt, 168
- copper, 169
- E_h-pH diagram, 349
- iron, 149
- lead, 171, 351
- manganese, 150
- mercury, 170
- nitrogen, 158
- sulfur, 145
- zinc, 172

equivalence (chemical, meq/L), 7
enthalpy, entropy, 59
errors, 313
- error balance calculation, 45, 314
- systematic and random, 313

Esters, 119

F

fugacity, 86
functional organic groups, 114, 117

G

gas, 30, 76
- atmospheric, 76, 160
- CH_4, 84, 238
- CO_2 system, 86, 125, 235, 238
- CO_2, O_2, N_2, 30, 77
- fugacity, 86
- gas activity coefficient (y), 86
- Henry's solubility and volatility, 112
- ideal solution, 80
- noble, 10, 159
- non-ideal solution, 86
- partial pressure, 79
- properties, 76
- temperature-dependent solubility, 30, 84
- universal gas constant, 79
- Van't Hoff equation, 82
- vapor pressure and density, 112

gas laws (Boyle's, Charles, Daltons Ideal, Raoults'), 78
- Henery's Law, 78, 112

Gibbs free energy, 29, 60
- mineral precipitation, 100

geothermal groundwater
- geothermal gradient, 219, 375
- stable isotopic shift, 220

geothermometer, 375
- adiabatic and conductive cooling, 375
- carbon isotope geothermometer, 390
- exchange geothermometer, 380
- gas geothermometer, 388
- geothermal gradient, 375
- Giggenbach Ternary geothermometer, 385
- hydrogen isotope geothermometer, 392
- isotopic geothermometer, 389
- K-Mg geothermometer, 384
- mixing geothermometer, 387
- Na-K-Ca geothermometer, 380
- oxygen isotope geothermometer, 392
- oxygen isotope shift, 378
- qualitative geothermometer, 376
- quantitative solute, 378
- silica, 377
- solubility geothermometer, 379
- sulfur isotope geothermometer, 392

global meteoric water line (GMWL), 200
Goldrich Weathering Series, 151
Graphical methods, 319
- Bar graph, 320
- Durov diagram, 328
- four corner diagrams, 325
- pie diagram, 320
- piper diagram, 329
- radial diagram, 325
- shape diagrams, 320
- stiff diagram, 326
- trilinear diagram, 328
- vector diagram, 320

Great Oxidation Event (GOE), 147
Great Salt Lake, Utah, 214, 320
groundwater dating, 272
- ^{14}C (carbon-14), 283
- ^{14}C activity, 283
- ^{3}H (tritium), 274
- $^{3}H/^{3}He$, 279
- absolute and relative age, 272, 283
- calculating groundwater age, 272, 288
- carbon isotopic exchange (ε), 286
- chlorine-36, 298
- chlorofluorocarbon, 300
- dating model instabilities and sensitivity, 295

dead carbon and live carbon, 285
half-life, 273
sulphur hexafluoride, 300
guyot layer, 32
gypsum, 142

H

$^3H/^3He$ dating, 279
half-life, 98, 273
Henry's law, 81
Hinkley, California, 167, 345
Hofmeister series, 35
Honey Lake, California - Nevada, 217, 345
humic substances, 44, 120
hydrocarbon, 114
formation, 238
functional groups – aliphatic and aromatic benzene ring, 114
halogenated, 118
hydrogeochemical facies, 331
hydrolysis, 22, 152, 286
hydronium ion, 19
hydroxyl group, 119

I

IAEA model, 294
Iceland, 375
Idaho Batholith, 221, 334, 376
industrial revolution, 281
inorganic chemical reactions, 17
acid-base reactions, 18
congruent and incongruent dissolution, 18, 50, 73, 151, 242, 358
hydrolysis, 22
mineral dissolution-precipitation, 18
reduction-oxidation (Redox), 25
ion
activity, 51
common ion effect, 72
exchange, 35
hydration, pairs, complexation, 36
ion activity product (IAP), 70
ionic strength effect, 72
ionic strength, 53
calculate, 53
definition, 53
effect, 51, 133
vs. activity coefficient, 54
ionization constant, 19
iron, 69, 148
isoelectric points, 34
isotherms, 33
isotope, 183
absolute abundance ratio, 183
abundances, 11, 183
atmospheric, 160
carbon, 229, 285
chlorine-36, 298
delta unit (δ), 183
dual isotope approach, 184
equilibrium, 190, 192
exchange, 185
global meteoric water line (GMWL), 200
hydrogen and oxygen, 190, 194, 199
nitrogen, 255
positive oxygen shift, 219, 378
radioactive (unstable isotope), 183, 193
reporting standards, 193, 281
stable, 183
standards, 193
sulfur-34, 246
tritium (^3H), 274
isotopic fractionation and exchange, 184
Fractionation factors (K, α), 186
isotopic enrichment factor (ε), 188
isotopic exchange, 185, 234
kinetic fractionation, 185
photosynthetic plants (C3, C4, CAM), 232
sulfur fractionation, 246
isotope reporting standards stable isotopes, 193
carbon-14 (pmc), 281
tritium unit (TU), 274

K

kinetic and potential energy, 59
kinetics, 90
activation energy, 92, 100
Arrhenius equation, 96
classification of reactions, 91
dissolution and precipitation mechanisms, 99
linear free energy (LFR), 91
molecularity, 93
ozone, 95
reaction rates, 94
transition state theory, 91

L

Lake Nyos, Cameroon, 112
law of mass action, 19, 49
lead, 166
Lewis acid, 18
Ligand, 37
Lithium, 165
LNAPL (Light nonaqueous phase liquid), 109

M

magmatic water, 222, 240
manganese, 69, 148
Mauna Loa Observatory, Hawaii, 77
McGil Hot Spring, Nevada, 70, 315
metamorphic carbon, 241
methane, 113
methanogenesis, 135, 238
mercury, 166
meteoric water isotopic trends, 204
air temperature effect, 204
altitude effect, 208
continental effect, 205
evaporation, 213
latitude, 206
paleoclimate, 215
rain-out effect, 205
seasonal trend, 211

mid-Atlantic ridge, 375
minerals (silicate, carbonate, chloride, fluorite, hydroxide and oxide, sulfide, 52
mineral dissolution
 common ion, ionic strength effects, 72
 congruent and incongruent dissolution, 73
 dissolution-precipitation mechanisms, 99
 ion activity product (IAP), 70
 order of encounter, 72
 relative solubility, 51
 saturation index (SI), 70
mineral surface reactions, 31
 adsorption and desorption, chemical substitution, electrostatic adsorption, specific adsorption, 32
 cation exchange capacity (CEC), 36
 electric double layer, 33
 Hofmeister series, 35
 ion exchange, 35
 isoelectric points, 34
 sorption isotherms, 33
Mook model, 292

N

Nerst equation, 350
NETPATH, 315
Niagara Dolomite, 218
nitrogen, 69, 156, 255
 ammonification, 158, 261
 biologic cycle (ammonification, nitrification and denitrification, 134, 138, 157, 261
 contamination (groundwater), 159
 cycle, 255
 groundwater, 262
 nitrate reduction, 134
 organic compounds, 119
 sources, 255
 terrestrial cycle, 157, 259
nitrogen isotopes, 255
 anthropogenic and natural), 255
 fractionation, 259
 nitrogen fractionation, 259
 nitrogen sources (values –
 reporting standard, 196
noble gas, 10, 159
 solubility, 162

O

octanol-water partition coefficient, 109
open and closed system, 129
order of encounter, 133
organic carbon
 functional groups, 117
 humic, 120
organic chemistry, 108
 carbon functional groups, 117
 diagenesis and lithification, 238
 dissolved organic carbon, 108
 DNAPL and LNAPPL, 109
 humic substances, 120
 hydrocarbon, 114
 nitrogen compounds, 119
 octanol-water partition coefficient, 109
 properties of organic compounds, 108
 trace metal complex, 44
 volatile organic compound (VOC), 112
organic matter (SOM, Humin), 44
oxidation state, 25
oxidizing environment, 29
 partial pressure, 31
 soil and atmosphere O_2 and CO_2, 31
oxygen and hydrogen isotopes, 199
 connate and magmatic water, 222
 deuterium excess, 192
 fractionation, 203
 global meteoric water line (GMWL), 200
 meteoric water evolution
  air temperature, 204
  amount, rainout and continental effects, 205
  evaporation, 213
  latitude effect, 206
  seasonal trends, 211
 paleoclimate, 215, 221
 Rayleigh distillation (ocean, atmosphere, evaporation), 201
 stable oxygen isotopic shift, 219
 standards, 194
 thermal groundwater, 219
ozone, 95

P

Pah Tempe, Utah, 88
paleoclimate, 215, 221
partial pressure fugacity diagrams, 365
PDB (VPDB), 195
p*E* - E_h, 66
Pearson model, 291
percent modern carbon (pmc), 196
pH, 19, 125
phase rule, 64
phosphorus, 24
PHREEQC, 235, 315
Pilot Valley, California-Nevada, 16, 214
Pitzer activity coefficient, 55
Piper diagram, 239
pressure
 atmospheric, hydrostatic and lithostatic, 5
 partial pressure, 7
 standard temperature and pressure, 5
Pourbaix diagram (E_h-pH)
 antimony, 174
 arsenic, 173
 chromium, 166
 cobalt, 168
 copper, 169
 iron, 149
 lead, 171
 manganese, 150
 mercury, 170
 nitrogen, 158
 sulfur, 145
 zinc, 172
Powder River basin, Wyoming, 239
Project design, 311

Q

quartz dissolution, 154

R

radioactive decay (half-life), 98, 193
radioactive isotopes, 193
rare earth elements (REE), 240
rate laws, 97
Rayleigh distillation, 190, 202
reducing environment, 31
reduction-oxidation (Redox), 25, 66
  E_h vs. p*E*, 67
  groundwater processes, 28
  half-reactions, 28
  inorganic and organic processes, 28
  oxidation states, 26
  oxidizing environments, 29
  reducing environments, 31
  rules for assigning oxidation state, 25
root zone processes, 133
rubidium, 165

S

salinity, 16
salting-out-effect, 87
Salton Sea, California, 375
San Andreas Fault, California, 375
San Luis Valley, Colorado, 110, 137, 223, 339, 345
saturation index (SI), 70
scatter plot, 333
  Bjerrum plot, 342
  dilution and mixing diagram, 336
  Schoeller diagram, 336
  Tie line, 334
second law of thermodynamics, 59
selenium, 172
Sichuan Basin, China, 263
silica, 150
  mineral weathering, 152
  tetrahedron, 150
SI units, 3
SMOW (VSMOW), 193
Snake River Plain, Idaho, 221
soil organic matter (SOM), 44
solubility (minerals), 51
solution (definition), 3
  non-ideal, 51, 86
  solute, 3
  solvent, 3
solute concentration units, 5
  formality, 6
  formula weight, 6
  gram formula weight, 6
  molality, 6
  molarity, 6
  mole, 6
  mole fraction, 7
solvent, 3
sorption isotherms, 33
speciation
  colloidal, gas, ion, ion pair, ion complex, 14
  complex ion, 40
  saline water, 42
  stability constants, 38
  trace metal complex, 44
specific conductance, 13, 38
spring leak, 275
stability constant, 38
stability diagrams, 358
  congruent solubility, 358
  incongruent dissolution, 360
  microcline-muscovite, 361
  silica system, 358
standard mean ocean chlorine (SMOC), 194
standard temperature and pressure (STP), 5
stiff diagram, 134, 327
storm tracks (Utah), 208
strontium, 165
Suess effect, 281
sulfate
  assimilatory-dissimilatory reduction, 144
  microbiologic conversion, 251
  oxidizing microorganisms (SOM), 143
  reducing microorganisms (SRM), 143
  reduction, 134, 242, 250
sulfide
  dissociation, 25
  sulfide minerals, 146
sulfur, 141
  anthropogenic, 253
  atmospheric, 142
  biogenic, 143
  carbonyl sulfide (COS), 247
  CAS (carbonate associated sulfur), 247
  dimethyl sulfide (DMS), 142, 247
  groundwater, 145
  isotope reporting standard, 196
  minerals, 142
  reduction, 134
  sulfur cycle, 141, 246, 250
  Terminal Electron Accepting Process (TEAP), 147
sulfur hexafluoride, 300
sulfur isotopes, 196, 246
  anthropogenic, 253
  atmospheric, 248
  calcium sulfate and calcium carbonate minerals, 249
  evaporite values over geologic time, 250
  inorganic and organic values, 246
  sulfate-sulfide biologic fractionation, 250
surface reactions, 17, 31
  adsorption – desorption, 32
  electrical double layer, 32
  guyot layer, 33
  mineral dissolution – precipitation, 18

T

tamers model, 291
  carbonyl sulfide (COS), 247
  CAS (carbonate associated sulfur), 247

temperature conversions, 5
terminal electron accepting process (TEAP), 147
Thermal decomposition (carbonate), 135, 241
thermal groundwater, 219
  positive oxygen isotope shift, 220
thermodynamics, 59
  enthalpy, entropy, 59
  Gibbs free energy, 60
  heat of reaction (ΔH), 60
  phase rule, 64
  reduction-oxidation, 66
  thermodynamics and kinetics, 91
total dissolved inorganic carbon, 272
total dissolved solids (TDS), 13, 15
trace
  elements, 163
  organic trace metal species and complexing, 43
transition metals, 166
trilinear diagram, 238
tritium, 274
  atmospheric, 274
tritium dating, 278
  $^3H/^3He$ dating, 279
  Piston flow model, 278
tritium units (TU), 274
  effects (continental, latitude, mountain, season), 275
  groundwater dating, 278
  spring leak, 275
  stratosphere-troposphere exchange, 275
Truesdell-Jones activity coefficient, 55

U

units, 3, 4
  Becquerel, 274
  meq/L, mg/L, ppm, 6
  mmole/L, 7
  SI system (System International), 3
  Temperature (°C, °F, °K), 5

V

valance, 13
vapor density, 112
viscosity, 76
volatile organic compound (VOC), 112
VSMOW, 194

W

Wasatch Range, Utah, 320
WATEQF, 235, 315
water
  connate water, 222, 242
  dissociation, pH, 19
  magmatic, 222, 240
  metamorphic carbon, 241
  meteoric, 201
  three phase system, 65
water density, temperature and viscosity, 12, 16
  electrolyte solutions, 13
  isotopic composition, 10
  molecule, 10
  nonelectrolyte, 13
  properties (physical and chemical), 10
  salinity, 15
  speciation, 14
  structure, bonding, dipole moment, 10
  TDS vs specific conductance, 13
weathering, 104
  aluminosilicate mineral, 21, 135, 150
  carbonate mineral, 285
  dissolution, 104
  Goldrich series, 151
  hydrolysis, 152, 286
  resistance to weathering, 153

Z

zeolite, 36
zinc, 166